THE GREENHOUSE GAMBIT:

Business and Investment Responses
to Climate Change

Douglas Cogan

Investor Responsibility Research Center
Washington, D.C.

The Investor Responsibility Research Center compiles and impartially analyzes information on the activities of business in society, on the activities of institutional investors, on efforts to influence such activities, and on related public policies. IRRC's publications and other services are available by subscription or individually. IRRC's work is financed primarily by annual subscription fees paid by some 400 investing institutions for the Environmental Information Service, the Social Issues Service, the Corporate Governance Service, the South Africa Review Service and the Global Shareholder Service. This report is a publication of the Environmental Information Service. The Center was founded in 1972 as an independent, not-for-profit corporation. It is governed by a 21-member board of directors.

Executive Director: Margaret Carroll
Director, Environmental Information Service: Scott Fenn

This report was prepared by the Investor Responsibility Research Center for educational purposes. While IRRC exercised due care in compiling the information in this report, from time to time errors do occur. IRRC makes no warranty, express or implied, as to the accuracy, completeness or usefulness of this information and does not assume any liability with respect to the consequences of the use of this information. Changing circumstances may cause the information to be obsolete.

ISBN 0-931035-86-4

Copyright 1992, Investor Responsibility Research Center Inc.
1755 Massachusetts Ave., N.W., Washington, D.C. 20036
Telephone: (202) 234-7500

Table of Contents

Acknowledgements iii
Introduction and Summary v

Chapter 1: A Climate of Uncertainty
    Introduction 3
    Has Global Warming Begun? 5
    Forecasting the Future 23
    Expecting the Unexpected 39
    Making Order Out of Chaos 51

Chapter 2: Agriculture
    Introduction 73
    Effects on Crops 77
    Effects on Water 106
    Responses of Farmers and Food Producers 123
    Conclusions 137

Chapter 3: Forest Products
    Introduction 157
    Effects on Forests 161
    Responses of the Forest Products Industry 183
    Carbon Sequestration Potential 205
    Conclusions 217

Chapter 4: Automobiles
    Introduction 235
    Fuel Efficiency Potential 249
    Alternative Transportation Fuels 281
    Conclusions 317

Chapter 5: Electric Power
    Introduction 347
    Effects on Power Generation 356
    Energy Efficiency Potential 375
    Alternative Generating Sources 390
    Conclusions 435

Chapter 6: Conclusions
    The Greenhouse Gambit 467

List of Tables

Chapter 1
Table 1: Greenhouse Gases: Their Sources
and Warming Potential 11
Table 2: 10 Warmest Years Since 1880 20
Table 3: Selected Major Conclusions of the IPCC Supplement 61

Chapter 2
Table 1: Climate Parameters of Selected Agricultural Regions .. 100
Table 2: Potential Regional Impacts of Climate Change
on Water Uses 117
Table 3: Genetically Engineered Species 129
Table 4: Major Net Cereal Exporters and Importers 139

Chapter 3
Table 1: Climate Parameters and Their Effects on Forests 167
Table 2: World's Top 15 Timber-Producing Countries 181
Table 3: Commercial Uses of U.S. Forests 187

Chapter 4
Table 1: Technologies to Improve Fuel Economy 263
Table 2: Vehicle Exhaust Emission Standards 286
Table 3: CO_2 Emissions from Alternatively Fueled Vehicles 308
Table 4: Alternative Transportation Fuels: Pros and Cons 324

Chapter 5
Table 1: Effects on Electricity Demand of Demand-Side
Mangement and Beneficial Electrification 358
Table 2: Major Assumptions of ICF Reference Scenarios
on CO_2 Reductions 368
Table 3: World Fossil Energy Reserves 396
Table 4: Nuclear Power Plants in Operation and Under
Construction 409
Table 5: Selected U.S. Mitigation Options 448

Acknowledgements

The Investor Responsibility Research Center is indebted to the Rockefeller Foundation for a generous grant that made this report possible. The grant was intended to support the Center's studies of corporations as they affect the global environment and the ways in which the environment and environment-related policies affect corporations in turn.

The author would like to thank many individuals who assisted in the preparation of this report. Within IRRC, Susan Brackett conducted indispensable research on the agriculture and forest products industries. Scott Fenn, Teresa Opheim and Carolyn Mathiasen reviewed preliminary drafts and offered many helpful comments. Margaret Carroll reviewed the final draft with unfailing attention to detail. Shirley Carpenter prepared this report for publication with utmost patience and care. Michael J. Davis of Davis Designs created the cover illustration.

Many other people reviewed preliminary drafts of this report. The author wishes to acknowledge them here for their wide-ranging comments, which added immeasurably to the accuracy, balance and completeness of the report. Any errors that remain are those of the author. Any opinions expressed herein are those of the author as well.

Climate chapter reviewers:
Anthony Del Genio, Goddard Institute for Space Studies, National Aeronautics and Space Administration.
William Kellogg, National Center for Atmospheric Research (retired).

Agriculture chapter reviewers:
Stanley Changnon, Illinois State Water Survey.
Pierre Crosson, Resources for the Future.
Thomas Hebert, Agriculture Committee, U.S. Senate.
Cynthia Rosenzweig, Goddard Institute for Space Studies, NASA.
Peter Weber, Worldwatch Institute.
Karl Weinkauff, Monsanto Agricultural Co.
Edward Whereat, University of Maryland (PhD candidate).
Carol Whitman, Global Change Program, U.S. Department of Agriculture.

Forest products chapter reviewers:
Scott Berg, American Forest Council.
Michael Coffman, Champion International Corp.
Dieter Deumling, Enforsys.
Peter Farnum, Weyerhaeuser Corp.
Thomas Hinckley, College of Forest Resources, University of Washington.
Alan Lucier, National Council of the Paper Industry for Air and Stream Improvement.
Neil Sampson, American Forestry Association.
Con Schallau, American Forest Resource Alliance.
John Turner, Georgia-Pacific Corp.
Ted Wolf, University of Washington (PhD candidate).

Automobile chapter reviewers:
Gordon Allardyce, Chrysler Corp.
John DeCicco, American Council for an Energy-Efficient Economy.
Thomas Gage, Chrysler Corp.
Toni Harrington, Honda North American Inc.
Samuel Leonard, General Motors Corp.
Alan Miller, Center for Global Climate Change, University of Maryland.
Steven Plotkin, U.S. Office of Technology Assessment.
Michael Schwarz, Ford Motor Co.
Albert Slechter, Chrysler Corp.

Electric power chapter reviewers:
William Fang, Edison Electric Institute.
Eric Hirst, Oak Ridge National Laboratory.
Alan Miller, Center for Global Climate Change, University of Maryland.
Evan Mills, Lawrence Berkeley Laboratory.
Ralph Perhac, Electric Power Research Institute (retired).
Nicholas Sundt, *Energy, Economics and Climate Change.*
Richard Tempchin, Edison Electric Insitute.
James Young, Southern California Edison Co.

Finally, the author wishes to express his heartfelt gratitude to family, friends and colleagues for their patience and support during the four-year gestation period of the report. You know who you are.

> Doug Cogan
> Manager, Global Issues
> IRRC Environmental Information Service
> August 1992

Introduction and Summary

When world leaders gathered in Rio de Janeiro in June 1992 for the "Earth Summit"—the largest international conference ever held—their first order of business was to sign a treaty to curtail a buildup of gaseous pollutants in the atmosphere. With more controversy than fanfare, industrialized nations agreed to try to reduce emissions of so-called greenhouse gases to 1990 levels by the year 2000, and developing nations pledged to try to limit their emissions as well. But at the insistence of the American delegation, led by President George Bush, the treaty signed in Rio commits no nation to achieving specific targets or timetables for controlling these emissions, which are feared will lead to a rapid warming of the Earth. Ironically, the world's richest nation voiced the same principal concern as the world's poorest nations: that measures taken to ameliorate the threat posed by global warming would create a serious impediment to economic growth.

Econometric studies of the possible effects of stabilizing or marginally reducing emissions of carbon dioxide, the main greenhouse gas, generally find that the U.S. gross national product would be reduced by 1 or 2 percent, or $60 billion to $120 billion a year. This does not mean that the economy would stop growing, however, only that the rate of growth would be suppressed. The level of national income that would have been reached in 2050, say, would be delayed until 2052, if 2 percent of GNP were invested in carbon abatement and the rate of real income growth were 1 percent a year. Some engineering studies suggest, however, that meaningful reductions in carbon emissions could be achieved at a lower net cost or possibly even at a net benefit to the economy. Real-cost pricing of energy in combination with regulatory reforms and market-based incentives could transform the economy and provide a new investment base for the 21st century.

On one point, virtually all who have studied the issue agree: Market forces alone will not solve the global warming problem, but without them, nothing else will. The findings of this study confirm that "business as usual" is bound to increase greenhouse gas emissions. Yet the institution of market-based pricing and trading schemes in place of traditional command and control laws may turn an environmental crisis into an economic opportunity.

A Climate of Uncertainty

While forecasts of the economic costs and benefits of ameliorating global warming remain highly speculative, so, too, are the climate change projections that gird the debate. The consensus among most of the scientific community is that a doubling of atmospheric carbon dioxide is likely to raise the mean global temperature of the Earth 3 to 8 degrees Fahrenheit by 2050. To put this forecast in some perspective, the planet was about 10 degrees cooler during the last Great Ice Age and about 10 degrees warmer during the Age of the Dinsosaurs. Accordingly, the warming could bring about dramatic changes in climate. But a prudent investor must be careful not to invest too much in pat assumptions about the greenhouse effect. The climate may have many surprises in store. Indeed, it has surprised climate forecasters already by not warming nearly as fast as their general circulation models have suggested it would.

Two years ago, a distinguished international panel of scientists, assembled under the auspices of the United Nations Environment Program and the World Meteorological Organization, issued a landmark report about what was "certain," "likely" and "to be determined" about climate change. The Intergovernmental Panel on Climate Change concluded it was "certain" that chlorofluorocarbons—the man-made chemical that is depleting the Earth's protective ozone layer—also plays an important role in raising global warming potential. Yet in a new assessment released only four months before the Earth Summit, the climate change panel disclosed that a counteractive effect of CFCs had been overlooked. During the 1980s, CFCs apparently did not contribute to global warming at all, since the loss of ozone in the upper atmosphere is having a chilling effect on the planet. So much for scientific "certainty."

Another startling revelation concerns the combustion of coal—the most carbon-rich of all fuels. Scientists have discovered that the scattering of sulfur aerosols from older coal-burning power plants creates a "parasol effect" in the atmosphere, much like a volcano that casts an umbrella of ash following a volcanic eruption. This process also has a cooling effect on parts of the globe that scientists did not fully appreciate until recently. With a degree of humility, the IPCC concluded in its 1992 update that "the unequivocal detection of the enhanced greenhouse effect from observations is not likely for a decade or more." Therefore, a climate of uncertainty will pervade policy decisions regarding the greenhouse effect for the remainder of this century and perhaps into the next one.

Nevertheless, general circulation models of the Earth's climate—the most sophisticated computer models ever created—continue to say that a warming *is* in the offing. The recent findings concerning CFCs and sulfur aerosols help to explain, in fact, why the climate to date has warmed only half as much (about 1 degree F in the last 100 years) as the models said it would. Still, the models have a tenuous grasp at best on how the real world behaves. In making their calculations, they reduce the planet to an abstract patchwork quilt, offering proxies for important climatic variables like clouds, while giving crude approximations of others like oceans. When climate modelers run their incredibly complex (yet highly simplified) computer programs, they cannot account for all of the surprises that a changing climate may have in store. Indeed, weather, as a chaotic system, is inherently unpredictable. Perhaps the only thing one can really expect is more of the unexpected.

Despite many remaining uncertainties, enough has been learned about humanity's effects on the climate that economic decisions that alter the production of greenhouse gases can no longer be regarded as unwitting acts. Rather, they must be viewed as conscious choices that could have far-reaching consequences for the globe. Whether governments and industries decide to act on premonitions about climate change or carry on with business as usual is almost beside the point. They have become active participants in the climate system, whether they choose to acknowledge their role or not.

This book examines four industries with the most at stake in the greenhouse debate: agriculture, forest products, automobiles and electric power. All of these industries essentially face two choices: Act now to blunt the possible momentum of climate change, or wait and see if the basic forecast is correct, accommodating any change as it occurs. These choices involve a trade-off between further information-gathering to ensure a proper course of action and implementing a strategy quickly to maximize its intended effect. Such a trade-off is the essence of risk, the stuff of investing. For the purposes of this book, it defines the "greenhouse gambit."

Agriculture

If global warming indeed emerges as a problem to be addressed, farmers and agribusinesses may find themselves in a Catch-22. Their success in boosting crop yields over the years has been largely the result of increased reliance on mechanized equipment and chemical inputs, which in turn has contributed a growing amount of greenhouse gases to the atmosphere. At the same time, farmers' transition away from diversified crop-livestock agriculture toward production-driven, single-crop cultivation has made them more dependent on cooperation from the weather—something that global climate change puts in doubt.

Pinpointing regional precipitation changes is the greatest unknown in farmers' greenhouse gambit. General circulation models offer a range of conflicting possibilities; accordingly, confidence in their projections is low. Largely because of these uncertainties, few signs point to agricultural interests factoring climate change into their strategic plans—even though the stakes in charting a proper future course for agriculture are higher than ever.

Global climate change would be likely to alter the geography of agricultural development around the world, creating new winners and losers in the production and distribution of food. The United States—as the world's largest agricultural producer and exporter—is the nation most vulnerable to a reduction in agricultural capacity that would compromise its geopolitical position. The agricultural capacity of high-latitude regions such as Scandinavia, northern Europe, the Ukraine, Russia, Canada and Japan could benefit from longer crop-growing seasons even as the productivity of American farmland—especially in the Great Plains and possibly the Southeast—declines. In the arid West, where growing conditions already are marginal and may get drier still, a day of reckoning may come for growers of especially water-intensive crops.

The prospect of climate change is not all bad for farmers, however. On the contrary, the greenhouse effect could provide the best of all worlds for many plants, concerns about heat-related stress notwithstanding. There may be fewer encounters with frost because of warmer nighttime temperatures, faster accumulation of biomass because of CO_2 enrichment and reduced vulnerability to drought because of greater water use efficiency. Farmers' adaptations and the advent of new agricultural technologies also must be taken into account. Continued introduction of drought-tolerant crops and advances in genetic engineering may improve the outlook for farmers even if the climate takes a decided turn for the worse. Genetic engineering will not be a panacea, however. Ultimately, the growth of plants always will depend on the availability of soil moisture, nutrients and suitable climate—things that genetic engineering cannot provide.

The fundamental challenge in agriculture is to develop means of food production that sustain the land as well as the population base over the years to come. The specter of global warming compounds this challenge. At the same time, it compels the consideration of policies to reduce agricultural greenhouse gas emissions while making the land more resilient to potential climate-related stresses. Low-input, sustainable agriculture offers such a coupling of benefits. It creates an agricultural land base where soils retain more water, carbon and nutrients through natural processes, and it reduces greenhouse gas emissions through scaled-back use of synthetic fertilizers, pesticides, heavy machinery and irrigation equipment. Yet it provides no more of a magic solution to farmers' problems than advances in genetic engineering. Decisions about what crops to plant and how to raise them will continue to have elements of risk, much as they always have.

Another critical, unanswered question is whether alternative agricultural methods—or even more intensive ones—will be able to sustain ongoing rapid population growth in an era of possibly unprecedented warming. Perhaps the toughest ethical issue is how global warming might affect the trade and distribution of food around the world. For those with grain available to sell, an ethical (as well as economic) question arises: Should the surplus grain be sold in the world market—driving up domestic prices—or should it be reserved for the domestic market—possibly at the expense of tens of millions of additional people going hungry abroad?

Forest Products

Perhaps more than any other industry, the forest products industry has the opportunity to profit from climate change—or lose because of it. In a direct sense, carbon dioxide enrichment should accelerate the growth of forests and bolster trees' water-use efficiency. Forests also might become more productive if the climate gets warmer and wetter. Yet some climate models raise a disturbing possibility that portions of the country could dry out and become less hospitable for forests. One such region is the American South, where the U.S. forest products industry has invested half of its assets and planted 80 percent of its new tree seedlings.

Global warming concerns have not caused forest products companies to amend their basic strategies to date. Seven of the 10 major forest products companies surveyed by the Investor Responsibility Research Center in 1990 agreed with the statement, "Greater scientific certainty is required before our company includes global warming in its strategic planning." The crystal balls of climate research apparently remain too clouded for these companies to act on premonitions about global change.

The climate models do suggest, however, that ambient temperatures could rise as much over the next 100 years as they did in six millennia—from 12,000 to 6,000 years ago. North Carolina's climate could reside in, say, New England after the year 2050. Such a climate shift would exceed forest communities' demonstrated rate of migration by 10, 30, even 50 times. Commercial species that grew well on company-owned plantations in the last rotation might not fare as well the next time around. Barring the successful introduction of new species—which remains a distinct possibility—tree-farming companies might become saddled with huge tracts of timberland no longer suited for commercial operations.

To make up for the potential shortfall of timber, the greenhouse effect may yet emerge as the industry's greatest ally for two reasons. First, carbon dioxide enrichment and climate change may stimulate the growth of trees in new locations and allow them to mature more quickly. Second, trees' ability to sequester carbon from the atmosphere promotes the concept that planting trees is good for the planet as well as for commerce—and that tree-planting efforts should be encouraged at every turn.

Moreover, wood is a renewable resource that distinguishes itself from most other materials with which it competes. Processes for manufacturing wood into finished products tend to be much less energy-intensive than those of alternatives. As a result, policies to address global warming could lead to an overall expansion of wood-products markets. It is even conceivable that other industries in need of carbon offsets might pay forest products companies to expand their tree-growing base.

A few energy companies already have volunteered to pay others to grow trees as a means of offsetting power plant emissions. Forest products companies could share in this bounty. Such a program would sequester more carbon in the short term, buy time for energy companies to switch to noncarbon alternatives over the medium term and help alleviate a possible timber supply shortage over the long term. Once these carbon-storing trees have served their initial purpose, they could be harvested and converted to wood products that continue to store the carbon for decades longer. Depending on the success of such reforestation efforts, the forest products industry could even make a push into biomass energy production, which would have a far greater effect in offsetting carbon emissions than the tree-planting efforts themselves.

Another way in which forest products companies could make an important contribution is by carrying out more environmentally sound timber-harvesting practices and limiting the removal of old-growth stands. While 90 percent of the Amazon rainforest remains intact, only 5 percent of American virgin forest groves are still standing. President Bush expressed his concern about the plight of the rainforests recently by pledging $150 million in additional assistance to help developing nations preserve their forests. Yet if American forest products companies were to favor the preservation of old-growth stands and encourage the expansion of agroforestry on erodible cropland in the United States, they would set an example for those who possess the real terrestrial lungs and gene pool of the planet—the tropical-forested nations—and make a contribution that is far greater than the president's monetary contribution and the industry's own comparatively limited ability to sequester carbon.

As long as forest products companies stake their reputation as "tree *growing*" companies—planting and nurturing new seedlings to replenish the harvest of mature stock—they will be play an important role in keeping global warming in check. Whether they are able to turn the greenhouse effect more to their advantage remains an open question. With thousands of people at work in the forests and hundreds more carrying on research in laboratories, the forest products industry has an effective early warning system in place to detect any adverse signs of climate change. The risk in waiting for the alarm bell to sound, however, is that it may become too late for them to do much of anything constructive to protect their long-term investment in the forests. The greater risk for this industry, therefore, may be to pass on the greenhouse gambit rather than to put it into play.

Automobiles

Among the industries considered in this book, the automobile industry is unique in that it is a major contributor of greenhouse gases yet relatively immune to the potential effects of climate change. Every gallon of gasoline burned in a car's engine releases nearly 20 pounds of carbon dioxide to the atmosphere. Altogether, about a billion tons of carbon dioxide are vented from U.S. cars and light-duty trucks each year; heavy trucks emit another 200 million tons of CO_2 annually.

Although the number of motor vehicles in the United States has nearly reached parity with the population, around the world only one car is available for every 12 people. The burgeoning market for motor vehicles suggests that the auto industry is a part of the potential global warming problem and must be a part of any eventual solution. Already, the industry's contribution of carbon dioxide into the atmosphere represents 14 percent of all such emissions from fossil fuels worldwide. Unless market demand for motor vehicles experiences a dramatic downturn, the auto industry is on course to becoming one of the very largest collective CO_2 emission sources by shortly after the turn of the century— rivaling emissions from fossil-fueled electric-generating plants.

To reverse the emissions trend within the auto industry, the onus will fall

Introduction and Summary

mainly on 15 large companies that command more than 85 percent of the world's motor vehicles market. Cars and trucks throughout the world are based on designs originated by these companies. In the future, these same companies will be counted on to offer alternative forms of transportation that do not compound the threat of global warming. This constitutes their greenhouse gambit.

Essentially, the auto makers have two options. In the near term, they can design engines and car bodies that result in far more efficient use of carbon-based fuels. A bit farther down the road, they can build cars and trucks to run on nonfossil-carbon energy sources, including electricity and hydrogen derived from nuclear and/or renewable sources. That way, no CO_2 would be released to the atmosphere regardless of the amount of fuel consumed.

The most fuel-thrifty cars now manufactured produce only one-third of a pound of CO_2 per vehicle mile traveled compared with today's average of one full pound per mile. The industry's greatest near-term potential to curb CO_2 emissions, therefore, is to shift the mix of autos toward the more fuel-economic models. But consumers have little interest in fuel economy when a gallon of gasoline sells for less than the price of bottled water. Moreover, U.S. auto makers have basically conceded their role to Japan in the small-car market. All of the 10 most economical cars now sold in America have engines designed, engineered and built by Japanese companies (although some are sold under American nameplates). Meanwhile, Japanese car makers are moving into the mainstays of the American market, proving that they can build cars every bit as big, powerful and fuel-consuming as their American counterparts.

Power is back in style. Car companies are building fewer models with four-cylinder engines and more with powerful V-6s and V-8s. In addition, they are adding more accessories and weight to their vehicles and emphasizing acceleration over fuel economy at practically every turn. The Big Three automotive fleet rolling off the assembly line in 1992 is rated at 27.5 miles per gallon—the most prodigal showing since 1984.

The fuel economy of new cars sold in the United States actually peaked in 1988 at 28.8 mpg. The fuel economy of the new car fleet was down 4 percent by the 1990 model-year, with the average car gaining more than 6 percent in weight and 10 percent in horsepower. The trend has continued since then, with even larger percentage fuel-economy declines among the Japanese auto makers. Yet the Japanese have proven themselves to be market innovators in such fuel-saving technologies as multivalve and lean-burn engines. The Big Three auto makers are several years behind the Japanese in putting these technologies to commercial use.

Auto manufacturers have to look well into the future to make their strategic planning and marketing decisions. Domestic manufacturers redesign their entire product line only every eight years or so, and many technologies require five years of lead time. Consequently, a new technology slated for development today would not be expected to reach the market until 1997, or to reach a high level of market penetration until after the year 2000. Moreover, high-sales-volume models require $1 billion in capital spending before the first car is rolled out of production.

If Congress were to raise fuel economy standards dramatically—as environmental and efficiency advocates have recommended—the production plans conceived by the auto makers in the wake of falling oil prices would have to be completely overhauled. A 40 percent improvement in the rated fuel economy of new cars by 2001 would mean, in effect, that the Big Three auto makers would have only a few years in which to institute design changes that would raise their fleet averages from 27.5 mpg to 38.5 mpg.

In the current recessionary environment, new government marching orders to address the greenhouse effect are the last thing the auto makers want. Nevertheless, a 40 percent gain in the rated fuel economy of new cars would take the nation a long way toward achieving a cap on its CO_2 emissions at 1990 levels—the target sought in the climate treaty signed at Rio. By some estimates, the fuel savings would represent nearly a 400-million-ton annual reduction in the nation's carbon dioxide emissions, equal to 7.7 percent of the present-day total. In addition, the amount of yearly fuel savings would nearly match the amount of oil that the United States imports from the Persian Gulf each year, or save as much oil in five years as is thought to reside in all of Alaska's Arctic National Wildlife Refuge.

The CO_2 benefit may be fleeting, however, because of the growing number of cars and trucks on the road, the greater number of miles they travel and the increasingly congested conditions in which they operate. One analysis concludes that even a 39-mpg standard achieved by the year 2000 would hold the nation's fuel consumption and transportation-related CO_2 emissions constant for only a decade, before they start to rise again. As a result, auto makers will have to turn to nonfossil energy sources eventually to achieve any permanent reduction in the sector's CO_2 emissions.

With the passage of the 1990 Clean Air Act and tougher clean-air standards in California, the first steps toward the development of alternative transportation fuels are happening now. Unfortunately, some of these steps may be in the wrong direction as far as the greenhouse effect is concerned. The auto industry would prefer to see the development of new fuels that require the fewest changes to conventional gasoline and diesel engines (and associated fuel delivery systems). By the same token, the oil industry would prefer to manufacture petroleum-derived products that make the greatest use of existing reserves and refining capacity. On that basis, reformulated gasolines are ideal from an industry standpoint, because they should enable most vehicles now on the road to emit fewer noxious pollutants, while maintaining the preeminence of oil. But gallon for gallon, these new formulas consume more crude oil during refining and may impose a slight fuel economy penalty. As a result, "cleaner burning" gasolines actually will increase carbon dioxide emissions rather than decrease them.

Methanol is the auto industry's next bet to comply with provisions of the Clean Air Act. In 1992, each of the Big Three auto makers introduced methanol-fueled vehicles in the California market. But methanol derived from natural gas reduces CO_2 emissions only a little, while increasing America's dependence on foreign supplies a lot. And methanol derived from coal would the worst possible choice in terms of global warming potential; it emits far more CO_2 per gallon than ordinary gasoline.

Ethanol made from corn increases CO_2 emissions as well. Moreover, global climate change may diminish the capacity to grow corn at a time when nutritional needs are rising. On the other hand, methanol or ethanol made from woody biomass could serve as environmentally sound transportation fuels in a greenhouse world. But there is no infrastructure or cost-effective means of producing such biomass-derived motor fuels at present.

Among the alternative transportation fuels, compressed natural gas (especially in fleet vehicles) appears to be the most promising near-term option from a greenhouse perspective. CNG vehicles could reduce tailpipe emissions of CO_2 by perhaps 15 to 30 percent relative to gasoline-powered vehicles (and reduce conventional auto pollutants as well). In the far-distant future, the installation of on-board hydrogen fuel cells or solar-powered photovoltaic systems may offer the greatest potential to reduce the automotive sector's greenhouse gas emissions overall.

Transportation policymakers may decide eventually that it is time to switch directions and emphasize the development of alternative forms of transportation—such as passenger buses, trolley cars and "bullet" trains. Considering the deteriorating infrastructure of roads and bridges—and the increasingly congested traffic conditions that motorists face—pursuing this option may make the most sense from a societal standpoint. Yet most observers put the least stock in this option. Since neither the government nor industry has developed a comprehensive transportation strategy to address the greenhouse effect, drivers will be left to their own devices. The family car is still likely to run on gasoline for a long time to come. But in a nation of two- and three-car families, the second car in the driveway may soon be powered by an alternative fuel. In any event, the automobile and its accoutrements have become such an integral part of modern-day living—and of the landscape itself—that it will remain with us for decades to come.

Electric Power

Energy policy throughout the Industrial Age has pursued a simple goal: to acquire fuel supplies at the lowest possible cost. Those nations able to acquire and consume the most fuel generally have had the fastest-growing economies, attained the highest standards of living and secured the brightest prospects for the future. The greenhouse effect may turn this conventional energy formula on its head: The faster fossil fuels are burned, the more the economy is taxed, the farther the standard of living declines and the dimmer the outlook for the future becomes.

We are at the dawning of a new Electronic Age. By the turn of the century, electricity is expected to account for 40 percent of U.S. primary energy demand, and by 2025 it may fulfill half of the nation's energy needs. Should electric vehicles enter the picture, electricity could represent more than two-thirds of U.S. primary energy demand eventually. Since generation of electricity derives simply from the flow of electrons—and no greenhouse gases need be emitted—concerns about global warming should not eclipse the coming Electronic Age.

The fuel supply remains a vestige of the Industrial Age, however. Oil, coal and natural gas provide 38, 27 and 19 percent of the world's primary energy requirements, respectively. In the United States, electric utilities account for 36 percent of the nation's CO_2 emissions (and 7.5 percent of the world's), because of heavy reliance on coal. With the expectation of further electrification of the U.S. economy—and coal leading the way—American utilities may emit 2.4 billion tons of carbon dioxide by 2005 and about 3.5 billion tons by 2015. Carbon dioxide emissions in the electricity sector could *double* in a quarter-century, in other words, if the current forecast holds up.

Because of this large—and growing—contribution, proponents of action on global warming believe American utilities are a requisite choice for a carbon abatement strategy. If this industry does not lead on combatting the greenhouse effect, no other industries are likely to follow. Yet a call for holding CO_2 emissions constant in the face of these energy projections seems daunting, and a recommendation for a 20 percent cut in CO_2 emissions is all the more ambitious. Nevertheless, the Intergovernmental Panel on Climate Change calculates that a two-thirds reduction in CO_2 emissions would be necessary to keep the atmospheric content of carbon dioxide from rising further.

In the view of the Bush administration, the time has come to launch a major new round of power plant construction. The administration's National Energy Strategy calls for doubling the nation's nuclear and renewable generating capacity over the next 40 years. More ominously for global warming, it also seeks a two-thirds increase in the nation's coal-fired capacity, even though coal already accounts for 85 percent of the utility industry's CO_2 emissions.

Policies instituted to combat the greenhouse effect would almost certainly alter future generating options. Since the cleanest fuels are ones that are never burned, conservation and efficiency programs—or "demand-side management" (DSM) in the industry's jargon—have become an integral part of utilities' business. And they are likely to take precedence in the future, shaping the supply requirements of all other fuels. At the same time, developers of low-carbon and noncarbon energy sources will jockey to win the hearts, minds and pocketbooks of energy consumers who at present remain heavily dependent on fossil fuels. If coal is the fossil fuel with the most to lose in case programs are enacted to ameliorate global warming, natural gas is the fossil energy source that stands the most to gain. As a relatively clean-burning fuel with half the carbon content of coal, natural gas may serve as a bridge fuel until the world crosses over to nonfossil energy sources.

Coal is too dominant a fuel source to be written off completely, however. In an effort to combat acid rain, new coal plants will feature technologies that reduce emissions of sulfur and nitrogen oxides substantially. Yet even these "clean-coal" plants would not reduce CO_2 emissions much, if at all. Moreover, some of these plants would increase emissions of nitrous oxide, a much more potent greeenhouse gas. It is even possible that clean-coal plants may increase the propensity for global warming, considering recent evidence about the atmospheric cooling effect of sulfate aerosols from coal plants lacking scrubbers. Clean-coal plants—by virtually eliminating sulfur emissions—would take away this counteractive effect and yet continue to release large amounts of carbon dioxide and other greenhouse gases to the atmosphere. Accordingly, the net effect of replacing "dirty," old coal plants with "clean" new ones may be to exacerbate the greenhouse effect.

By contrast, a new generation of "passively safe" nuclear reactors offers a more attractive generating option, since atomic energy plants emit no greenhouse gases. In the United States, however, political and economic pressures have kept a de facto moratorium on new reactor orders for 14 years. In fact, all U.S. reactors ordered since 1974 subsequently have been canceled. Nuclear construction programs around the world are winding down as well, with the possible exception of programs in France and Japan.

Congress's recent streamlining of licensing rules has brightened prospects for a restoration of nuclear plant orders in the United States. With Wall Street still reluctant to bankroll new projects, consortia of equipment firms, construction companies and utilities may work together to build new nuclear power stations as means of spreading the financial risks. Toward that end, Congress also has approved regulatory changes that would allow utility and nonutility companies to gain access to utility transmission lines, become wholesale generators and operate power plants in more than one state without being subject to restrictive utility regulations.

If the nation decided to rely on nuclear power exclusively to meet unconstrained electricity demand and to control greenhouse gas emissions, a mammoth construction program would be required. One 600-megawatt reactor would have to be completed almost every week between 2000 and 2015 to supply the amount of growth forecast in the National Energy Strategy. The cost of such a Herculean program could easily top $1 trillion.

Beyond the financial constraints of nuclear power, other technological and socio-political problems remain. With demonstration of advanced reactor designs not expected until after the turn of the century, their performance and safety characteristics are uncertain. In addition, no permanent repositories for high-level waste have been located anywhere in the world (although the United States hopes to have such a site in Nevada by 2010). Moreover, global uranium supplies are expected to last only 100 years without fuel reprocessing. While breeder reactors could extract 100 times as much energy from the same amount of uranium as conventional light-water reactors—extending the fuel supply—nuclear fuel reprocessing also raises the troubling possibility of diversion of bomb-grade plutonium.

Alternatively, the nation could choose to accelerate the development of renewable energy sources that, by definition, are practically limitless in supply. Most renewable sources do not emit any greenhouse gases (biomass being the major exception). Yet they account for less than one-fifth of the world's primary energy demand at present—and an even smaller fraction of electricity demand.

The early 1990s has been a time of retrenchment and consolidation for renewable power producers. The world's largest photovoltaics company, Arco Solar, was sold to a German firm, Siemens AG, in 1990. The world's largest solar thermal company, Luz International, filed for bankruptcy protection in 1991. America's wind energy developers—the fastest-growing suppliers of renewable power during the 1980s—plan to erect no new major windfarms in 1992. And the nation's geothermal industry, which added more megawatts of generating capacity during the 1980s than any of the other renewable industries, has brought on-line fewer than 100 megawatts since 1990. Meanwhile, the Bush administration has ranked renewable energy at the bottom of its list of federal energy funding priorities.

Despite adverse political and economic trends and lack of interest among utilities, renewable power developers still can point to steady advances in their technologies in recent years. If progress continues—and concerns about global warming mount—utilities may take another look at renewables and decide the time has come to diversify their coal-dominated fuel mix. Volume orders from utilities could enable manufacturers of renewable power systems to achieve better economies of scale and lower the cost of financing installations.

Still, to provide 400,000 megawatts of new capacity requirements by 2015, as forecast in the National Energy Strategy, 15 times as many renewable energy installations would be required each year than has occurred in recent years. Lead times for most renewable generating facilities are quite short, rarely exceeding a year or two. But the independent power industry would have to mature very fast or utilities would have to take a much greater interest in renewables for these sources to fulfill the expected generating requirements of the early 21st century.

While past energy policies put an emphasis on securing plentiful, low-cost fuel supplies, future strategies must stress the value of using those stocks wisely. The future clearly belongs to the efficient. Decoupling energy demand from economic growth complicates the task of setting new energy priorities, however. The difference between 1 percent growth and 2 percent growth in annual energy demand—compounded over 50 years—amounts to more energy than is presently consumed. Therefore, not only are future sources of energy now in question, so is the amount of power ultimately required.

The National Energy Strategy's demand forecast may turn out to be much higher than is necessary to meet the nation's future generating requirements. A burgeoning number of utilities have begun to turn to "demand-side management" to satisfy their customers' energy-related needs. The greater the role these demand-side programs play in utilities' business plans, the more the industry can say that electrification of the economy is helping to alleviate the greenhouse effect rather than exacerbating it.

In essence, DSM has become a way for utilities to increase their overall share of the energy market by emphasizing the efficiency advantages that electricity has to offer. While the net result is that electricity sales increase relative to the amount of generation that would have occurred without customers switching away from other carbon-rich fuels, DSM programs at least hold the increase in power generation to more manageable levels. More important from a greenhouse standpoint, the combination of industrial fuel switching and utility DSM programs is able to shrink primary energy demand—and, hence, carbon dioxide emissions—relative to baseline projections, making the strategy suitable as a greenhouse gambit.

In 1991, three electric utilities announced that they would take steps to reduce greenhouse gas emissions in an effort to appease the global warming threat. As such, they became the first—and so far the only—major American companies to make such publicly stated commitments. Each of the companies expects to offset 20 percent of its CO_2 emissions by 2010 without raising electricity prices appreciably or lowering the quality of service. Conservation and efficiency programs, increased reliance on natural gas and expedited use of renewable energy sources figure prominently in each of these plans.

In the final analysis, no one energy source seems capable of supplying all future generating requirements. Nor does it seem likely that efficiency improvements will obviate the need for new power plants. As long as there is a growing economy in an Electronic Age, demand for electricity will persist. The question is whether future generating requirements will be close to today's level—or much larger. If conservation and efficiency measures hold future electricity demand growth to a minimum, the industry can afford to be choosy. If consumer and business demand grows rapidly, however, then the industry will be inclined to tap energy sources wherever it can develop them.

The 'Greenhouse Gambit'

As America's military actions in the Persian Gulf demonstrate, the U.S. government finds the money and the muster to act when it perceives a threat to one of the nation's vital interests. If global warming were regarded in the same way, there is little doubt that the government would resolve to act once again. The real question is whether energy policy will be transformed in the absence of a clear and present climatic danger.

The "greenhouse gambit," as defined here, represents a strategic decision to devitalize the importance of fossil fuels before global warming leaves decisionmakers with no other alternative. The greenhouse gambit is further designed as a "no-regrets" policy, since benefits would accrue even if the climate does not change. It would provide a rationale for the government to reduce or eliminate taxpayer subsidies for fossil energy development, for example, saving billions of dollars a year; it would also give the government more discretion in the use of military force, saving lives as well as money.

Oil and coal interests would be the apparent losers in such a gambit. Ironically, however, the long-term interests of these fossil energy producers could be served if a deliberate approach is taken to the development of alternative fuels and energy-saving technologies. The combination of more availability of noncarbon sources and slower energy demand growth would make it easier to stay within CO_2 emissions limits without requiring wholesale shifts away from oil and coal to natural gas. In effect, oil and coal producers would secure a place for themselves by acknowledging a larger role for competing sources, even though their first choice would be a future in which global warming presented no encumbrance.

The greenhouse gambit offers other potential benefits. Since the U.S. economy has more energy efficiency gains still to reap than the economies of most of its trading partners, it should be able to spend fewer dollars per ton of greenhouse gas abatement and realize a comparative advantage. Rather than being hindered, then, the U.S. economy should become more competitive in the world marketplace. Moreover, the greenhouse gambit would create additional jobs in the energy sector even as it shrinks the required size of energy-related investments. The trade associations representing the nation's natural gas, solar energy and energy services industries estimate that 215,000 new jobs would be created if market forces were used to promote the growth of these industries. At the same time, only 44,000 jobs would be lost in the coal industry.

A narrow focus on the energy-related impacts of CO_2 abatement also overlooks the interaction of capital markets. America's energy industries are extremely capital-intensive. As reduced growth in energy demand reduces the size of investments in new energy facilities, more capital would be available to invest in new manufacturing plants, machinery and the nation's infrastructure, creating more jobs and further bolstering U.S. competitiveness. As an added bonus, the nation could reduce its $450 billion annual energy bill through investments in energy efficiency, putting more spending money in consumers' pockets.

Employed to its fullest logical extent, the greenhouse gambit also would make use of carbon taxes to incorporate the environmental costs of energy consumption. Several studies show that carbon taxes could provide $80 billion to $140 billion in annual revenues. While fossil energy producers may regard such a tax as a loss of income, other sectors of the economy could view it as a transfer of income. Indeed, the macroeconomic effect of such a tax hinges much more on how the tax proceeds are allocated than on the size of the tax itself.

The most politically palatable option may be to offer a combination of investment tax credits for energy-intensive industries and personal income tax reductions with the revenues generated by a carbon tax. The investment tax credit would ease the financial burden on carbon-intensive industries and could be targeted for investments in energy efficiency and diversification into noncarbon energy sources. In effect, such a fiscal program would bring about an ecological redistribution of the tax burden, from those who emit more CO_2 in the atmosphere as they go about their business to those who emit less. Several European nations are trying this approach already. Recent studies sponsored by the Environmental Protection Agency suggest that such a coordinated approach with carbon taxes may make it possible to achieve a substantial reduction in the nation's CO_2 emissions at no net cost to the economy. Even without the institution of carbon taxes, however, a number of authoritative studies suggest greenhouse gas emissions could be reduced at a low net cost—or possibly even at a net benefit— to the economy. These studies buck the conventional wisdom that a carbon abatement program necessarily would bring harm to the economy.

Eventually, a completely market-based trading approach may be used to reduce greenhouse gas emissions. Forest interests could be paid to plant trees in Brazil to offset the cost of utility CO_2 emissions in America. Natural gas developers in Eastern Europe could be paid to plug methane leaks in their pipelines. Coal companies in West Virginia could be paid to trap coal-seam methane emissions. Even junkyard owners could be paid to capture methane and CFCs that otherwise would escape from garbage, old refrigerators and abandoned automobiles. Through market trading, a least-cost means of reducing greenhouse gas emissions would be realized.

The first chapter of this book describes how "chaos" may reign in the global climate as climate patterns shift from one equilibrium state to another. A similar thoery could apply to the global economy. The global economy today is in a state of flux as the Industrial Age gives way to the new Electronic Age. During this transitional phase, technology, energy and investment are working largely at cross-purposes, amplifying economic disruptions as well as compounding the threat to the globe. Absent a strategic plan to bring energy and environmental priorities into balance, chaos lurks in both the climate *and* the economy.

But the world has an alternative: to develop strategies for the sustainable use of resources and more efficient use of fuels. The sooner this is accomplished, the smoother the transition to the new era will be, the more the quality of life will be enhanced and the greater the likelihood that global warming will be held in check. This is a new economic formula for the Electronic Age. It is the essence of the greenhouse gambit.

Chapter 1
A Climate of Uncertainty

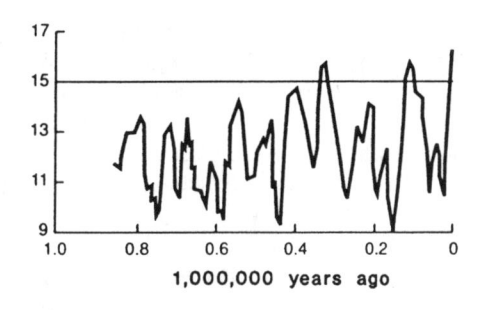

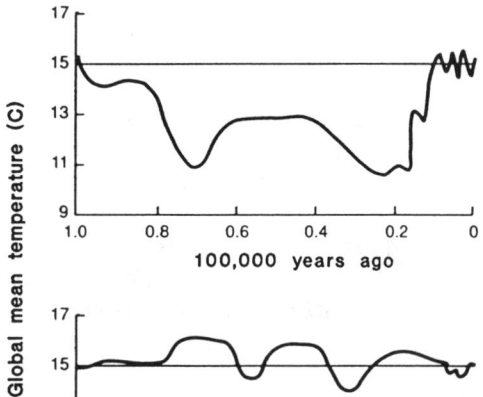

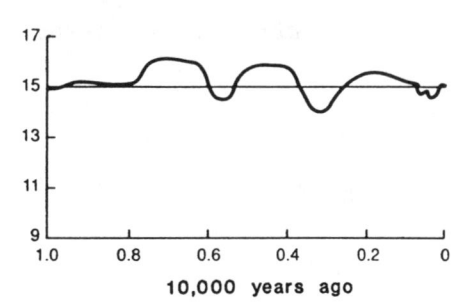

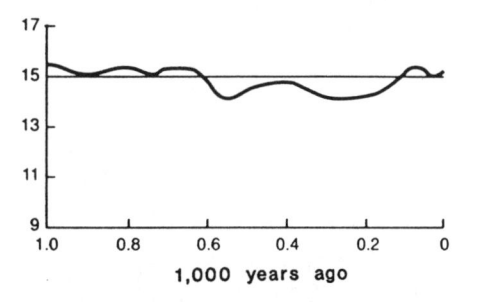

Contents of Chapter 1

Introduction 3

Has Global Warming Begun?
    The Greenhouse Effect: A Century-old Theory 5
    The Coming Ice Age 12
    'The Greenhouse Effect Is Here' 16
    The Drought of '88—and Beyond 19
    Search for a Smoking Gun 22

Forecasting the Future
    The Primitive Art of Weather Prediction 23
    Lost in Space 25
    Chaos 27
    General Circulation Models 29
    Clouds in the Forecast 32
    Motion in the Oceans 35
    Transient Behavior 37

Expecting the Unexpected
    Climatic Surprises 39
    Recent Weather Anomalies 42

Making Order Out of Chaos
    Climate's 'Strange Attractor' 51
    An Ensemble Forecast 54
    Rolling the Dice 58

Notes 63

Introduction

Throughout recorded history, no matter of science has commanded more of mankind's attention than weather. The earliest book of the Bible, the book of Job, credited God for weather in its ever-changing forms. "Great things doeth God, which we cannot comprehend," a humbled Job wrote. "For he saith to the snow, fall thou on the Earth; likewise to the shower of rain, and to the showers of His mighty rain. Out of the chamber of the south cometh the storm; and cold out of the north. By the breath of God, ice is given and the breadth of the waters is congealed." To this day the most famous of all weather forecasts has assumed such biblical proportions: 40 days and 40 nights of rain that would flood even the highest mountains—and prompt Noah to build an Ark.

In modern times, humanity has embraced a more secular, scientific approach to weather prediction. Orbiting satellites transmit pictures of the Earth from space. Meteorologists dutifully interpret these swirling images, and reams of ground-based data, for an inquisitive television audience. And yet for all this high-tech wizardry, weather's secrets have not been revealed completely. Weathermen still profess doubts about their forecasts, conceding a 20 percent chance of snow here, an 80 percent probability of rain there, with skies becoming partly sunny—or is that mostly cloudy? Viewers still must make up their own minds about how to prepare when venturing out into the elements.

Now another ominous weather prediction has entered the long-range forecast. A warming trend could raise the planet's temperature over the next 50 years as much as it has climbed in the history of human civilization, dating back more than a hundred centuries. Storms of unprecedented ferocity might wreak havoc on some parts of the globe, while others could be parched by interminable drought. And once again the seas may rise—not as they did during Noah's Flood—but enough to inundate coastal population centers and cover surrounding farmland.

Once such a prediction would have been attributed to an angry God. But this epic forecast arises from humanity itself. Propagation of the species and rapid industrialization of the globe are changing the composition of gases in the atmosphere and may be giving weather a human face. This startling possibility has touched off an international debate about what society can and should do to try to change the climatic forecast. This book examines the options of four industries with the most at stake in this debate: agriculture, forest products, automobiles and electric power.

All of these industries—and the people they serve—essentially face two choices: Act now to blunt the possible momentum of climate change, or wait and see if the basic forecast is correct, accommodating any change as it occurs. These choices involve an axiomatic trade-off between further information-gathering to ensure a proper course of action—"look before you leap"—and implementing a strategy quickly to maximize its intended effect—"he who hesitates is lost." Such a trade-off is the essence of risk, the stuff of investing.

For the purposes of this book, the latter strategy—acting in the face of uncertainty—defines the "greenhouse gambit."

Given the perplexities of weather forecasting, it is hard not to be skeptical of long-range climate predictions. Weather is an inherently chaotic system that is full of surprises. It defies accurate prediction. And so it is that any forecast concerning global warming will be shrouded in a climate of uncertainty.

Even if the whims of weather remain unfathomable, however, it may be possible to discern long-term climatic trends with help from some of the most sophisticated mathematical formulas and the most powerful computers ever created. Such "general circulation models" alter the boundary conditions of the climate system, rather than attempting to predict—futilely—specific weather-related events into the far-distant future. By raising the forecasted amount of greenhouse gases, the models offer a glimpse of how existing weather regimes may adjust over time, until descriptions of "average weather," or climate, are changed the world over.

At the moment, general circulation models have only a vague sense of the convergent forces at work in nature—with little resolution at the regional level. Therefore, predictions about future climate are no more iron-clad (and probably even less so) than the five-day weather forecasts of television weathermen. In a decade, maybe longer, the models may be able to clarify how global warming would affect particular regions of the globe. But for now, humanity is left like Job to ponder things "which we cannot comprehend." Are existing climate anomalies—droughts, heat waves, powerful storms and other freakish weather occurrences—early manifestations of a long-term warming trend, or are they natural perturbations of an always-erratic climate system?

The answers may be less important for the time being than the fact that the questions are being raised. They posit the idea that humanity—and its attendant industries—may assume some responsibility for shaping the future evolution of the climate. Even if society's role is not yet completely understood, weather's manifestations are no longer to be considered solely "great things that doeth God." From now on, economic decisions that alter the production of greenhouse gases must be regarded as conscious ploys—not unwitting acts—that could influence the climate's temperament. Whether governments and industries decide to act on premonitions about climate change or carry on with business as usual is almost besides the point. They have become active participants in the climate system, whether they choose to acknowledge their role or not.

The four industry chapters of this book lay out options to defend against the global warming threat (if indeed the climate turns adversarial) as well as ways for industries to go on the offensive and beat back the threat. Policy responses, ultimately, must be coordinated at the international level. The conclusions describe the greenhouse gambit in this context. This introductory chapter sets the stage for the rest, offering a historical, scientific and conceptual framework for thinking about the Earth's climate, the greenhouse effect and the possibly turbulent forecast in the making.

Given the prevailing climate of uncertainty, the best advice may be the simplest: Prepare for change, expect the unexpected, and stay tuned for further developments.

Has Global Warming Begun?

Viewed from the ground, the sky appears as a vast expanse, with virtually limitless capacity to store the gaseous byproducts of human and industrial activity. Yet seen from space, the atmosphere is revealed as something completely different: a thin membrane that shrouds the planet's immense surface. If one could get far enough away from Earth so that its image was that of a big blue marble, say, 80 inches around, the atmosphere would appear only an eighth-of-an-inch thick—barely visible to the naked eye. Such an outer space perspective illustrates how emissions from the Earth's surface can accumulate readily in the atmosphere and alter its gaseous composition.

The Greenhouse Effect: A Century-old Theory

Long before there were spaceships, however, clever scientists already had begun to unravel the mysteries of the sky.[1] Antoine Lavoisier, a French chemist, determined two centuries ago that the Earth's atmosphere consists of 78 percent nitrogen, 21 percent oxygen and only a "trace" amount of carbon dioxide—0.027 percent at the time he was alive. (Argon and other "trace gases" account for the remaining 0.9 percent.) A contemporary of Lavoisier's, Jean-Baptiste Joseph Fourier, postulated that some of these gases warm the atmosphere by trapping energy from the Sun in the form of infrared heat. Fourier even used the metaphor of glass in a "hothouse" to describe this warming effect: The gases allow shortwave visible sunlight to strike the Earth's surface, where it is converted to longwave infrared heat, but do not allow all of the radiated heat to escape back into space. Finally, in the mid-19th century, Englishman John Tyndall measured the actual heat-absorbing properties of carbon dioxide and water vapor and concluded that the Earth's surface temperature could be altered by changing the atmospheric concentration of these gases.

Thus, the groundwork for the greenhouse effect theory was laid well over 100 years ago. As implausible as it may seem, the basis for global warming stems from a likely increase in the atmospheric content of carbon dioxide from 27-thousandths-of-a-percent to 54-thousandths-of-a-percent or even 75-thousandths-of-a-percent of all gases present in the atmosphere. (The water vapor content of the atmosphere, about two-tenths-of-a-percent, would rise as well.) These increases would be very small in absolute terms, yet perhaps large enough to render the climate to behave in ways very different from the way it acts now.

Despite more than a century of research, the public's interest in global warming was not piqued until 1988. That summer, a National Aeronautics and Space Administration scientist announced—in the midst of a torrid Washing-

Figure 1

The Greenhouse Effect

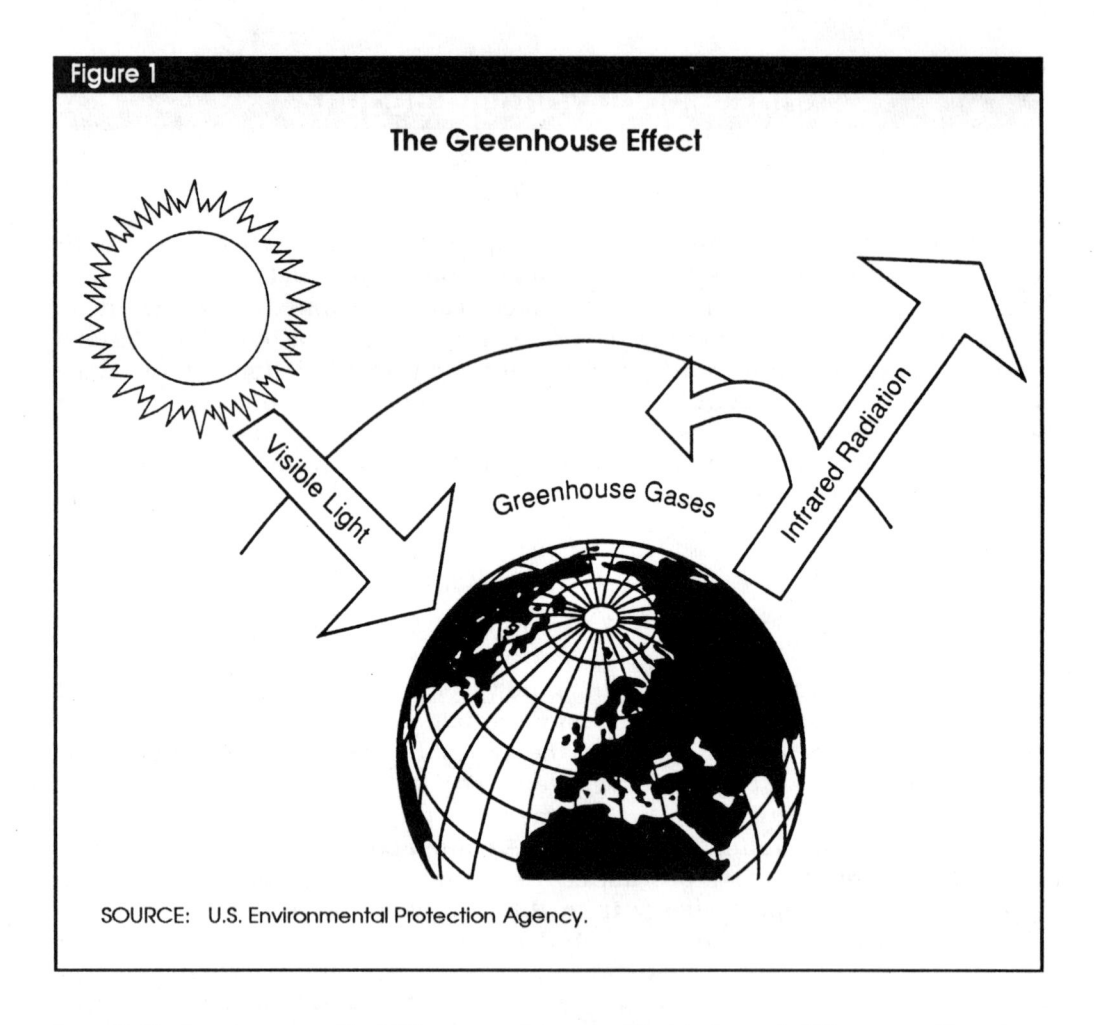

SOURCE: U.S. Environmental Protection Agency.

ton, D.C., heat wave—that "the greenhouse effect is here." Of course, Fourier, Tyndall and other atmospheric scientists knew long ago that the greenhouse effect did not happen onto the scene just like that. The presence of water vapor and carbon dioxide in the atmosphere always has kept the Earth nearly 60 degrees Fahrenheit warmer than it would otherwise be. A pre-existing greenhouse effect, in fact, makes the difference between our inhabitable planet and a frozen planet like Mars, which lacks the greenhouse gases necessary to hold radiated heat in its atmosphere. For that matter, it also makes the difference between the Earth and Venus, which has an atmosphere consisting of 97 percent carbon dioxide and a surface temperature of nearly 900 degrees F.

The real import of the announcement made four years ago was that it seemed to confirm a century-old prediction about an *enhanced* greenhouse effect. That prophecy was made by yet another atmospheric scientist named Svante Arrhenius, a Swede who was one of the first scientists to win a Nobel prize for chemistry. In the 1890s, Arrhenius recognized the fact that society was burning coal at a rapidly escalating rate—"evaporating our coal mines into the air."[2] He

Figure 2

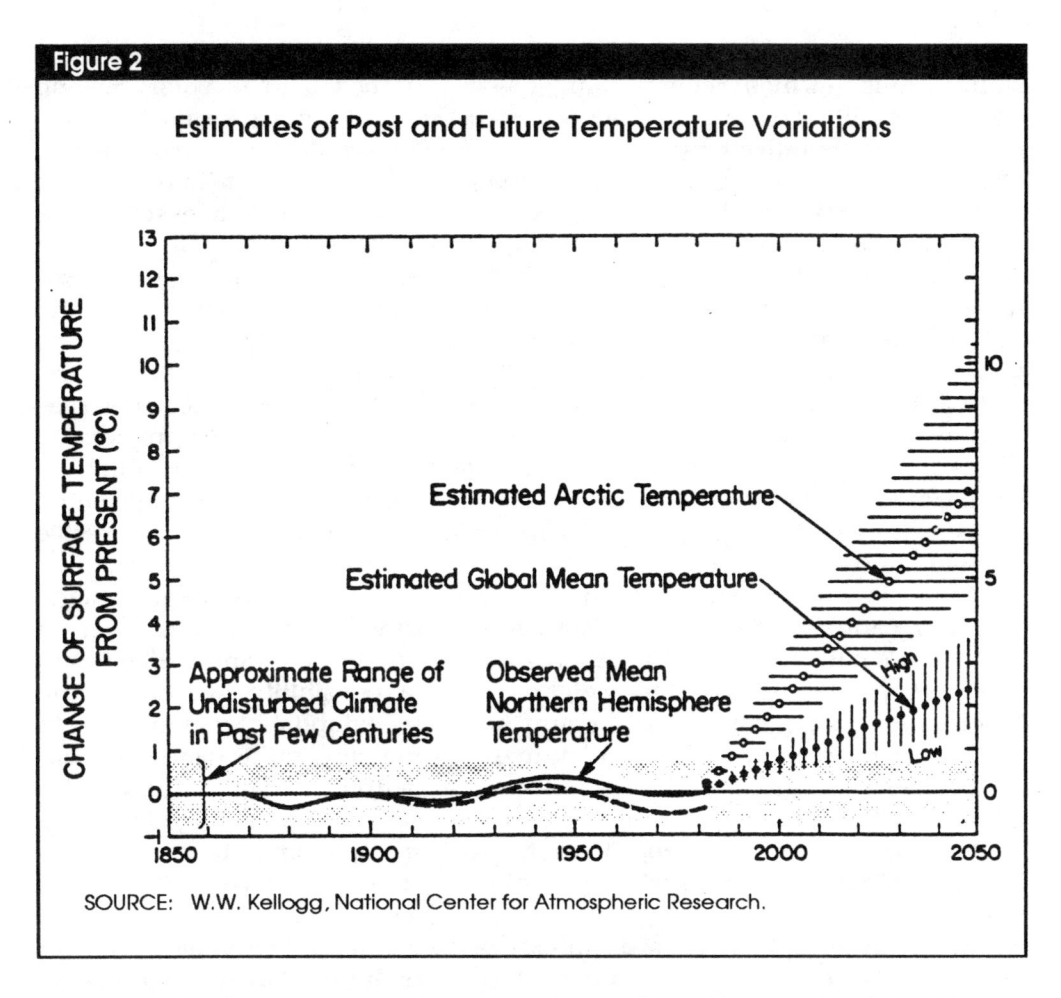

Estimates of Past and Future Temperature Variations

SOURCE: W.W. Kellogg, National Center for Atmospheric Research.

wondered what would happen to the Earth's surface temperature if the burning of coal and other fossil fuels continued to the point where the carbon dioxide content of the atmosphere doubled.

To answer that question, Arrhenius considered a primary and second-order warming effect on the planet. The primary effect was the added carbon dioxide, which would keep more infrared heat from escaping into space, thus causing surface temperatures to rise. That increase, in turn, would enable the atmosphere to hold more water vapor—a vitally important second-order effect because water vapor has an even greater capacity to retain infrared heat than CO_2. (Today, scientists call such an indirect effect a "positive feedback.") Arrhenius concluded originally that a world with twice as much atmospheric CO_2 would be about 10 degrees F warmer than the one he lived in; he later revised this estimate downward to about 8 degrees F.[3]

No one paid much attention to Arrhenius's projection at the time, but he established some key principles about global warming that remain valid today. Arrhenius figured that the temperature swing would be exaggerated at the poles, since dry polar air would become relatively more humid and retain more infrared heat emanating from the Earth's surface. Moreover, he predicted that white sea ice—which reflects sunlight—would melt as the polar regions warmed, revealing more of the dark oceans below. In technical terms, the polar region's sunlight reflectivity quotient—or "albedo"—would decrease as a result of the oceans' absorbing more of the sunlight and converting it to heat. This positive feedback would exacerbate the warming, although to a lesser degree than the rising water vapor content of the atmosphere would.

In many respects, what climate modelers know today about the greenhouse effect is little more refined than what Arrhenius mused about before the turn of the century. The Intergovernmental Panel on Climate Change, a distinguished group of 300 scientists from 25 nations assembled by the United Nations Environment Programme and the World Meteorological Organization, still uses a doubled-CO_2 atmosphere as a milestone in its projections of global warming. And remarkably, the panel's consensus forecast continues to suggest that the greenhouse effect should lead to a rise in average global temperatures of 3 to 8 degrees F—with perhaps four times as much relative warming toward the poles.[4]

Most climate modelers also hold to a prediction that Arrhenius made nearly a century ago: that cold-weather nations of the globe, such as Arrhenius's native Sweden, have more to gain from the greenhouse effect than any others.[5] Arrhenius first weighed such a possibility in 1896. The "enormous combustion of coal by our industrial establishments suffices to increase the percent of carbon dioxide in the air to a perceptible degree" so that far-northern countries "may hope to enjoy ages with more equable and better climates," he wrote.[6]

The most significant change in Arrhenius's forecast since 1896, in fact, is how long it would take for such "enjoyable ages" to arrive. In the last hundred years, worldwide industrial production has grown fiftyfold. From 1900 to 1973, use of fossil fuels grew at an exponential rate of 4 percent a year, an astounding rate of increase that even Arrhenius did not foresee. Fossil-fuel combustion now releases approximately 22 billion metric tons of carbon dioxide into the atmosphere a year (equivalent to 6 billion metric tons of carbon).[7] Consequently, the atmospheric concentration of carbon dioxide in 1992 stands at 356 parts per million, compared with 275 parts per million at the start of the Industrial Age, representing a 29 percent increase. Moreover, the atmospheric CO_2 concentration is rising rapidly—by 1.5 parts per million a year—compared with only 0.8 parts per million a year as recently as the 1960s. Therefore, the doubled-CO_2 milestone that Arrhenius thought would take many centuries to arrive now is likely to be reached sometime during the next century.

At the time Arrhenius made his original calculations, he also did not factor in other impacts of a growing human population, which has tripled during this century alone. To make way for crops, livestock and human settlements, forest cover has shrunk one-fifth since pre-agricultural times. Cultivated fields retain only 5 percent as much carbon as temperate forests and forest soils. Tropical deforestation has accelerated this carbon loss, as tropical forests possess three to five times as much carbon as temperate forests. The net result is that deforestation, especially in the tropics, releases another 800 million tons to 2 billion tons of carbon into the atmosphere annually—in addition to the 6 billion tons resulting from combustion of fossil fuels.[8]

Arrhenius also did not account for the rise in atmospheric methane in his global warming forecast. Although methane's concentration in the atmosphere amounts to only 1.75 parts per million, each methane molecule has roughly 20 times the heat-trapping potential of carbon dioxide. Since the start of the Industrial Age, methane emissions have more than doubled and now are increasing at a rate of 0.01 parts per million a year. About 20 percent of methane emissions are associated with production of coal, oil and natural gas.[9] (Methane is the principal component of natural gas.)

The agricultural sector is an even larger source of methane emissions. The burgeoning number of mouths to feed, especially in Asian nations, has resulted in a huge increase in rice production. Rice plants grown in flooded paddies tap methane harbored in oxygen-deficient mud and vent it to the air. The livestock population also has more than quadrupled in the last century. Anaerobic bacteria found in the stomachs of grass-eating livestock digest the cellulose in ruminants' cud and pass methane from the animals in the form of burps and flatulence.[10] (Methane-producing termites do the same with wood.) Putting all of these sources together—and adding in methane emissions from landfills—atmospheric

Figure 3

Atmospheric Carbon Dioxide Concentration
(parts per million by volume)

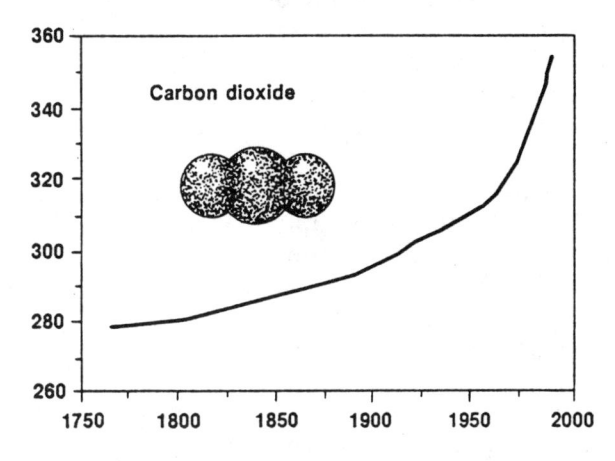

Carbon dioxide

SOURCE: Office of Technology Assessment, 1991, from IPCC, 1990.

Figure 4

Atmospheric Methane Concentration
(parts per billion by volume)

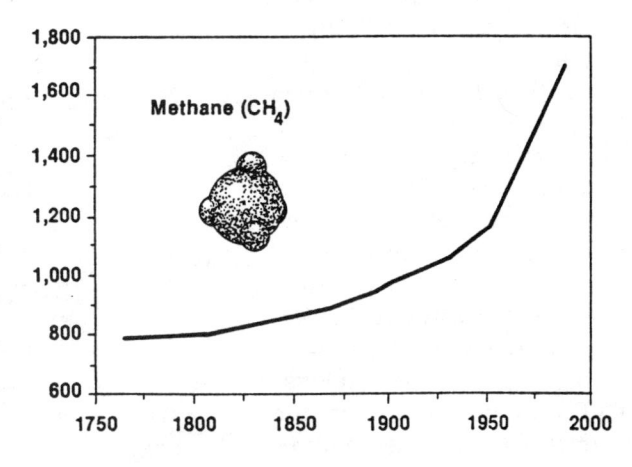

Methane (CH$_4$)

SOURCE: Office of Technology Assessment, 1991, from IPCC, 1990.

concentrations of methane are rising about 0.6 percent a year, or slightly faster than the growth rate of CO_2.[11]

Yet another greenhouse gas that Arrhenius overlooked in his calculations is nitrous oxide. Nitrous oxide is an especially potent greenhouse gas, with 270 times the global warming potential of carbon dioxide, molecule for molecule. Nitrous oxide's presence in the atmosphere is quite small, however, at 0.31 parts per million, and its concentration is rising slowly—about 0.25 percent a year. The sources of nitrous oxide emissions have been hard to pin down. Nitrogen fertilizers are thought to account for more than half of N_2O emissions. Adipic acid (nylon) and nitric acid production as well as operation of automobile catalytic converters are other suspected sources of N_2O emissions.[12]

Finally, in 1896 Arrhenius could not have known that man-made chemicals called chlorofluorocarbons would be invented some 30 years later. CFCs are used as refrigerants, foam-blowing agents, solvents, sterilants and for many other industrial purposes. In terms of global warming potential, CFC molecules are up to 7,100 times as efficient as CO_2 molecules in trapping infrared heat. Because CFCs are depleting the Earth's protective ozone layer, however, they are being phased out of production through the end of 1995. (In 1991, worldwide production of CFCs already was 46 percent below 1986 production levels.) Nevertheless, CFCs will continue to factor into the greenhouse debate

(continued on p. 12)

Figure 5

Atmospheric Nitrous Oxide Concentration
(parts per billion by volume)

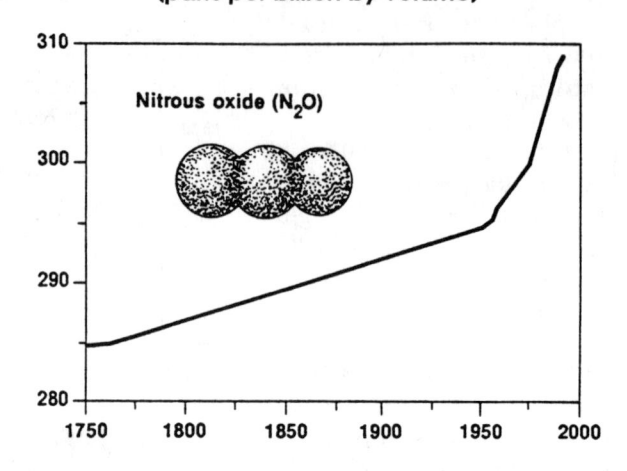

SOURCE: Office of Technology Assessment, 1991, from IPCC, 1990.

Figure 6

Atmospheric Concentration of CFC-11
(parts per billion by volume)

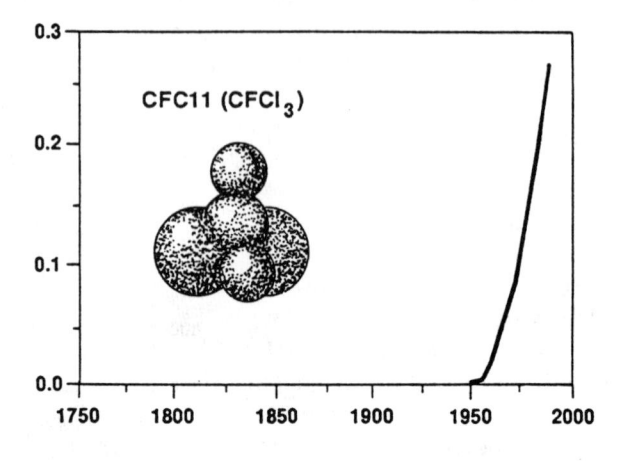

SOURCE: Office of Technology Assessment, 1991, from IPCC, 1990.

Table 1

Greenhouse Gas	Pre-Industrial Concentration (parts per billion)	1992 Concentration (ppb)	Annual Rate of Increase	Direct Global Warming Potential Index/ Feedback Effect*
Carbon Dioxide	275,000	356,000	0.5%	1/none
Methane	800	1,750	0.6%	11/positive
Nitrous Oxide	288	312	0.25%	270/uncertain
CFC-11	0	0.290	3%, decreasing	3,400/negative
CFC-12	0	0.500	3%, decreasing	7,100/negative

Greenhouse Gas**	Major Anthropogenic Emissions Sources
Carbon Dioxide 61%	• Combuston of fossil fuels (coal, oil, natural gas) for electricity generation and for residential, commercial and industrial uses. • Combustion of fossil-based fuels (gasoline, diesel) for transportation. • Deforestation.
Methane 15%	• Fossil fuel production, transportation and combustion, including: coal mining; exploration, production and transportation of oil and natural gas; and incomplete combustion of natural gas. • Agricultural sources such as livestock and rice paddies, biomass burning and degradation of wetlands for development. • Decomposition of waste in landfills.
Nitrous Oxide 4%	• Agricultural fertilizer use, biomass burning and fossil fuel combustion, among other sources.
CFCs 11.5%***	• Industrial processes, such as refrigeration, air conditioning, foam blowing and solvents.
Tropospheric Ozone 8.5%	• Vehicle exhausts and industrial emissions of nitrogen oxides, volatile organic compounds and carbon monoxide, which form ozone in the presence of sunlight.

*   Global warming index calculated as a weight basis over 100 years. Feedback effect denotes indirect effects of the greenhouse gas. A positive feedback increases the total warming relative to the index.
** Percentage denotes relative cumulative effect of 1990 manmade emissions over next 100 years.
*** Does not include negative indirect effect of ozone depletion.

SOURCE: Intergovernmental Panel on Climate Change, *1992 IPCC Assessment*, and Investor Responsibility Research Center.

for many decades to come, since their residence times in the atmosphere range up to 130 years.[13]

Recent evidence suggests that CFCs counteract most of their global warming potential by depleting ozone, which itself is a greenhouse gas.[14] Many CFC substitutes deplete ozone to a much smaller degree (and in some cases not at all), but they remain potent greenhouse gases—with more than 1,000 times the molecular global warming potential of carbon dioxide in some cases. The net result may be that CFC substitutes will do more to warm the Earth's atmosphere than CFCs themselves.

When the heat-trapping effects of methane, nitrous oxide, CFCs (and their substitutes) and ozone are pooled together and converted into units of CO_2-equivalent, it is as if an additional 70 parts per million of CO_2 has been added to the atmosphere—on top of the 81 parts per million increase in CO_2 since the start of the Industrial Age.[15] All told, the atmosphere now contains more than 425 parts per million of CO_2-equivalent. Accordingly, the doubled-CO_2 milestone that Svante Arrhenius thought would take centuries to reach now appears to be only several decades away.

The Coming Ice Age

If Arrhenius's now-amended forecast is to remain credible, the onset of global warming should be detectable long before the doubled-CO_2-equivalent milestone is reached. Today, a spirited scientific debate centers on whether the "signal" of global warming has emerged from the random "noise" of natural climate variability. Approximately 1 degree F of global warming has in fact been detected since Arrhenius's time, after adjustment for the so-called "urban heat island effect." (This effect raises temperature readings in cities because asphalt, bricks and other man-made surfaces absorb greater amounts of sunlight and radiate more heat than surrounding natural areas.)[16] Such warming could be within the bounds of natural climatic ups and downs. But the climate record dating back to the last Ice Age, some 18,000 years ago, reveals that temperature swings of this magnitude are more likely to occur over several millennia rather than within a single century.

More perplexing to climate modelers is that the present degree of warming is less than half what they would have expected by now, given the amount of greenhouse gases that have accumulated in the atmosphere. Several factors may account for this disparity. The most important of these is "thermal lag" caused by the Earth's oceans. The oceans act as a giant heat sink that slows the rate of atmospheric temperature changes. Constant circulation causes the oceans' heat-absorbing (and CO_2-rich) top layers to sink to the sea floor, where they may remain for hundreds of years before resurfacing. The oceans are a giant sink for carbon dioxide as well, as they store 60 times more CO_2 than the atmosphere. Because of the oceans' influence on the atmosphere, scientists believe that the amount of warming now detected is only one-half to four-fifths of the total warming built into the climate system.[17] In other words, even if the amount of greenhouse gases in the atmosphere were to level off tomorrow,

Figure 7

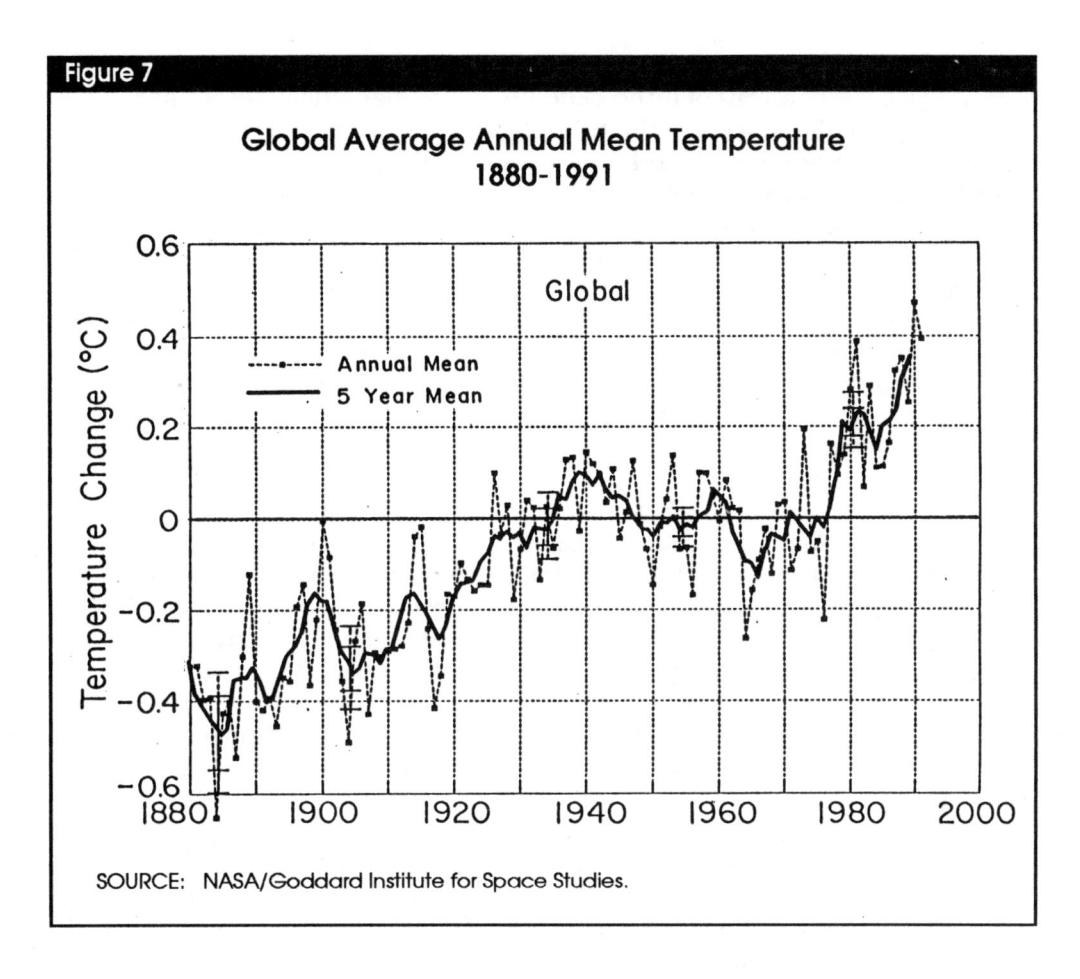

Global Average Annual Mean Temperature 1880-1991

SOURCE: NASA/Goddard Institute for Space Studies.

global temperatures would continue to rise up to another full degree F as the oceans assimilate the changes in atmospheric conditions.

Other factors contributing to the disparity between actual temperature trends and the models' projections have become apparent only recently. These factors include the cooling effects of stratospheric ozone depletion and the presence of sulfate aerosols in the atmosphere. Sulfate aerosols—mainly associated with burning coal—diffuse sunlight and possibly seed clouds. As these and other factors are accounted for in the climate models, the degree of forecasted warming appears more consistent with the observed temperature trend of the last 100 years.

Except for rapid warming during the 1980s, however, the observed temperature record over the last hundred years has been highly erratic and inconclusive in showing a warming trend. From the 1880s through the 1930s, average temperatures did rise distinctly in most parts of the world. But then temperatures dipped a bit from the 1940s through the 1970s—especially in the Northern Hemisphere—even as the atmospheric levels of carbon dioxide, methane and CFCs soared. This temperature decline prompted speculation that maybe the Earth was in the midst of a long-term cooling trend instead of a warm-up.

Until the 1980s, in fact, many knowledgeable scientists—if asked to predict a coming ice age or an era of unprecedented warming—chose the ice age.[18] Ice-age theorists had a heyday in the mid-1970s after a series of especially severe winters. Numerous books and articles cropped up with ominous-sounding titles like "Ice," "The Weather Conspiracy: The New Coming Ice Age" and "The Cooling: Has the Next Ice Age Already Begun? Can We Survive It?"[19] Even the highly respected National Academy of Sciences warned that the beginning of a new ice age might be felt within 100 years. The Central Intelligence Agency weighed in with its own report in 1974, offering a Machiavellian assessment: "In a cooler and therefore hungrier world, the U.S.'s near monopoly as a food exporter...could give the U.S. a measure of power it never had before—possibly an economic and political dominance greater than that of the immediate post-World War II years."[20]

While the ice-age hysteria has since receded, the underlying concerns about a coming ice age are not to be dismissed entirely. Although ice ages have been rare during the Earth's 4.5 billion year history, the climate has been locked in an Ice Epoch for most of the last 3 million years. Continental drift, of all things, has been an important factor in this icy era. Antarctica has positioned itself over the South Pole, providing a base for falling snow on which a tremendous ice sheet has built up. At the same time, the northern continents have formed a ring around the Arctic Ocean, blocking the flow of warm, equatorial water toward the North Pole, allowing a northern icecap to form. Consequently, ice now covers both poles—a condition that may never have existed before in the planet's history.[21]

In the last million years, the Earth has experienced eight ice ages. Each has lasted 100,000 years or so, only to be interrupted briefly by interglacial periods. The present interglacial is about 13,000 years old, and its passing is overdue, since no previous interglacial has persisted for more than 12,000 years. For the last 9,000 years, summertime solar radiation has been decreasing in the far north as a result of long-term, periodic changes in the tilt, rotation and orbit of the Earth. Eventually, cooler summers and snowier winters will prevail in far northern latitudes, and the excess snow slowly will be compressed into giant sheets of ice, marking the beginning of a new ice age. In perhaps 10,000 years, the ice sheets will become massive enough to launch another inexorable advance to the south.[22]

To put this geologic clock in perspective, human civilization has existed for a "split second" in the 24 "hours" since the Earth was formed—and by the time this second is finished, the Earth probably will have slipped into another ice age. The entire history of civilization, dating back to the rudimentary society in Mesopotamia, is in fact briefer than the time that has passed since the peak of the last great Ice Age. During the present interglacial, temperatures have risen more than 10 degrees F. In North America, the warming has melted a mile-thick sheet of ice that once covered the land as far south as Chicago and Long Island. The receding glaciers left behind the Great Lakes—holding 20 percent of the world's fresh water supply—and deposited in the Great Plains some of the most fertile topsoil in the world. Given these geologic circumstances, it is little wonder that North America has emerged as one of the most agriculturally productive regions in the world. All the same, astrophysicists insist that the

A Climate of Uncertainty

Figure 8

Glacial-Interglacial Temperature Changes

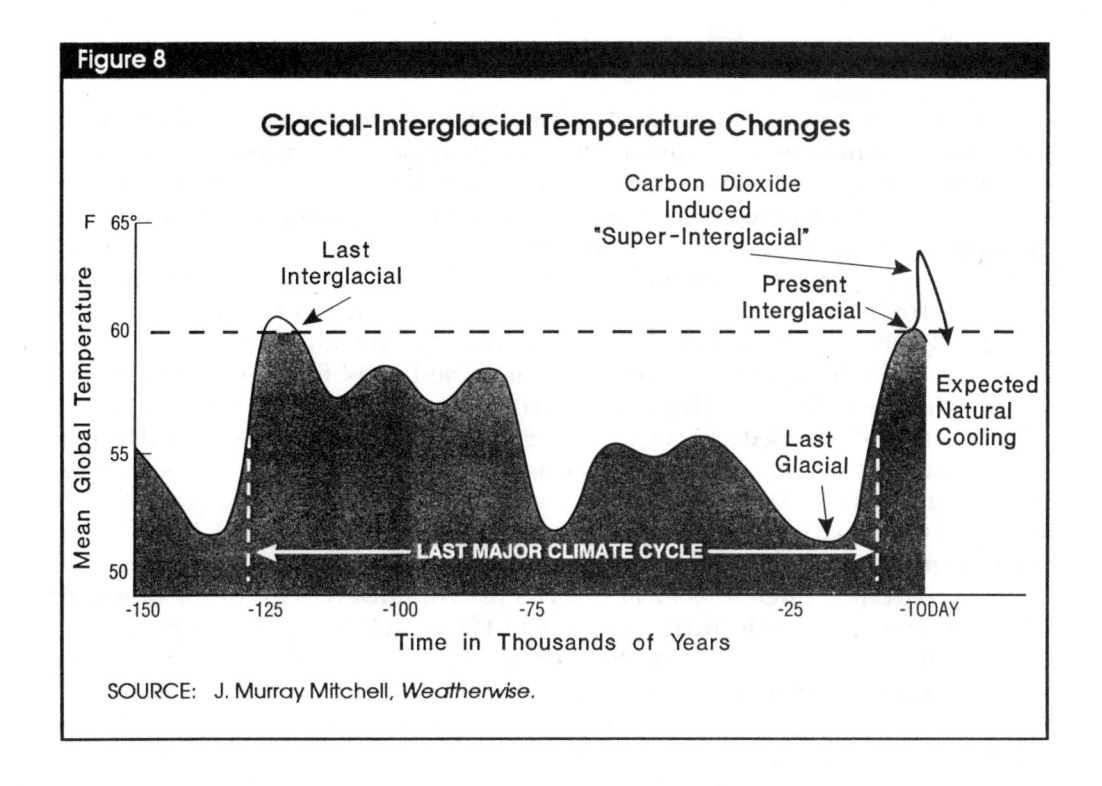

SOURCE: J. Murray Mitchell, *Weatherwise*.

Earth's wobbly path around the sun will spawn another ice age within the next millennium or two, and the continent's cornucopia will come to an end.

To add insult to injury, some scientists now are saying that global warming might accelerate the onset of the next great ice age. The continuing decline in solar radiation in northerly latitudes, combined with the greenhouse forecast for warmer, wetter winters in polar regions, could lead once again to more snowfall in winter and less melting in summer—triggering the advance of the ice sheets.[23] Whether global warming would actually hasten its own demise is a matter of some debate. In any event, few reputable scientists dispute that the next ice age is inevitable—and coming soon, geologically speaking—regardless of a warming trend in the 21st century.

Why, then, does hardly anyone seem alarmed by the fact that the planet is now at least two decades closer to the next ice age than it was during the ice-age scare of the 1970s? The answer appears to be in the second-hand of the geologic clock. General circulation models of global climate suggest that global temperatures could rise as much in the next century as they have cumulatively in the last 10 millennia. In a tick of geologic time, the planet's atmosphere could swing from about 50 degrees F (during the last great Ice Age) to about 70 degrees F—its warmest average in all of human civilization. Not since the age of the dinosaurs, some 100 million years ago, might the Earth have endured temperatures any warmer than what may now be in store. The Earth is indeed on the verge of another ice age, but a "super-interglacial" period may precede it—fire coming before ice.[24]

'The Greenhouse Effect Is Here'

Several times since Svante Arrhenius's original pronouncements about an amplified greenhouse effect have climatologists sounded public alarm over global warming—each time with greater amplification. In the 1930s, British physicist G.S. Callendar revived Arrhenius's theory, pointing out the warming trend over the previous 50 years. But like Arrhenius, Callendar speculated that a continuation of the trend would be mainly beneficial "for the northern margin of agriculture" and for stimulated growth of plants. "In any case the return of the deadly glaciers should be delayed indefinitely," he surmised.[25]

When the warming trend reversed itself in the 1940s, fears of an impending ice age returned. Partly for this reason, articles written in the early 1950s about global warming sounded a decidedly neutral tone. Then hurricanes Carol, Edna and Diane successively pounded New England in 1954 and 1955. Because this region rarely faces the wrath of hurricanes—and since global warming is expected to raise the temperature of ocean surface waters on which hurricanes feed—some viewed these storms as conclusive evidence of a change in global climate patterns. (Meteorological records reveal, however, that New England also suffered bouts with hurricanes in the 1800s and in the 1930s.)[26]

In 1957, two scientists at the Scripps Institution of Oceanography in California published an article in the journal *Tellus* about another aspect of the oceans' relationship to the atmosphere.[27] Roger Revelle and Hans Suess reported that while upper layers of the oceans were readily able to absorb a portion of the excess heat building up in the atmosphere—delaying a greenhouse warming—the oceans would be able to soak up less of the excess carbon dioxide accumulating in the atmosphere than was commonly believed at the time. Given this new constraint, Revelle and Suess halved the prevailing estimate of the oceans' capacity to store the excess CO_2, and they went on to make a statement that has been quoted perhaps more often than any other about the greenhouse effect: "Human beings are now carrying out a large-scale geophysical experiment that could not have happened in the past or be repeated in the future." Such a premise, "if adequately documented, may yield a far-reaching insight into the processes of weather and climate."

Also in 1957, the Scripps Institution hired Charles David Keeling, then a young researcher, to set up CO_2 monitoring stations at the South Pole and atop the Mauna Loa volcano in Hawaii. One year later, Keeling had the documentation that Revelle and Suess were seeking: The atmosphere at Mauna Loa contained 315 parts per million of carbon dioxide, about one-sixth more CO_2 than ice core data suggested was in the atmosphere at the start of the Industrial Revolution around 1750.

While the concentration of CO_2 in the atmosphere rose steadily after 1958—another 40 parts per million at last count—the weather during the 1960s and 1970s confounded predictions about global warming. A White House report in 1965 did acknowledge for the first time that human activities could bring about changes in climate that would have important consequences for world commerce in the decades ahead.[28] But the same report, issued by President Johnson's Science Advisory Committee, addressed many other air and water

Figure 9

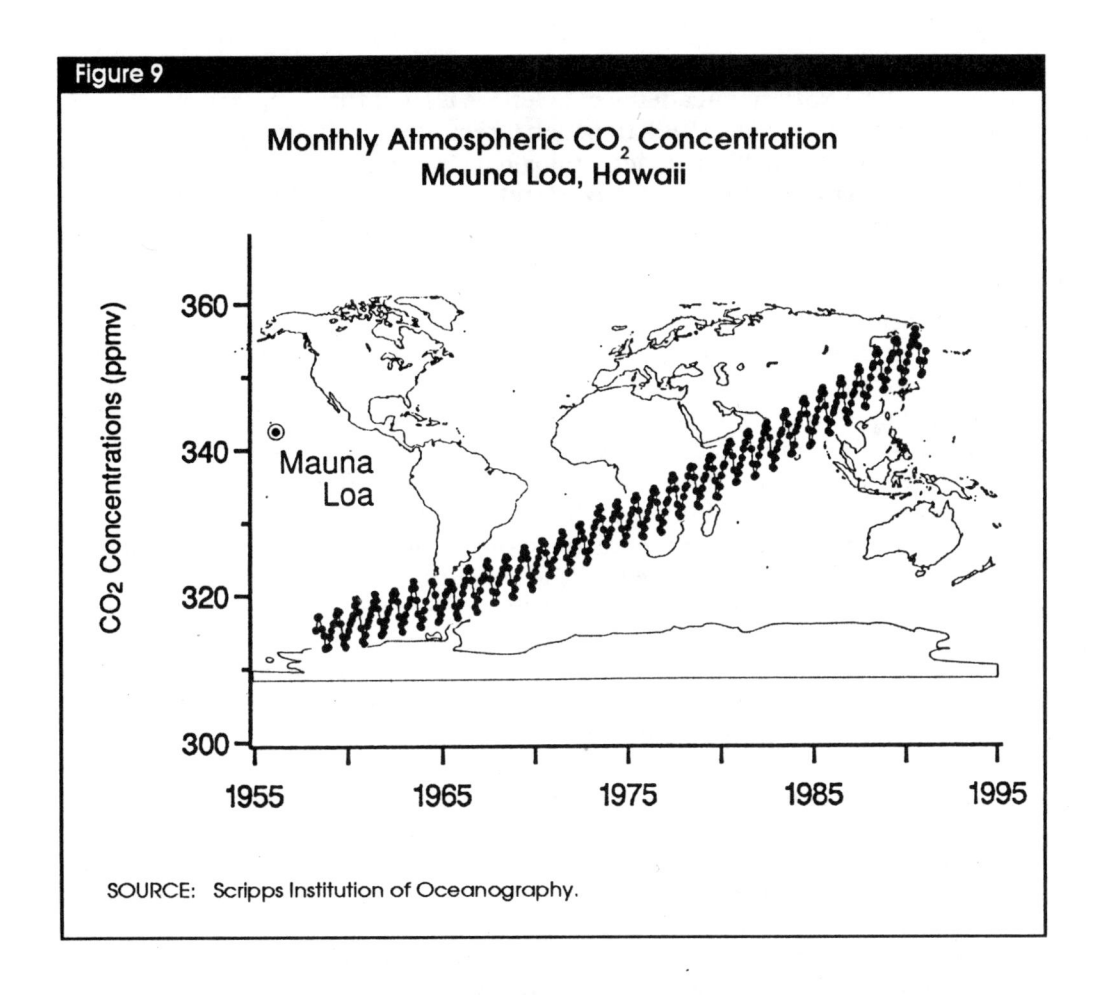

**Monthly Atmospheric CO_2 Concentration
Mauna Loa, Hawaii**

SOURCE: Scripps Institution of Oceanography.

pollution problems that commanded more immediate attention. By comparison, global climate change remained a far-off problem (or, as some preferred, a "far out" theory).

By 1970, climate modelers fell loosely into two camps. "Climate coolers" believed that industrial and agricultural pollutants—the likes of which were addressed in the earlier White House report—would reduce the amount of sunlight reaching the Earth's surface and therefore have a net cooling effect on the planet. "Climate warmers," on the other hand, believed that carbon dioxide and other greenhouse gases would trap enough heat radiating from the Earth's surface to overwhelm the cooling effect of the sunlight-blocking aerosols. The dispute seemed to be settled in favor of the climate warmers in 1975, when scientists reported at an international symposium that particles of dust and smoke absorb more sunlight than was previously thought (and reflect light less into space) and do not result in cooling of the lower atmosphere, at least over land.[29] This was by no means the final word on aerosol cooling (as we shall see later), but the findings persuaded most atmospheric scientists to join the global warming camp.

Important developments in the late 1970s and early 1980s built on this consensus. The National Academy of Sciences issued three reports on climatic change between 1979 and 1986, in which it recanted its earlier warnings about a coming ice age.[30] The World Meteorological Organization sponsored a World Climate Conference in Geneva, Switzerland, in February 1979, from which the following warning was issued: "It is possible that some effects on a regional and global scale may be detectable before the end of this century and become significant before the middle of the next century."[31]

The World Meteorological Organization then sponsored a second conference on global climate change in Villach, Austria, in October 1985. The statement from this conference was even more urgent, noting that "some warming of climate now appears inevitable due to past actions." In an effort to prompt international policy discussions, it added, "the rate and degree of future warming could be profoundly affected by government policies on energy conservation, use of fossil fuels and the emission of some greenhouse gases."[32]

While each of these announcements caught the attention of the press (and to some degree the public), none had the sensationalized effect of a statement uttered by a NASA scientist on Capitol Hill on June 23, 1988. On a day when the temperature in Washington, D.C., was approaching 100 degrees F, James Hansen told a packed Senate hearing room: "It is time to stop waffling so much and say that the evidence is pretty strong that the greenhouse effect is here."[33]

Hansen's declaration was not meant as a comment on the searing heat wave that had descended on the Nation's Capital that June (although the sweltering weather surely added to the dramatic effect). Rather, it was based on a calculation that global temperatures had risen a full degree F since 1900, when the random fluctuation in temperature over a century's time normally is only a third of a degree F. Combining this long-term temperature change—three times the standard deviation—with the fact that the five warmest years on record (dating back to 1880) all had occurred in the 1980s, Hansen concluded that it was "99 percent certain" that the recent warming trend was not because of a natural variation in climate, but because of an enhanced greenhouse effect.

Hansen's testimony transformed the 100-year-old debate on global warming. His proclamation led off the television news that evening and made the front page of next morning's papers. *Time* and *Newsweek* magazines subsequently ran cover stories about the greenhouse effect. Serendipitously, Hansen seemed to be answering a question that had been on everyone's mind for weeks: "Is something weird about the weather?" That June, not only Washington but much of the nation (indeed the Northern Hemisphere) was suffering through a scorching early-summer heat wave. The newspapers had been full of stories about a drought ravaging crops in the Midwest and barges grounded by record-low water levels on the Mississippi River. Moreover, the public was still digesting other news from NASA that those man-made chemicals invented back in the 1930s—CFCs—had somehow ripped a hole in the Earth's ozone layer. If a NASA scientist now claimed the greenhouse effect was to blame for that summer's drought and the heat wave, the public was inclined to believe him.

Of course, Hansen did not say he could prove that global warming triggered the specific events of that summer, only that they were indicative of the kind of weather phenomenon that would occur more frequently now that "the greenhouse effect is here." In hindsight, there is a much more mundane explanation for the heat wave and drought during the summer of 1988—a stationary weather pattern that developed over the American heartland early in the year, which, come summer, ushered in a lot of heat but kept rainstorms out of the forecast. Crops shriveled in the sun-baked soil and huge agricultural losses followed. Whatever the cause of that oppressive weather, it provided ideal tinder for Hansen's incendiary testimony.

The Drought of '88—and Beyond

While climatologists continue to debate whether the summer of '88 is a telltale sign of global warming or a fluke of Mother Nature, they know now what specific meteorological events caused that "greenhouse summer" to occur. In February 1988, a ridge of high pressure—towering five miles into the atmosphere—anchored over the Great Plains and refused to budge. Over the next

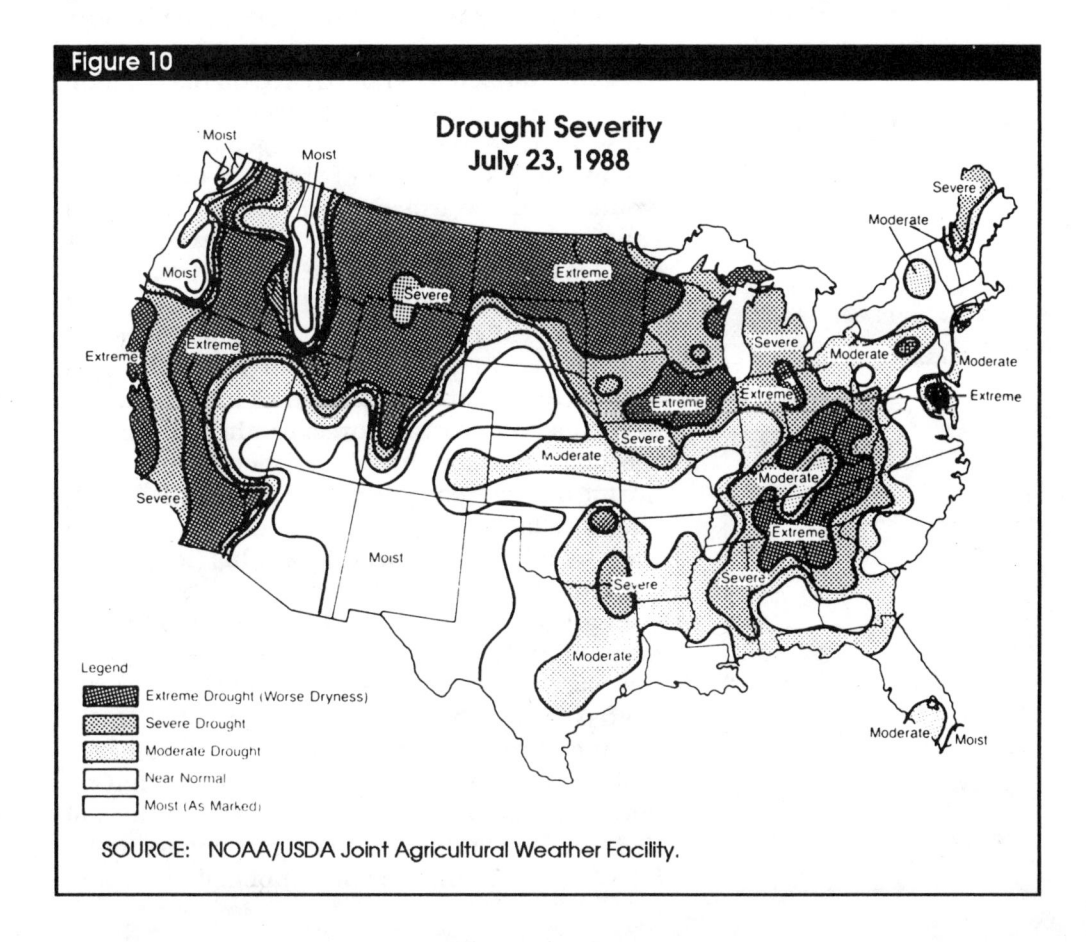

Figure 10

**Drought Severity
July 23, 1988**

Legend:
- Extreme Drought (Worse Dryness)
- Severe Drought
- Moderate Drought
- Near Normal
- Moist (As Marked)

SOURCE: NOAA/USDA Joint Agricultural Weather Facility.

five months, the system ballooned in size; by June it covered more than half of the United States and Canada. Inside this dome of high pressure, temperatures soared six to 12 degrees F above normal. The North American jet stream—with its parade of rainstorms—steered around the dome. Precipitation over much of the affected area was less than 40 percent of normal.[34]

When the bubble finally burst in late July, storms penetrated the Midcontinent region, and rainfall returned to near-normal levels, despite persistent heat. But the damage had been done. America's corn harvest fell 34 percent in 1988, while soybean production dropped 21 percent and the wheat harvest slipped 14 percent. All told, it was the worst year-to-year drop in agricultural performance since the Dust Bowl years of the 1930s. (Canada's harvest was equally devastated.) Congress subsequently approved $4 billion for what the President's Interagency Drought Policy Committee described as "the largest disaster relief measure in U.S. history."[35]

Memories of that fateful summer have been etched into the minds of farmers, property insurers, even congressmen. Meanwhile, the controversy over the appropriateness of Jim Hansen's testimony before the Senate that June is still simmering. Many in the scientific community have criticized Hansen for playing fast and loose with the laws of statistics, which make it extremely difficult to prove anything with "99 percent certainty"—especially something as incorrigible as the weather.

Still, Hansen has not backed off his position that global warming has begun. On the contrary, in 1989 he issued a wager to his fellow climatologists that one of the next three years would be the warmest on record. His payoff came only one year later. At an average 59.81 degrees F, 1990 has been the warmest year yet.[36] (Second place belongs to 1991, ranking just ahead of 1981 and 1988.) In fact, the eight warmest years in the 110-year record of land surface temperature readings all have occurred in the last 12 years. Moreover, the warming of the last 25 years has been more rapid than during any comparable period on record.[37] With statistics like these, it is no wonder that Hansen was the odds-on-favorite to win his bet.

The global warming trend may be a more telling sign of the greenhouse effect than the drought and heat wave of '88. But the drama of that summer sparked a burst of activity in the United States that has not been rivaled since. Dozens of major corporations—the likes of Atlantic Richfield, Bechtel, Ford and Pacific Gas and Electric—assembled ad hoc committees of their staffs or commissioned white papers in the fall of that year on the subject of global climate

Table 2

10 Warmest Years Since 1880

The 10 warmest years since 1880 and average temperatures in Fahrenheit.

Year	Temp
1990	59.81
1991	59.67
1981, 1988	59.64
1987	59.56
1980, 1983	59.51
1989	59.45
1973	59.31
1977, 1986	59.30

SOURCE: NASA/Goddard Institute for Space Studies.

change. Some in the agricultural business went even further, enacting plans to diversify their land holdings and supply contracts to make sure they would not get caught short again as they were in 1988.

Pioneer Hi-Bred International, the world's leading producer of seed corn, decided, for example, to increase its winter seed-growing acreage in the southern United States as well as in South America to guard against potential summer drought.[38] Archer-Daniels-Midland purchased a 16 percent stake in the Illinois Central Railroad to improve its capacity to move grain and derivative products in case low-water levels once again blocked barge shipments on the Mississippi and Ohio Rivers.[39] And two breweries that once boasted of all-American barley, hops, corn and rice in their beers—Miller Brewing and Anheuser-Busch—reluctantly decided it was time to line up potential importers for some of their ingredients, from France and elsewhere.[40]

The greenhouse summer of '88 also mobilized the environmental community. The majority of the so-called "Group of 10" placed the greenhouse effect on their list of top priorities for action in 1989, as they have done every year since. Congress also got into the act by holding dozens of hearings on climate change and introducing scores of greenhouse-related bills. But not much in the way of substance has become law since, nor has the Bush administration advocated that the international community commit to targets and timetables to reduce specific greenhouse gases.

The fact is, the sense of urgency that gripped the United States after the drought of '88 has since let go—and three relatively good years of crop production may have had something to do with it. While 1990 and 1991 posted even warmer global temperatures than the previous record set in 1988, drought has not been a persistent problem in any region of the United States except California, where most crops are irrigated anyway. Without a dramatic weather event to galvanize the public's attention, global warming has faded from the American limelight.

By contrast, the Europeans—who have been the most forceful advocates of curbing greenhouse gas emissions—have been dumbfounded by the weather's behavior in recent years. A localized drought starting in pockets (particularly the Alps) in the mid-1980s spread to several European nations in 1988, and then encompassed virtually the entire continent in the summer of 1990. Temperatures ran more than 10 degrees F above normal for weeks at a time. While the southeast of England basked in weather more typical of the Mediterranean, southern Europe endured almost sub-Saharan heat and dryness in 1990.[41]

Moreover, the 1990 European drought came on the heels of three hurricane-force winter storms that tore through England and the continent, killing 140 people and causing billions of dollars in property damage. The ferocity of these gales—matched only by one other storm, in 1976—prompted the same kind of speculation that followed the 1950s hurricanes that battered New England: Perhaps the greenhouse effect was to blame.[42]

Search for a Smoking Gun

The tradition in policy debates concerning regional and global environmental problems has been to search for some kind of smoking gun. The appearance of "dead" lakes in the Adirondacks and dying trees along the ridgetops of the Appalachian Mountains became a focal point in Congress's decision to enact acid rain legislation, for example. Similarly, the development of an "ozone hole" over Antarctica transformed the international debate on banning CFCs. With global warming, however, there are many clues—yet no smoking gun—to resolve doubts about the long-range forecast.

The climate-change clues come in many forms: The oceans are getting warmer, coral reefs are blanching and the sea level is rising. Average sea level since 1900 has risen four to eight inches; now it may be rising at a rate of an inch per decade.[43] The extent of polar sea ice has shrunk about 6 percent globally over the last 15 years.[44] The Rhone Glacier in Switzerland, the Grinnell Glacier in Montana's Glacier National Park, the Mt. Quelccyaya Ice Cap in Peru and the Speke Glacier in the Ruwenzori Range of Uganda are among dozens of famous glaciers around the world that are receding at record rates.[45] Snow cover throughout the Northern Hemisphere has shrunk 8 percent since 1973—to its lowest amount in the 19 years that such records have been kept.[46] At the same time, precipitation has increased globally, especially in the Southern Hemisphere, consistent with general circulation model forecasts.[47] And of course there is the global temperature record dating back more than a century, in which the 1980s stand out as the warmest decade by far.

Despite this circumstantial evidence, the American public has yet to be persuaded that global warming is a serious problem. A 1990 survey of 1,413 Americans conducted by the Roper polling service found global warming ranking 19th among 29 environmental problems listed. Leaking underground storage tanks, water pollution from sewer plants, and solid waste and litter were among other issues that respondents ranked as being of greater concern.[48]

Ultimately, people would be inclined to take global warming more seriously if there were a sudden barrage of aberrant weather events that made news headlines: more wildfires like the ones that razed 2,700 dwellings in Oakland in 1991[49] and burned a quarter of Yellowstone National Park in 1988;[50] more killer hurricanes along the lines of Hurricane Hugo in 1989[51] and Hurricane Gilbert in 1988;[52] more summertime heat waves persisting in large population centers such as New York[53] and Washington;[54] and drought that reaches beyond California[55] and localized pockets of the Farm Belt.[56]

This is not to suggest that Americans have lost interest in the greenhouse effect—or the weather for that matter. Local news broadcasts in the United States devote far more time to the weather than news programs practically anywhere else in the world. (There is even a 24-hour weather channel on cable television.) But absent a smoking gun on global warming, Americans will continue to doubt climatologists' fuzzy forecasts of climate change just as they discount the accuracy of predictions made by their own television weathermen. If the weather outlook for the coming week is subject to change, why should people trust a global climate forecast issued for the next 50 years?

Forecasting the Future

The livelihood of tens of millions of people around the globe depends on the weather. It is not just farmers toiling in their fields; it is also those who work in construction, transportation, telecommunications and the electric power industries, to name but a few. They all have to take account of the weather in creating their schedules, mapping out their routes, choosing the proper equipment and making other important planning decisions. Because these industries have so much riding on the elements, they have come to depend on accurate weather forecasts. In the United States, forecasts are issued by the National Weather Service and about 100 private weather forecasting companies that constitute a $200-million-a-year industry.

Some may wonder whether this money is well spent. By reputation, weather forecasters are about as trustworthy as used car salesmen and politicians running for office. In reality, meteorologists have made great strides in improving the accuracy of their 6-, 12- and 24-hour weather forecasts. With help from satellites and radar, they also are doing a much better job of pinpointing the movement of tornadoes and hurricanes, saving countless lives.

Be that as it may, a much greater investment will be required to replace a now-aging network of computers, radar, satellites and ground instruments. The National Weather Service, for one, plans to spend more than $1 billion over the next five to 10 years to further improve its forecasting capabilities—particularly of isolated, short-lived storms, such as tornadoes, flash floods and hurricanes.[57] (The present system for isolated severe storms signals a false alarm about 60 percent of the time.) But such improvements will do little to resolve uncertainties about the global warming forecast.

The Primitive Art of Weather Prediction

While climate forecasts are put together in a different way than weather forecasts, any meteorological forecast at its root has to solve a problem in fluid dynamics. The key is to measure the important variables—temperature, humidity, barometric pressure, wind velocity and so on—and ferret out the relationships between these variables. Long-range weather forecasts (seven to 14 days) encompass dozens—even hundreds—of equations, calculated on a computer over and over again. The thinking goes: If one has enough data on the important variables, if the models describing the variables' interrelationships are sound and one has a big enough computer, it should be possible to forecast the weather as far in advance as is desired.[58]

Unfortunately, this theory has four problems. There are never enough weather observation points, the models do not replicate the way the real world operates, the computers are still too small, and turbulence in the atmosphere imposes an outward limit of about two weeks in predicting specific weather-related events. After that, it becomes anybody's guess what will happen to the weather.

Today's weather forecasts incorporate 10,000 observations made every six hours by weather stations, balloons, aircraft, radar and satellites across the globe. That may seem like a lot of data points until one considers that the surface of the Earth measures almost 200 million square miles and the active weather portion of the atmosphere is more than 10 miles high. What is more, observations from certain locations around the planet are quite spotty, especially over the oceans, which cover 70 percent of the planet.

Then there is the problem of "sensitive dependence on initial conditions." Since the 10,000 observation points must serve as proxies for all points in between, this biases the data set's snapshot of the atmosphere's initial conditions—even before the computer model starts to run. Moreover, if the measuring devices are not calibrated perfectly, they introduce errors of their own into the data set. As the computer churns through the calculations that go into making a weather forecast, such errors are compounded. By six or seven days into the model's forecast, the errors usually have made such a mockery of the initial conditions that any predictive capability in the forecast is lost.[59]

Forecasts looking 90 days into the future have an especially abysmal track record. Faced with the seemingly simple question, "Will the next three months be warmer or cooler than normal?" forecasts from the National Weather Service provide a correct answer fewer than six times out of 10. When asked if the next 90 days will be wetter or drier than normal, the chance of the forecast being right is barely better than 50-50. For all the trouble that meteorologists go through in making long-range forecasts, in other words, they could do nearly as well by flipping a coin.[60]

At a time when climatologists are trying to determine how changes in climate may affect the seasonal weather patterns three, four, even five *decades* into the future, it is disheartening to learn that a recent National Weather Service 90-day weather forecast may have been the least accurate it has ever issued. It predicted—with a degree of confidence uncharacteristic of the forecasting profession—that the winter of 1990-91 would be colder than average in the eastern United States, and warmer in the West. In retrospect, there was not much cold weather to be found anywhere in the country that winter. To the extent it could be found, it was in the West, not in the East.[61]

It is instructive to see what went wrong with that 90-day outlook, because it points out one of many problems that bedevil the weather forecasting business—both short- and long-term. In this instance, it was the ungainly relationship between the oceans and the atmosphere that spoiled the forecast. Out in the Pacific, meteorologists keep a watchful eye on something called the Southern Oscillation—a weakening of east-to-west trade winds that affects the movement of a warm surface layer of the ocean. While the movement of this warm-water body is highly unpredictable, usually it trudges back and forth across the equatorial Pacific every three to seven years. When the trade winds weaken and the warm water sloshes to the east, the event is known as an "El Niño/Southern Oscillation." El Niño's trigger dramatic weather changes around the globe. Desert portions of Peru and Ecuador are deluged with rain;

normally moist parts of Brazil, southeastern Africa, Australia and the Philippines are struck by drought; and in Asia, the summer monsoons weaken.

During the winter of 1990-91, the U.S. Weather Service predicted the onset of an El Niño in the eastern Pacific. Strong storms were supposed to form in the equatorial Pacific and put a giant kink in the jet stream. As it entered North America, the jet stream was expected to careen into western Canada, scooping up cold air as it went. Then it would plunge back into the United States, ensuring a frosty winter in the East.

But the climatic trigger was never pulled. The body of warm water, instead of moving east, stayed put way out in the Pacific. With the jet stream maintaining an orderly course, another 90-day forecast by the National Weather Service was spoiled—only this time the outlook was off by 180 degrees. Defying the forecast, the West was colder than normal when it was supposed to be warmer, and 13 eastern states enjoyed one of their 10 mild winters on record—despite the prediction of an especially cold winter.[62] As a small consolation for the National Weather Service, an El Niño did finally develop in the fall of 1991—one year overdue.[63]

Deciphering the intentions of El Niño is not essential to making a proper global warming forecast. El Niño, after all, represents one periodic fluctuation of one component of the climate system, whereas general circulation models probe the sensitivity of the entire system to an external force, such as increased radiative forcing by greenhouse gases. But the apparent inability to anticipate the behavior of the oceans poses a tremendous problem to meteorologists and climatologists alike (as we shall see later). Just as El Niño does not always conform to meteorologists' expectations, general ocean circulation patterns may not behave as climatologists expect them to as they make their climate change forecasts. The result is that very unexpected things can happen.

Lost in Space

Satellites are one of the most effective tools that scientists have to monitor the Earth's climate system. But the situation here, too, is far from ideal. For the last three years, the National Oceanic and Atmospheric Administration has been relying on just one geostationary operational environmental satellite (GOES) to make meteorological observations over the United States. This satellite, GOES-7, transmits the swirling cloud images seen on television plus data on temperature-humidity profiles across the nation. When the imager on GOES-6 failed in 1989, GOES-7 was repositioned over the center of the country to watch out for Pacific storms bound for North America as well as hurricanes spinning west across the Atlantic. But GOES-7 is running out of fuel and soon will start to drift out of position. Should it fail unexpectedly, the blinds would be shut on America's lone orbiting window.[64]

As a precaution, NOAA has borrowed the European Meteosat-3 satellite stationed over South America and moved it westward to give better coverage of storms brewing in the Atlantic. In its present position, however, Meteosat-3 sees only the eastern half of the United States, and it, too, will begin drifting as

Figure 11

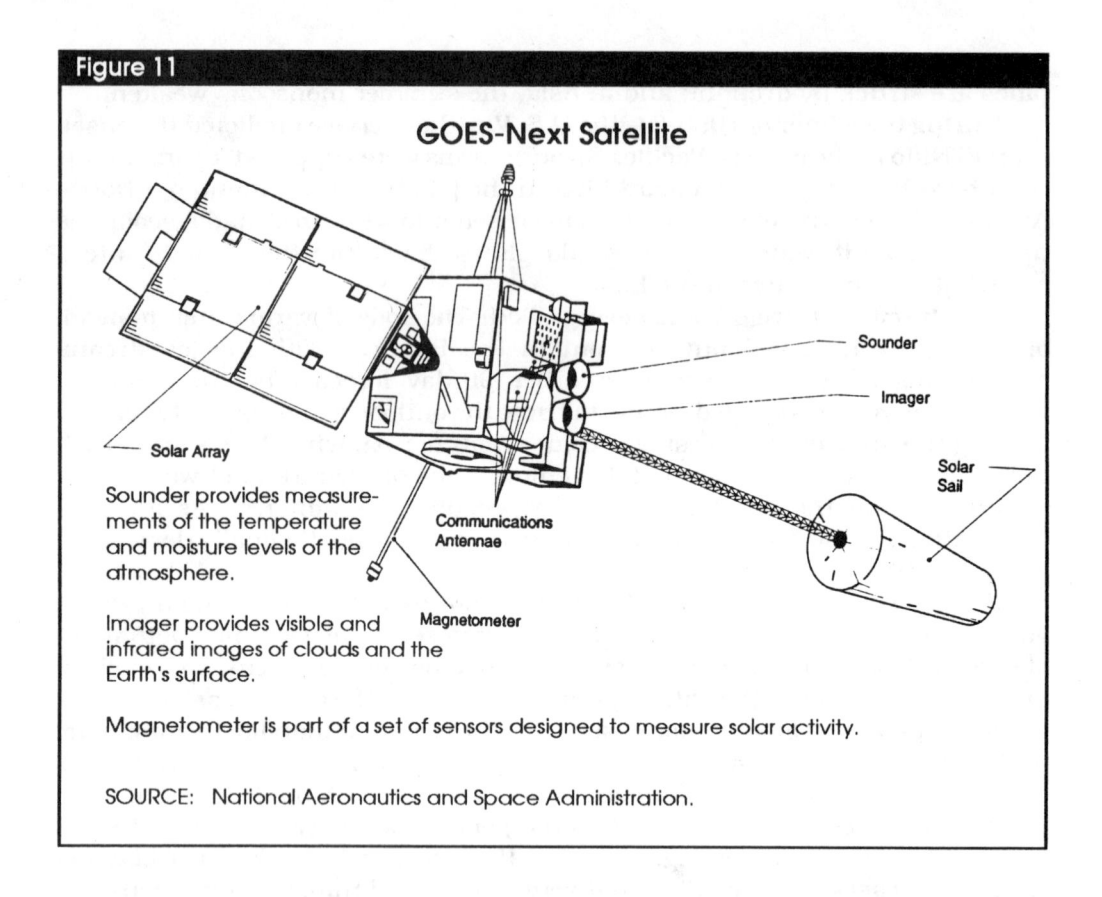

GOES-Next Satellite

Sounder provides measurements of the temperature and moisture levels of the atmosphere.

Imager provides visible and infrared images of clouds and the Earth's surface.

Magnetometer is part of a set of sensors designed to measure solar activity.

SOURCE: National Aeronautics and Space Administration.

it runs out of fuel at the end of 1993. Any further slippage in the launch date of the next GOES satellite—originally scheduled for 1989 and now not expected before early 1993—could leave American meteorologists lost in space.[65]

NASA's "Mission to Planet Earth" also is in trouble. As originally proposed, NASA was going to spend more than $30 billion over the next 20 years to launch and operate earth observing satellites (EOS) to study global change. But considering the additional $37 billion that NASA wants to spend on Space Station Freedom, Congress has instructed the space agency to trim its sails. In 1991, Congress lopped $5 billion from the $16 billion budgeted for EOS between now and the end of the century. That reduction now makes it unlikely that the first of two 13-ton EOS platforms will get off the ground before 2000.[66] (EOS-A originally was scheduled for lift-off in 1998, and EOS-B several years later.) As a result, U.S. climatologists—like American meteorologists—could be in the dark when it comes to observing the Earth's systems from space.

There may be a light on the horizon, however. Recalling the space shuttle Challenger accident and the equipment malfunctions now plaguing the Hubble space telescope and the Galileo probe, an engineering review panel has suggested that it might not be such a good idea to load dozens of earth sensors on just two satellites anyway. A better solution, the panel recommended, would be to launch smaller satellites carrying only those instruments that take simultaneous measurements. Sensors that provide the most useful information to resolve questions about climate change (such as the relationship between clouds and solar radiation or oceans and heat transport) would be launched first, perhaps as early as the mid-1990s. Eventually, a squadron of satellites would fly in close formation, allowing near-simultaneous measurements, even if not quite as precise as EOS had intended originally.[67]

Chaos

There are some problems in meteorology that money and even the most sophisticated measuring instruments cannot solve. In this age of orbiting satellites, sophisticated modeling techniques and powerful supercomputers, a typical weather outlook is still good for only seven days—a two-day improvement over a decade ago. More than a week into the future, no one can say for sure what the weather has in store. In a word, the problem is "chaos." There is so much going on simultaneously in the atmosphere—and in relation to the oceans and land below—that it is impossible for forecasters to make sense of it all.

Edward Lorenz, a theoretical meteorologist at the Massachusetts Institute of Technology, discovered the weather-chaos problem purely by chance in 1961. While checking out his weather forecasting model on a new computer, Lorenz decided to run the model twice using the same set of data. Only to save time, he dropped a decimal point's worth of data the second time he ran the model. Much to his amazement, he found that the tiny change in the data—his initial set of conditions—snowballed to the point where the forecast in his second run eventually bore no resemblance to the one in his first run. Chaos had taken over. The upshot of Lorenz's discovery came to be known as "the butterfly effect": A butterfly flitting its wings in Tokyo theoretically could make the difference between whether Miami basks in sunshine or is struck by a hurricane a month later.[68]

Long-range climate forecasts presumably avoid the chaos problem by not attempting to describe the evolution of weather from one day to the next. But it is possible that chaos may lurk in the climate system nonetheless. Changing the boundary conditions of the climate triggers another kind of chain reaction that must work its way through the entire system. While the climate could make a smooth and orderly transition to a new equilibrium state, it is also possible that it may gyrate wildly for a time, yielding unexpected results. It all depends on how the climate variables interact as buildup of greenhouse gases ensues.

As with so many other fields of science, chaos is changing the way meteorologists think about their profession. One of the newest methods of weather forecasting, in fact, uses the same technique that Lorenz applied by accident in 1963. It is called ensemble forecasting. Meteorologists run up to 10 nearly identical sets of climatic data through the same computer model. The almost imperceptible changes in the data are intended to mimic real-life errors

Figure 12

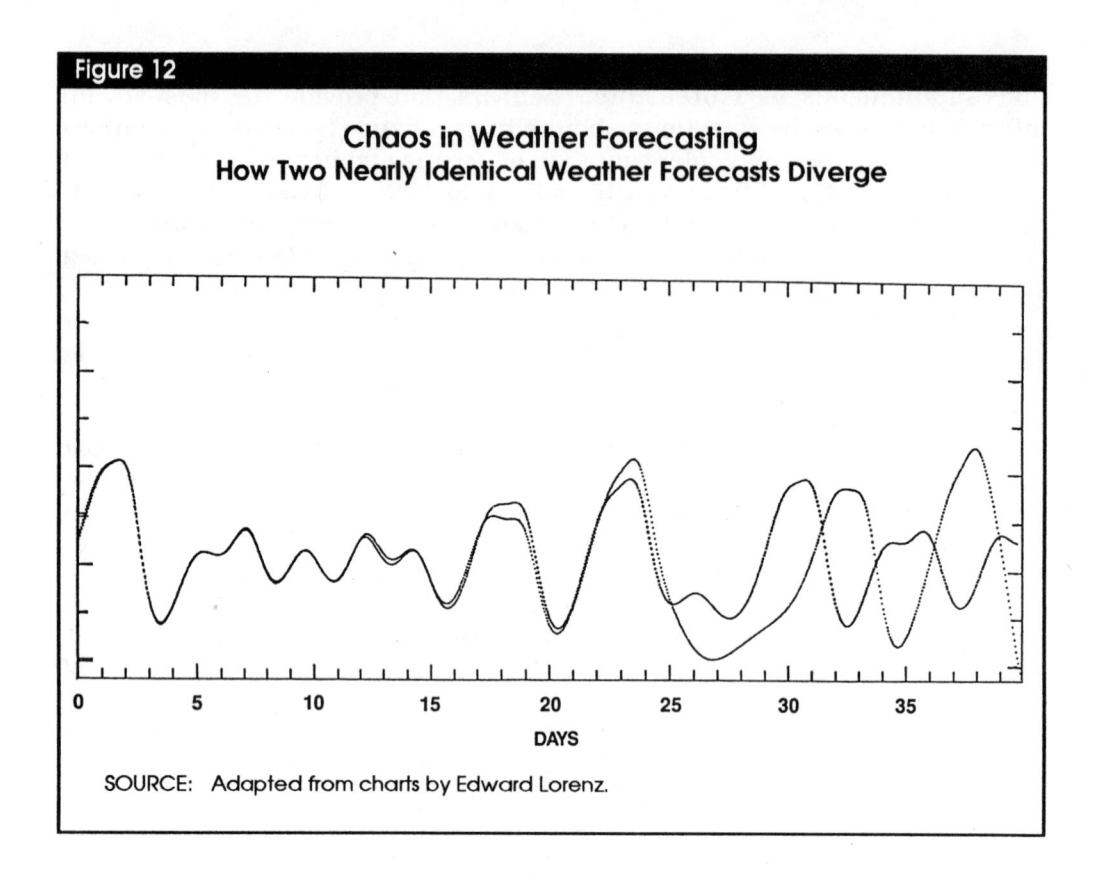

Chaos in Weather Forecasting
How Two Nearly Identical Weather Forecasts Diverge

SOURCE: Adapted from charts by Edward Lorenz.

in observing the weather's initial conditions. If, say, five days into a 10-day forecast the runs are showing completely different kinds of weather patterns emerging, meteorologists can assume that the atmosphere is in a highly volatile state and that they should assign a low level of probability to any forecast they make. But if the runs show an absence of major changes—i.e., differences come about only slowly—then they can be more confident that their forecast is going to hold up throughout the period.

As a real world example, one might recall the dome of high pressure that built up over the North American continent during the drought summer of 1988. It represented a kind of dominant weather pattern that (much to the lament of Midwestern farmers) was not about to give way to any pesky storm fronts. Forecasters could lay odds that the drought and heat wave would persist. But had they been modeling a spring or fall weather pattern—when the storm track is much more active—they would have been less inclined to bet whether it would rain or shine in Des Moines 10 days down the road.

Meteorologists have given ensemble forecasting a nickname befitting of gamblers: They call it the "Monte Carlo" approach.

General Circulation Models

With the greenhouse effect, the stakes of meteorological forecasting have never been higher. Trillions of investment dollars hinge on the outcome of global climate change, depending on whether the change is slow and relatively benign, impetuous and swift—or whether the change occurs at all. Soon, ground-based observations and forecasting techniques may reveal a clear climatic trend. Or perhaps orbiting satellites will discover a smoking gun, as they did with the ozone hole. But until one of these things happens, much of what is "known" about the greenhouse effect will remain in the theoretical realm of general circulation models. Thus, climatologists will continue to operate under some of the same constraints that apply to short-term meteorological forecasting, despite a raising of the ante. Moreover, chaos may still hold the trump cards.

When climatologists set up their computerized general circulation models, the first thing they have to do is to input the variables that constitute the initial conditions of the Earth's climate: sunlight, temperature, air pressure, wind, humidity and the like. Next they have to plug these variables into functions that govern the basic laws of nature: Newton's laws of motion, the conservation of energy, the Ideal gas law and so forth. Then it is nearly time to run the model—but not quite. Because it would be impossible to build a model that replicates real world conditions completely, climatologists must find short cuts. Like the weatherman who cannot possibly enter weather data from every point on Earth into his meteorological model, the climate modeler cannot represent the entire planet in infinite climatic detail in his general circulation model.[69]

As a result, the Earth gets divided into a gridwork of rectangular boxes. Typically 3,300 boxes cover the surface of the globe, with each box measuring about 300 square miles—roughly the size of Colorado. In this schematized view of the world, Florida does not exist, nor does Panama, Italy or the Great Lakes. Each box representing a piece of land is assigned a certain kind of topography. The land may have mountains, be covered with snow and possess up to six types of vegetation. The oceans also get divided into boxes, and the ocean depths may have up to 12 layers underneath each box. In the most sophisticated models, each layer of ocean is assigned a particular current and salinity level. The atmosphere over each box also is sliced into as many as 12 layers, each signifying a given altitude above the Earth's surface.

Climate phenomena smaller than the size of a box are represented collectively rather than individually—a process called "parameterization." Clouds, for example, are typically shown as a parameter of temperature and humidity. A cloud appears when enough water saturates the atmosphere, and it disappears when the theoretical conditions are ripe for rain or snow. In some general circulation models, all clouds are the size of Colorado. In others, they cover only a fraction of a grid box. The clouds move from box to box as weather systems pass over the landscape.[70]

When the general circulation model is finally set into motion, each box on the Earth's surface—as well as those in the atmosphere above and the ocean

Figure 13

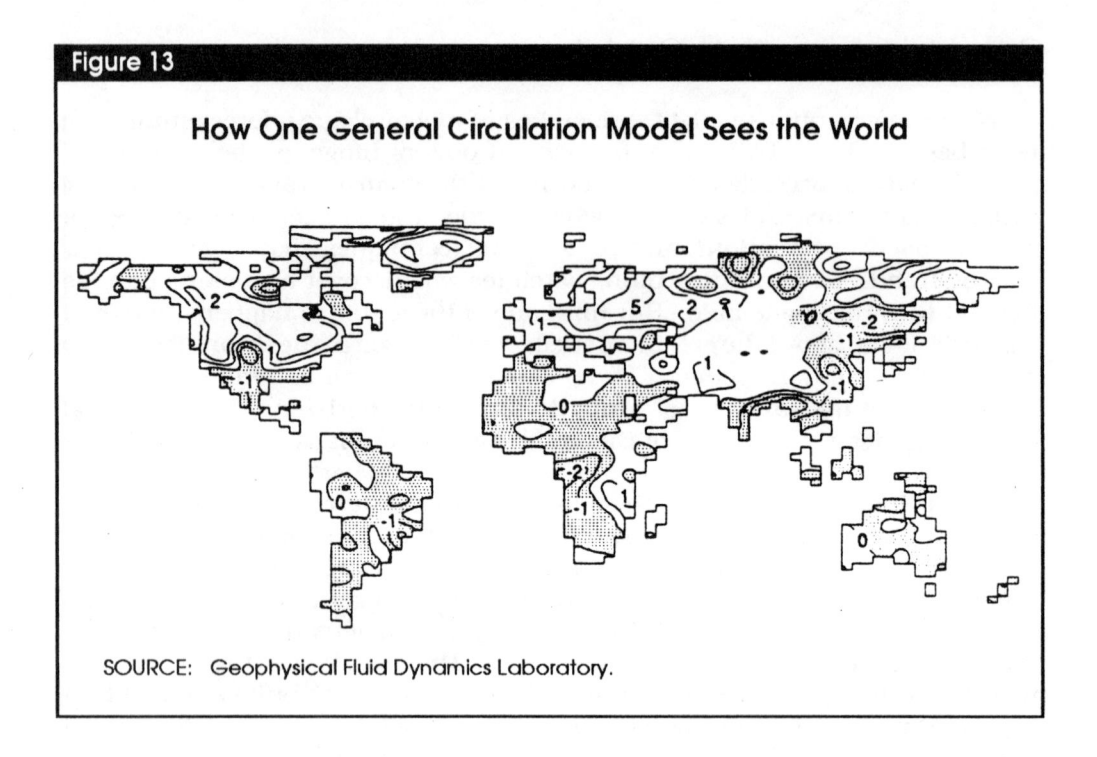

How One General Circulation Model Sees the World

SOURCE: Geophysical Fluid Dynamics Laboratory.

depths below—is treated as a separate entity. The mathematical formulas describing the laws of nature are applied to the variables within each box and are used to calculate how the conditions in one box affect the variables in each adjacent box. The calculations are done again and again until the effects have rippled through all of the boxes and connecting layers. Then the computer reperforms all of the calculations as if an interval of real time—anywhere from 10 minutes to an hour—had passed. A supercomputer typically takes 40 seconds to forecast a day's worth of weather changes.[71]

As the modeled days pass into months, the sunlight variable in each of the grid boxes is altered to reflect the changing of the seasons. As the seasons pass into years, an image of the Earth's climate pattern begins to emerge. Given all of the short cuts and all of the potential to magnify errors as the stream of calculations is made, it is remarkable that general circulation models work at all. Nevertheless, when control runs of the Earth's present climate are performed on the computer, familiar seasonal weather patterns show up in the results. It snows in the wintertime in the Northern Hemisphere; seasonal monsoons appear in the tropics; and desert regions of the globe rarely receive rain. In some of the models with a sophisticated treatment of oceans, even the periodic movement of the Southern Oscillation in the Pacific Ocean is picked up.[72] The models also have been manipulated successfully to replicate conditions that existed on Earth during the last Ice Age or that prevail now on other planets such as Venus and Mars.

Once climatologists are satisfied that they have incorporated all of the important climate forcing variables into their models and have completed

A Climate of Uncertainty

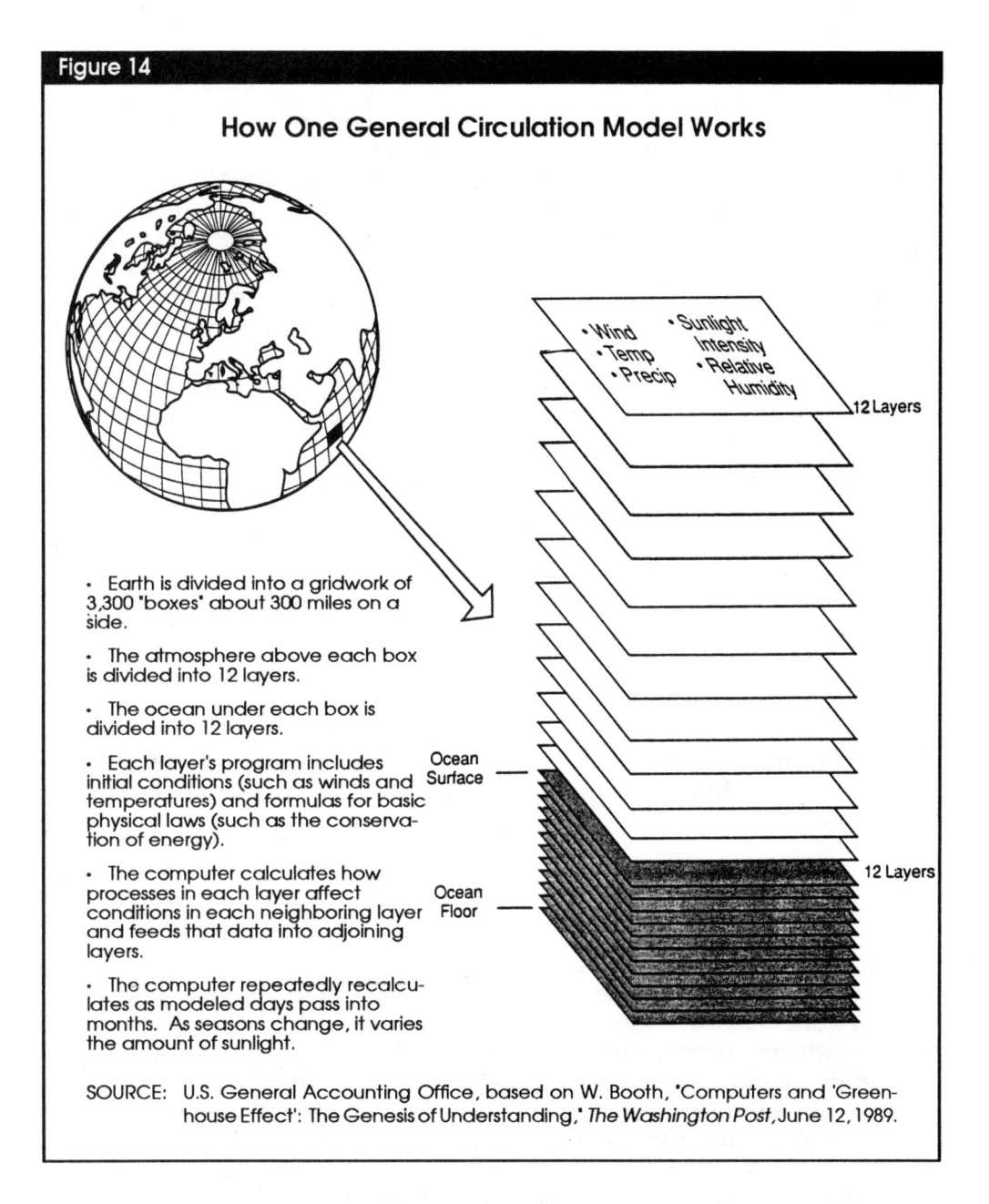

> **Figure 14**
>
> ## How One General Circulation Model Works
>
> - Earth is divided into a gridwork of 3,300 "boxes" about 300 miles on a side.
>
> - The atmosphere above each box is divided into 12 layers.
>
> - The ocean under each box is divided into 12 layers.
>
> - Each layer's program includes initial conditions (such as winds and temperatures) and formulas for basic physical laws (such as the conservation of energy).
>
> - The computer calculates how processes in each layer affect conditions in each neighboring layer and feeds that data into adjoining layers.
>
> - The computer repeatedly recalculates as modeled days pass into months. As seasons change, it varies the amount of sunlight.
>
> SOURCE: U.S. General Accounting Office, based on W. Booth, "Computers and 'Greenhouse Effect': The Genesis of Understanding," *The Washington Post*, June 12, 1989.

control runs that replicate the existing climate, they typically double the amount of carbon dioxide in their equations and see what happens. The added CO_2 effectively changes the model's energy budget, permitting the same amount of energy into the hypothetical climate system while letting less energy back out. The model is then allowed to run until the system achieves a new state of thermal equilibrium. On a powerful YM-P supercomputer, built by Cray Research, nearly three weeks of continuous operation is required before such a steady state is achieved.[73] In terms of model time, however, a century may have passed.

Through this painstaking process, climate modelers have come to verify Svante Arrhenius's century-old postulations about an enhanced greenhouse effect. According to the Intergovernmental Panel on Climate Change, which has conducted the most thorough review of general circulation models to date, a doubling of CO_2-equivalent in the atmosphere would be likely to cause the Earth's surface temperature to rise 3.5 degrees F by 2025 (compared with pre-industrial times) and perhaps by as much as 10 degrees F before a new thermal equilibrium is reached. Warming in higher northern latitudes would be greater than the global mean. And it would be possible that faster warming of land surfaces in relation to the oceans would lead to a greater occurrence of drought in mid-continental areas and more intense monsoons around the equator.[74]

But the IPCC has hedged its global warming forecast—just as any responsible meteorologist does when it comes to predicting the weather. It concluded in 1990:

> There are many uncertainties in our predictions with regard to the timing, magnitude, and regional patterns of climate change, especially changes in precipitation. This is due to our incomplete understanding of sources and sinks of greenhouse gases and the responses of clouds, oceans and polar ice sheets to a change of the radiative forcing caused by increasing greenhouse gas concentrations.[75]

In other words, there is a 20 percent chance of drought here, an 80 percent probability of flooding there, with skies becoming partly sunny—or is that mostly cloudy?

Clouds in the Forecast

The reason climatologists sometimes doubt their own general circulation models is that they recognize the inability of the models to account fully for the forces at work in nature. Clouds and oceans play especially important roles in driving weather patterns. But these variables are ones that climatologists tend to generalize and parameterize most in their models. It concerns them that when they tweak their models—changing the parameters for clouds or giving a more sophisticated treatment of oceans—ofttimes their results change dramatically. It suggests that their models have yet to account fully for all of the major factors influencing global climate change.

Take the case of clouds. Clouds are like curtains for the planet, shading about half the Earth's surface at any given time. Generally speaking, clouds reflect more solar energy off their tops than would be the case if sunlight passed through unobstructed to the Earth's surface and then reflected back into a clear sky. Consequently, clouds tend cool the planet, especially over the oceans at mid-latitudes.[76] Yet certain kinds of clouds are effective at trapping infrared heat radiating from the Earth's surface, producing a net warming effect.

Therefore, it is important to know how climate change might influence average cloud cover and the types of clouds that form.

Most climate models predict that a warmer world would have fewer clouds, even though the water vapor content of the atmosphere would increase. Since most clouds exhibit a cooling effect at present, a decrease in average cloud cover presumably would warm the planet, with the increase in available sunshine heating up the Earth's surface. If the climate models' predictions are wrong, however, and cloud cover increases in a greenhouse world, more of the Earth's surface would be shaded from the sun's rays. Then the net effect might be to keep daytime high temperatures down, while reducing radiational cooling at night.

Meteorological records compiled by Thomas Karl of the National Climatic Data Center in Asheville, N.C., indicate that average nighttime temperatures in the United States, China and the former Soviet Union have indeed been rising since 1950. At the same time, daytime high temperatures hardly have increased at all, especially in the United States and China. This day-night temperature disparity may be the result of an increase in average cloudiness in these countries, which cover about 15 percent of the Earth's surface. (The disparity also could be partly the result of rising sulfur emissions, which block sunlight effectively during the day but have no ability to retain heat in the atmosphere at night.) If a trend toward cloudy skies were to continue, longer crop-growing seasons might be in store, since killing frosts usually occur at night. Moreover, concerns about summer drought might be lessened, since warmer nights do not enhance soil evaporation nearly as much as if the warming occurred during daylight hours.[77]

There are other factors to consider with clouds besides whether their numbers are increasing or decreasing, however. Because clouds come in many shapes and sizes, they exhibit different radiational tendencies. Bright, puffy cumulus clouds block incoming sunlight very effectively, for example. Wispy cirrus clouds, on the other hand, allow some shortwave sunlight to pass right through them—but then they trap the longwave infrared heat bouncing back off the Earth's surface. It becomes important to determine, therefore, how the brightness, thickness and location of clouds might change in response to global warming. Satellite observations indicate that high clouds in the tropics, such as cirrus and cumulonimbus, have increased over the last 30 years, while lower and middle clouds, such as cumulus and stratus, have decreased.[78] These trends in cloudiness augur a positive feedback—a warming.

Not all climatologists agree on what types of clouds would become more prevalent as a result of global warming, however. One theory posits that convective air currents should intensify as sea surface temperatures rise near the equator, allowing cumulus clouds to tower higher into the troposphere. Since the moisture that goes up eventually must come down, the net effect of all this roiling could be to trap a layer of moist air near the surface—with dry, detrained air piled on top of it. This in turn would diminish the greenhouse effect, since more sunlight would bounce off the tops of the cumulus clouds and less water vapor would be present in the middle troposphere to absorb infrared heat radiating from the Earth's surface. One greenhouse skeptic, Richard

Lindzen of the Massachusetts Institute of Technology, has argued that general circulation models are overestimating global warming potential by 20 to 50 percent because they fail to account for this phenomenon.[79]

Others are skeptical of the skeptics' arguments. They point out that the satellite observations tend to confirm climate model forecasts of a positive cloud-water vapor-feedback effect. That is to say, rising temperatures create more of the types of clouds that trap longwave infrared heat near the surface rather than those that block shortwave sunlight from reaching the surface in the first place. Moreover, the satellite data and the climate models indicate that the water vapor content of the atmosphere increases substantially as the atmosphere warms, especially in the upper troposphere, near where Lindzen says the most drying would occur.[80] While a negative feedback is possible in some tropical zones if cumulus clouds become more prevalent, the appearance of other types of clouds in parts of the world that are relatively cloud-free at present could create a countervailing positive feedback. The exaggerated warming away from the equator might induce an even greater regional disparity in climate change than is now forecast in most general circulation models.[81]

In any event, clouds bear watching. The United Kingdom Meteorological Office—considered to have one of the most sophisticated climate models in the world—created a stir a few years ago when it scrapped its old parameter for clouds in its model and replaced it with a new, more complex one. The British team decided to make a distinction between clouds consisting mainly of ice particles at below-freezing temperatures and other warmer clouds consisting mainly of water droplets. With this amendment, the model's global warming prediction dropped nearly in half—from a 9 degree F increase to only a 5 degree F increase. The reason for the dramatic change, according to the modelers, is that a slight warming of the lower troposphere would convert ice particles to water droplets in many clouds hovering around the freezing point. Since water droplets fall out of clouds less readily than ice particles, these "wetter" clouds would tend to hang around longer in the atmosphere—and block sunlight more of the time.[82]

To further dramatize the role that clouds play in climate model forecasts, the U.S. Department of Energy coordinated a study of 19 general circulation models in 1990. When the models' assumptions about clouds were suppressed—projecting the Earth with a clear sky—the models showed remarkable agreement in "climate sensitivity" (a proxy for global temperature change caused by direct radiative forcing of greenhouse gases). But when the models' cloud-feedback effect was taken into account, a threefold variation in climate sensitivity emerged. While nearly all of the models assigned a positive value to cloud feedback—suggesting an enhanced warming effect—the wide range of values "emphasizes the need for improvements in the treatment of clouds if they are ultimately to be used as climatic predictors," the study concluded.[83]

Motion in the Oceans

As unpredictable as clouds are in general circulation models, most climatologists consider oceans to be the real "wild card" of climate prediction. Until the mid-1980s, many general circulation models treated oceans like giant ponds—motionless slabs of water that warmed and cooled seasonally to a fixed depth of 150 or 200 feet. Of course, oceans are much more dynamic than that. Their temperatures and salinity vary considerably, and ocean currents transport heat from warm tropical waters toward the poles—exerting a tremendous influence on climate. Were it not for the Atlantic Gulf Stream, for example, annual temperatures in Europe would be 12 or 13 degrees F colder on average.[84]

As noted earlier, the El Niño/Southern Oscillation in the Pacific Ocean also plays a major role in driving the world's weather patterns. Influencing the movement of the warmest waters of the Pacific, the Southern Oscillation acts like a giant heat pump, distributing heat from these balmy equatorial waters to higher latitudes, aided by powerful atmospheric storms brewed in the western Pacific. When this warm water spreads to the eastern Pacific—an El Niño event—it usually sets off a climatic chain reaction around the world. Even global temperatures tend to rise during an El Niño.[85] When the water ricochets back to the west, in what has become known as a La Niña event, the climatic effects are generally reversed. The Indian and Australian summer monsoons intensify, while the deserts of Peru along with the western and southern coasts of the United States tend to dry out. Some believe a La Niña event was in fact the principal cause of the North American drought during the "greenhouse summer" of 1988.[87]

Given the ways in which the oceans and the atmosphere interact to influence weather, most general circulation models now attempt to couple their behavior, but the relationship remains fairly crude. Many of the models assign variable sea surface temperatures and fixed ocean currents to simulate ocean heat transport, but they do not allow the heat transport variable to evolve as the climate warms. Some models have assumed away this important relationship altogether by setting sea surface temperatures according to historical analogs and by ignoring the horizontal and vertical transport of heat in ocean currents.

That is precisely what climate modelers did for 20 years at the Geophysical Fluid Dynamics Laboratory in Princeton, N.J., for example. Considered pioneers in the effort to model a coupled ocean-atmospheric system, the GFDL modelers were unable to synchronize the rise in sea surface temperatures with the rise in atmospheric temperatures as hypothetical global warming ensued. Such a correlation quickly fell apart in their early computer programs; temperature disparities of up to 7.5 degrees F cropped up after a century of projected warming.[87] Finally, in 1989, the GFDL team arrived at a satisfactory model that factored in the poleward transport of heat along the oceans' western sides as well as the return flow of cooler water toward the equator on the oceans' eastern sides. When the GFDL modelers ran this more sophisticated program, they were in for a big surprise: The simulation predicted that only half the planet would feel the heat of global warming during the next century.

Climate modelers have long taken for granted that the effects of global warming would be muted in the Southern Hemisphere because of the frozen Antarctic continent and a paucity of land masses south of the equator. But the coupled ocean-atmosphere GFDL model projected that, after a short spurt of warming early in the 21st century, the southern half of the globe actually would *cool off* around 2070. In some high latitude regions near Antarctica, the cooling might amount to as much as 7 degrees F—precisely where conventional greenhouse theory suggests some of the greatest warming should take place.[88]

The inclusion of ocean currents in GFDL's general circulation model made the difference in its conclusions. Several ocean currents converge around Antarctica, especially in the Weddell Sea, where the world's coldest—and therefore densest—water is formed. This heavy water sinks to the bottom of the ocean and remains isolated from the atmosphere for 500 to 2,000 years. Consequently, when warm surface water is circulated into these depths, it has less chance to heat the surrounding layers of the atmosphere. Simultaneously, natural upwelling of frigid waters in the southern Pacific Ocean cools the atmosphere and prevents much of the sea ice around Antarctica from melting.

In the 21st century, especially cold water from the "Little Ice Age" is due to return to the surface. (The Little Ice Age was a period between 1450 and 1850 A.D. when global temperatures dropped off by 2 degrees F.) The reappearance of these elderly waters may cause atmospheric temperatures to drop in portions

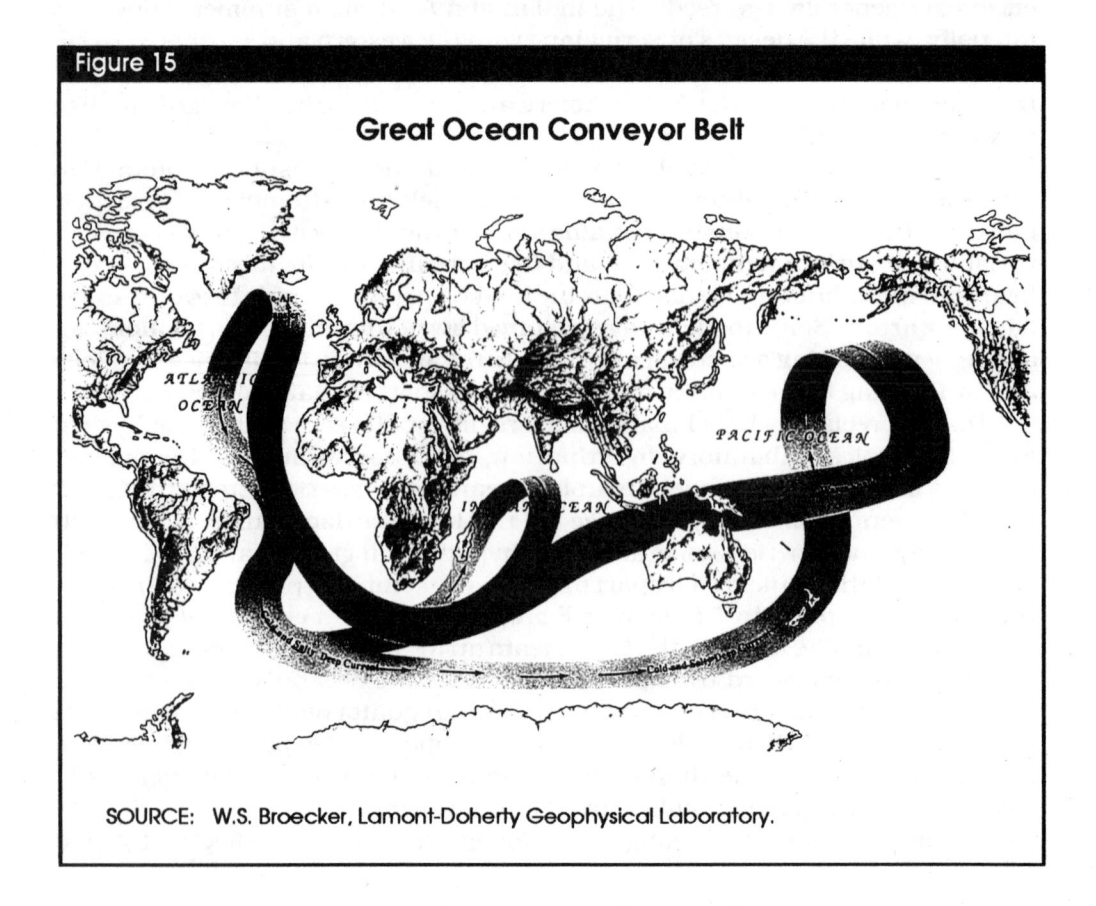

Figure 15

Great Ocean Conveyor Belt

SOURCE: W.S. Broecker, Lamont-Doherty Geophysical Laboratory.

of the Southern Hemisphere in the latter half of the 21st century, even as average temperatures in the Northern Hemisphere rise 3 to 7 degrees F, according to the coupled GFDL model. The findings are significant not only because they suggest that the Northern Hemisphere will bear the brunt of global warming during the next century; they also indicate that melting of sea ice around Antarctica may occur at a slower rate than previously thought, lessening the prospect for a rapid rise in sea level that would threaten low-lying coastal areas around the globe.[89]

Transient Behavior

One other important feature of the GFDL model and other dynamic general circulation models is that they offer a glimpse of how the climate may respond over time to an incremental accumulation of greenhouse gases in the atmosphere. Until two years ago, all but one of the climate models doubled the amount of carbon-dioxide equivalent all at once and then let the computers churn until the climate system arrived at a new state of thermal equilibrium. Such a method permits theoretical comparisons of the climate before and after global warming but fails to capture the time in between when the buildup of greenhouse gases is taking place.

It is vital to consider how the climate system might respond step by step to global warming because some parts of the system will take longer than others to assimilate the influx of gases. Oceans are a prime example, as the coupled GFDL model illustrates. While surface layers are exposed to the atmospheric warming as it happens, the ocean depths are insulated from such transient effects. The mixing of the oceans' layers will, in fact, carry on the planet's response to global warming for many centuries after the amount of greenhouse gases in the atmosphere has leveled off—and long after land surfaces have achieved a new state of thermal equilibrium. As a result, significant climatic effects will transpire long before the ripples of change propagate through the entire system and before the climate finds a new rhythm.

"Hence during the transient phase the warming and other climatic effects induced by the enhanced greenhouse effect could well display worldwide patterns significantly different from ones inferred on the basis of equilibrium simulations," writes Stephen Schneider of the National Center for Atmospheric Research, who was one of the first to develop a model showing the importance of the transient phase of global warming. As a cautionary note, Schneider adds, "Furthermore, the social impact of climatic changes would probably be greatest rather early, before equilibrium has been reached and before human beings have had a chance to adapt to their new environment."[90]

One historical example helps to illustrate this point—which is to expect the unexpected. At the end of the last great ice age, the planet was thawing out and average global temperatures were rebounding to their highest levels in more than a million years. Then a remarkable climatic event put a large portion of the Northern Hemisphere back into an icy grip for hundreds of years—even as the rest of the planet continued to warm.

At the time, about 10,700 years ago, the huge North American ice sheet had been melting for several thousand years, and its runoff flowed through the Mississippi basin into the Gulf of Mexico. When the ice sheet retreated far enough to the north, its passage to the Gulf of Mexico became blocked and the runoff found a new route to the sea. Instead of depositing an Amazon River's-worth of cold freshwater into the warm Gulf, the flow apparently was diverted into the St. Lawrence Seaway, which empties into the especially salty waters of the North Atlantic. This massive infusion of freshwater diluted the denser, salt water that normally causes the Gulf Stream to sink as it reaches the North Atlantic. In fact, the dilution may have stopped the Gulf Stream from flowing altogether.[91]

Without the Gulf Stream's transport of heat from the tropics, maritime Canada, Greenland, Iceland and northern Europe suddenly reverted to an ice-age climate. Forests that had dominated parts of northern Europe for 2,000 years succumbed during the first few decades of renewed cold, and Arctic flora that had retreated far to the north re-staked their claim in the south. The cold spell lasted for about a millennium. Eventually, remnants of the Laurentide Ice Sheet, covering the general area that is now Hudson Bay, diminished to the point where not much meltwater remained. As the high salt content of the North Atlantic was restored, the Gulf Stream began to flow again and the localized ice age ended. Only then did northern North America and northern Europe resume the warmup that had been going on elsewhere around the world for several thousand years.

Expecting the Unexpected

The interruption of the North Atlantic Gulf Stream nearly 11,000 years ago has come to be known as the "Younger Dryas" event. The name derives from an arctic flower that flourished during the Ice Age and made a brief resurgence during the time the Gulf Stream was cut off. The Younger Dryas is a small symbol of something far more significant, however: the ability of climate to exhibit completely unpredictable behavior—in this case, a sudden regional cooling on the way to a far more predictable rise in global temperatures.

Climatic Surprises

Today, a jolt to the Earth's climate system—say, a major volcanic eruption or a change in ocean currents—could cause regional or global temperatures to drop just as suddenly despite an underlying warming trend. Indeed, the massive eruption of Mount Pinatubo in June 1991—the largest volcanic eruption since Indonesia's Krakatua in 1883—is having such a cooling effect on the planet now. The 25 million tons of sulfur aerosols lofted high into the atmosphere by Mount Pinatubo is expected to blot out 2 percent of the incoming sunlight and lower the Earth's temperature by 1 degree F over the next few years, before the aerosols settle out of the sky. (As of April 1992, satellite soundings already had detected a cooling of about 0.7 degrees F around the globe.)[92]

Meanwhile, researchers at the Lamont-Doherty Geological Observatory at Columbia University have come across another surprising subterranean development. The amount of cold, dense seawater that forms near Greenland and sinks to the bottom of the Atlantic Ocean has decreased 80 percent since 1980, apparently because of decreased salinity of the surface layer.[93] (The colder the water and the greater its salt content, the denser it is and the more it tends to sink.) This dilution may have been caused by an increase in freshwater runoff, melting sea ice, rising ocean temperatures or a combination of factors.

Whether global warming ultimately is responsible for the dilution of saltwater near Greenland is purely speculative. However, a coupled ocean-atmosphere model developed at the National Center for Atmospheric Research has forecast a greater flow of low-salinity water into the East Greenland Sea during the third decade of a hypothetical warming.[94] In the model run, the Atlantic Ocean "conveyor belt" that brings warm Gulf Stream water to the coast of Europe becomes obstructed for a time because of a change of prevailing winds in the North Atlantic from the west to the northwest. This wind change has the effect of pushing the warm Gulf Stream waters to the south and introducing colder air to the region, cooling off the ocean's surface layers. If in fact the deep-water circulation process in the Greenland Sea were to be interrupted—as the

NCAR model and Woods Hole data suggest—Europe could enter another mini-ice age just as it did during the Younger Dryas event.

While this is a counter-intuitive example of how global warming could affect regional climate, there are other forces that could just as easily shift the Earth into a rapid warming mode. That same interruption of ocean circulation, for example, could inhibit the transfer of carbon dioxide-rich surface waters into the ocean depths. Bearing in mind that the oceans store at least 60 times more CO_2 than the atmosphere, the loss of this huge sink (at least temporarily) could cause excess CO_2 to build up much more quickly in the atmosphere. Indeed, reduced downwelling of ocean water in the Greenland Sea may be a factor in the recent increased rate of CO_2 accumulation in the atmosphere. According to measurements by Charles Keeling of the Scripps Oceanographic Institute, the annual rate of increase has jumped from 0.3 percent to 0.5 percent since the mid-1980s.[95]

Accelerated respiration and decay of plants also may be contributing to the increased rate of atmospheric CO_2 accumulation. Since 1958, when Keeling began recording fluctuations in the levels of carbon dioxide in the atmosphere, global temperatures and naturally occurring concentrations of atmospheric CO_2 usually have changed in tandem. But when the contribution of fossil-fuel burning and the effect of seasonal variation is eliminated from the data, temperature changes appear to precede changes in atmospheric CO_2 levels by a matter of months.[96] The suggestion is that higher temperatures lead to greater CO_2 concentrations in the atmosphere in a positive feedback effect.

The broader significance of this finding is that it may answer a chicken-and-egg question about fluctuating CO_2 and methane levels during the ice ages and previous interglacials. At the end of the last great Ice Age 12,000 years ago, for example, CO_2 levels rose from fewer than 200 parts per million to approximately 260 parts per million in a matter of a few thousand years. Similarly, the atmospheric concentration of methane jumped from 300 parts per billion to more than 500 parts per billion during that period.[97] The unresolved question is whether the rise in atmospheric temperatures led to an increase in CO_2 and methane emissions—causing the warming to build on itself—or whether the increase in greenhouse gas emissions came first. The new data seem to suggest that rising temperatures spur the release of additional greenhouse gases and exacerbate the warming, but this finding is by no means conclusive.

Another important means by which global warming could spur a natural increase in greenhouse gas emissions is by thawing permafrost in high-latitude regions. Permafrost is thought to contain a high concentration of methane gas in ice-like compounds, known as methane clathrates. These clathrates could be released to the atmosphere as a thawing took place. By some estimates, a sudden increase in methane emissions from this permafrost might contribute about as much to global warming potential as the loss of ice and snow in polar regions. In technical terms, the positive feedback from the release of methane clathrates would affect the radiative budget as much as the reduction in planetary albedo from the loss of polar reflective surfaces—doubling the cumulative effect.[98] The bottom line: The world would get warmer faster.

Figure 16

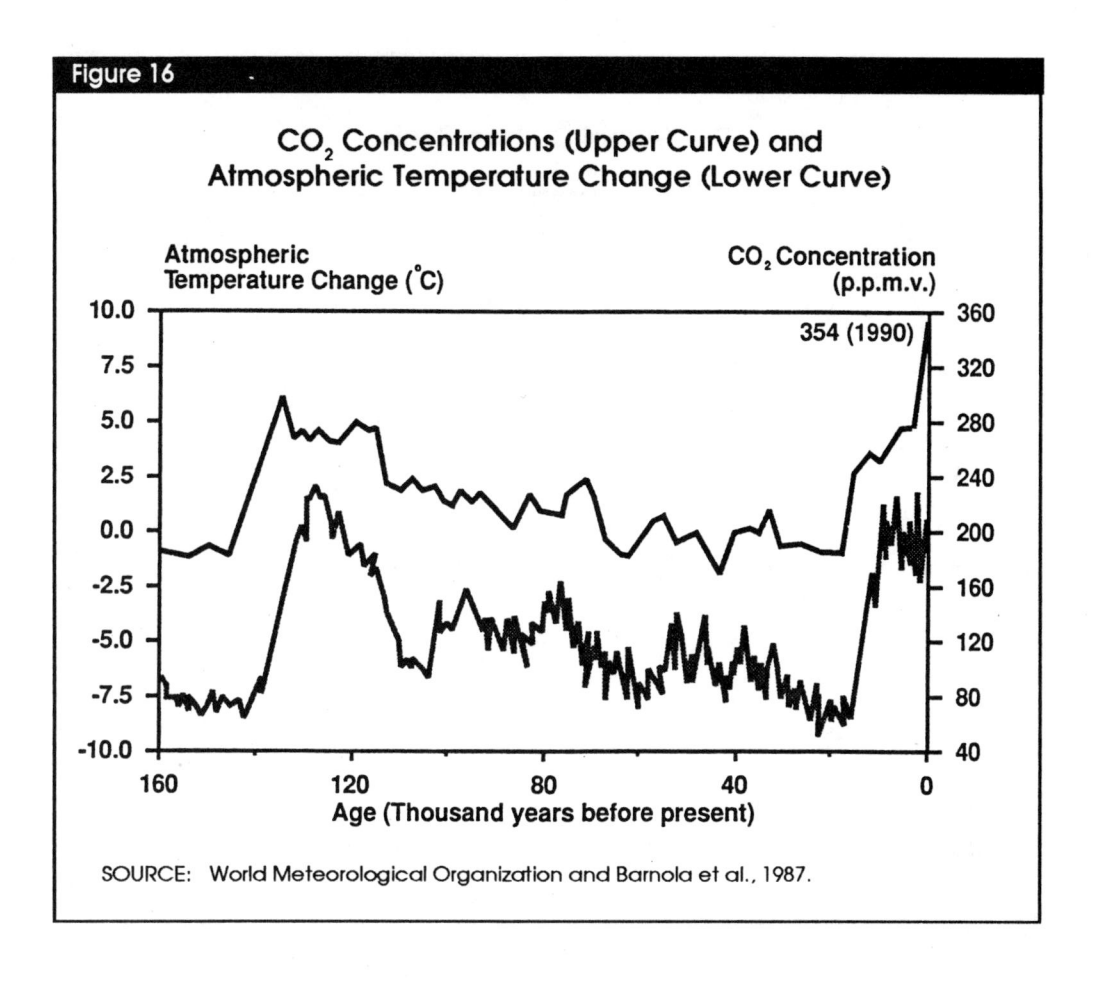

CO$_2$ Concentrations (Upper Curve) and Atmospheric Temperature Change (Lower Curve)

SOURCE: World Meteorological Organization and Barnola et al., 1987.

One of the greatest mysteries of all concerning the greenhouse effect is that more than a billion tons of carbon in the annual carbon budget is unaccounted for. Scientists are fairly confident that carbon emissions from combustion of fossil fuels and burning of tropical forests totals approximately 7 billion tons a year—half of which remains in the atmosphere. The other half apparently is swallowed by oceans and sequestered by plants and trees, but the numbers on this end of the carbon cycle do not add up. Oceanographers generally agree that the oceans are capable of taking up about 2 billion tons of carbon annually, which leaves 1 billion tons to be taken up on land. Terrestrial ecologists have little evidence to show that the biosphere is sequestering more carbon dioxide than it is releasing, however. It may be possible that temperate forests, rebounding from the severe cuts of the 18th and 19th centuries, are absorbing the extra carbon. Another possibility is that a "fertilization effect" brought about by the rise in atmospheric carbon dioxide may enable plants and trees to store more carbon through photosynthesis. In any event, one must hope that this mysterious carbon sink keeps working. Without it, the amount of carbon dioxide added to the atmosphere would increase by 30 percent a year, increasing the chances of a severe warming.[99]

Figure 17

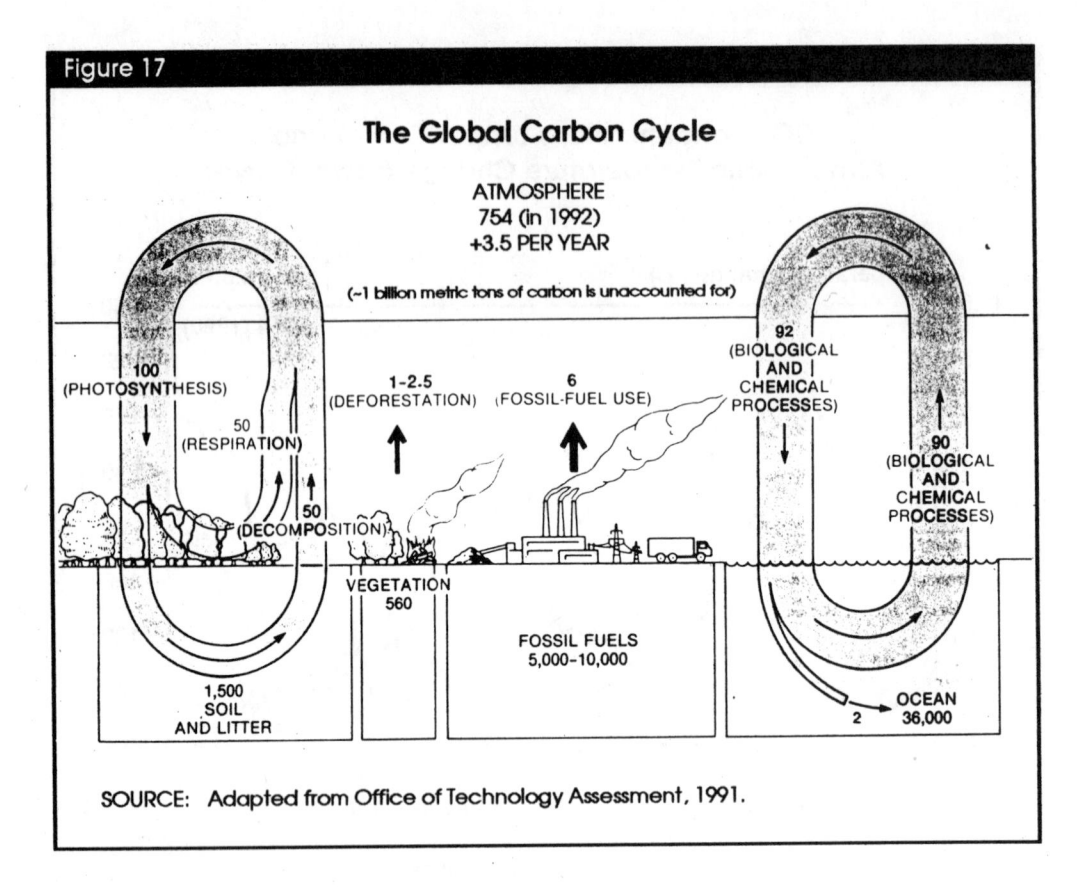

The Global Carbon Cycle

ATMOSPHERE
754 (in 1992)
+3.5 PER YEAR

(~1 billion metric tons of carbon is unaccounted for)

100 (PHOTOSYNTHESIS)
50 (RESPIRATION)
50 (DECOMPOSITION)
1-2.5 (DEFORESTATION)
6 (FOSSIL-FUEL USE)
92 (BIOLOGICAL AND CHEMICAL PROCESSES)
90 (BIOLOGICAL AND CHEMICAL PROCESSES)

VEGETATION 560
FOSSIL FUELS 5,000-10,000
1,500 SOIL AND LITTER
OCEAN 36,000
2

SOURCE: Adapted from Office of Technology Assessment, 1991.

Recent Weather Anomalies

Given the many surprises that global warming may have in store, it is worth remembering that weather always behaves in chaotic fashion, even when no new forces are working their way into the climate system. While loose predictions can be made about the likely behavior of weather over time, meteorologists are not yet able to predict the exact nature and timing of future weather events, and given the limitations of forecasting, they probably never will. Long-term climate analysis becomes especially dicey when the entire planetary system is being subjected to a new stress—in this case an influx of greenhouse gases.

General circulation models seem to assure climatologists that the Earth's climate system eventually will settle into predictable behavioral patterns after the influx stabilizes and the climate system achieves a new state of thermal equilibrium. But chaos theory suggests that as the climate makes the necessary adjustments—the transient phase of global warming—weather patterns may become especially erratic and unpredictable. This is a problem that general circulation models cannot readily resolve.

A simple analogy to the present situation is trying to predict when water will drip from a faucet. When the flow of water through the spout is slow and regular, one can observe the pattern of drips and easily predict when the next drop will fall. If the flow is then increased sufficiently and maintained at that rate, one can similarly forecast that the drips will bead together and form a continuous stream. The system has shifted from one steady state to another. But if the flow of water is increased gradually, there will be a period when the drips come two or three in a row, then pause for an instant, followed by five drips in rapid succession, and so on; the pattern becomes erratic and highly unpredictable.[100] This unsteady phase is the part of global warming that the planet may now be experiencing, as the flow of greenhouse gases entering the atmosphere slowly increases.

Is there any real-world evidence to suggest that the influx of greenhouse gases is causing the Earth's climate to behave erratically? Such a proposition is difficult to test because "normal" weather patterns are themselves so unpredictable. Nevertheless, it is interesting to consider some of the unusual climatic events that have taken place since the summer of 1988—when climatologist Jim Hansen announced to the world that the "greenhouse effect is here."

Record cold and warmth: If global warming has indeed begun, the nation's weather data for December 1989 certainly would not reveal it. That December was the coldest on record for 15 U.S. states east of the Missisippi River—and the fourth coldest December for the nation as a whole. (Weather records date back about 100 years.) However, the following month was the warmest January on record for the entire country; maybe global warming had begun after all. But at the end of 1990 it was the West's turn to shiver through *its* coldest December on record. More than 260 weather stations set daily low-temperature minimums, and a few stations even experienced all-time record cold (i.e., the coldest temperature measured at the station on any date).[101]

Even though 1990 was the warmest year on record for the planet overall, that came as cold comfort to California's citrus growers, who sustained three-quarters of a billion dollars in crop damage during the December 1990 deep freeze.[102] Meanwhile, back in the East, 160 weather stations broke daily high-temperature records in December 1990. Cities from Portland, Maine, to Philadelphia, Pa., posted their highest average temperatures ever for the month of December, after experiencing their coldest December ever just the year before.[103]

The winter of 1991-92 (or lack thereof) was more consistent across the United States. It was the warmest winter in the 97 years that the National Weather Service has been keeping records, with temperatures running 4 degrees F above normal nationwide, and as much as 11 degrees F above normal in the Upper Midwest.[104] The lack of any appreciable snowfall outside of mountainous regions led many to wonder if global warming was to blame. But to confound the pundits, a series of spring snowstorms struck the Southeast, including a freak storm in May that provided northern Georgia with its latest snowfall on record. Other parts of the world that rarely get snow also had plenty of it during the winter of 1991-92. Jerusalem received nearly two feet of snow. Shanghai, China, had a rare snow shortly after Christmas. It snowed on Cyprus for the first time in 40 years, in Beirut for the first time in 55 years, and in Eilat,

Figure 18

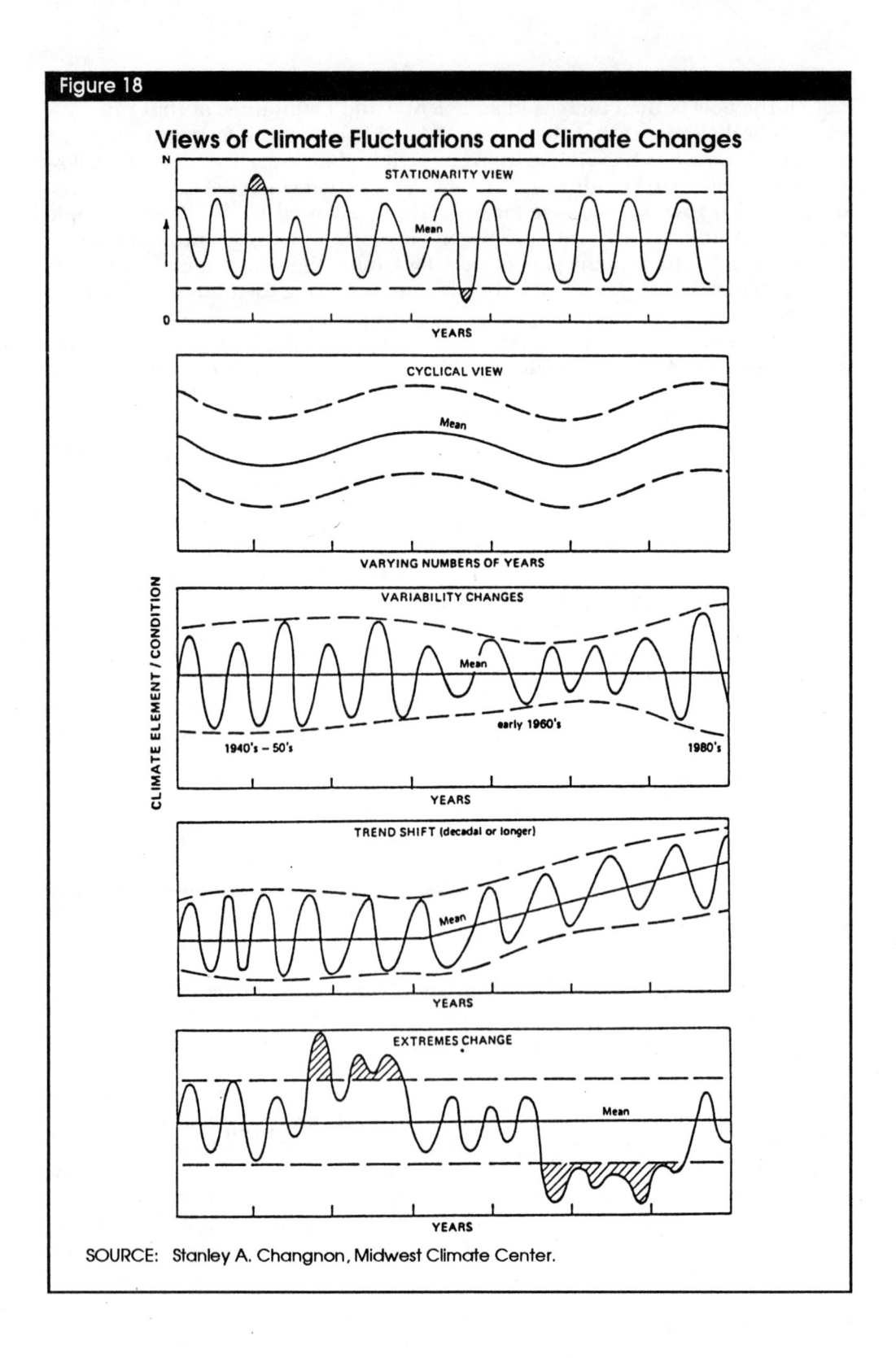

Views of Climate Fluctuations and Climate Changes

SOURCE: Stanley A. Changnon, Midwest Climate Center.

Israel, for the first time ever (in recorded history).[105]

What is one to make of this jumble of record highs and lows, snows and no-snows? It is hard to build a convincing case that the "greenhouse effect is here" when both the East and the West endured record-breaking winter cold in 1989 and 1990, respectively, and snow fell in normally mild places during the winter of 1991-92. But the fact that many of these same regions broke warm-weather records over the period may be indicative of the kind of climatic instability—chaos—that is caused by the ongoing build-up of greenhouse gases in the atmosphere.

Record barometric pressure: Two other examples of recent weather extremes involve measurements of barometric pressure, one of the most basic tools in weather forecasting. Normally, barometric pressure slowly vacillates within a narrow band between 29 and 31 inches of mercury. Masses of cold, dense air push the barometer up; warm, unstable air causes the mercury to fall. In a five-month period between late 1988 and early 1989, however, both the all-time-high and the all-time-low barometric pressure records for North America were shattered. The all-time low reading of 26.22 inches was set on Sept. 13, 1988, in the eye wall of Hurricane Gilbert, which cut a path of devastation through the Caribbean and Mexico.[106] With sustained winds of 175 miles an hour, Gilbert is the strongest hurricane ever measured in the Atlantic (although Hurricane Hugo, the tenth strongest on record, reached greater sustained winds of 200 miles an hour one year later).

Just five months after Hurricane Gilbert set the low-pressure record, a frigid air mass on Jan. 31, 1989, broke the high-pressure mark with a reading of 31.85 inches at Northway, Alaska.[107] This Siberian Express subsequently headed south, creating the most extensive cold-weather outbreak in the continental United States of the 20th century. What made this arctic blast especially notorious is that it followed on the heels of a record warm spell. On Feb. 1, 1989, 54 American cities matched or broke daily high temperature records; only five days later, many of these same cities (and 41 U.S. cities in all) set or equaled daily low temperature records. These weather anomalies broke through formidable barriers on opposite ends of the barometric spectrum—and did so in rapid succession.

Atmospheric 'hiccups': Also in late 1989 and early 1990, meteorologists at the Midwestern Climate Center in Champaign, Ill., observed another unusual weather phenomenon when four mysterious atmospheric pressure waves rippled across the lower atmosphere of the Midwest. These so-called "solitary waves" are like atmospheric hiccups. Although they had been observed before in localized areas, these waves were of unprecedented size and duration. Each wave measured several hundred miles in length and up to 60 miles in width, and towered up to four miles in the atmosphere. At the front of each wave, the atmospheric pressure dropped suddenly—the hiccup—followed by a burst of rain. Then as the wave crested about an hour later, the atmospheric pressure surged to its previous level, accompanied by gusty winds and cessation of the rain. Between Dec. 21, 1989, and Jan. 14, 1990, four of these giant waves washed across the Midwest—from central Missouri to eastern Ohio—spanning two states (from north to south) as they went. Needless to say, aircraft were advised not to "ride" these strange and turbulent waves as they rolled across the country.[108]

Record tornado outbreaks: Recent years also have been notable for another kind of turbulent weather, namely tornadoes, which have struck all but six American states. The 1990 tornado season was the most active ever—1,115 twisters in all—exceeding the prior record of 1,102 set in 1973. The preliminary total for 1991 is 1,141 tornadoes, perhaps another record.[109] (Figures date back to 1953. Improved monitoring equipment may be a factor in the reported increase of tornado sitings.) The severity of tornadoes also appears to be on the upswing. In 1990, 14 tornadoes were in the violent "F4" and "F5" categories (on the Fujita scale), including three tornadoes with estimated wind speeds of 260 miles an hour.[110] Not since 1976 had three F5 tornadoes struck the nation in a single year.

Record hurricanes and tropical cyclones: Severe storms are not confined to the land, of course. With global warming, there is an even greater concern about ocean storms, which draw their strength from the heated surface layers of the sea: The warmer the surface water becomes, and the deeper this warm layer extends beneath the surface, the more ferocious the storms will be. Hurricane Gilbert now ranks as the most intense hurricane ever measured, approaching the upper theoretical limit of a "Class 5" storm. (Hurricanes are ranked according to wind speed on a scale of 1 to 5.) Yet if sea surface temperatures were to rise another 3 or 4 degrees, it would be possible for barometric pressure in the eye of a hurricane to drop below 24 inches of mercury and for its destructive potential to rise another 40 to 50 percent. Meteorologists would have to add a Class 6—and perhaps a Class 7—ranking to their repertoire.

While Hurricane Gilbert ranks as the strongest Atlantic hurricane on record, a tropical cyclone that struck Bangladesh in April 1991 now rates as the most powerful storm ever to ply the Bay of Bengal. Known simply as "Cyclone 2B," it reached sustained winds of 160 to 200 miles an hour shortly before landfall. A 20-foot tidal crest ushered the storm ashore, which killed 140,000 people in cyclone-prone Bangladesh.[111]

Record gales: Far to the north, Europe also has braced for record-breaking ocean storms. Three successive gales struck Great Britain and the European continent in late January and early February 1990 with sustained winds reaching 100 miles an hour. Such a series of severe storms—unprecedented in modern European history—led to popular speculation that the greenhouse effect was responsible for them.[112] While that cannot be proven, it is known that the storms fed on a clash of unusually cold air embedded to the north and especially warm tropical air to the south, making the gales much more powerful than usual. The contrast in temperatures between the Arctic and the equator also caused the jet stream—that fast-moving river of air that propels surface storms across the Northern Hemisphere—to move at breakneck speed. In January 1990, the velocity of the jet stream accelerated to 230 miles an hour in the North Atlantic—as fast as it has ever been measured there.[113] Ocean storms propelled along this fast track gained full intensity just as they reached Europe, rather than losing most of their punch over open water as they normally do.

Near-record drought and wetness: The gales of 1990 were just parts of an extended series of bizarre weather anomalies to strike Europe in recent years. Before the gales, drought was the predominant concern over much of the continent and in the United Kingdom. The water table as measured in West Sussex, England—the longest continuous observation site in the world—was within an inch or two of the lowest level ever recorded as of mid-December 1989. But the winter of 1989-90 turned out to be the wettest winter in more than 70 years in Great Britain. February 1990 was in fact the wettest February ever. The water table at the West Sussex borehole surged 125 feet in just eight weeks, an unprecedented rise in the 150-year history of the site.[114]

That was not the only strange thing about Britain's weather in 1990. Scotland registered its wettest year on record, but for the rest of the United Kingdom it was the second driest year of this century—despite that record-breaking precipitation in February.[115] March through September was the driest spring and summer period for England and Wales in a rainfall series dating back more than 220 years. "The juxtaposition of drought, widespread flooding and drought again is not without precedent," two British hydrologists noted in the *Weather* journal of the British Meteorological Society, "but there is no modern parallel to the remarkable swings in water resources recently experienced. Certainly there was little hint in the decade up to 1988 of the erratic conditions to come." In a cautionary note about global climate change, the hydrologists continued: "Whether the recent wild swings in weather patterns represent the extreme limit of normal variability or are illustrative of conditions which may be expected with greater frequency over ensuing decades remains to be determined."[116]

Record European heat: It is not just unusual rainfall patterns that have the British wondering about climate change. The temperature record also is intriguing. Three consecutive winters, beginning with 1987-88, constitute the mildest "triplet" of winters in 332 years of record-keeping (as measured at four sites that make up the Central England temperature record). In 1989 and 1990, the temperature dropped below the freezing mark on only one day each year. Almost every month during 1989 and 1990 was warmer than normal, making it the warmest two-year period on record, dating back to 1659. On Aug. 3, 1990, two locations in England approached the 100-degree mark for the first time, establishing a new all-time-record high for the country, adding to speculation about the greenhouse effect.[117] "Over the year ending in September 1990," the British hydrologists noted, "...our climate took on a complexion more normally associated with, say, northern Portugal—a comparison made more compelling by the very mild winter and hot dry summer."[118]

Such comparisons are hard to make, however, when the climate is also acting up elsewhere. In Portugal, for example, temperatures climbed to 108 degrees F during the summer 1990 heat wave, while the mercury reached 110 degrees F in southern Spain, and 104 degrees F in northern Spain and central France.[119] Elsewhere across southern Europe, temperatures ran 7 to 12 degrees above normal for much of the summer.

Greece suffered through its worst drought in more than a century. France endured its second worst drought of this century. Half of Italy's olive harvest was ruined by drought in 1990. In some areas of Istanbul, Turkey, where drought had persisted for more than two years, the only water available for washing was in public baths; drinking supplies were limited to four hours a day.[120] Japan also experienced an extended drought and heat wave in 1990, stretching its electric generating supply to the limit.[121]

Chaos in 1991: Nineteen ninety-one also was a chaotic year. Reports from around the world noted dramatic weather anomalies, particularly with respect to rainfall. In eastern Asia, it was exceptionally dry to start the year, except in Japan and eastern China, where it was exceptionally wet. Then the pattern reversed itself for a time; it became wet where it had been dry and dry where it had been wet. Finally, a barrage of tropical typhoons, heavy rains and flooding soaked Japan, the Philippines, most of southern China and the Indochina Peninsula. Seven typhoons took direct aim at Japan in 1991, including "super typhoon" Mireille, which killed 50 people and caused more damage to insured property than any other storm in the nation's history.[122] Super typhoon Ruth struck the Philippines four weeks later and was an even more powerful storm, packing winds of 195 miles an hour before making landfall in Luzon.[123] However, it was tropical storm Thelma in early November that caused the most human devastation, killing 5,000 in the central Philippines, making it the worst natural disaster to strike that island nation since 1976, when a tsunami (tidal wave) claimed 8,000 lives.[124]

The international weather summary for 1991 issued by the Climate Analysis Center in Asheville, N.C., was full of other headlines indicative of capricious weather patterns: Northern and Eastern Australia—"Wet Early in the Year, Then Very Dry;" Europe and the Middle East—"Dry Winter, Wet Spring;" West-Central Africa and the Sahel—"Wet, Then Dry"; Central and Eastern South America—"A Variety of Conditions First Half of Year"; and so on.[125]

In the United States in 1991, much of the "lower 48" experienced balmy weather in mid-winter that was more typical of mid-spring. In just one week in February, more than 170 daily record high temperatures were tied or broken in 28 states, stretching from the Far West to the Northeast. But while Washington, D.C., was basking in 70-degree-heat, London, England, was freezing at 12 degrees F, and 10-foot snowdrifts covered much of Yorkshire. England's string of mild winters had come to an abrupt end. France shivered through its coldest 48-hour period in recent memory. Even normally mild Athens watched the thermometer sink to a near-record-low of 25 degrees. For one of the few times this century, it snowed in the Riviera. And for the first time ever, snow accumulated on the island of Crete.[126] Is this further evidence of global warming? It would appear not. Of chaos? Perhaps so.

In March 1991, the weather on both sides of the Atlantic did a complete flip-flop. Temperatures returned to seasonal levels in the United States and it started to rain in places that had been in a drought situation for months or even years. In California, they called it the "Miracle March." Parts of southern California that had received only 1 percent of normal rainfall during January and February 1991 got eight inches of rain in the first week of March alone—making it the wettest California week in five years.[127] For the entire month, two to seven times the normal amount of precipitation fell in the southern third of

A Climate of Uncertainty

the state. Then it stopped raining, just as quickly as it had started. Some parts of southern California saw no more appreciable rainfall for the next eight months.

Meanwhile, back in Europe, the winter cold snap ended just as abruptly as California's rains began. Abnormally mild air spread over much of Europe in March, with average temperatures soaring as much as 11 degrees F above normal. In French wine country, many vines sprouted grape buds early. But Mother Nature was only fooling about the onset of spring. In the fourth week of April—fully one month into the new season—a cold spell returned to the continent, and temperatures dropped below the freezing point almost all the way to the Mediterranean coast. France's flowering grapes were devastated. In the Bordeaux, Medoc and white wine Sauternes regions of France, some vineyards reported losses of 80 to 100 percent of the young crop. It was not just the freakish cold snap in April that caused the damage; it was the fact that it followed the equally strange warm spell in March, which tricked the buds into flowering early.[128] Chaos triumphed again.

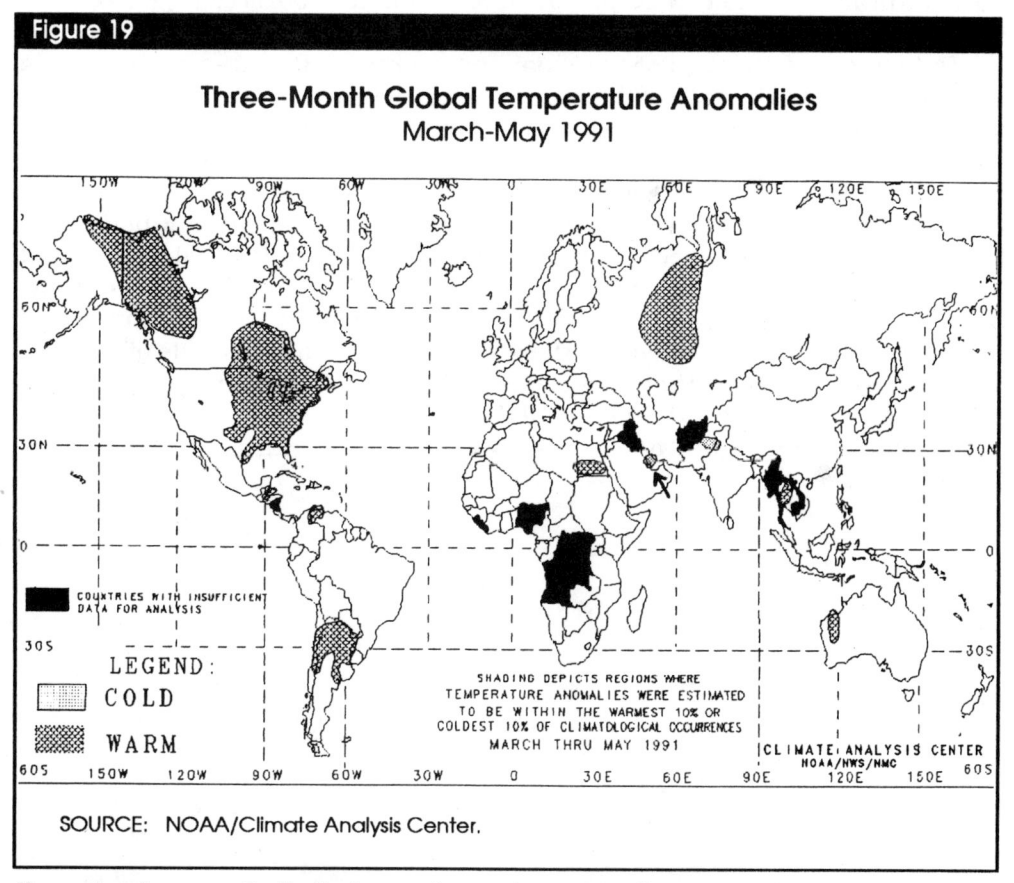

Figure 19

Three-Month Global Temperature Anomalies
March-May 1991

SHADING DEPICTS REGIONS WHERE TEMPERATURE ANOMALIES WERE ESTIMATED TO BE WITHIN THE WARMEST 10% OR COLDEST 10% OF CLIMATOLOGICAL OCCURRENCES MARCH THRU MAY 1991

LEGEND:
COLD
WARM
COUNTRIES WITH INSUFFICIENT DATA FOR ANALYSIS

CLIMATE ANALYSIS CENTER
NOAA/NWS/NMC

SOURCE: NOAA/Climate Analysis Center.

The only place on Earth that experienced a colder-than-normal spring in 1991 was immediately downwind of the Kuwaiti oil fires. (See arrow.)

It was also a confusing year for American plants. The record warm spell in February was a harbinger of unusually hot weather for much of the eastern half of the nation. Thirteen eastern states experienced their first- or second-warmest start (January through May) ever, dating back to 1895. In some states, temperature readings in May were more typical of August. Botanists were astounded at plants' response to the weather. At the New York Botanical Garden in the Bronx, daffodils, tulips and lilacs all bloomed in early April—instead of several weeks apart, lasting into May. Many backyard gardeners observed a similar flowering pattern. In New England, the blueberries ripened in early July—right along with the strawberries—instead of waiting until August. In Maryland, chrysanthemums flowered in the summer instead of the fall.[129]

Then an even more remarkable thing happened. In September, the magnolias bloomed—again—at the New York Botanical Garden. A few weeks later, the lilacs blossomed—again—on Martha's Vineyard. And in Washington, D.C., rhododendrons, dogwoods and pear trees bloomed—again—at the National Arboretum. "It's like the spring of '92 is peaking in the fall of '91," one horticulturalist for the U.S. Department of Agriculture observed. "It's eerie. I don't remember a season like this in over 50 years."[130]

By themselves, these highly unusual—often unprecedented—weather events prove nothing. In the case of the bizarre flowering pattern of the plants of '91, a combination of early season heat, mid-summer drought and late season rain caused the flowers to bloom twice.[131] As for global warming, it is hard to convince skeptics that the "greenhouse effect is here" when one considers all of the cold-weather anomalies since 1988. Yet there can be little doubt that the last few years have set a record for breaking many kinds of records—high and low recorded temperatures, high and low precipitation totals, high and low barometric pressure readings, most severe hurricanes, record number of tornadoes, etc. In terms of chaos, these events represent the kinds of climatic shudders and gyrations that may come along more often as the planet works through a period of transitional instability.

Perhaps the only thing one can really expect in the years ahead is more of the unexpected.

Making Order Out of Chaos

How is a farmer, a businessman or an investor to make sense of climate change when the weather has become perhaps more unpredictable than ever? Chaos theory offers two possibilities. One approach is to analyze weather trends with the help of something called the "strange attractor." The other option is to apply ensemble forecasting—the "Monte Carlo" approach of weather prediction—to general circulation models and see if they have come to any firm conclusions about future climate. Neither approach can divine the future, as if through a crystal ball, but for now they are the best tools we have.

Climate's 'Strange Attractor'

To see the weather through the eyes of chaos, one must come to know the "strange attractor." The strange attractor "pulls" seemingly random sequences of events into patterns that start to look familiar over time. Meteorologists go through this exercise when they characterize a region's climate. Although the weather is far too unpredictable for them to forecast what the weather will be like on any given day, they know from the weather's past behavioral patterns generally what to expect in the future.

Consider a corn farmer near Des Moines, Iowa, with access to a century's worth of weather records. Since his livelihood depends on the way temperature, precipitation and sunlight interact in the weather systems passing over his fields, he and his forebears have made observations of these climatic variables to indicate when is a good time to plant, fertilize and harvest the crop. For the month of June, the weather records in Des Moines show that daily temperature values have ranged from a record low of 37 degrees F to a record high of 103 degrees F, but the daily average temperature hovers right around 70 degrees F. Similarly, the precipitation variable has a wide range of values, but it clusters around an average of four inches of rain for the month of June. Finally, the values for available sunlight are locked in a tighter range and are "attracted" to a 68 percent average.

A strange attractor arranges these variables in a matrix showing all of the combinations of weather that have occurred in Des Moines over the period. (Since there are three variables to consider in this example, the matrix has three dimensions.) If every hour of every day in June were the same—70 degrees F, a few sprinkles of rain and 68 percent available sunshine—the weather in Des Moines could be described by one dot in the matrix that represents the intersection of all three variables. But since the weather changes, the matrix is arranged in a broader array of dots. Most of the dots cluster around the average values for the variables, but there are also outliers representing the cloudless days when the temperature soared above 95 degrees F and the rainy days when the temperature never got above 60 degrees F.

Figure 20

The Strange Attractor

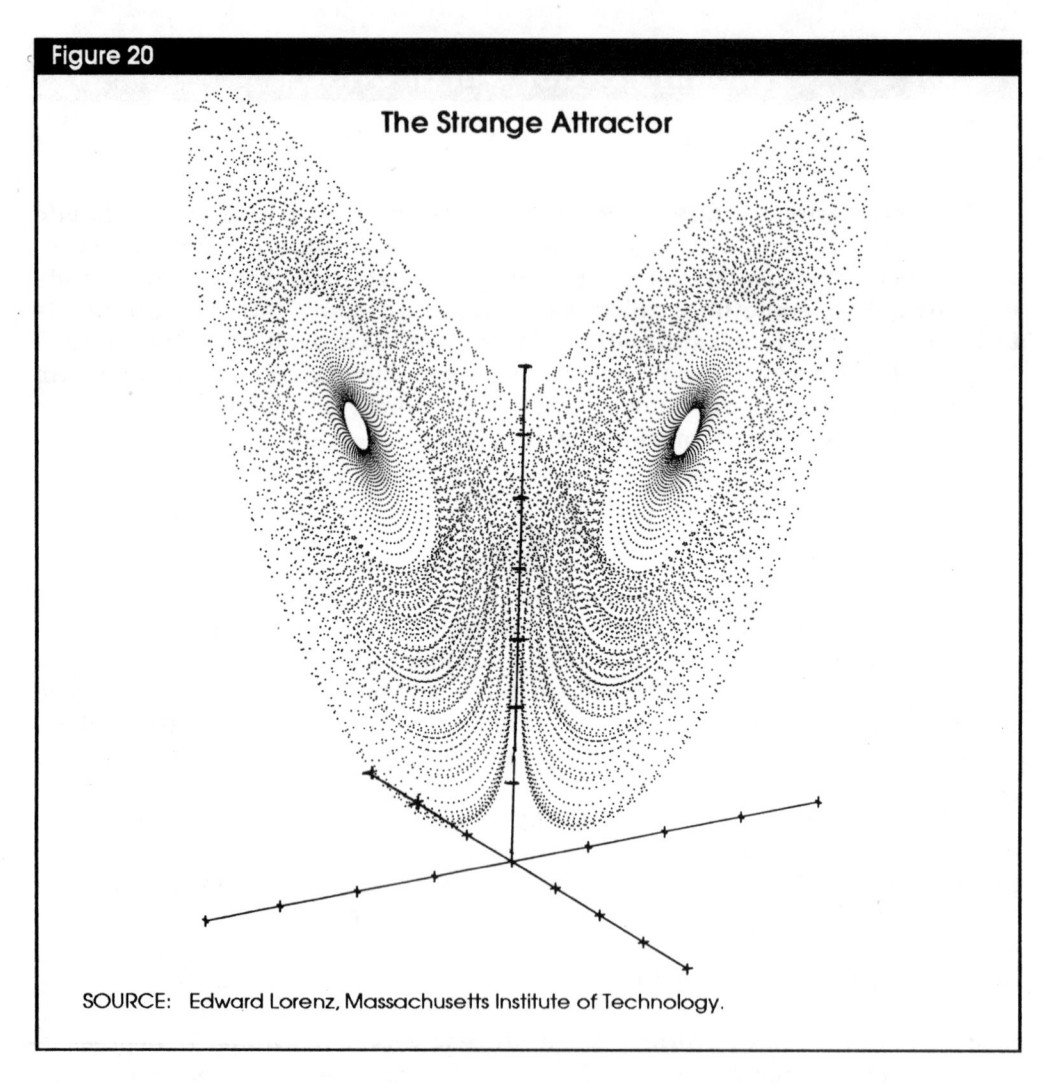

SOURCE: Edward Lorenz, Massachusetts Institute of Technology.

All assembled in the strange attractor, Des Moines's climate begins to take on a distinctive appearance, displaying the right combination of warmth, rain and sunshine for growing corn in June. But perhaps the most important climate information is that which lies outside the strange attractor. Since there are no dots representing days in June when temperatures have dipped below freezing, that means corn planted in May should not be nipped by frost as it emerges from the ground in June. (Had this example featured California citrus growers, however, a new dot would have appeared on the December plot as a result of the record cold in 1990.)

The strange attractor for Des Moines does reveal something more disturbing for Iowa corn farmers, however. There are several dots representing days in June when daily high temperatures exceeded 100 degrees F, such as during the 1988 heat wave. Even more troubling in terms of the attractor, dry weather and abundant sunshine are associated with this especially hot portion of the climate matrix. The image conveyed here is one of drought. Accordingly, the attractor is alerting the farmer to the possibility that the corn he plants in May may succumb to hot, dry weather in June—even if frost is not a concern. But there is comfort in this image as well. Since the "drought dots" are on the perimeter of the attractor and are few in number, drought does not appear to be a consistent part of the weather pattern. If the Des Moines farmer loses one crop out of 10, it is an average he can live with.

Of course, farmers do not have to build a strange attractor to make judgments about the suitability of the climate for growing corn. They have an intuitive sense based on their own experiences and generations' worth of weather records. What is vital to remember about climate change, however, is that the weather's past behavior may no longer serve as a reliable guide to the future. In terms of chaos, it means that the strange attractor may begin to alter its appearance—or the climate may choose a new strange attractor altogether. Climatic events that once were outside the attractor—say, a blistering-hot 110-degree F day—could start to show up on the fringes of the attractor. And climatic events that once had been on the periphery—the 103-degree temperature record, for instance—might work closer toward the center.

In statistical terms, the odds that had been working in the farmer's favor might begin to turn against him. A small relative change in one climatic variable opens the door to a much greater change in the frequency of extreme events, which are so critical to determining a farmer's fortunes. Using Des Moines's June rainfall totals as an example, a 10 percent drop in the four-inch monthly average would increase by 50 percent the number of Junes when the crop receives only two inches of rain or less—half the "normal" average.[132]

Turning to the temperature variable, a 3 degree F increase in average daytime highs would result in a three-fold increase in the average number of heat waves, when maximum temperatures topped the 95-degree mark for five consecutive days or more. If such a heat wave bore down on Des Moines when the corn crop was in the tassle-forming stage in late June, the excessive heat would prevent proper kernel formation and ruin the crop. Thus, a relatively small change in the local climate—a 10 percent drop in precipitation or a 3 degree F increase in temperature—could have a dramatic effect on the shape of the strange attractor and the assumed frequency of extreme events. A devastating drought or heat wave that used to appear once a decade now might come along every five years or so. Moreover, the odds of such extreme events coming in successive years, which had been one in a hundred, would rise to one in 25—a four-fold increase—based on the law of averages.[133] Suddenly, the fearsome prospect of back-to-back crop failures—which could wipe out a farmer's savings—falls well within the span of his farming career.

Of course, this is not the only scenario that could be borne out by the strange attractor. If the climate in Des Moines becomes cloudier, those days representing a combination of less available sunshine, higher minimum temperatures, lower maximum temperatures and (perhaps) greater rainfall might gravitate toward the center of the plot. This would suggest to the corn farmer that extreme heat waves that sterilize his crop would become less of a concern after climate change. Better still, he could look forward to greater yields, since corn appears to grow best when average nighttime temperatures increase but the overall temperature range declines.[134] Even if the strange attractor for June showed signs of increasing heat and drought, an astute farmer might take a look at the temperature plot for May. If he found that the chance of frost was becoming more remote in May as a result of global warming, he could plant his corn crop earlier in the season. That way, the crop would be more likely to pollinate before the onset of hot, dry weather, and be harvested before pests or hail damage the crop and reduce his yield.

The farmer can never be too sure of himself, however, lest the weather remind him of its inherent unpredictability. While the strange attractor helps to reveal the frequency of extreme events, it cannot say when the next climate anomaly will enter the forecast or what it will consist of. Moreover, there remains a possibility that the climate itself will shift so dramatically (as it did during the Younger Dryas event) that one strange attractor will be replaced by another that looks completely different. This is perhaps the farmer's—and the climatologist's—worst nightmare: that global warming will not cause subtle, orderly, predictable shifts in climate, but rapid, vacillating, unexpected changes that make weather-related planning decisions even more of a gamble than they are today.

An Ensemble Forecast

While the strange attractor offers a structured (albeit limited) way to think about the randomness of weather, general circulation models are still the best tool available to ponder the long-term future of climate. The problem with these models, as we have seen, is that they build themselves upon precarious notions about how the natural world behaves. They reduce the planet to a patchwork of squares. They offer proxies for some important climatic variables such as clouds and give crude approximations of others like oceans. When climate modelers run these incredibly complex (yet highly simplified) computer programs, they cannot account for all of the surprises that a changing climate may have in store. Indeed, they have surprised themselves many times by making seemingly minor amendments to their programs only to witness major changes in the outcome of their modeled results.

Nevertheless, some residual value may be found in general circulation models by combining their results in an ensemble forecast, similar to the way some meteorologists forecast short-term weather patterns. Areas of agreement in their results are signs that they have probably modeled contingent variables and feedback processes adequately—even if some minor differences remain in their findings. Areas where they flatly disagree point to places where certain variables may have been miscalculated or feedback processes overlooked, requiring additional refinement before the modeled results are regarded as sound.

Ideally, an ensemble forecast for global climate change would involve at least 10 general circulation models, but not that many advanced models exist. Four such models have been mentioned earlier: NASA's model at the Goddard Institute for Space Studies (GISS), NOAA's model at the Geophysical Fluid

Dynamics Laboratory (GFDL), the National Center for Atmospheric Research's (NCAR) model in Boulder, Colo., and the United Kindgom Meteorological Office's (UKMO) model in Sussex, England. A fifth leading model was developed at Oregon State University (OSU) in Eugene and now resides at the University of Illinois, Champaign-Urbana. (Other important climate change experiments are being conducted using general circulation models created by the Canadian Climate Center, the Laboratory of Dynamic Meteorology in France and the Meteorological Institute at the University of Hamburg in Germany.) Although each of these models takes a slightly different approach and varies in its base assumptions, they all strive to compare the Earth's climate before and after a doubling of carbon-dioxide-equivalent in the atmosphere. Critical determinations must be made about changes in temperature, precipitation and soil moisture.

One important point of agreement in all of the models is that a greenhouse world would indeed appear to be a warmer one. In the United States during summertime, for example, the models provide average temperature increases ranging from 5.4 degrees F in the NCAR model to 10.1 degrees F in the GFDL model. The GISS model projects a 6.8 degree F summertime increase, and the OSU model, 6.3 degrees F. The UKMO model, before modifying its cloud parameterization a few years ago, tended to resemble the GFDL model's results. Now it more closely resembles the NCAR model with regard to temperature increases. This basic agreement on warming is critical because increased summertime heat would raise the rate of water evaporation (particularly as daytime temperature maximums increase). Therefore, land areas would be expected to dry out—unless precipitation also increases sufficiently to offset the added evaporation.

And here is where the ensemble forecast begins to fall apart. While the general circulation models suggest that precipitation would increase as a global average, their forecasts have great disparities about which land areas would become wetter or drier. Using the United States again as an example, results of three of the models—GFDL, OSU and UKMO—indicate that virtually the entire North American continent would dry out during summertime, the key agricultural season. The GISS model results indicate, however, that only the southern Great Plains and the Northeast would become drier in the summertime. Meanwhile, the NCAR model suggests that the Southeast and virtually all of the Great Plains would dry out.

When the U.S. regional results of the five models are compared, in fact, they find themselves in agreement on only 3 percent of the surface area of the continental United States—one small portion of the Great Plains, covering western Nebraska, Kansas and eastern Colorado, where it would become drier.[135] This seems to suggest that agricultural interests in the High Plains region are highly vulnerable to an increase in summertime drought. On the other hand, the modeled results merely could be a statistical fluke. Even if the models had no predictive capability—choosing dry regions at random—all five would still be expected to agree on 3 percent of the surface area examined, based on the law of averages.[136]

The ensemble forecast thus leaves the regional precipitation question

Figure 21

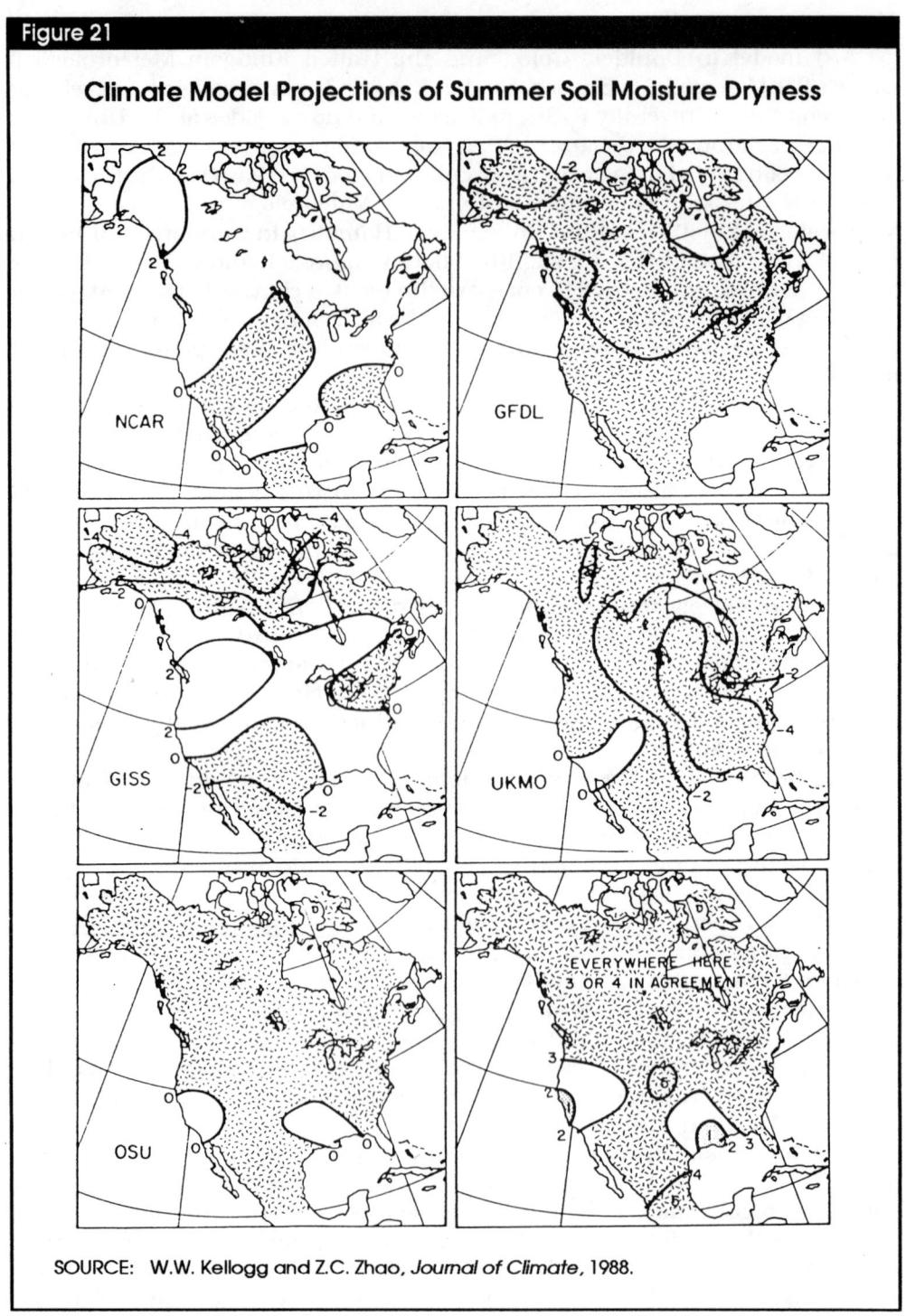

Climate Model Projections of Summer Soil Moisture Dryness

SOURCE: W.W. Kellogg and Z.C. Zhao, *Journal of Climate*, 1988.

Shaded area denotes change to a drier condition. Clear area denotes change to a moister condition. Final box denotes areas of agreement among the five models.

unresolved. But when the general circulation models are set in motion, they do agree on the general principle that the interiors of continents would warm faster than the oceans, since water takes longer to assimilate added heat than land. In North America, a more rapidly warming continent could lead to the positioning of stationary high pressure systems off the Atlantic and Pacific coasts. Since winds flow clockwise around these fair-weather systems, the Atlantic high would channel warm, humid air from equatorial regions into the Southeast and Midwest, while similar winds around the Pacific high would pull cool Arctic air toward the West Coast.[137]

Now the question becomes whether this modified air flow would lead to persistent dryness or wetness of soils in the Mid-continent, and here again the ensemble forecast is of little help. Soils that hold moisture well in the spring and early summer enable more evaporation to occur during the warmest months of the year. This evaporation enhances low-level cloudiness, which in turn inhibits the tendency of the sun to dry out the soil. Thus, the models' assumptions about soil moisture become critical to forecasting drought. If soils are assumed to absorb a significant portion of early-season rainfall, rather than allowing a high percentage to run off into rivers and streams, the onset of summer drought becomes less likely—and some of the worst fears about global warming vanish into thin air.

Accordingly, the soil moisture parameterization in the general circulation models has an important bearing on the severity of climate change. The GISS model is considered to have the most elaborate parameterization for soil moisture.[138] It divides the soil into two distinct layers—each with its own absorptive capacity depending on the type of soil—creating a wide distribution of values for soil moisture across the country. Significantly, GISS is not one of the models predicting an increase in summertime dryness over virtually the entire continent, although it forecasts that much of the Farm Belt could be adversely affected.

On the other hand, the model that has predicted the most dryness, GFDL, factors in a projected increase in bare ground during winter that would lead to deeper frosts in the spring. Since water runs off frozen ground more readily than thawed-out ground, more early-season spring rains are assumed to run off in the GFDL model.[139] As a result, some runs of the GFDL model have garnered a reputation as the "doomsday scenario." With a projected 10 degree F summertime temperature increase and a 25 percent projected decline in precipitation nationwide, the doomsday scenario finds parts of the Midwest, the Southeast and most of the Great Plains inheriting an arid, steppe-like climate—not unlike what now exists in the Mediterranean and sub-tropical Africa and Asia. (More recent runs of GFDL with a coupled ocean-atmosphere system produce a less severe outcome, however.) Some runs of the GFDL model also suggest that a similar fate may befall other middle- and high-latitude regions of the globe—including much of Canada, western Europe and Siberia—where Svante Arrhenius and others thought that "equable and better climates" would develop as a result of global climate change.[140]

Which of the general circulation models offers the most realistic view of the future? No one can say for sure. Perhaps none of them does. But at least the ensemble forecast finds agreement on a global warming trend. When will the differences between the models on precipitation and soil moisture be resolved? Probably not for another 10 to 20 years. According to a report by the Intergovernmental Panel on Climate Change, reviewed by Stephen Schneider:

> [I]t will take at least five years to build the high-resolution coupled atmosphere, ocean, biosphere, land-surface, sea-ice, and chemistry submodels that are needed if scientists are to have any hope of predicting the evolving regional climatic changes that are so important to impact assessment. [The IPCC report] notes that five to 10 years will be necessary to get computers large enough to run such models routinely to determine the quality of the forecasts. Also, some five to 10 years will pass before major data-gathering projects begin to provide data to validate the various subcomponents of such high-resolution coupled modeling, let alone coupled model performance. Thus, 10 to 20 years is suggested as the time required for everything to come together and for detailed predictive skill to become credible.[141]

In short, a climate of uncertainty will continue to pervade policy decisions regarding the greenhouse effect for the remainder of the 1990s and into the 21st century.

Rolling the Dice

Global climate change may take the planet into uncharted territory in the 21st century, but the matter of what to do about it poses a familiar question: Do we know enough to act in the face of uncertainty, or should we wait until more facts become available? Increasingly, this question has become a political one rather than a scientific one.

A turning point in the policy debate came, of course, when NASA scientist James Hansen declared before Congress in 1988 that "the greenhouse effect is here." Hansen returned to Capitol Hill with more political dynamite one year later. This time he said that the GISS model had become reliable enough for him to conclude that rapid strengthening of the greenhouse effect would lead to "drought intensification at most middle- and low-latitude land areas" over the next 10 to 20 years—and that the southern Great Plains and Southeast of the United States would be among the first to feel the ill-effects. Over Hansen's objections, the President's Office of Management and Budget attached a caveat at the end of his written testimony, explaining that "these changes should be viewed as estimates from evolving computer models and not as reliable predictions." Hansen publicly took issue with the disclaimer and once again found himself leading off the evening news. This time, the media charged the White House with censorship of Hansen's "impartial" scientific research.[142]

While many climate modelers took umbrage at the White House's tactics, few faulted the language of the disclaimer itself. "Where Jim has some problems with his friends, and I count myself as one," Stephen Schneider subsequently remarked, "is when he says that the location of specific areas of drought in his model are robust." Among other things, the GISS model is "not running a realistic ocean. You really don't know what [the ocean] is going to do," Schneider said. It is a problem that continues to plague all of the models. Yet Schneider concedes that Hansen is "probably right anyway. The odds are better than 50:50 that the drought areas are robust."[143]

Hansen has his own odds-making terminology to describe the situation: "The dice are becoming loaded," he says. Before the onset of global warming, the six faces of a forecasting die provided equal representation of warmer-than-average, colder-than-average and merely average summers, so when the die was rolled, any given summer would offer a one-in-three chance of being warmer than normal. But in the 1980s, Hansen contends, the dice became loaded; one "cold" face on the die was replaced with another "warm" face. Now the die, with three warm faces, has a 50-50 chance of rolling a warmer-than-average summer—even if climatologists cannot predict which summers within the period. By the turn of the century, the GISS model forecasts that the die will have four "warm" faces, so that two out of three summers will be warmer than the historical average, and the hottest summers will shatter all records.[144]

If Hansen's forecast proves to be correct, the policy debate over whether to respond quickly to global warming may become a no-brainer—the smoking gun will have been revealed. But the debate could get more complicated if the change does not develop in line with modelers' expectations. For all the circumstantial evidence pointing to a warming trend and increasingly chaotic weather behavior, the fact remains that global temperatures have not risen as much to date as modelers thought they would. As long as this disparity persists, the greenhouse theory will continue to have a battery of skeptics.

Climate modelers may have underestimated the thermal lagtimes built into the coupled ocean-atmosphere system. Perhaps chaos in the climate system itself is throwing unexpected curves in the underlying buildup of greenhouse gases. Or maybe periodic changes in the output of the sun have had a greater influence on the climate than modelers now comprehend. (But if this were the case, it would seem to defy physical law, since the climate would have to be fairly unresponsive to major changes in the radiative forcing of greenhouse gases yet highly responsive to comparatively minor changes in the output—or radiative forcing—of the sun.)[145]

A more compelling explanation for the climate's delayed response to the influx of greenhouse gases is just now emerging. In February 1992, the Intergovernmental Panel on Climate Change came forward with a remarkable explanation that the temperature disparity is caused not so much by flaws within the general circulation models as it is by other environmental problems that have had a distinct cooling effect on the planet.[146] Apparently, ozone depletion and acid rain—two "smoking guns" that nations now are taking steps to control—have been prime factors in holding back global warming. When the climate-interactions of these phenomena are plugged into the general circulation models, forecasts of global warming fall more in line with observed temperature trends. If these mitigating factors are removed from the picture, however, climate change resumes a course that more closely resembles the original (warmer) forecast. Thus, the atmosphere appears to have developed a precarious system of checks and balances, in which policymakers are "damned if they do, and damned if they don't" when formulating response strategies for these global environmental problems.

As for ozone depletion, climate modelers have determined that damage to the ozone layer is cooling the atmosphere, principally because ozone itself is a greenhouse gas. Moreover, the ozone layer acts like a warm blanket wrapped around the planet; its temperatures range higher than those found on the Earth's surface. (This is largely because ozone molecules create heat when they are split apart by ultraviolet light.) As CFCs and other halogenated compounds remove ozone from the ozone layer, it is as if the threads of this planetary blanket are being stripped away. The depleted ozone layer generates less heat (because there is less ozone for ultraviolet light to split apart) and more infrared heat is able to pass through to the stratosphere and, ultimately, into outer space. The net effect is that ozone depletion cancels out the radiative forcing potential of CFCs and halons, even though they are potent greenhouse gases.[147]

Alarming new evidence about depletion of the ozone layer over densely populated regions of the Northern Hemisphere has prompted the international community to press forward with plans to ban most ozone-depleting substances by the end of 1995.[148] The Bush administration favors this expedited schedule, partly because it is consistent with a "no regrets" strategy on global warming.[149] But the new data call this premise into question. Ironically, the phase-out may increase global warming over time by promoting a shift to alternative compounds that are potent greenhouse gases in their own right but which lack the ability to counteract global warming through ozone destruction. This suggests that the onus will have to fall even more on CO_2 reductions to redress global warming.

Then there is the acid rain phenomenon. The United States has adopted a plan to cut emissions of sulfur dioxide, a precursor of acid rain, by 50 percent over the next decade as part of the 1990 amendments to the Clean Air Act. The hope is that damage to crops, trees, monuments and human health (both realized and feared) will be minimized by the reduction in sulfurous emissions, which come mainly from coal-burning power plants lacking scrubbers. Once again, this has been considered a "no regrets" strategy from a greenhouse point of view, since CO_2 emissions are likely to fall as power generation from these older coal plants is reduced.

But new scientific evidence is reviving an old debate. As the sulfur dioxide emissions from these older coal plants (and to a lesser extent from metal smelting plants and automobiles) waft into the atmosphere, they turn into tiny sulfuric acid aerosols, which can be either droplets or particles. These aerosols reflect sunlight back into space—even when the sky is "clear"—in much the same manner put forward by the "climate coolers" of the early 1970s. Moreover, for the few days they reside in the atmosphere, sulfur aerosols drift along the air currents and may actually seed clouds. (The tiny sulfate particles serve as "cloud condensation nuclei" that attract moisture and form water droplets. These droplets in turn produce clouds that hold on to moisture quite well.) The result appears to be an increase in persistent low clouds that shade parts of the planet from the sun.

> **Table 3**
>
> ## Selected Major Conclusions of the 1992 IPCC Supplement
>
> - Global mean surface air temperature has increased by 0.3° to 0.6° C (0.5 to 1.1 F) over the last 100 years.
>
> - The unequivocal detection of the enhanced greenhouse effect from observations is not likely for a decade or more.
>
> - The evidence from the modeling studies, from observations and the sensitivity analyses indicate that the sensitivity of global mean surface temperature to doubling CO_2 is unlikely to lie outside the range of 1.5° to 4.5° C (2.7 to 8.1 F).
>
> - There are many uncertainties in our predictions, particularly with regard to the timing, magnitude and regional patterns of climate change due to our incomplete understanding.
>
> - Depletion of ozone in the lower stratosphere in the middle and high latitudes results in a decrease in radiative forcing which is believed to be comparable in magnitude to the radiative forcing contribution of chlorofluorocarbons (globally averaged) over the last decade or so.
>
> - The cooling effect of aerosols resulting from sulfur emissions may have offset a significant part of the greenhouse warming in the Northern Hemisphere during the past several decades.
>
> - The consistency between observations of global temperature changes over the past century and model simulations of the warming due to greenhouse gases over the same period is improved if allowance is made for the increasing evidence of a cooling effect due to sulfate aerosols and stratospheric ozone depletion.
>
> SOURCE: Intergovernmental Panel on Climate Change, *IPCC 1992 Supplement*, February 1992.

When the direct scattering of sunlight by the aerosols is combined with the reflection off these cloud tops, the result is an increase in planetary albedo that offsets a large portion of the anthropogenic increase in greenhouse gases. To be precise, the man-made aerosols are reducing sunshine on portions of the Earth's surface—especially in the Northern Hemisphere—by one to two watts per square meter. This shading is comparable in magnitude to the increase in radiative forcing potential caused by the buildup of greenhouse gases—but opposite (negative) in sign.[150] As control measures are enacted to limit sulfur dioxide emissions from coal-burning power plants, this shading effect will be reduced, while carbon dioxide emissions continue unabated. Here again, a policy is in motion to reduce a negative feedback effect even as the atmospheric loading of greenhouse gases continues.

All of this suggests that a plan to ameliorate global warming inevitably must focus on abatement of carbon dioxide—the principal greenhouse gas that John

Tyndall and Svante Arrhenius first studied a century ago. While Arrhenius chose a doubled-CO_2 atmosphere as the arbitrary cut-off point in his analysis—a milestone that climatologists still use today—the reality is that carbon dioxide will continue to accumulate in the atmosphere as long as population growth, economic development and government policies permit it to. Even when its concentration does level off, global warming will continue for perhaps hundreds of years as the climate system works to achieve a new state of thermal equilibrium.[151] Accordingly, decisions made in our time will do more to affect the course of future generations than our own, just as any perturbed climatic effects experienced today mainly reflect what previous generations have done before us.

Given the climate of uncertainty that now prevails, policymakers may be inclined to wait a while longer before taking action on the greenhouse effect. But even then, not all outstanding questions are likely to be resolved. Chaos will remain a formidable adversary, as always. What we have learned from a century of research is that society has gained the ability to influence the weather, but not the power to control it. Indeed, what Roger Revelle and Hans Suess remarked in 1957 remains true today: "Human beings are now carrying out a large-scale geophysical experiment that could not have happened in the past or be repeated in the future." The question is whether society is willing to make a move—one that is sure to involve risks and trade-offs—in order to gain a possible long-term advantage in its relationship with climate.

This is the greenhouse gambit society has to consider.

Notes

1. For a definitive history, see William W. Kellogg, "Mankind's Impact on Climate: The Evolution of an Awareness," *Climatic Change*, April 1987.
2. Svante Arrhenius, "On the Influence of Carbonic Acid in the Air Upon the Temperature of the Ground," *Philosophical Magazine*, 1896.
3. See note 2, and William W. Kellogg and Robert W. Schware, "Society, Science and Climate Change," *Foreign Affairs*, Summer 1982.
4. Intergovernmental Panel on Climate Change, Working Group I, *Policymakers' Summary of the Scientific Assessment of Climate Change*, World Meteorological Organization and United Nations Environment Programme, Geneva, Switzerland, 1990.
5. See, for example, Martin Parry, *Climate Change and World Agriculture*, Earthscan Publications Ltd., London, England, 1990. Parry led a team investigating the agricultural impacts of climate change for the IPCC.
6. See note 2, and Michael Weisskopf, "'Greenhouse Effect' Fueling Policy Makers," *The Washington Post*, Aug. 15, 1988.
7. Intergovernmental Panel on Climate Change, *1992 IPCC Supplement*, World Meteorological Organization and United Nations Environment Programme, Geneva, Switzerland, February 1992.
8. *Ibid.*, and R. Neil Sampson and Dwight Hair, editors, *Forests and Global Warming*, American Forestry Association, Washington, D.C., May 1991.
9. See note 7.
10. U.S. Environmental Protection Agency, Office of Air and Radiation, *Reducing Methane Emissions from Livestock: Opportunities and Issues*, EPA 400/1-89/002, Washington, D.C., August 1989.
11. See note 7. Methane emissions had been increasing at a rate of more than 1 percent annually in the late 1970s, but for reasons that are not completely understood, the rate of emissions has dropped since then.
12. *Ibid.*
13. Douglas G. Cogan, *Stones in a Glass House: CFCs and Ozone Depletion*, Investor Responsibility Research Center, Washington, D.C., July 1988.
14. See note 7.
15. See note 4. This calculation does not factor in the counteractive effect of stratospheric cooling resulting from depletion of the ozone layer by CFCs.
16. Philip D. Jones and Tom M. L. Wigley, "Global Warming Trends," *Scientific American*, August 1990.
17. See note 4.
18. See second citation listed in note 3.
19. Michael H. Glantz, "Politics and the Air Around Us: International Policy Action on Atmospheric Pollution by Trace Gases," *Societal Responses to Regional Climatic Change*, Michael H. Glantz, editor, Westview Press, Boulder, Colo., 1988; and Michael J. McCarthy, "Is the Crazy Weather Still Another Sign of a Climate Shift?" *The Wall Street Journal*, Aug. 8, 1988.
20. Quotes appear in the second citation of note 19.
21. John Gribbin, "Climate Now," *New Scientist*, Inside Science Number 44, March 16, 1991.
22. Wallace S. Broecker and George H. Denton, "What Drives Glacial Cycles?" *Scientific American*, January 1990.

23. William K. Stevens, "Scientists Suggest Global Warming Could Hasten the Next Ice Age," *The New York Times*, Jan. 21, 1992. The article cites a study in a recent issue of the British journal *Nature* by Gifford Miller of the University of Colorado and Anne de Vernal of the University of Quebec at Montreal.
24. J. Murray Mitchell, "Carbon Dioxide and Future Climate," *Weatherwise*, 1977, reprinted August/September 1991. Mitchell credited Wallace Broecker of the Lamont-Doherty Geological Observatory of Columbia University for the "super-interglacial" analogy.
25. G.S. Callendar, "The Artificial Production of Carbon Dioxide and Its Influence on Temperature," *Quarterly Journal of the Royal Meteorological Society*, Fall 1938.
26. See second citation of note 19.
27. Roger Revelle and Hans E. Suess, "Carbon Dioxide Exchange Between Atmosphere and Ocean and the Question of an Increase of Atmospheric CO_2 During the Past Decades," *Tellus*, January 1957.
28. President's Science Advisory Committee, *Restoring the Quality of Our Environment, Report of the Environmental Pollution Panel*, The White House, Washington, D.C., 1965.
29. See note 1.
30. National Academy of Sciences, *Carbon Dioxide and Climate: A Scientific Assessment*, Climate Research Board, Washington, D.C., 1979; National Research Council, *Changing Climate: Report of the Carbon Dioxide Assessment Committee*, Board on Atmospheric Sciences and Climate, Washington, D.C., 1983; and National Research Council, *Global Change and Geosphere-Biosphere: Initial Priorities for an IGBP*, Washington, D.C., 1986.
31. World Meteorological Organization, *Proceedings of the World Climate Conference*, Report No. 537, Geneva, Switzerland, 1979.
32. United Nations Environment Programme, World Meteorological Organization and International Council of Scientific Unions, *International Assessment of the Role of Carbon Dioxide and of Other Greenhouse Gases in Climate Variations and Associated Impacts: Statement of the Villach Conference*, Villach, Austria, October 1985.
33. Philip Shabecoff, "Global Warming Has Begun, Expert Tells Senate," *The New York Times*, June 24, 1988.
34. Harry Lins, Eric T. Sundquist and Thomas A. Ager, *Information on Selected Climate and Climate Change Issues*, U.S. Geological Survey, Open File Report 88-718, Reston, Va., 1988.
35. "1988 Drought Termed Nation's Worst Disaster," Associated Press wire story, Jan. 19, 1989.
36. William K. Stevens, "Separate Studies Rank '90 As World's Warmest Year," *The New York Times*, Jan. 10, 1991.
37. Richard A. Kerr, "1991: Warmth, Chill May Follow," *Science*, Jan. 17, 1992.
38. Bruce Ingersoll, "How Much to Plant After the Drought?" *The Wall Street Journal*, Sept. 20, 1988.
39. Stanley A. Changnon, Midwest Climate Center, presentation in *The Greenhouse Effect: Investment Implications and Opportunities*, proceedings of an Oct. 4, 1989, conference sponsored by the Investor Responsibility Research Center and the World Resources Institute, Douglas G. Cogan, editor, Investor Responsibility Research Center, Washington, D.C., 1990.
40. See note 38, and Richard B. Cogan, Anheuser-Busch Co., personal communication, Oct. 12, 1988.

41. T.J. Marsh and R.A. Monkhouse, "A Year of Hydrological Extremes-1990," *Weather*, British Meteorological Society, December 1991; and Mick Hamer, "The Year the Taps Ran Dry," *New Scientist*, Aug. 18, 1990.
42. Edwin Unsworth, "European Storms Heighten Fears of 'Greenhouse,'" *The Journal of Commerce*, Feb. 27, 1990; and William K. Stevens, "Europe's Wild Weather: By Air Mail from U.S.," *The New York Times*, March 5, 1990.
43. A.E. Strong, "Greater Global Warming Revealed by Satellite-derived Sea-surface-temperature Trends," *Nature*, April 20, 1989; and Thomas H. Maugh II, "Ocean Swell From Greenhouse Effect," *The Los Angeles Times*, April 20, 1989.
44. Richard Monastersky, "Signs of Global Warming Found in Ice," *Science News*, March 7, 1992; and Richard A. Houghton and George M. Woodwell, "Global Climatic Change," *Scientific American*, April 1989.
45. Richard Monastersky, "Shrinking Ice May Mean Warmer Earth," *Science News*, Oct. 8, 1988.
46. William K. Stevens, "Northern Hemisphere Snow Cover Found to Be Shrinking," *The New York Times*, Oct. 30, 1990.
47. Richard Monastersky, "Recent Decades Saw Wetter Continents," *Science News*, Jan. 28, 1989.
48. David Stipp, "EPA, Public Differ Over Major Risks," *The Wall Street Journal*, Oct. 1, 1990; and Leslie Roberts, "Counting on Science at EPA," *Science*, Aug. 10, 1990.
49. Ralph T. King Jr., "Insurers' Losses On Oakland Fire May Be $1.5 Billion," *The Wall Street Journal*, Oct. 29, 1991.
50. David S. Wilson, "Worst Fire Season Appears to Be at an End," *The New York Times*, Nov. 10, 1988.
51. Beatrice E. Garcia, "Hugo to Be Costliest Storm for Insurers But No Early Rate Increase Is Expected," *The Wall Street Journal*, Sept. 27, 1989.
52. Miles B. Lawrence, "The Weather of '88: Return of the Hurricanes," *Weatherwise*, February 1989.
53. "Hot Time, Summer in the City," *The New York Times*, Oct. 20, 1991. In 1991, the maximum temperature in Central Park reached 90 degrees F or above on 38 days, breaking a record set in 1944. The 10-year average for 90-degree-plus readings is 17 days per summer.
54. Kevin Sullivan, "For Those Who Sweat the Details: Summer Was 2nd Hottest," *The Washington Post*, Sept. 23, 1991. In 1991, the maximum temperature at National Airport reached 90 degrees F or above on 59 days. The 10-year average for 90-degree-plus readings is 40 days.
55. Jane Gross, "California Painfully Faces Grim Truth of Drought," *The New York Times*, Feb. 26, 1991.
56. Michael J. McCarthy, "Southeast May Face Repeat of 1986's Severe Drought," *The Wall Street Journal*, July 11, 1990; and Bruce Ingersoll and Scott Kilman, "Drought Damage to Winter Wheat Crop to Exceed Expectations, Forecast Says," *The Wall Street Journal*, June 13, 1989.
57. William K. Stevens, "The Outlook for Weather Forecasting," *The New York Times*, July 30, 1989.
58. For a summary of weather forecasting methodology, see Richard A. Kerr, "Is Something Strange About the Weather?" *Science*, March 10, 1989.
59. American Meteorological Society, "Weather Forecasting: Policy Statement," *Bulletin of the American Meteorological Society*, August 1991; and Richard Monastersky, "Forecasting into Chaos," *Science News*, May 5, 1990.
60. See note 58, and Richard A. Kerr, "Telling Weathermen Where to Worry," *Science*, June 9, 1989.

61. "United States Seasonal Climate Highlights: Winter 1990-91," *Weekly Climate Bulletin*, National Weather Service, March 16, 1991.
62. William Booth, "Prediction of Cold Winter Was Only 180 Degrees Off," *The Washington Post*, March 5, 1991.
63. Richard Monastersky, "'Tis the Season for an El Niño Warming," *Science News*, Dec. 14, 1991.
64. "GOES-NEXT to Wait Out Next Round," *Science*, Oct. 4, 1991.
65. *Ibid.*, and Peter Aldhous, "Weather Satellite Woes," *Nature*, July 11, 1991; and R. Cowen, "Launch Delays Jeopardize Weather Forecasts," *Science News*, July 6, 1991.
66. Richard A. Kerr, "Why Bigger Isn't Better in Earth Observation," *Science*, Sept. 27, 1991.
67. Richard Monastersky, "NASA Inches Toward Smaller Satellites," *Science News*, *Sept. 28, 1991.*
68. James Gleick, *Chaos: Making a New Science*, Penguin Books, New York, 1987.
69. Stephen H. Schneider, "Climate Modeling," *Scientific American*, May 1987.
70. Anthony D. Del Genio, Goddard Institute for Space Studies, personal communication, March 11, 1992; and William Booth, "Computers and the 'Greenhouse Effect': The Genesis of Understanding," *The Washington Post*, June 12, 1989.
71. See note 69.
72. Gerry Mahlman, Geophysical Fluid Dynamics Laboratory, verbal presentation at "Global Change and Our Common Future," a conference sponsored by the National Research Council, Washington, D.C., May 2, 1989.
73. See note 70, and Frank Edward Allen, "Labs Rely on Computer Models To Predict Changes in Climate," *The Wall Street Journal*, June 3, 1992.
74. See note 4.
75. *Ibid.*
76. V. Ramanathan et al., "Cloud Radiative Forcing and Climate: Results from the Earth Radiation Budget Experiment," *Science*, Jan. 6, 1989.
77. Tim Beardsley, "Night Heat," *Scientific American*, February 1992. Thomas Karl's original study was published in the December 1991 issue of *Geophysical Research Letters*.
78. William W. Kellogg, "Response to Skeptics on Global Warming," *Bulletin of the American Meteorological Society*, April 1991.
79. Richard S. Lindzen, "Some Coolness Concerning Global Warming," *Bulletin of the American Meteorological Society*, March 1990.
80. Anthony D. Del Genio, Andrew A. Lacis and Reto A. Ruedy, "Simulations of the Effect of a Warmer Climate on Atmospheric Humidity," *Nature*, May 30, 1991.
81. Stephen H. Schneider, "A Follow-up Response to 'The Global Warming Debate Heats Up: Perspective and Analysis,'" *Bulletin of the American Meteorological Society*, July 1991. In his article cited in note 78, Kellogg notes that the water vapor content of the middle troposphere actually has increased 20 to 30 percent during the last several decades, casting further doubt on Lindzen's hypothesis.
82. As reported by William Booth, "Climate Study Halves Estimate of Global Warming," *The Washington Post*, Sept. 14, 1989; and Richard Monastersky, "Warmer Clouds Could Keep Earth Cooler," *Science News*, Sept. 23, 1989.
83. R.D. Cess et al., "Interpretation of Cloud-Climate Feedback as Produced by 14 Atmospheric General Circulation Models," *Science*, Aug. 4, 1989.
84. Wallace C. Broecker, "The Biggest Chill," *Global Climate Change Linkages*, edited by James C. White, Elsevier Science Publishing Co., New York, 1989.
85. Kevin E. Trenberth, Grant W. Branstator and Phillip A. Arkin, "Origins of the 1988 North American Drought," *Science*, Dec. 23, 1988.

86. K.M. Lau and P.J. Sheu, "Annual Cycle, Quasi-Biennial Oscillation, and Southern Oscillation in Global Precipitation," *Journal of Geophysical Research*, Sept. 20, 1989.
87. See note 72.
88. R.J. Stouffer, S. Manabe and K. Bryan, "Interhemispheric Asymmetry in Climate Response to a Gradual Increase of Atmospheric CO_2," *Nature*, Dec. 7, 1989.
89. *Ibid.*, and William Booth, "Computer Predictions of 'Greenhouse Effect' Have a Northern Accent," *The Washington Post*, Dec. 7, 1989.
90. See note 69.
91. See note 84.
92. Boyse Rensberger, "Volcano Reverses Global Warming," *The Washington Post*, May 19, 1992; and Constance Holden, "Post-Pinatubo Cooling on Target," *Science*, May 29, 1992.
93. Peter Schlosser et al., "Reduction of Deepwater Formation in the Greenland Sea During the 1980s: Evidence of Tracer Data," *Science*, March 1, 1991.
94. G.A. Meehl, "Seasonal Cycle Forcing of El-Niño-Southern Oscillation in a Global, Couple Ocean-Atmosphere GCM," *Journal of Climate*, 1990.
95. William W. Kellogg, "Overview of Global Environmental Change: The Science and Social Science Issues," *MTS Journal*, 1990. This article cites personal communication between Kellogg and Keeling.
96. William K. Stevens, "Warming of Globe Could Build on Itself, Some Scientists Say," *The New York Times*, Feb. 19, 1991. The original study appeared in the journal *Nature*.
97. See note 44.
98. Gordon J. MacDonald, "Climate Impacts of Methane Clathrates," and Daniel A. Lashof, "The Dynamic Greenhouse: Feedback Processes that Can Influence Global Warming," both in *Proceedings of the Second North American Conference on Preparing for Climate Change*, The Climate Institute, Washington, D.C., 1989.
99. See note 7; P.D. Quay, B. Tilbrook and C.S. Wong, "Oceanic Uptake of Fossil CO_2: Carbon-13 Evidence," *Science*, April 3, 1992; and Richard A. Kerr, "Fugitive Carbon Dioxide: It's Not Hiding in the Ocean," *Science*, April 3, 1992.
100. Drip analogy from Nova broadcast #1603, "The Strange New Science of Chaos," Jan. 31, 1989.
101. Statistics derived from various issues of the *Weekly Climate Bulletin*, National Weather Service, Washington, D.C.
102. "California Growers Lost Over $780 Million in December's Freeze," *The Wall Street Journal*, Jan. 14, 1991.
103. "1990 United States Climate Summary," *Weekly Climate Bulletin*, National Weather Service, Washington, D.C., Jan. 12, 1991.
104. "United States Seasonal Climate Summary: Winter 1991-1992," *Weekly Climate Bulletin*, National Weather Service, Washington, D.C., March 14, 1992.
105. "Significant Global Climate Highlights, Events and Anomalies: October 1991-mid-March 1992," *Weekly Climate Bulletin*, National Weather Service, Washington, D.C., March 14, 1992.
106. Douglas LeComte, "Weather Highlights of 1988," *Weatherwise*, February 1989.
107. Douglas LeComte, "Weather Highlights of 1989," *Weatherwise*, February 1990.
108. "Strange Pressure Waves Crossed Midwest in 1990," *The Wall Street Journal*, Jan. 15, 1991.
109. "1991 United States Climate Summary," *Weekly Climate Bulletin*, National Weather Service, Washington, D.C., Jan. 11, 1992.
110. Edward Ferguson and Frederick Ostby, "Tornadoes of 1990: An All-Time Record Year," *Weatherwise*, April 1991.

111. "Cyclone 2B," *Weekly Climate Bulletin*, National Weather Service, Washington, D.C., May 4, 1991.
112. M. Hulme and P.D. Jones, "Temperatures and Windiness Over the United Kindgom During the Winters of 1988/89 and 1989/90 Compared with Previous Years," *Weather*, British Metiological Society, May 1991.
113. See second citation of note 42.
114. See first citation of note 41.
115. *Ibid.*
116. *Ibid.*
117. "Global Climate Highlights," *Weekly Climate Bulletin*, National Weather Service, Washington, D.C., Aug. 4, 1990.
118. See note 112.
119. See note 101.
120. "Short and Long-Term Dryness Afflict the Mediterranean Basin," *Weekly Climate Bulletin*, National Weather Service, Washington, D.C., July 28, 1990.
121. Jacob M. Schlesinger and Quentin Hardy, "Japan Faces Possible Summer Electricity Crunch," *The Wall Street Journal*, April 4, 1991.
122. "Global Climate Highlights," *Weekly Climate Bulletin*, National Weather Service, Washington, D.C., Sept. 28, 1991.
123. "Global Climate Highlights," *Weekly Climate Bulletin*, National Weather Service, Washington, D.C., Oct. 26, 1991.
124. "Global Climate Highlights," *Weekly Climate Bulletin*, National Weather Service, Washington, D.C., Nov. 9, 1991.
125. "Annual Climate Summary: Major Climatic Events and Anomalies Around the World During 1991," *Weekly Climate Bulletin*, National Weather Service, Washington, D.C., Jan. 25, 1992.
126. "Global Climate Highlights," *Weekly Climate Bulletin*, National Weather Service, Washington, D.C., Feb. 9, 1991.
127. "Global Climate Highlights," *Weekly Climate Bulletin*, National Weather Service, Washington, D.C., March 2, 1991.
128. "Global Climate Highlights," *Weekly Climate Bulletin*, National Weather Service, Washington, D.C., April 27, 1991.
129. Anne Raver, "With Autumn Showers Come, Yes, April's Flowers," *The New York Times*, Oct. 20, 1991.
130. *Ibid.*
131. *Ibid.* According to this article, "The heat and drought sent some plants into early dormancy. Then a mild rainy spell in August or September caused them to break bud, or flower, as if it were spring." Plants produce flower buds for next year's growth during the previous summer.
132. Example derived from Paul E. Waggoner and Roger R. Revelle, "Summary of Climate Change and the Planning and Management of U.S. Water Resources," American Association for the Advancement of Science, Sept. 27, 1988.
133. See note 5.
134. David Stooksbury, "Is Climate Change Already Giving Us Greater Maize Yields?" *New Scientist*, Nov. 23, 1991.
135. William W. Kellogg and Zong-Ci Zhao, "Sensitivity of Soil Moisture to Doubling of Carbon Dioxide in Climate Model Experiments. Part I: North America," *Journal of Climate*, April 1988.
136. Stephen H. Schneider, *Global Warming: Are We Entering the Greenhouse Century?*, Sierra Club Books, San Francisco, Calif., 1989.

137. James Hansen et al., "Global Climate Changes as Forecast by Goddard Institute for Space Studies Three-Dimensional Model," *Journal for Geophysical Research*, Aug. 20, 1988.
138. See note 135.
139. *Ibid.*
140. Fred Pearce, "High and Dry in the Global Greenhouse," *The New Scientist*, Nov. 10, 1990.
141. Stephen H. Schneider, "Three Reports of the Intergovernmental Panel on Climate Change: A Review," *Environment*, January/February 1991.
142. Philip Shabecoff, "White House Admits Censoring Testimony," *The New York Times*, May 9, 1989.
143. Richard A. Kerr, "Hansen vs. the World on the Greenhouse Threat," *Science*, June 2, 1989.
144. John Noble Wilford, "His Bold Statement Transforms the Debate on the Greenhouse Effect," *The New York Times*, Aug. 23, 1988.
145. E. Friis-Christensen and K. Lassen, "Length of the Solar Cycle: An Indicator of Solar Activity Closely Associated with the Climate," *Science*, Nov. 1, 1991; and Richard A. Kerr, "Could the Sun Be Warming the Climate?" *Science*, Nov. 1, 1991.
146. See note 7.
147. Guy Brasseur and Michael H. Hitchman, "Stratospheric Response to Trace Gas Perturbations: Changes in Ozone and Temperature Distributions," *Science*, April 29, 1988.
148. Bob Davis and Barbara Rosewicz, "Panel Sees Ozone Thinning, Intensifying Political Heat," *The Wall Street Journal*, Oct. 23, 1991.
149. Keith Schneider, "Bush Orders End to Manufacture of Ozone-Harming Agents by '96," *The New York Times*, Feb. 13, 1992.
150. R.J. Charlson et al., "Climate Forcing by Anthropogenic Aerosols," *Science*, Jan. 24, 1992.
151. See note 24.

Chapter 2
Agriculture

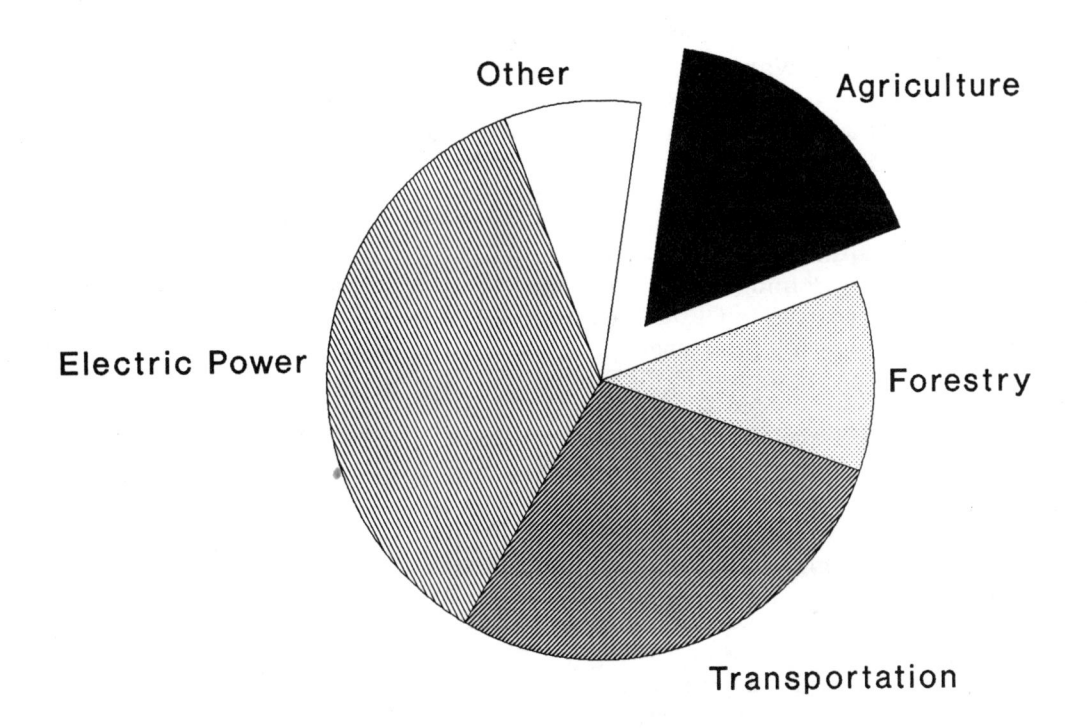

Relative contribution to global warming potential.

Contents of Chapter 2

Introduction 73

Effects on Crops
    The World As It Might Be 77
    Winners and Losers 78
    Feeding 6 Billion 80
    The Green Revolution 82
    Agriculture's Role in Global Warming 85
    Effects of Carbon Dioxide Enrichment 89
    Global Warming and American Agriculture 94
    Global Warming and Canadian Agriculture 97
    Uncertain Costs of Global Warming 99
    Constraints on Northern Agriculture 103

Effects on Water
    Liquid Gold 106
    The Coming Water Shortage 110
    Factoring in Climate Change 113
    Irrigating the Desert 116
    Making a Market for Water 119

Responses of Farmers and Food Producers
    Tapping Water Conservation 123
    Growing Drought-Tolerant Crops 125
    Engineering the Foods of the Future 128
    Designing Plants for Climatic Stress 130
    The Return of Sustainable Agriculture 132

Conclusions
    Feeding 12 Billion 137
    Global Warming and Food Security 138
    Farming in a Climate of Uncertainty 141

Boxes
    2-A: The Business of Agriculture 76
    2-B: Agriculture, Food Safety and the Environment 86
    2-C: Pests, Disease and Global Warming 92
    2-D: Weather's Role in Agriculture 102
    2-E: The Ogallala Aquifer: America's Sixth Great Lake 108
    2-F: Farming on Water Subsidies 120
    2-G: Federal Policy and Alternative Agriculture 135

Notes 145

Introduction

Each year at planting time, American farmers place their bets on the weather and the commodities markets. They wager that the soil and the climate will nurture their seeds into commercially salable crops—and that grains merchants will pay them enough to sustain their livelihood for another year. It is a profession wrought with anxiety. Farmers cannot dictate when the rains will fall, whether pests will invade their fields or how well their harvests will stack up against those of their competitors, some of whom are halfway around the world.

Now the possibility of climate change raises the stakes for American farmers and threatens to take away their comparative advantage in the global marketplace. Yet farmers have ways to defy the odds. They share an intimate relationship with the land, choosing crops to match the soil and nutrient characteristics of particular fields. They can anticipate many of nature's challenges by spraying their fields to protect against insects and disease and irrigating selected crops to overcome local water shortages. American farmers also have the U.S. government on their side, which offers price supports for surplus crops and which stands ready to offer remuneration in case natural disaster strikes.

Such a combination of favorable climate, fertile soils, agricultural technology and government backing has made the U.S. agricultural system one of the most productive on Earth. The 2 percent of the American population living on farms not only provides enough food for the rest of the country but also serves as a breadbasket for much of the world. As the global leader in agriculture, the United States supplies half the world's total exports of wheat and coarse grain—corn, barley, oats, rye and sorghum. Accordingly, any adverse effects of global warming felt in the American heartland would have repercussions around the world.

The sheer size and scope of American agriculture makes it big business. Broadly defined, the food sector is the nation's single largest industry—second only in size to the entire U.S. manufacturing sector—and it is the largest and most consistent source of U.S. trade. The traditional family farm, however—a bucolic enterprise of crops and livestock sharing a quarter-section of land—is fading from the American landscape. Today's typical farm is highly specialized, with three times the acreage of one in 1940. One-seventh of the nation's 2.2 million farms, in fact, now account for three-quarters of all U.S. farm sales, and many are incorporated.[1]

While farmers make up just 2 percent of the gross national product, those who provide the capital, supply the seeds, manufacture the fertilizer, store and ship the grain, and process the food for sale to consumers account for another 14 percent of GNP—nearly $800 billion of economic activity every year.[2] The specter of climate change looms over each of these agricultural concerns, and it threatens to disrupt the food distribution network on which much of the world depends. Under the gloomiest forecasts, a climate reminiscent of the Dust Bowl years of the 1930s—or the more recent drought of 1988—could return with a vengeance. But the prevailing sentiment among the experts is not that dire.

"The overall outlook for climate change is not catastrophic," surmises Cynthia Rosenzweig, a Columbia University agronomist who is the co-author of a major study on global warming and U.S. agriculture prepared for the U.S. Environmental Protection Agency. "In general," she finds, American "consumers may have to pay a small to moderate amount more for their food, while food producers either gain or lose depending on the severity of the climate change scenario."[3]

At present, EPA's guess is that the U.S. agricultural sector actually might make $3.5 billion *more* a year than the $200 billion or so it already earns, since a climate-induced drop in the food supply would increase retail prices.[4] "None of the scenarios predict mass starvation," reassures another co-author of the EPA study, Texas A&M University professor Bruce McCarl. Although regional effects of climate change in the United States could be profound, "Food security is not an issue," McCarl insists.[5]

A broader view of the world situation offers a less sanguine outlook, however. While Americans spend only about 15 percent of their personal disposable income on food, more than a billion people in less-developed countries spend 50 to 80 percent or more of their income on food.[6] For them, a moderate increase in the price of food might not lead to further belt-tightening; it could be life-threatening.

Meeting the nutritional needs of a world population expected to double in less than a century is perhaps the most formidable challenge facing humanity today. At the same time, the prospect of global warming is raising questions about the fortitude of the existing agricultural system. Ironically, production methods developed during this century to make agriculture more bountiful and less susceptible to nature's elements may increase the system's vulnerability to climate change in the long run.

Modern crop-production methods are generally rigid, monocultural and very energy-intensive. High-yielding crops may wither if the generous supply of water and nutrients on which they depend is cut short. So, too, may the herds of livestock that live off the fat of the land. As for energy requirements, nearly 10 calories of fossil fuels are now burned for each calorie of food Americans ingest. Put another way, when all of the agrichemicals, nutrients and fuels are combined, the energy equivalent of nearly six barrels of oil is consumed each year just to put food on each of our plates.[7] Such resource-intensive methods may be increasingly difficult to sustain as the world has less fuel to burn and many more mouths to feed in the years ahead.

On a brighter note, carbon dioxide enrichment may stimulate the growth of plants even if global temperatures warm and local precipitation falls. This so-called "fertilization effect" is a potentially important mitigating factor of climate change on agriculture. But the long-term effect of CO_2 enrichment on crop productivity remains about as uncertain as the effect of the CO_2 buildup (and

Figure 1

World Grain Exports
(percentage by weight, 1988)

- United States 50%
- Canada 15%
- Western Europe 12%
- Australia 8%
- Argentina 5%
- Others 10%

SOURCE: U.S. Department of Agriculture.

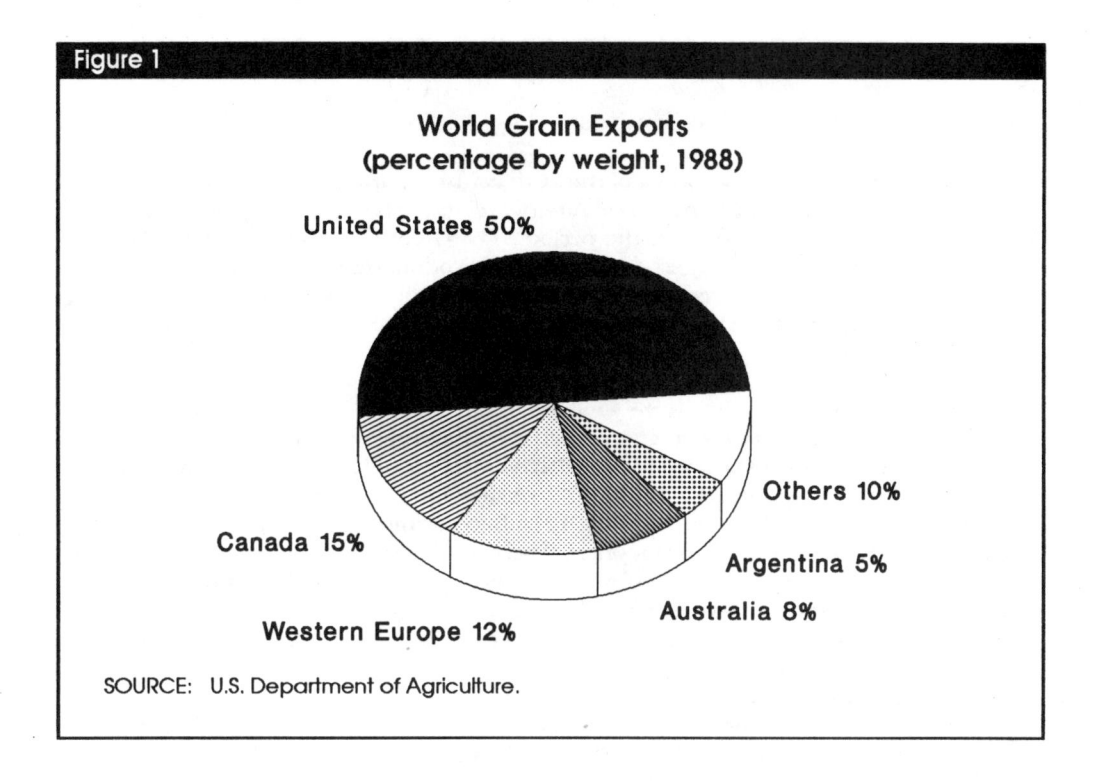

other greenhouse gases) on the behavior of climate itself. Therefore, farmers cannot bank on the assumption that CO_2 enrichment will provide a ready antidote to global warming.

In any event, the future of agriculture—even absent climate change—is pointing in two new directions. One path is toward a Brave New World: genetically altered plants (maybe even animals) created by firms that shepherd products from the farm all the way to the supermarket. (If this sounds far-fetched, think of Frank Perdue's success with breeding chickens.) The other path is back-to-basics: growing food with an eye toward nature more as an ally than an adversary. (If this smacks of '60s nostalgia, consider the rise of integrated pest management methods and consumers' desire for organically grown foods.)

These two emerging trends for agriculture do not exclude one another, nor do they portend the elimination of conventional growing techniques. Yet the specter of climate change makes it harder to chart a proper future course for agriculture. In this uncertain climate, planting decisions and soil- and water-management strategies that farmers employ over time will come to define their greenhouse gambit. The remainder of this chapter explores the potential effects of climate change on agriculture, the range of responses available to farming enterprises and the possible implications for world food security.

Box 2-A

The Business of Agriculture

Year in and year out for most of the 20th century, farmers have planted crops on approximately 350 million acres of American land. That has been one of the few constants in rural America over the period. The Great Depression and the Dust Bowl years of the 1930s triggered a population exodus from the Great Plains that has never ceased. The percentage of people now tilling the soil is lower than ever before—only 2 percent of the American population. Yet crop yields on the 2.2 million American farms that remain have never been higher.

Huge investments in land, mechanized equipment, irrigation systems, chemical fertilizers and the like have enabled large farming enterprises to realize economies of scale and become more profitable than the rest. From 1983 to 1987, for instance, the 300,000 or so farms with sales of more than $100,000 annually accounted for three-quarters of total farm sales and nearly seven-eighths of net farm income. Meanwhile, the smallest 1.6 million farms, with annual sales of less than $40,000 each, accounted for only 10 percent of total farm sales and even less of net income over the period.[1a] As small farms consolidate into bigger ones, some are assuming corporate ownership. The two largest agricultural states, California and Texas, allow corporate ownership, as do Florida, Indiana and Illinois, among others. Several key farm states still do not permit corporate farm ownership, however, including Kansas, Nebraska, the Dakotas, Iowa and Oklahoma.

Institutional investors have turned increasingly to farm investments over the last 20 years as a hedge against inflation and to seek capital growth. Equitable Agri-Business (a subsidiary of Equitable Life Assurance Society), for one, had about $2.2 billion invested in 12,000 farms and ranches as of mid-1990, mostly through mortgages. Aetna and Prudential Insurance, Batterymarch Financial and Morgan Stanley Asset Management are among the other institutional investors with large farmland interests. Equitable also has created a $100 million farm mutual fund to allow smaller investors to add agricultural holdings to their portfolios.[2a]

Buoyed by a boom in world food trade and generous rainfall, farm income and farm land prices soared in the 1960s, 1970s and early 1980s. But the boom ended in 1982—with a drought year to follow. As farmers' dreams of acquiring ever-more-valuable land on credit dried up along with their fields in 1983, farm debt rose to a record $192 billion. Since then, nearly 300,000 farmers have left the profession, many of them penniless. The shakeout has yielded some positive benefits, however. Farm debt has fallen one-third since its peak in 1983. Agricultural banks are now considered to have the strongest balance sheets within the banking industry. And in 1990 the U.S. farm sector posted a record net income of $47 billion.[3a]

Land prices have been slower to recover. After peaking at $823 an acre in 1982, average prices dropped to below $640 an acre in 1986 and then crept back up to just under $700 an acre at the beginning of 1990. The rebound in land prices has varied considerably by region. The southwestern states—where oil and gas companies acquired huge tracts during the oil boom years—have had the slowest price improvements. The western ranch states have experienced some of the steepest price rises, since cattle raising has become more profitable in recent years and ranchers have been willing to pay a premium to acquire more land.

In the years ahead, global warming could alter the productivity of these agricultural regions, some for better and some for worse. But the specter of climate change has yet to emerge as a factor in appraising farmland values.

Effects on Crops

The geographical extent of agricultural development today is largely determined by temperature gradients and rainfall amounts around the globe. Near the equator, where the world's lush tropical rain forests are found, the climate is hot and rainfall is heavy; here much of the soil nutrients necessary for the growth of crops are simply washed away. From the Tropics of Cancer and Capricorn halfway to the poles, temperatures remain high but rainfall is relatively sparse. With high rates of evaporation, this is where the world's great land deserts are found. Farther to the north and south, roughly between 45 and 65 degrees of latitude, precipitation picks up again and temperatures become more variable. The world's highly productive agricultural zones are contained within this temperate band, with summer growing seasons bracketed by killing frosts in the spring and fall. Finally, from the Arctic Circle to the North Pole, as well as in Antarctica, there is little in the way of precipitation or warmth. These polar regions essentially are frozen deserts.[8]

The World As It Might Be

Global climate change poses many implications for world agriculture—and many are positive. An increase in global temperatures, for example, seems poised to benefit agriculture in northerly temperate zones, since most crops consist of 80 to 90 percent water and are prone to freezing at temperatures below 32 degrees Fahrenheit (F). Corn, soybeans, wheat and other annual crops presumably would encounter fewer days of frost and enjoy longer growing seasons, although more frequent summer heat waves could impair their seed yield as well. Similarly, citrus trees and other perennial crops might be subject to fewer devastating cold snaps that in recent years have ravaged groves in California, Texas and Florida—and nipped the buds of grape vineyards in France. (Of course, crops everywhere would remain susceptible to chaotic temperature swings, especially if the climate does not adjust smoothly to the influx of greenhouse gases in the atmosphere.)

Some crops also are likely to benefit from the increase in atmospheric carbon dioxide levels that would accompany warmer temperatures. The photosynthetic optimum for many species would rise by as much as 10 degrees F with a doubling of ambient CO_2.[9] A CO_2-enriched atmosphere, in turn, would spur plants' production of tissues, since carbon dioxide is the fuel that regulates their rate of photosynthesis. A 25 percent rise in ambient CO_2 levels since 1800, in fact, may already have boosted the average seed yield of selected crops, such as soy beans, according to U.S. Department of Agriculture researchers.[10] A doubling of CO_2 could increase the average yield of such crops considerably more.

As an added bonus, an increase in photosynthetic activity by most CO_2-enriched plants does not require a concomitant increase in water intake; hence, the plants would become more efficient users of water. Concerns about heat-related stress notwithstanding, global warming could provide the best of all worlds for many plants: fewer encounters with frost because of warmer temperatures, faster accumulation of biomass because of CO_2 enrichment and reduced vulnerability to drought because of greater water use efficiency.

Given these positive factors, precipitation and soil moisture holds the key to determining whether global warming would benefit agriculture overall. To put this vital question in perspective, one can imagine what would happen if a gardener failed to water plants grown in a greenhouse. Despite optimal growing temperatures and enriched levels of CO_2, the plants' productivity would start to decline when the amount of water in their tissues fell by just 5 percent. If the water content dropped by another 10 percent, photosynthesis in most irrigated plants would stop altogether. Then, by casting off their leaves in a process called abscission, the plants would try to curb any additional water loss, retaining enough moisture in their tissues to regenerate leaves when water was resupplied. But if the gardener failed to water in time, the plants inevitably would die of dehydration.[11]

In a greenhouse world, climate modelers must gauge how much rain would fall and whether it would come at regular enough intervals to keep soils moist and plants thriving. Here again, many of the first-order indications are positive. Although higher temperatures would cause more water to evaporate, the warmer atmosphere would hold on to more of this moisture. To maintain equilibrium in the hydrological cycle, greater precipitation must balance out the accelerated evaporation. The net result: the enhanced greenhouse effect would increase projected global rainfall totals by 7 to 15 percent.[12]

The precipitation increase would not be uniform across the globe, however. Some regions would give up more in the way of evaporation than they would get back as precipitation. Soil moisture consequently would decline in those areas, and plants' ability to grow would be adversely affected. Pinpointing such regional precipitation changes is the greatest unknown in farmers' greenhouse gambit. General circulation models of global climate change offer a range of possibilities, but their projections often conflict, so confidence in their results is low. In any event, global climate change would appear likely to alter the geography of agricultural development more than it would reduce total output around the globe. These geographical shifts would create new winners and losers in the production and distribution of food.

Winners and Losers

With a rise in global temperatures, the climate zones forming the latitudinal rings around the planet would likely expand toward the poles. Abundant precipitation at the equator might extend toward subtropical areas, benefiting these water-deficient zones. At the same time, some of the world's deserts—the Sahara, the Mojave, the Gobi, the Great Victoria Desert of Australia and others—might encroach on lands that produce some of the world's most bountiful grain harvests—the Mediterranean, the North American Great Plains, the fertile river valleys of China and the plains of western Australia. Summer rainfall totals in these agriculture-intensive zones could decline by 15 or 20 percent, having a disproportionate, negative effect on global food production.[13]

Fortunately, other regions of the globe would likely become warmer and wetter as a result of climate change, enhancing their own crop-producing capabilities. Northern Scandanavia, for example, "stands to gain more from global warming than perhaps any other region of the world," according to Martin Parry, the lead author of a landmark international study of climate change and agriculture conducted by the International Institute for Applied Systems Analysis.[14] Similarly, much of the British Isles, Denmark and the Low Countries of Europe might experience stimulated growth of grasses, potatoes and other plants. Simultaneously, the Ukraine could benefit from a significant jump in wheat production, while northern Japan's Honshu and Hokkaido provinces might enjoy increases in the yields of rice, corn and soybeans. Within North America—the world's greatest corn-growing region—production of corn in Canada could rise to offset possible losses in the United States.[15] Of course, all of these projections remain highly speculative, pending further refinement of general circulation and crop response models.

What about developing nations closer to the equator? The projected effects of global warming are less pronounced in these countries because temperatures would be expected to rise far more toward the poles of the Earth than in the middle. On the other hand, crops grown at low latitudes are subject to considerable heat stress already. Future warming might push some of these crops beyond their limits of temperature tolerance. One recent study led by Rosenzweig and Parry found that climate change could put an additional 63 million to 369 million people at risk of hunger in developing nations (depending on which general circulation model is used.)[16] Especially vulnerable regions include Mexico, Central America, western South America, as well as most of India, parts of Southeast Asia and virtually all of sub-Saharan Africa. Some fear that food shortages in these regions could lead to mass human migrations and associated refugee problems. (In parts of Africa, these problems exist already.)

In equatorial nations, the current growing season is dictated largely by seasonal monsoons: giant sea breezes that blow from cold continents to warm oceans in the winter and then reverse direction in the summer as the continents grow relatively warmer than the oceans. If global warming were to increase the seasonal temperature disparity between land and sea, monsoons in Africa and Asia might grow more intense and penetrate further poleward. This could benefit drought-prone regions such as the African Sahel and northwestern India.[17] But it could also exacerbate problems in flood-prone regions such as Bangladesh, where well over a half-million lives have been lost to floods since 1970.

A final consideration of global warming's effects on agriculture is that it could lead to a rise in sea level that inundates low-lying, highly productive farmland (to say nothing of sopping coastal infrastructure). Bangladesh, Egypt, Indonesia and Pakistan—nations where arable land already is at a premium—would be among those most vulnerable to encroachment by the sea. Bangladesh, for one, has embarked on a $10 billion project to shore up its defenses against

Figure 2

World Population Growth 1750-2000

6 billion in 2000?

North America & Oceania
Latin America
Africa
Europe & USSR
Asia

Lifespan of the Rev. Robert Malthus

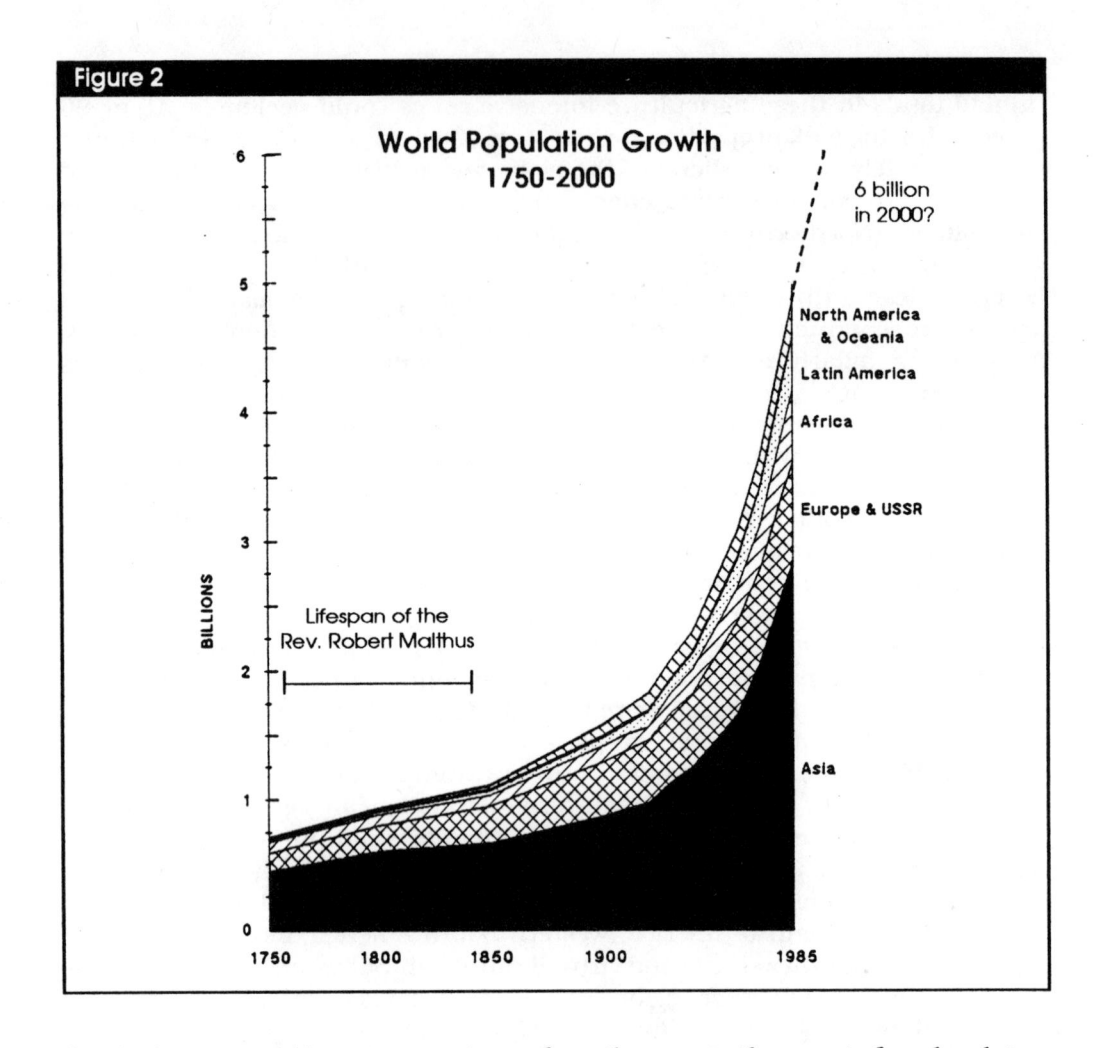

BILLIONS

floods, but it would have to invest significantly more in the event of sea level rise. The world's master dike-builders, the Dutch, calculate that up to $10 billion would be required to improve Holland's own system to hold back the sea if the ocean rises by three feet or more.[18]

Feeding 6 Billion

While the regional effects of climate change remain highly speculative, the global distribution of the food supply would be almost certain to change in the event of global warming. Hard-hit agricultural regions would have to develop new trading arrangements with the emerging breadbaskets of the world. If such markets did not develop properly, or if there simply were not enough food to go around, the human population could crash in resource-stricken, rapidly growing regions.

The Rev. Robert Malthus conjured such an image of the human population

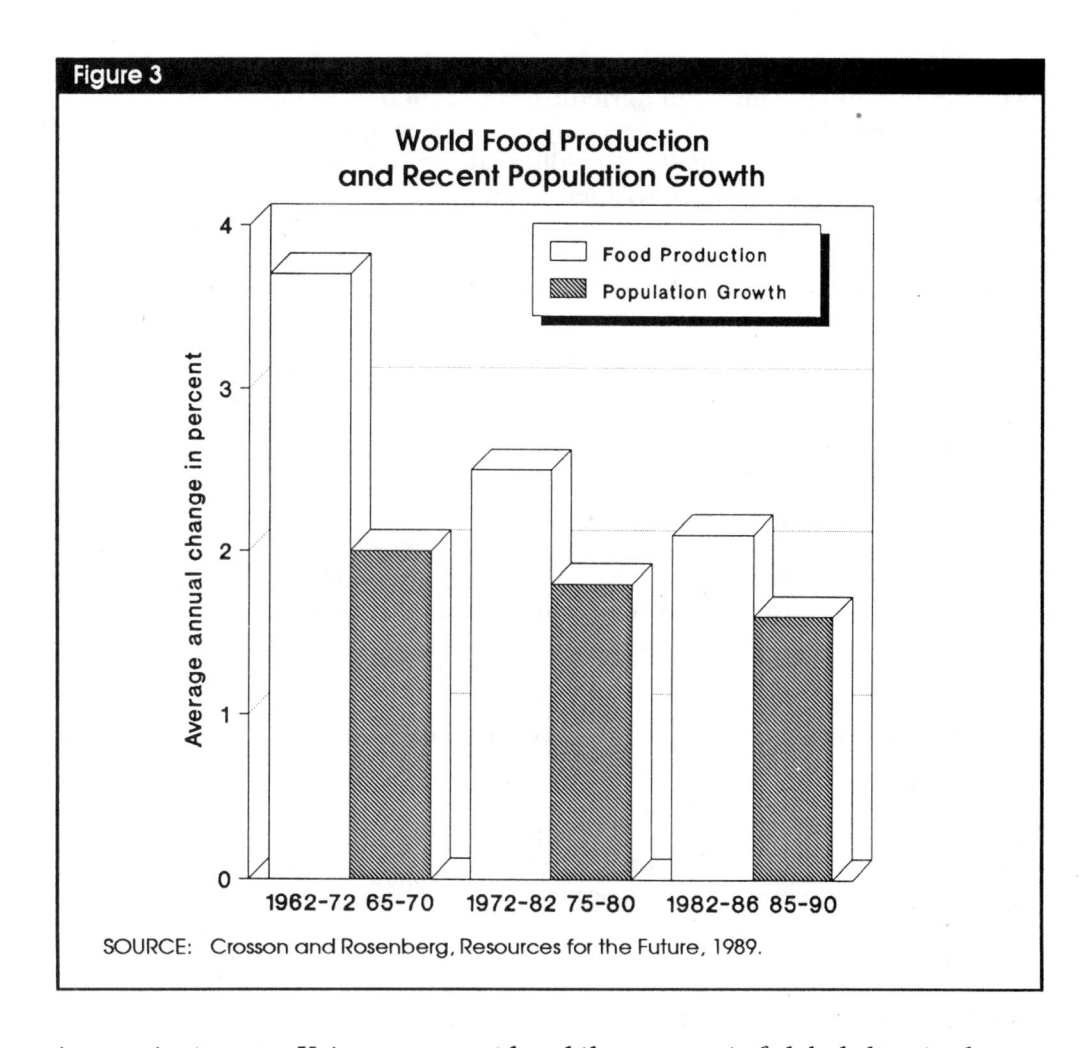

Figure 3

World Food Production and Recent Population Growth

Average annual change in percent

Period	Food Production	Population Growth
1962-72 / 65-70	~3.7	~2.0
1972-82 / 75-80	~2.5	~1.8
1982-86 / 85-90	~2.0	~1.6

SOURCE: Crosson and Rosenberg, Resources for the Future, 1989.

two centuries ago. Yet no one considered the prospect of global climate change in his time. Besides, only a billion people inhabited the planet back then, well within the Earth's carrying capacity. A century after Malthus's death, however, the world's population reached 2 billion. It doubled again—to 4 billion—between 1930 and 1975. The 5 billion mark was passed in 1986. By 2000, 6 billion humans are expected to be living on the planet.[19]

To sustain the human population explosion that is expected to continue well into the 21st century, the world's food supply must keep growing at least 1.7 percent a year, estimates Pierre Crosson, a senior fellow at Resources for the Future. "If history is a reliable guide," he says, "world agricultural capacity [absent climate change] could accommodate the prospective increase at acceptable economic and environmental costs, provided that farmers continue to apply new knowledge about how to manage agricultural systems."[20]

"But," Crosson warns, "history may not be a reliable guide for two reasons. One is that the expansion in capacity in recent decades undoubtedly entailed a variety of environmental costs that are not reflected in world market prices of food and fiber. The second reason is that the quantities of good land and water available to supply additional agricultural production are more limited now than they were 35 years ago, especially in developing countries."[21]

It was only 50 years ago that traditional farming methods prevailed the world over. Farmers in India and Argentina were able to achieve about the same crop yields as those in Indiana and Arkansas.[22] Since World War II, however, those who have planted high-yielding crop varieties and applied greater doses of water, fertilizer and pesticides have boosted crop yields tremendously. In the United States, average corn yields per acre have tripled, per-acre yields for wheat have more than doubled, and yields for soybeans have increased by more than 40 percent.[23] At the same time, the amount of labor required to farm an acre of U.S. land has declined 75 percent, even as farm output per acre has doubled. That means today's American farmer—outfitted with modern equipment—is eight times more productive than one in 1940.[24]

Such energy- and capital-intensive farming techniques pioneered in the United States have since spread to the far corners of the globe. World food production has increased by about 2.5 percent a year since 1950, outpacing annual population growth by nearly a full percentage point. This added margin has allowed food prices to fall (adjusted for inflation) even as the population has doubled.[25]

An acre of arable land today produces an average of slightly more than a ton of grain, whereas less than a half-ton was the norm only 40 years ago.[26] Yet despite the tremendous gains in crop yield, the average amount of food produced *per person* has been declining since 1984.[27] As a result, one out of ten human beings—some 550 million people—is now malnourished.[28] This implies that the added production share is being used not to feed hungry people more, but to feed more hungry people. Global warming could compound this challenge.

The Green Revolution

To make room for the expanding human population, 500 million acres of mostly tropical forest—an area about the size of the United States east of the Mississippi River—has been felled in the last 20 years. Such deforestation has contributed a substantial amount of greenhouse gases (principally carbon dioxide and methane) to the atmosphere. Yet even more land would have been cleared were it not for the Green Revolution. This radical transformation in farming methods is responsible for about 80 percent of the rise in world crop production since World War II, whereas clearing more arable land accounts for only 20 percent of the gain.[29] Should agricultural productivity be maintained at current levels, the area of tropical agriculture would have to expand by another 60 percent to meet projected food demand in 2025.[30] Accordingly, the Green Revolution must carry on if the world is to preserve remnants of its tropical rainforests.

The Green Revolution is showing signs of aging, however, making it increasingly difficult to replicate past productivity gains. Moreover, the Green

Agriculture

Revolution is contributing greenhouse gases of its own to the atmosphere because of the energy- and and chemical-intensive practices involved. Therefore, farmers have fewer high-benefit/low-cost options than they did in the decades immediately following World War II to sustain a burgeoning human population.

Hybrid corn and semi-dwarf varieties of wheat and rice gave huge boosts to food production in the 1960s, for example, as these high-yielding crops became staples in developing countries. At the same time, these new crops led to rising demand for irrigation, fertilizers and pesticides—and the energy they all require. World fertilizer use climbed tenfold between 1950 and 1990, from 14 million metric tons a year to an estimated 145 million tons, largely because hybrids respond so well to fertilizers.[31] Meanwhile, irrigated acreage nearly tripled worldwide—mainly as a result of expansion of agriculture to semi-arid regions—but also because of greater water requirements among heavily fertilized crops. According to the World Bank, one-half to three-fifths of the massive increase in agricultural output in developing nations between 1960 and 1980

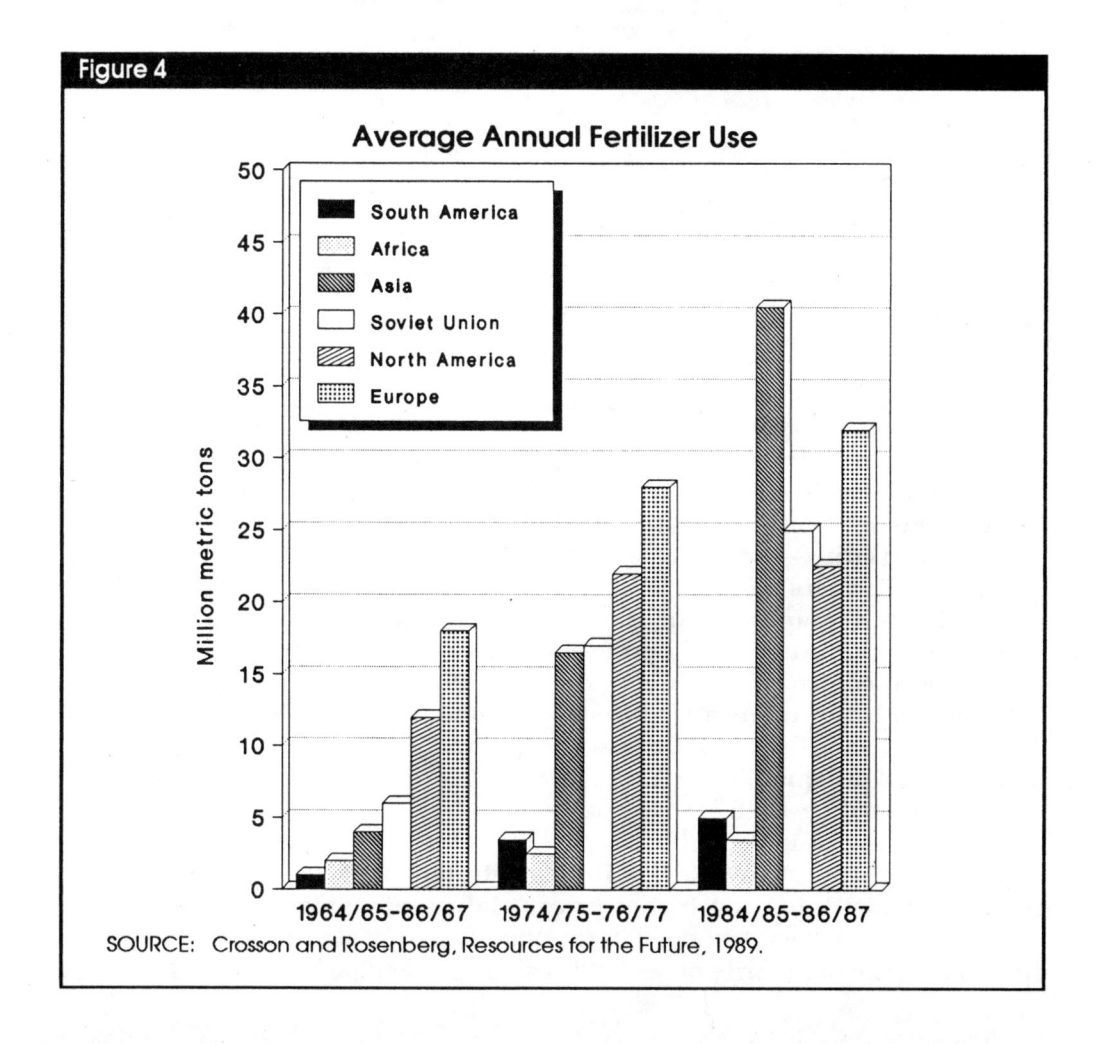

Figure 4

Average Annual Fertilizer Use

SOURCE: Crosson and Rosenberg, Resources for the Future, 1989.

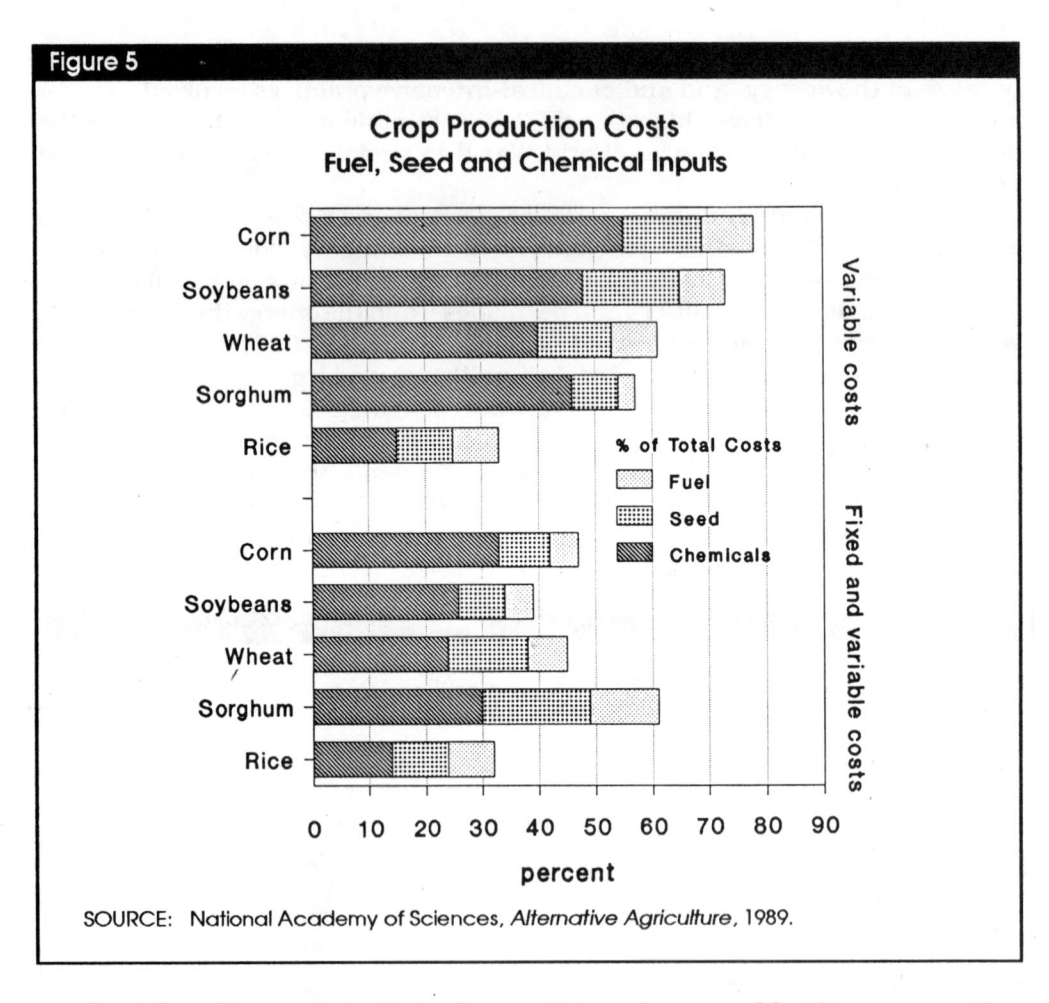

Figure 5

Crop Production Costs
Fuel, Seed and Chemical Inputs

% of Total Costs
- Fuel
- Seed
- Chemicals

Variable costs: Corn, Soybeans, Wheat, Sorghum, Rice

Fixed and variable costs: Corn, Soybeans, Wheat, Sorghum, Rice

percent

SOURCE: National Academy of Sciences, *Alternative Agriculture*, 1989.

resulted from the spread of irrigation, with greater use of fertilizer accounting for most of the rest of the gain.[32]

A market for synthetic pesticides also emerged after World War II. Before the war, most farms—including American ones—were diversified crop-livestock operations. Farmers provided forage and feed grains for their own livestock and returned animal manure to the land. As such, crop rotations were longer and had less need for purchased inputs, especially nitrogen fertilizers. Unwanted insects were controlled (with varying degrees of success) using a range of cultural and biological means. Weeds were kept down mainly by turning over the soil with a mull-bore plow, a practice that tended to dry out the soil and led to erosion. Consequently, far fewer insecticides—and almost no herbicides—were applied before 1945.[33]

The successful introduction of pesticides such as DDT, 2,4-D and BHC during and shortly after World War II demonstrated that lower-priced chemicals could be substituted for higher-priced, labor-intensive weed and insect control methods. As American farms increased in size and became more specialized, the popularity of pesticides soared. Pesticide use since 1945 has increased more than tenfold in the United States, peaking at 500 million pounds of active ingredients in 1982.[34] While as recently as 1970 only two-fifths of U.S. corn and soybean acreage was treated with herbicides, by 1985 the fraction had risen to 95 percent. Today, corn, soybeans, wheat and cotton—the nation's four largest cash crops—account for about 90 percent of the volume of herbicides and insecticides applied domestically.[35] Worldwide, the use of pesticides has more than doubled in the last two decades, reaching 4 billion pounds in 1989.[36]

Farmers' growing use of chemical inputs has been a boon to manufacturers of these products—mainly chemical and petroleum companies. The total dollar value of the domestic fertilizer market is now estimated at $6.5 billion a year, while domestic pesticide sales exceed $4 billion.[37] The international market provides more than $25 billion in additional sales each year. The market has become so lucrative, in fact, that for many major commodities, fertilizer and pesticide expenses now exceed other variable costs—such as seeds and fuel—by a wide margin. Fertilizers and pesticides represented more than one-half of the variable costs and one-third of the total costs of growing corn in the United States in 1986, for example. For soybeans, chemical inputs represented almost half of the variable costs and one-quarter of the total production costs; and for wheat, they accounted for two-fifths of the variable costs and nearly a quarter of the total costs on a nationwide average.[38]

Agriculture's Role in Global Warming

Given the large amounts of chemicals, nutrients and fuel now required to maintain high crop yields—and the sheer number of mouths to feed—agriculture has become an important contributor of greenhouse gases around the world. When the global food sector is considered in its entirety, it accounted for 16 percent of the total release of greenhouse gases during the 1980s, the congressional Office of Technology Assessment estimates.[39] Up to one-fifth of net global carbon dioxide emissions, 15 percent of chlorofluorocarbon emissions, one-third of global methane emissions and anywhere from one-tenth to one-half of nitrous oxide emissions stem from agriculture and food-related activities.

Half of the food sector's contribution—about 8 percent—is not related to modern agricultural practices, however. Rather, it results from clearing and burning of tropical forests to make way for crop production, and from the use of traditional biomass fuels for cooking in many developing countries. Food refrigeration contributes another 2 percent of greenhouse gas emissions through the release of CFCs, which are used as a cooling agent.[40] (This analysis does not factor in the counteractive effect of ozone depletion on global warming potential, however.)

Most of the remaining contribution of greenhouse gases in the food sector is related to modern farming practices. Consumption of fossil fuels in heavy farm machinery, irrigation pumping equipment and fertilizer manufacturing

(continued on p. 88)

Box 2-B

Critics of the Green Revolution

The Green Revolution can be credited for tremendous gains in food production since World War II. It has enabled farmers to feed billions of people affordably while slowing the rate of deforestation in nations with rapidly expanding populations. Yet some critics are ambivalent about the modern trend in agriculture. They argue that farmers have become dependent on institutionalized processes that squeeze their profits, degrade the environment and contribute excessively to global warming potential. The result, they say, is an agricultural system that emphasizes short-term production gains over long-term sustainability—and farmers that are less able to cope with potential climate-related stresses on top of the economic and environmental problems they already face.

Critics like Jim Hightower, the former agriculture commissioner of Texas, are as outspoken as they are controversial. "The great decline in American family agriculture is the direct result of a system that benefits the corporate farm at the expense of the family farm," Hightower contends. It "breeds dependence on expensive synthetic chemicals rather than the replenishment of natural resources....It is the result of a system that locks farmers into debilitating pesticide and fertilizer cycles that deplete the real value of American farms. The land itself becomes addicted to chemical fertilizer, becoming less productive and losing its value." In the end, Hightower says, "Farmers find themselves spending more in order to put more chemicals into the land each year while getting ever diminishing returns."[1b] Defenders of modern farming are quick to point out, however, that farmland values and farmers' profits are still rising.

Some advocates of low-input, sustainable farming methods believe that agribusiness has shaped modern farming to serve its parochial interests. The plethora of chemicals used to stimulate the soil and kill off pests suggests that "Nature is [something] to be subdued or ignored," maintains Wes Jackson of the The Land Institute in Salina, Kan. "The purpose of agricultural research and farming has been and still is to increase production," Jackson says; "agriculture is to serve as an instrument for the advance of industry."[2b] The defenders respond, "Aren't these supposed to be the dual objectives of agribusiness?"

Even the prestigious National Academy of Sciences has weighed in with a controversial assessment of American agriculture that is critical of some current farming practices. It places a large share of the blame on the rigid policies instituted by the federal government. "As a whole, federal policies work against environmentally benign practices and the adoption of alternative agricultural systems," the academy wrote in a landmark report, *Alternative Agriculture*.[3b] The 1989 report concluded that such policies tend to discourage "crop rotations, certain soil conservation practices, reductions in pesticide use, and increased use of biological and cultural means of pest control. These policies have generally made a plentiful food supply a higher priority than protection of the resource base."

"Pesticides can also cause crop losses," the academy added in its report. "This can occur when the usual dosages of pesticides are applied improperly; when herbicides drift from a treated crop to nearby, susceptible crops; when herbicide residues prevent chemical-sensitive crops from being planted in rotation or inhibit the growth of subsequent crops; and when excessive residues of pesticides accumulate on crops, causing the harvested products to be destroyed or devalued in the marketplace."[4b]

Four billion pounds of active pesticide ingredients are applied routinely around the world each year. Farm workers and wildlife come in contact with these chemicals as well as the targeted pests. Indiscriminate use of pesticides also kills soil organisms that naturally enrich the soil. With a reduced capacity to store and absorb water, the soil becomes more susceptible to drought. Pesticide residues also leach into groundwater, risking contamination of drinking water supplies. Sometimes residues persist in the food chain, raising highly publicized questions about food safety.

Although federal regulators consider the vast majority of pesticides to be safe, many chemicals have not been adequately tested. Partly for this reason, American consumers continue to put pesticide residues at the top of their list of food safety concerns.[5b] And partly in response to consumers' anxiety, manufacturers have begun to develop new products that contain much-reduced levels of active pesticide ingredients.[6b] Some manufacturers are even introducing natural insecticides as an alternative to synthetic bug killers for the billion-dollar U.S. lawn care and home gardening markets.[7b] Such developments are not all bad for pesticide manufacturers. Despite a 14 percent drop in production of active pesticide ingredients between 1979 and 1988, manufacturers' sales revenue grew 18 percent to $4.35 billion.[8b]

Nitrate and phosphate fertilizer applications pose a similar dilemma, since they raise crop yields, greenhouse gas emissions and water quality problems all at the same time. Fertilizers collecting in surface waters cause eutrophication of lakes and suffocating algal blooms. High nitrate concentrations in groundwater also interfere with the oxygen-carrying capacity of red blood cells in infants—known as "blue baby syndrome." More than half of the wells tested in a recent EPA survey of drinking water supplies found detectable levels of nitrates; 1.2 percent of the urban and suburban wells and 2.4 percent of the rural wells contained levels that EPA considers unsafe.

A final concern about modern farming methods is their contribution to topsoil erosion. Lester Brown of the Worldwatch Institute, another critic of some modern farming practices, estimates that farmers lost five pounds of topsoil to wind and water erosion for every pound of grain they produced during the early 1980s—a yearly topsoil loss of 1.6 billion tons.[9b] (Others insist that while the soil may have moved downwind or been deposited along fence rows, it was not "lost" altogether.) A federal program enacted by Congress in 1985 to take highly erodible farmland out of production—the Conservation Reserve Program—has since reduced topsoil losses to about 1 billion tons annually. But this erosion is still quite serious, Brown and others believe, since it takes roughly 100 years to build up one inch of topsoil naturally. And from a greenhouse perspective, erosion takes much of the carbon bound in the soil and releases it to the atmosphere.

While a continuation of intensive farming methods associated with the Green Revolution will be necessary to satisfy the world's growing appetite for food, critics insist that the rising ecological toll associated with these methods also must be taken into account. Over time, they say, institution of more environmentally benign management practices may serve three purposes. First, they would address ongoing food safety, water quality and soil erosion concerns. Second, they would create an agricultural land base that is more resilient to climate-related stresses—as soils retain more carbon, water and nutrients through natural processes. Finally, the agricultural sector's greenhouse gas emissions would fall as a result of the scaled-back use of synthetic fertilizers, pesticides, heavy machinery and irrigation equipment.

The unanswered question is whether such alternative agricultural methods—or even the more intensive ones—will be able to sustain ongoing rapid population growth in an era of possibly unprecedented warming.

operations contributes about 1 percent to global warming potential (since these fuels emit carbon dioxide as they are burned). Fertilizer production, in particular, is extremely energy-intensive, requiring large amounts of natural gas as a feedstock.[41]

Nitrogen fertilizers are a major source of nitrous oxide emissions as well. While the sources of this potent greenhouse gas are not well understood (soil erosion and microbial processes are thought to account for a large portion of N_2O emissions), nitrous oxide is estimated to have contributed about 6 percent of the total weighted greenhouse gas emissions during the 1980s. Current estimates of N_2O releases to the atmosphere range from 1.5 million tons to 2.25 million tons a year, according to the Intergovernmental Panel on Climate Change. Of this amount, up to 1.25 million tons is thought to derive from applications of nitrogen fertilizer. Over the next 35 years, nitrogen fertilizer applications may increase from 40 million tons to 60 million tons a year. Therefore, N_2O's contribution to global warming potential is expected to increase as well, with nitrogen fertilizer applications accounting for perhaps 2 million tons of N_2O emissions by 2025.[42]

Agriculture is a significant source of yet another growing greenhouse gas—methane. This gas is produced when bacteria decompose organic material in oxygen-deficient (anaerobic) environments. During the 1980s, methane was thought to be responsible for between 15 and 19 percent of the accumulation of greenhouse gases in the atmosphere. Ruminant digestion in livestock accounted for 4 to 7 percent of the contribution to global warming potential from all methane sources, the Office of Technology Assessment estimates, and releases of methane from the bottom of flooded rice paddies contributed another 3 to 6 percent. Methane emissions from the food sector represent about three-fifths of all methane emissions from man-made sources. Natural sources, such as termite mounds and wetlands, account for perhaps two-thirds of methane's total contribution to global warming potential.[43]

The booming livestock population is expected to further raise the agricultural sector's contribution of greenhouse gases in the years ahead. One analysis by EPA researchers finds that methane emissions from livestock could increase by two-thirds over the next 35 years.[44] Methane-producing bacteria in the stomachs of these animals is not the only greenhouse consideration, however; there is also the fact that they consume so much food in the first place. To sustain a population of a billion cows and 3 billion other ruminants—goats, pigs, horses, sheep and the like—nearly 50 percent of the world's production of coarse grains, and more than two-thirds of that grown in the United States, now is used as animal feed. The net effect is that farmers have nearly 10 billion mouths to feed—not just 5.5 billion human ones—and this increases demand for agricultural chemicals, nutrients and fuel in turn. What is more, if human and animal population trends continue, four new livestock will be appear at the trough for every five additional humans taking a place at the dinner table.

Effects of Carbon Dioxide Enrichment

Because agricultural activities release such a wide range of greenhouse gases, it is difficult to figure the food sector's exact contribution to global warming potential. Harder still is forecasting how the atmospheric buildup of these gases might affect the climate and the general conditions for growing food. The airborne concentration of greenhouse gases, however, is one thing that scientists can measure with great precision. They know that the amount of carbon dioxide in the atmosphere is 356 parts per million, and by sampling ice core data, they have determined that its concentration has risen approximately 80 parts per million since 1750.

The level of carbon dioxide in the atmosphere continues to grow by 1.5 parts per million a year. Whether or not this enhanced greenhouse effect will result in global warming, it means that plants *will* be exposed to a CO_2-enriched environment for many decades to come. Therefore, consideration must be given to how plants will respond to this changing atmosphere, absent any other effects of climate change.

Although the jury is still out on carbon dioxide enrichment, it may well stimulate the growth of 95 percent of the world's plants—including 12 of the world's 15 major crops. Wheat, soybeans, rice, barley and oats are among these so-called "C3" plants, whose photosynthetic rates could increase 10 to 50 percent with a doubling of ambient CO_2.[45] As these C3 plants devote more energy to building tissues, they should also lose less water through photorespiration. In effect, these plants should not have to work as hard to turn carbon dioxide into carbohydrates, a process known as carboxylation. The principle is the same as equipping a mountain climber with an oxygen mask: The ultimate objective becomes easier to achieve.

Some plants already are efficient absorbers of carbon dioxide, however. These C4 plants have a built-in "pump" that concentrates CO_2 near the site of carboxylation. Since they would get less of a lift from a doubling of ambient CO_2, their photosynthetic rates are expected to increase by only 0 to 10 percent.[46] Although C4 plants account for just 5 percent of the world's living biomass, they represent a disproportionate 20 percent of the world's food supply. Sorghum, sugar cane, millet—and most importantly corn—are among the world's major C4 crops. Many pasture and forage grasses, including the prairie forbs of North America, are C4 plants as well.

Where farmers can choose among C3 and C4 crops to grow, CO_2 enrichment bolsters the case for planting C3s. In India, for example, the prevailing trend toward planting wheat, rice and barley in place of corn and millets could be accelerated in an effort to boost harvest yields in a CO_2-enriched world.[47] By the same token, farmers in parts of the United States could switch from corn silage to rape seed (canola) as a forage crop.

The effects of carbon dioxide enrichment must be evaluated further before farmers are sure they are making the right moves, however. To date, most CO_2 enrichment studies have taken place in growth chambers and greenhouses that are ideally suited for the growth of plants. Out in the field, plants are subject to many other environmental stresses that could affect their ability to respond

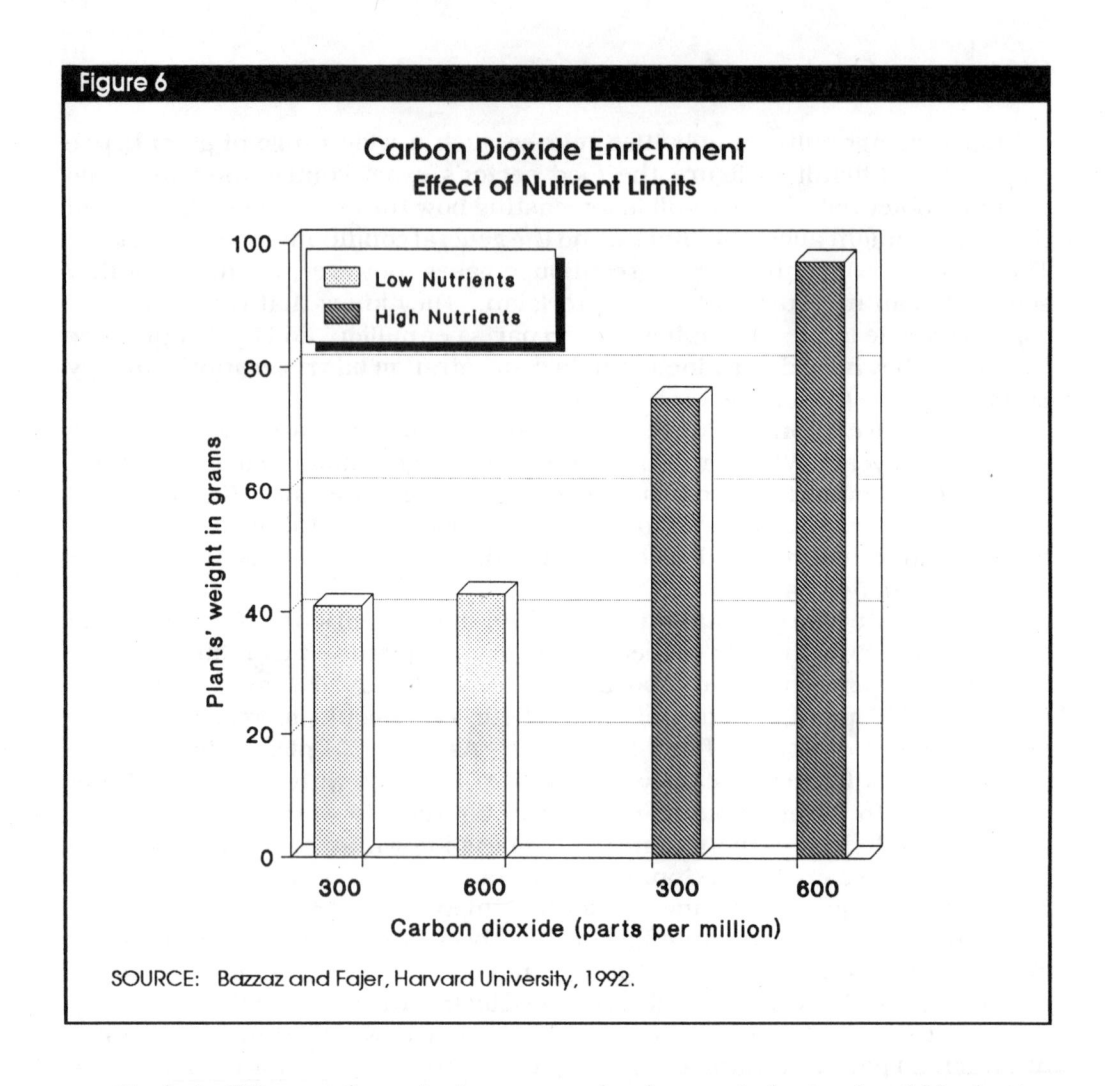

Figure 6

Carbon Dioxide Enrichment
Effect of Nutrient Limits

[Bar chart showing Plants' weight in grams vs Carbon dioxide (parts per million). Low Nutrients: ~40g at 300 ppm, ~42g at 600 ppm. High Nutrients: ~75g at 300 ppm, ~97g at 600 ppm.]

SOURCE: Bazzaz and Fajer, Harvard University, 1992.

positively to CO_2 enrichment. As a general rule, most plants should be better able to overcome growth-limiting factors in a CO_2-enriched environment. Yet some recent studies suggest that the fertilization effect itself is muted when light, water or nutrient levels are constrained. Moreover, some plants may partition their added growth to less nutritious roots and stems—not just to leaves, flowers and seeds—so the increase in yield from the fertilization effect may not be as large as expected.[48]

Despite these caveats, CO_2 enrichment may have important implications for agricultural interests besides farmers and seed suppliers. Fertilizer manufacturers could be affected in a number of ways. First, more frequent rotations of nitrogen-fixing, C3 legumes (such as soybeans) might reduce the need to apply nitrogen fertilizer to the soil. Second, the added CO_2 itself might act as "free" fertilizer, allowing farmers to grow their plants to the same size without having to apply as much fertilizer to the soil.

Farmers seeking to maximize the potential productivity gains offered by CO_2

enrichment, on the other hand, might increase their fertilizer applications to make plants grow larger and faster. The stimulated growth of plants would consume additional quantities of nutrients such as nitrogen, phosphorous and potassium. Since the amount of nutrients in the soil is fixed, all of the increase in nutrients would have to come from fertilizer applications in order to keep the supply of soil nutrients stable. On that basis, a one-third gain in crop growth would require a two-thirds gain in fertilizer applications to prevent depletion of the soil nutrient supply.[49]

More use of fertilizers and pesticides also might be necessary in areas where climate change leads to greater precipitation, since rainwater runoff leaches nutrients and chemicals from the soil. One environmental consequence of this would be localized increases in groundwater contamination. An analysis of climate models by EPA suggests that cotton-growing areas in the southern United States might see an increase in fertilizer and pesticide runoff, while spring wheat and corn-growing regions farther north would have less runoff because of diminished rainfall and greater evaporation during the growing season.[50] (Encroachment by migrating insects could lead farmers to increase insecticide spraying in all regions, however, and a change in rainfall patterns different from those forecast here would pose other consequences for agricultural runoff.)

Carbon dioxide enrichment may affect demand for herbicides in yet another way. Fourteen of the world's 17 most troublesome terrestrial weeds are C4 species that grow among C3 crops. With CO_2 enrichment, these C4 weeds should be less able to compete with C3 crops, thereby reducing demand for herbicides. Conversely, where C3 weeds compete with C4 crops, use of more herbicides may be required to fend off the invigorated C3 weeds. Drought is potentially the great equalizer in this forecast, however. With fewer weeds to kill and fewer plants to protect in drought situations, demand for herbicides almost always falls.[51]

Meanwhile, out on the prairie, weedy grasses that predispose the land to burning may encroach on moister range grasses because of rising ambient CO_2 levels. That would increase the potential for more wildfires, regardless of whether the climate itself becomes more drought-prone.[52]

Much more must be learned about carbon dioxide enrichment before any firm conclusions are possible, however. One recent article by Fakhri Bazzaz and Eric Fajer in *Scientific American* warns that the perceived benefits of CO_2 enrichment on plants may be overstated. "Studies have shown that an isolated case of a plant's positive response to increased CO_2 levels does not necessarily translate into increased growth for entire plant communities," the Harvard University researchers noted. "Even the notion that plants will serve as sinks to absorb ever mounting levels of carbon dioxide is questionable."[53]

Many scientists believe that trees and plants will sequester more carbon dioxide in their tissues as the concentration of CO_2 in the atmosphere rises. If true, this uptake represents an important negative feedback for global warming: More CO_2 going into the atmosphere induces the Earth's biomass to take more out through the fertilization effect, slowing the overall rate of accumulation of CO_2 in the atmosphere. Yet Bazzaz and Fajer caution that plants may adjust

> **Box 2-C**
>
> ### Pests, Disease and Global Warming
>
> Much to the chagrin of farmers, plant-eating insects have proven to be persistent adversaries over the years, and global warming could make them even more of a problem. Since World War II, farmers have relied on chemical insecticides in their war on bugs. But insects' resistance to them has become 64 times as great over the last half-century. At least 500 different kinds of bugs have developed immunities to one or more insecticides. The U.S. Department of Agriculture estimates that crop losses to insects actually increased from 7 percent in 1945 to 13 percent in the early 1980s, despite an order of magnitude increase in insecticide spraying over the period.[1c]
>
> Carbon dioxide enrichment may cause leaf-eating insect populations to decline because of a loss of protein in plant tissues. Even so, higher temperatures brought on by the atmospheric rise in greenhouse gases might enable remaining insect populations to multiply faster and spread to new territory. For example, the European corn borer is a major pest of corn grain that can produce up to four generations a year in many parts of the world. Each 1 degree F increase in temperature could shift its northern limit by 100 to 300 miles.[2c] Similarly, the overwintering range of the potato leafhopper, a serious pest of soybeans and other crops in the United States, could double or triple from its present narrow range along the coast of the Gulf of Mexico.
>
> Recent climatic events have already illustrated how hot weather can promote the spread of insects. During the "greenhouse summer" of 1988, the spider mite population spread rapidly in America, ravaging soybean crops and causing losses of 15 to 20 percent in some areas.[3c] Anomalously warm weather conditions in Europe during the last half of the 1980s also enabled locust swarms to reach new northward limits in the continent.[4c] Populations of aphids and many other insects also tend to grow faster with higher temperatures. Accordingly, demand for insecticides may well increase in an era of global warming among farmers who do not employ other pest management techniques.
>
> Pest invasions can be especially damaging to crops early in the growing season, when the young plants are at a highly vulnerable stage of development. Increased air temperatures might permit the corn earworm moth to infest soybeans earlier in the season, for instance, since warm weather is conducive to the flight of moths.

to their CO_2-enriched environment over time and scale back their rate of photosynthesis, leaving the net assimilation rate unchanged.

Another possibility is that plant respiration and biomass decomposition rates will accelerate if global temperatures rise. Such developments might lead to a greater release of carbon dioxide from plants (and methane from soils, especially in high-latitude tundra regions) than plants themselves are able to sequester from the atmosphere—despite their stimulated growth. In this case, the net effect of carbon dioxide enrichment—if combined with global warming—would be a positive feedback: The increased release of greenhouse gases from photorespiration and organic decomposition would effectively shrink the size of the terrestrial carbon sink harbored in plants and soils. The Intergovernmental Panel on Climate

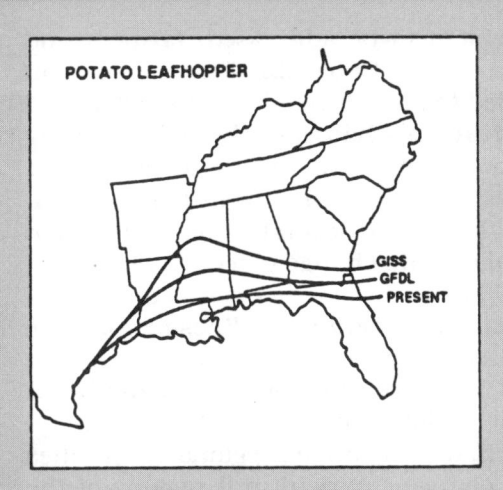

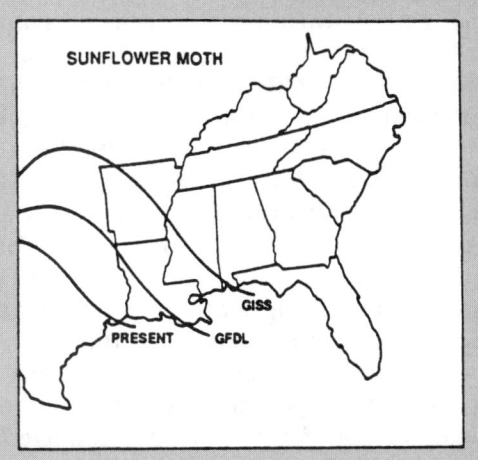

Such a development would add considerably to losses from corn earworm moths, which already amount to hundreds of millions of dollars a year.[5c]

Livestock dieseases might spread as well. Higher temperatures would allow a northward shift in the range of some tropical infections, enabling these diseases to reach the southern United States for the first time. Rift Valley fever, transmitted principally by mosquitoes, might become more of a threat as rising winter temperatures promote an increase in the mosquito population. African Swine fever, now limited to tropical countries, also could spread to the United States.[6c] The range and season could expand as well for other insects already posing a threat to domesticated livestock in the United States. The active season for horn flies—which cause nearly $750 million of losses annually in the beef, dairy and cattle industries—could lengthen by eight to 10 weeks under a warmer climate.[7c] The distribution of blue tongue, caused by a virus, also could spread northward and eastward, as could anaplasmosis, a rickettsial infection of ruminants that is the second most widespread disease among American cattle.[8c]

A final consideration for livestock is that higher temperatures could diminish their reproductive capabilities. On the other hand, fewer livestock would be expected to succumb to the cold during the winter.

Change estimates that decomposition of soil organic matter could lead to a soil carbon loss of 60 billion tons over the next 60 years, which equals the amount of carbon released from fossil fuels (at current rates) over 10 years.[54]

One final concern about carbon dioxide enrichment is that it may make plants less nutritious to eat. As more carbon is fixed in plants, the nitrogen content and, hence, protein in plant leaves is diluted. As a result, plant-eating insects might have to eat more leaves to get their fill or else experience a drop in the size of their populations. Neither option is particularly palatable from an ecological point of view. Insects would have to consume more plants, or fewer insects would themselves be available for consumption by others. Either way, the food chain would be affected.[55]

Global Warming and American Agriculture

If global warming emerges as a problem to be addressed, farmers and agribusinesses may find themselves in a bit of a Catch-22. Their success in boosting crop yields over the years has led to greater reliance on mechanized equipment and chemical inputs, which in turn has contributed a growing amount of greenhouse gases to the atmosphere. At the same time, their transition away from diversified crop-livestock agriculture toward production-driven, single-crop cultivation has made them more dependent on cooperation from the weather—something that global climate change places in doubt.

When EPA reported to Congress in 1989 about the possible effects of global warming on U.S. agriculture, it warned that considerable declines in domestic crop production are possible. Based on its review of three general circulation models, EPA concluded that the largest likely drops in crop yield would be for sorghum, a 20 percent projected decline; followed by corn, 13 percent; and wheat, 11 percent.[56] EPA also reported that tremendous regional disparities would be likely in the event of climate change. Agricultural regions of the country that might experience the greatest reductions in total crop acreage are the Southeast, the southern Great Plains and Appalachia, where projected acreage losses range from 5 to 25 percent. Conversely, the crop acreage projections in the Pacific Northwest, the Great Lakes region and the northern Great Plains show an increase ranging from 5 to 17 percent, according to EPA's analysis of the climate models.

The agricultural fortunes of farmers in the Great Plains are especially tied to nature's good graces. Encompassing 20 percent of the land area of the lower 48 states—yet home to only 2.5 percent of the nation's residents—some consider the Great Plains to be the "American Outback" because of its sparse population and paucity of rainfall. Even so, in a wet year the Great Plains is capable of producing bumper crops of small grains—adding billions of dollars to the nation's export trade balance. In fact, if the Great Plains were treated as a separate nation, it would be the world's fourth largest producer of wheat, on average, and one of the world's top 10 producers of several other grains.[57]

The principal Great Plains states are North Dakota, South Dakota, Nebraska, Kansas, Oklahoma and Texas, as well as the eastern portions of Montana, Wyoming and Colorado. The nation's breadbasket extends eastward from there to states bordering on or east of the Mississippi River. Altogether, 12 midwestern states—from Nebraska in the west to Ohio in the east—comprise 55 percent of the nation's cropland and 44 percent of its agricultural output.[58] On average, these states account for 75 percent of the nation's corn production, 70 percent of its soybeans production and 57 percent of its wheat production. Rainfall amounts typically are greater on the eastern side of the breadbasket than on the western side.

Regrettably, it is in this heartland where the general circulation models suggest that global warming may hit first—and hardest. While the regional details of these models remain quite sketchy and are subject to change, they indicate that the continental interiors of the Earth's surface may tend to dry out relative to other regions as global temperatures rise. The Midwest is one such

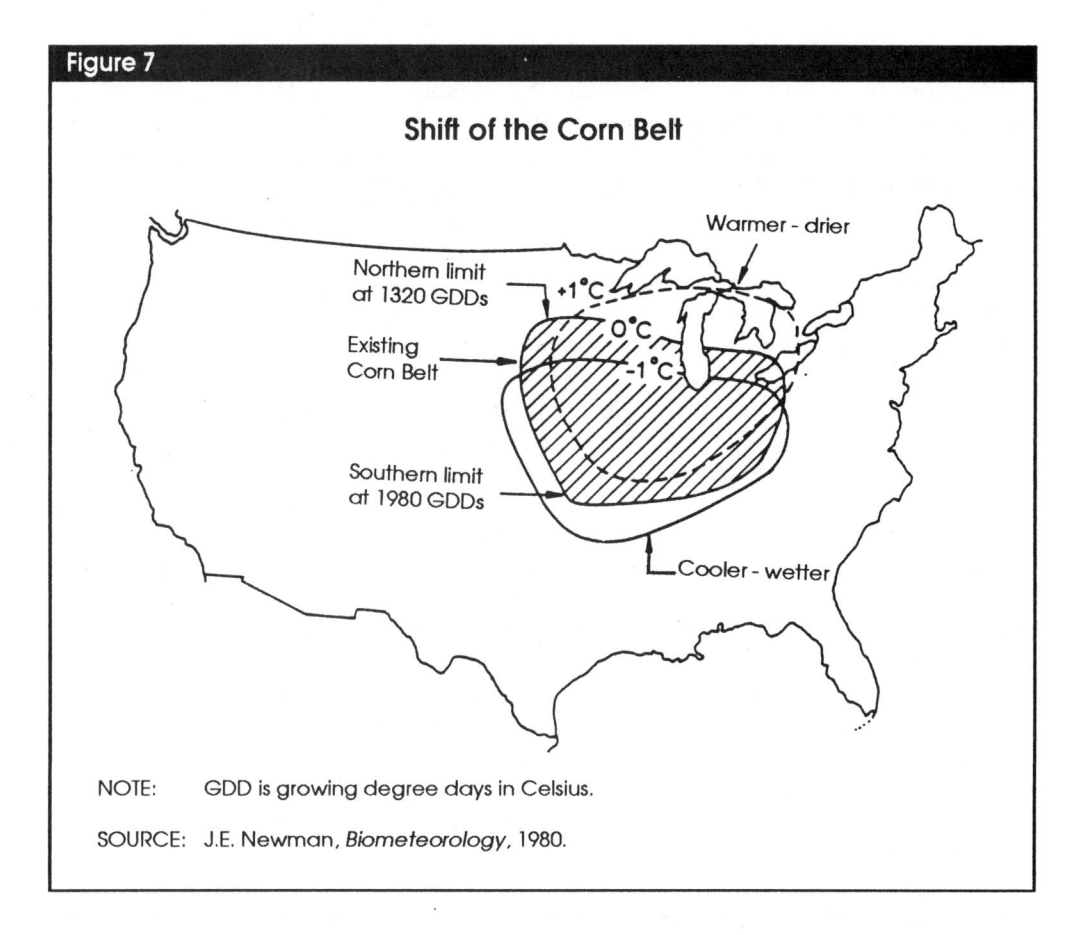

> **Figure 7**
>
> **Shift of the Corn Belt**
>
> [Map of the United States showing the existing Corn Belt with Northern limit at 1320 GDDs and Southern limit at 1980 GDDs, with shifts indicated at +1°C, 0°C, and -1°C. Arrows indicate "Warmer - drier" to the north and "Cooler - wetter" to the south.]
>
> NOTE: GDD is growing degree days in Celsius.
>
> SOURCE: J.E. Newman, *Biometeorology*, 1980.

interior place. As a result, the wheat-producing lands in the western Great Plains could revert to rangeland. At the same time, states in the Mississippi River valley—Iowa, Missouri, Arkansas and Illinois—might inherit the existing semiarid climate of the Great Plains, making it the preferred place to grow wheat and other drought-tolerant crops like sorghum.[59]

The Corn Belt also could experience important shifts in production as a result of climate change. Its present boundaries are set by lack of soil moisture in the west and lack of warmth in the north. Each degree F of climatic warming could therefore displace the Corn Belt by approximately 60 miles in a north-by-northeast direction.[60] Unfortunately, the Great Lakes lie in the path of this directional shift, covering millions of acres of what otherwise might become prime corn-growing country. EPA's analysis concludes that a temperature increase of 3 degrees F or more could lead to an eventual decline in corn production throughout the United States. The decline in corn production in the Great Plains might range from 16 to 25 percent; in the Southeast, from 5 to 14 percent; and in California, from 4 to 17 percent.[61] These hypothetical losses would come despite increased crop irrigation.

Of course, farmers are not likely to sit on their hands if these changes were to come to pass. Rather than abandoning their fields, they would seek out other drought-hardy and heat-resistant crop varieties to plant—and employ new cultivation techniques. An analysis by Resources for the Future, which considers the possible responses of farmers to climate change, finds that crop yields and acreage losses could be held to a minimum—contrary to the projections made by EPA in its report to Congress.[62]

Moreover, climate may not evolve in the manner now suggested by general circulation models. If, for example, more warming occurred at night than during the daytime, the effect on crops might be relatively benign—or even favorable. Nighttime frost and daytime heat dehydration would be reduced, thereby reducing the likelihood of crop damage. When combined with the possible fertilization effect of CO_2 enrichment, crops could thrive in a greenhouse world.

If a geographical shift in corn and wheat production does occur, however, it may cause "further weakening of an economic base already under pressure from long-term structural changes in U.S. agriculture," warns Columbia University agronomist Cynthia Rosenzweig.[63] A few demographers have even gone so far as to suggest that portions of the western Great Plains revert to a "Buffalo Commons," whereby the government would buy back vast tracts of farmland and restore them as grasslands—much as they were before their conversion to agriculture.[61] While few take this grand scheme seriously, climate change might leave some already-struggling farm communities with few other alternatives.

The regional effects of climate change as projected by the climate models are not all doom and gloom, however. Agricultural production could increase around the Great Lakes, for example, especially in its northern tier. Higher temperatures that extend the length of the growing season could increase the yields of unirrigated corn and soybeans by as much as 50 to 100 percent for farmers in Minnesota.[65]

Farmers in the Dakotas also could reap gains through a switch from spring wheat to winter wheat—the principal grain used to make bread flour. Winter wheat is planted in the fall and has a few months to become established in the soil before winter sets in. After it lies dormant over the winter season, its growth resumes in the early spring, with harvest in the early summer—usually before any hot, dry weather arrives. By contrast, spring wheat has a highly variable planting date, since the soil must thaw and dry sufficiently before this spring crop is sown. Moreover, it has to survive a long, hot summer—the potential for which increases with global warming—before it is harvested in the fall. Winter wheat generally yields 40 bushels per acre, compared with 30 bushels per acre for spring wheat.

With the development of new seed varieties, the growing area for winter wheat already has expanded tremendously during this century. Its range now extends from central Texas to the southern half of the prairie provinces of Canada. With global warming, farmers throughout the provinces of Alberta and Saskatchewan might be able to grow winter wheat in place of spring wheat—if soil conditions permit.[66]

Despite the apparent advantages of growing winter wheat, it is not immune to weather damage from drought, winterkill and wind erosion. Recently, a dry, relatively snowless winter in the major bread-wheat states of Kansas, Oklahoma, Texas, Colorado and Nebraska reduced the winter wheat harvest to 1.5 billion bushels in 1989—the poorest crop in 11 years.[67] (Kansas was so dry that

Figure 8

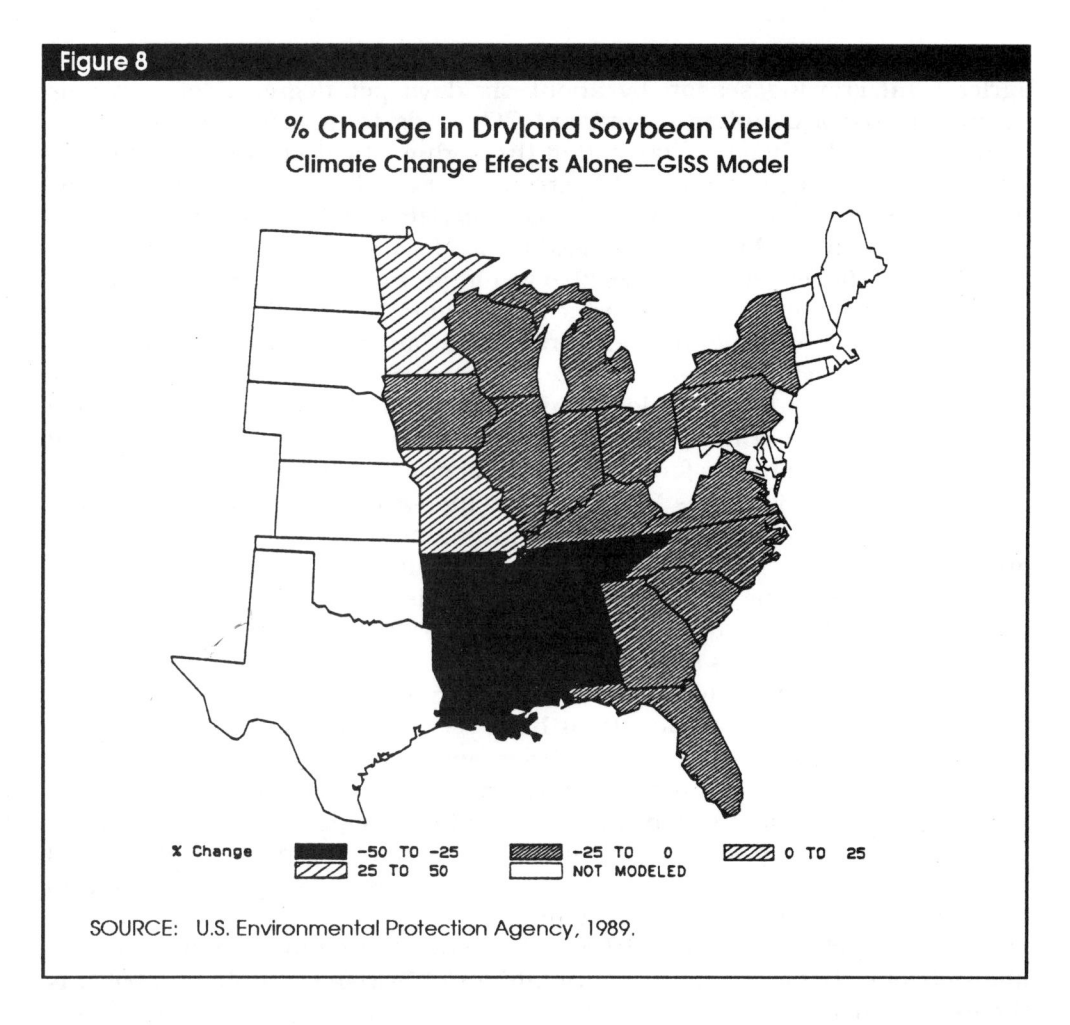

% Change in Dryland Soybean Yield
Climate Change Effects Alone—GISS Model

% Change: -50 TO -25 | 25 TO 50 | -25 TO 0 | NOT MODELED | 0 TO 25

SOURCE: U.S. Environmental Protection Agency, 1989.

it even had to battle a series of extensive brush fires that winter.) And in the winter of 1991-92, an early bitter cold snap followed by unusually warm weather damaged portions of the winter wheat crop and spread the growth of a fungus that feeds on injured plant tissue. At the same time, repeated freezing and thawing of the ground heaved the soil and impeded the development of roots in the young plants.[68] More winters reminscent of 1989 and 1992 might be possible in winter wheat-growing states in the event of global warming or chaotic, amplified temperature swings.

Global Warming and Canadian Agriculture

The outlook for Canadian agriculture in the event of global warming also is decidedly mixed. Canada usually is the world's second-largest exporter of wheat after the United States. Spring wheat is Canada's main wheat crop. Canadian farmers planted 29 million acres of spring wheat in 1991.[69]

In the Canadian prairies, global warming could extend the length of the agricultural growing season by about six days per degree F of warming. Assuming that a doubling of ambient CO_2 leads to at least 3 degrees F of warming, the likelihood of harvesting the spring wheat crop before the first autumn frost would increase by 10 percent or more. The trade-off for less frost damage, however, would be lower average crop yields, since the maturation time for spring wheat would be compressed by 1.5 fewer days for each degree F of warming.[70] (Higher temperatures also would promote faster crop growth and reduced per-acre yields of dry matter in the United States.)

As noted earlier, higher temperatures in the Canadian prairies might enable farmers to grow winter wheat in place of spring wheat. But a more ominous question is whether there would be sufficient rainfall for either crop to grow in the event of global warming. The climate model developed by NASA's Goddard Institute for Space Studies forecasts that temperatures in Saskatchewan could rise by more than 6 degrees F without any increase in precipitation. Such a climate change would raise wind-erosion potential 25 percent and the frequency of drought thirteen-fold. The net result would be that an average weather-year after the warming would resemble those of the most extreme period on record—the Dust Bowl years of 1933-37.[71]

Moreover, "Anomalies such as the dry year of 1961"—the driest year of the century—"would be likely to occur more frequently and quite possibly be surpassed by even drier years than have recently been experienced in the region," reports Martin Parry.[72] On average, spring wheat production would fall by more than 25 percent if the GISS scenario played itself out in the Canadian prairies. "Since Saskatchewan at present produces 18 percent of all the world's traded wheat," Parry observes, "such a reduction could well have global implications."[73] (Parry's analysis leaves out the possibility of switching from spring wheat to winter wheat in the region.)

Saskatchewan is not the only Canadian province that might suffer from the ill-effects of global warming. The dry interior of British Columbia probably is the most vulnerable region in all of Canada, where even greater precipitation deficits are projected. Elsewhere around the country, yields of grain corn, barley, soybeans and hay might decline in all provinces except the northern part of Ontario, where production now is limited by inadequate warmth. Despite a projected 50 percent increase in growing season precipitation, increased evapotranspiration could lead to drier conditions in most of Ontario. As a result, corn and soybean production might become especially precarious in the southern part of the province. Grain yields could improve in Manitoba, however, since that province is projected to be the least deficient in soil moisture, according to the GISS climate model.[74]

Of course, there can be no assurance that the GISS model's projections—more than those of any other climate model—will even remotely resemble what the climate has in store over the next 50 years. The analysis for Canada is based merely on interpretations of the GISS model's recent projections and may be subject to change as the model is refined and updated.

Uncertain Costs of Global Warming

It bears repeating that the effects of global warming on agriculture are purely speculative. The assumed physiological response of plants hinges on assumptions made about climate change at the regional level *and* the effects of CO_2 enrichment—two subject areas where great uncertainties remain. Most studies do not even attempt to go the next step and consider the ways in which farmers may respond to changing environmental conditions. Thus, those who attempt to digest such information are wise to add a large grain of salt.

The uncertain effects of climate change on American agriculture were revealed in the results of a multidisciplinary study performed by an esteemed group of agronomists, economists and climatologists in 1990.[75] Despite a thorough examination of the potential economic consequences of climate change, they could not determine the likely *sign* of the economic consequences, let alone the actual value of the costs or savings involved. What really mattered is which climate model they used and their built-in assumptions about the CO_2 fertilization effect. One model yielded a net economic *benefit* of nearly $11 billion a year (expressed in 1982 dollars) when the aggregate effects of climate change and CO_2 enrichment were tabulated. The other model forecast a net loss of more than $10 billion a year, despite using the same favorable assumptions about the fertilization effect. (Response strategies of farmers were not considered in either case.)

While the conclusions could not have been more contradictory, the study is instructive for review purposes because (believe it or not) it represents the most sophisticated economic analysis performed to date on the possible effects of climate change on U.S. agriculture. For the study, the researchers employed two climate models—the GISS model and the Geophysical Fluid Dynamics Laboratory model (GFDL), developed by researchers at the National Oceanographic and Atmospheric Administration. Both models indicated that the United States would become 7.5 to 9 degrees F warmer over the next 50 to 75 years in a doubled CO_2 environment, and that annual precipitation would increase as well. But their projections about the critical summer growing season varied considerably. The GISS model found that average summertime temperatures would rise by only 6.3 degrees F, whereas the GFDL model forecast a rise of 8.9 degrees F. More importantly, the GISS model projected a summertime increase in precipitation, whereas the GFDL model forecast a decrease. (More recent runs of the GFDL model have scaled back the expected severity of summer dryness in the United States, although GFDL still forecasts more summer dryness than GISS.)

The GISS model presents a favorable picture of climate change on U.S. agriculture. It projects that wheat crop yields in the eastern half of the Great Plains and throughout the Great Lakes states would rise modestly and that yields elsewhere would decrease only slightly. At the same time, soybean yields would rise 30 to 40 percent in the Corn Belt, and leap 40 to 50 percent in the eastern Great Plains and throughout the Great Lakes states. Corn yields also would increase everywhere except the Northeast and Southeast, the GISS model projects, and especially in the northern plains and Great Lakes, where

Table 1

Climate Parameters of Selected Agricultural Regions as Predicted by GISS and GFDL

Agricultural Region	Evaporation ratio* GISS	GFDL	Precipitation ratio** GISS	GFDL	Average annual temperature increase (°F) GISS	GFDL
Southeast	1.08	0.93	1.11	0.92	6.3	8.8
Delta	1.02	1.02	1.02	1.00	7.1	7.9
Northern Plains	1.09	0.99	1.07	0.97	8.5	10.6
Southern Plains	0.99	1.02	0.92	1.00	7.9	8.1
Mountain	1.10	1.06	1.11	0.99	8.8	7.1
Pacific	1.11	1.03	1.15	1.02	8.5	8.5

* Evaporation ratio is the doubled-CO_2 forecast relative to current evaporation.
** Precipitation ratio is the doubled-CO_2 forecast relative to current precipitation.

SOURCE: Richard Adams et al., *Nature*, 1990.

yield increases of up to 50 percent would be possible.

The hot, dry GFDL summertime forecast, on the other hand, envisions a very gloomy future for American agriculture. Wheat crop yields would likely fall 20 to 50 percent throughout the United States, according to this scenario, and soybean yields would drop everywhere except in the northern Great Plains and northern Great Lakes states, where yields would increase modestly. Meanwhile, corn yields would drop 10 to 20 percent in the Corn Belt and Great Plains and increase 0 to 20 percent in the Southeast, based on the early-run GFDL model.

The next step in this study was to plug the crop yield estimates into an agro-economic model that forecasts food price impacts. Then the discrepancies between the two climate models could be counted in dollars and cents. Under the GISS doubled-CO_2 scenario, the yield of eight major field crops and four livestock crops would rise by a collective 17 percent and consumer food prices would drop by 16 percent. On that basis, the nation's consumer food bill would fall $9.3 billion a year (in 1982 dollars). Producers also would reap a $1.6 billion annual gain in profits, bringing the total net benefit to $10.9 billion a year.

The GFDL scenario leads to a much more discouraging result. With doubled CO_2, crop yields would drop by 20 percent and food prices would rise by 34 percent as a result of adverse climate changes. Consumers' food bill would *rise* by $13.8 billion a year, and because of the higher prices, producers' profits would also rise by $3.5 billion annually. The net annual loss in total income would be $10.3 billion—with about half of the loss borne by foreign importers of American grain.

To put the projected economic cost of the GFDL scenario into perspective, one can compare its costs with those imposed by other environmental problems. Tropospheric ozone pollution is blamed for annual crop losses of $2 billion to $3 billion (in 1982 dollars), for example. A hypothetical 15 percent depletion of the Earth's protective ozone layer would tack on another $2.5 billion in estimated annual losses.[76] Economic losses forecast by the GFDL model, in other words, are four times as great as those stemming from either of these other potentially serious environmental problems.

But there may be even more to these losses than meets the eye. The crop-response models used in this analysis made highly favorable assumptions about the likely response of crops to CO_2 enrichment—offsetting some or all of the negative effects on crop yields projected by climate change alone. The crop-response models assume, for instance, that the photosynthetic rates for soybeans, wheat and corn would rise 35, 25 and 10 percent, respectively, under doubled-CO_2 conditions. Without this fertilization effect, the projected economic loss in the GFDL model would have been *three* times greater than the $10.3 billion annual loss already forecast.[77] Even the GISS model forecast would have shown an overall economic loss—rather than a large gain—were it not for the assumed fertilization effect.

Several other caveats are in order regarding assumptions made in this study. First, the crop-response models assume that sufficient soil nutrients will remain to support the fertilization effect in all geographical regions. Second, they assume that pests—insects, weeds and diseases—will pose no greater threat to crops than they do now. Third, the general circulation models assume there will be no change in climate variability—the frequency of extreme events. Under actual conditions of climate change, it is possible that shortages of soil nutrients would develop in some locations; pests may indeed spread; and drought, flooding and extreme heat waves could occur more often.

> **Box 2-D**
>
> ## Weather's Role in Agriculture
>
> To appreciate the role that weather plays in shaping the fortunes of agricultural interests, one need look no further than the nation's grain exchanges, such as the Chicago Board of Trade, where millions of dollars are made and lost daily according to the latest meteorological forecast. Commodities brokers know that weather can raise or lower U.S. agricultural production up to 15 percent in any given year.[1d] They also know that heavy machinery enables farmers to gear their planting schedules around the vagaries of the weather and to sow huge amounts of crops in only a matter of days. Therefore, the crop outlook for an entire growing season can change practically overnight.
>
> Spring planting time is an especially anxious period in the fields and the trading pits. Enough rain must fall to nurture the seeds once they are in the ground, but not so much as to turn the fields into mud, which would make planting impossible. Accordingly, too much rain as well as too little can send futures traders into a tizzy. As spring wears on and crucial deadlines pass, farmers and grain merchants keep a wary eye on the weather. In May and June, some farmers harvest the winter wheat crop, sown the previous autumn, while others plant the spring wheat crop, to be harvested in the fall. By Memorial Day, most of the corn and cotton crops should be planted, and by early June most of the soybean crop should be in as well.
>
> In late June, farmers and brokers must determine whether there is sufficient subsoil moisture to sustain the growth of summer crops through their critical fertile periods. During the drought summer of 1988—when the soil was dry and the long-term outlook for rain in the Corn and Wheat Belts was especially bleak—futures prices for some grains soared as much as 65 percent in an emotion-filled, mid-summer rally. The price surge came despite storage bins that were bursting from bumper crops harvested the previous year, and that ultimately held the rise in retail food prices to only 3 to 5 percent after the 1988 drought was over.[2d]
>
> In July, farmers and grain merchants have new weather worries. The corn crop should have completed its pollination process by the middle of the month. But an ill-timed heat wave can hinder kernel formation during a critical 10-day period. If the strands of silk emerging from corn cobs are seared by excessive heat, or if the heat sterilizes the pollen, the pollination process can be thwarted—leaving holes where kernels of corn were supposed to form.[3d] Summertime drought also can promote the spread of a fungus that contaminates the corn with aflatoxin. After the 1988 drought, aflatoxin was detected in unsafe levels in milled corn and milk; trace amounts were even found in popcorn, cereal and chips. Guatemala and India reported receiving at least 24 million contaminated bushels of corn under U.S. aid programs.[4d]
>
> August is the "make or break" time for soybeans, when they must complete their own flowering and pod-filling process. Compared with corn, soybeans are a durable crop. They can go into a semi-dormant state to protect themselves during early summer heat waves. But soybeans, too, are vulnerable to excessive heat if it comes in August. Even when summer growing conditions have been ideal, farmers know that a barrage of hail from a passing thunderstorm can damage a field or sometimes wipe out an entire crop. After the peak of the growing season has passed, one final weather hurdle is a fairly dry and frost-free early fall to assure the harvest of undamaged crops. During winter, farmers count their blessings and, if they are lucky, their money.

In many respects, then, this multidisciplinary analysis represents a best-case scenario under the assumed conditions of climate change. But since the analysis also leaves out the potential response of farmers to such changes, it cannot be considered a best-case scenario in all respects. Farmers' adaptations and the introduction of new agricultural technologies may improve the outlook for U.S. agriculture even if the climate itself takes a turn for the worse. Only time will tell.

Constraints on Northern Agriculture

Farmers have good reason to believe that use of modern equipment and CO_2-enriched air may meet the challenges presented by the weather in the years ahead. But they also know that their livelihood depends, ultimately, on things largely beyond their ability to control, such as receiving the right amount of sunlight, rainfall, nutrients and warmth. In a greenhouse world, such factors would continue to predominate.

Sunlight: Sunlight is a case in point. Cotton is grown in the South, for example, not just because it is warmer there, but also because it needs a daylength regime that is less variable than that found in the North. Most crops grown at high latitudes, in fact, are bred specifically for short growing seasons with long photoperiods. While substitution of crop varieties now grown at lower latitudes might make sense from a climatic standpoint in the event of global warming, such species might not be suited for the long daylengths of the growing season in their new northern habitat.[78]

The *kind* of light to which crops are exposed is another important factor, especially considering the depletion of the stratospheric ozone layer. This depletion, caused by chlorofluorocarbons and other halogenated compounds, is most acute in high-latitude regions, although recent evidence also points to a serious decline in mid-latitudes.[79] Studies by Alan Teramura at the University of Maryland indicate that a 25 percent thinning of the ozone layer (which is considerably greater than the loss now occurring) could reduce photosynthetic activity of wheat, rice and soybeans in a range of 5 to 10 percent. Moreover, the added exposure to UV-B light might inhibit the photosynthetic boost from carbon dioxide enrichment, limiting the gains of these crops to only a fraction of what would have occurred without the added UV-B exposure.[80]

Soil nutrients: Crop productivity constraints come from beneath the earth as well as above the sky. Soils nutrients are essential to the growth of plants. In northern Minnesota, Wisconsin and Michigan—where crop-simulation models indicate some of the biggest crop-yield gains might occur—leaching of carbonates of calcium and magnesium has created acid soils. Moreover, these podzol soils have a thin organic-mineral layer and a limited ability to retain water, making them better suited for pasture and growing hay than for growing crops. This largely explains why large cut-over forest tracts in the northern Great Lakes have never been cultivated for agriculture—and perhaps never will be.[81]

Soil nutrients are a limiting factor in expanding agriculture to northern Canada as well. During the last great ice age, glaciers scraped away most of the topsoil from the northern portions of Alberta, Saskatchewan, Manitoba, Ontario and Quebec and deposited it at points far to the south. (This is why the soils of the Great Plains are in fact so fertile.) Now only thin soils remain within the glaciated shield. Accordingly, farmers who try to establish crops in the northern halves of these provinces may have difficulty gaining a foothold—even if temperature improvements otherwise make it possible.

Temperature requirements: As a general rule, higher temperatures should stimulate the growth of most crops, but there are limits to the amount of heat they can withstand. One analysis by Rosenzweig and Parry found that crop yields for soybeans, wheat, rice and corn would be expected to increase in response to an average temperature rise of 3.6 degrees F during the growing season. But the yields of these crops would be likely to fall if the average temperature rise were 7.2 degrees F instead. Even greater losses are projected with a 9 degree F temperature rise. (This analysis assumes no change in precipitation and an increase in plants' water use efficiency because of CO_2 enrichment.)[82]

At the other end of the temperature spectrum, species like winter wheat and Douglas-fir trees require a *cold* period to ensure proper seed germination. Deciduous fruit trees also enter a rest period during cold-weather months during which no growth or injury occurs. Therefore, a rise in minimum temperatures also could be harmful to the growth of selected species.

Root crops such as sugar beets and potatoes should benefit from moderation of cold winter weather, however, since their growth habits depend not so much on the air temperature as on whether the ground is thawed. Subtropical fruits also would benefit from milder winters, since they are susceptible to damage from frost. At present, the only place in America that is free from the threat of frost is the Florida Keys.

Higher temperatures also would affect the yields of crops harvested after a single growing season, since the development phase of these determinate crops would shrink. With their leaves extended and intercepting light over a shorter period, cereal crops in particular (such as barley, corn, rice, rye and wheat) would have less time to produce biomass through photosynthesis. Accordingly, more acres would have to come into cultivation to produce the same output as before the warming occurred—unless the CO_2 fertilization effect is able to make up the difference. One analysis of the GISS model concluded that acreage under cultivation would indeed contract because of CO_2-enriched yields. A similar analysis of the GFDL model also found crop acreage shrinking—but for a different reason. The climate would no longer be suited for agriculture in some U.S. locations; therefore, crop acreage would have to expand elsewhere around the world.[83]

From an ecological perspective, what matters most is not the net number of acres remaining under cultivation, but the gross changes that would sweep across the landscape as climate change occurs. Presumably, agriculture would be abandoned in some areas and introduced in others. Many of the newly affected areas presently serve as breeding grounds and foraging areas for

Figure 9

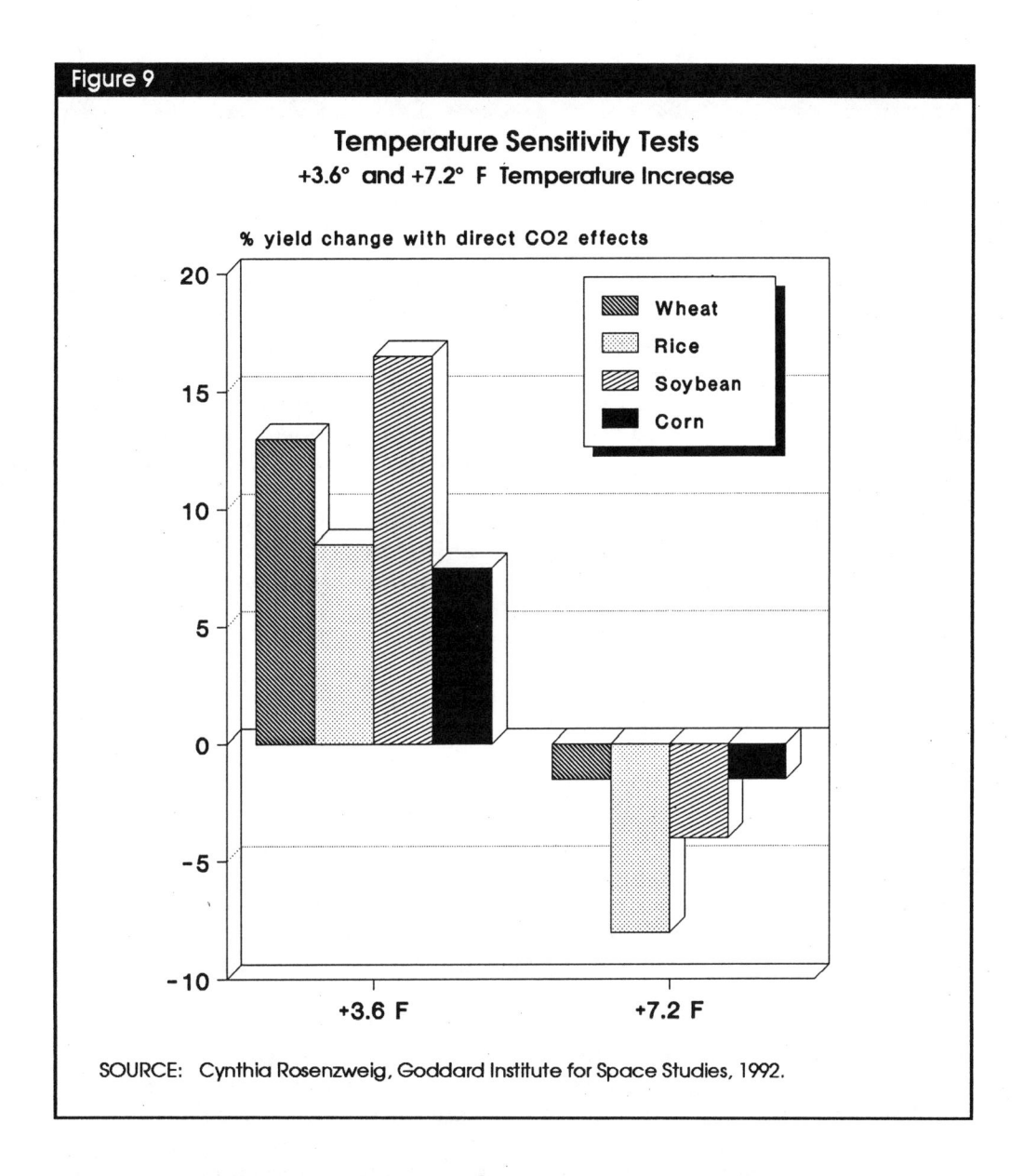

Temperature Sensitivity Tests
+3.6° and +7.2° F Temperature Increase

% yield change with direct CO2 effects

Legend: Wheat, Rice, Soybean, Corn

SOURCE: Cynthia Rosenzweig, Goddard Institute for Space Studies, 1992.

waterfowl and wildlife. Cultivating this land for crops and grazing of domesticated livestock almost certainly would lead to the replacement of natural perennial climax species. The soil would be exposed to more wind and water erosion, and natural processes for nutrient cycling would be disrupted. Considering all that modern farming entails—synthetic fertilizers, pesticides, heavy machinery and the like—the overall quality of the natural environment would seem almost certain to diminish, regardless of whether changes in the climate itself are benevolent or cruel.[84]

Effects on Water

While many issues would have a bearing on agriculture in the event of climate change, none is more important than the continued availability of water. Water is the lifeblood of agriculture; for farmers, it is "liquid gold." It takes three gallons of water to produce a gallon of milk, 26 gallons to grow an ear of corn and 430 gallons to raise a pound of pork.[85] Water used for irrigation purposes makes crops grow relatively impervious to arid climates and allows fields to become more productive in comparison with ordinary dryland farming. In the United States, only 13 percent of the nation's cropland is irrigated, yet those fields account for fully 30 percent of the total value of the nation's crops harvested each year.[85]

One might assume that agriculture would wither in portions of the globe that dry out because of climate change. But irrigation has proven throughout the ages that even deserts can be turned into green oases. As long as farmers are able to tap into surface water and groundwater supplies, they need not worry much about rainfall descending from the heavens. Only the continuity in the hydrological cycle must be maintained. What falls from above, over time, has to replenish what is drawn from below.

Liquid Gold

Today, U.S. agricultural interests draw an estimated 140 billion gallons of water daily from rivers, streams and aquifers, 97 percent of which is used for irrigation. Agriculture accounts for more than two-fifths of the nation's total water withdrawals, and in terms of water actually consumed (i.e., what is lost to the atmosphere through evaporation from soils and evapotranspiration through plants), agriculture's role is even more dominant—accounting for more than four-fifths of the nation's total water consumption.[87] The agricultural sector's water consumption is greatest west of the Mississippi River, since that is where all but 5 million acres of the nation's 45 million acres of irrigated cropland is found. Eighty percent of the water withdrawals and 90 percent of the total water consumption in the West are for agricultural purposes.[88]

Regrettably, evidence is mounting that groundwater consumption by the U.S. agricultural sector is exceeding the long-term rate of replenishment. Of the 31 million acres irrigated with groundwater across the nation, about 14 million acres, or 45 percent, are in areas where groundwater tables now are falling by a foot or more a year. California, Kansas and Nebraska, for example, each have more than 2 million acres where groundwater is declining at that rate; Texas has more than than 4 million such acres.[89] Much of the affected land is devoted to growing especially water-intensive crops, such as cotton, rice and assorted small grains. If global warming were to lead to even less rainfall in these areas, groundwater and surface water withdrawals would have to increase in order to

Figure 10

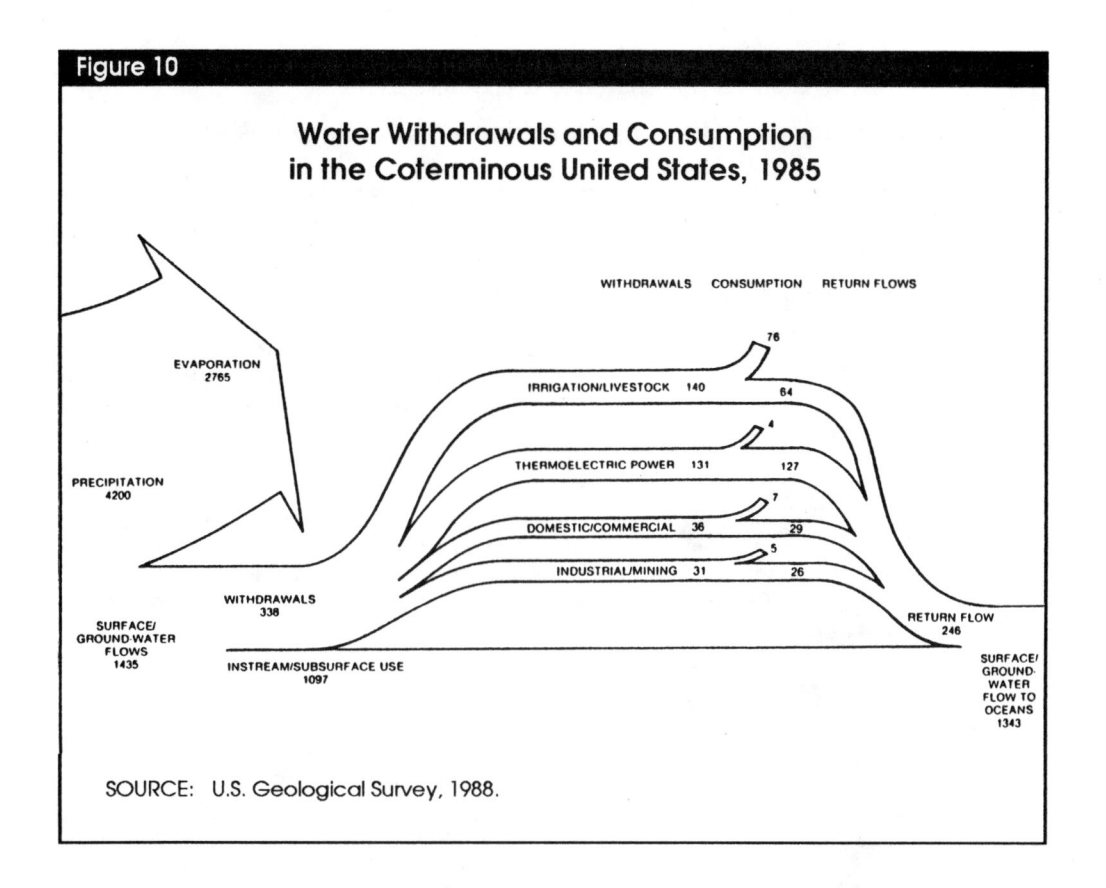

Water Withdrawals and Consumption in the Coterminous United States, 1985

WITHDRAWALS   CONSUMPTION   RETURN FLOWS

EVAPORATION 2765

IRRIGATION/LIVESTOCK   140   64   76

THERMOELECTRIC POWER   131   4   127

PRECIPITATION 4200

DOMESTIC/COMMERCIAL   36   7   29

INDUSTRIAL/MINING   31   5   26

WITHDRAWALS 338

SURFACE/GROUND-WATER FLOWS 1435

INSTREAM/SUBSURFACE USE 1097

RETURN FLOW 246

SURFACE/GROUND-WATER FLOW TO OCEANS 1343

SOURCE: U.S. Geological Survey, 1988.

---

preserve these crops—causing groundwater tables to fall even faster.

Irrigation offers diminishing returns for farmers when the water table is falling. Farmers in the Great Plains, for example, reduced their groundwater withdrawals from the giant Ogallala aquifer by nearly 20 percent during the first half of the 1980s.[90] The combination of having to pump water from deeper under ground and paying more for fuel to run the pumps rendered irrigated farming uneconomic on the fringes of this great aquifer, which underlies much of the western Great Plains. Throughout the western half of the nation, in fact, irrigated acreage has been in retreat for more than a decade. Meanwhile, in the East, irrigated acreage has continued to grow, since the added value to crops has covered the extra expenses associated with irrigation.[91]

Global warming appears poised to increase farmers' demand for irrigation and amplify these regional trends. Rising temperatures would increase the rate of evaporation from soils and through plants, causing proportionately more water to return to the atmosphere—even in some places where total precipitation increases.[92] That would leave less available water on the ground and in the plants themselves, so that irrigation would be called upon to make up the difference. Where irrigation water is available and affordably priced, farmers would be expected to avail themselves of more of it. Where irrigation water already is in short supply, however, farmers' fortunes could begin to dry up.

> **Box 2-E**
>
> ## The Ogallala Aquifer: America's Sixth Great Lake
>
> While most people do not realize it, North America actually contains six Great Lakes. The Ogallala Aquifer—hidden underground—contains 3.3 billion acre-feet of freshwater, enough to fill all of Lake Huron plus one-fifth of Lake Ontario.[1c] Technically known as the High Plains Aquifer, this groundwater supply system is an important reason why farms are so highly productive in the Great Plains, even though the region receives a sparse 16 to 28 inches of natural rainfall a year. In the event of global warming, the aquifer could be vital to sustaining agriculture in this semiarid region—but more so in some states than others.
>
> The High Plains Aquifer extends all the way from South Dakota to Texas, with parts of its reserves lying underneath eastern Wyoming, Colorado and New Mexico. It has accumulated drips of rainwater (and to a lesser extent surface water) for millions of years, in much the same way that water collects after it passes through a filter. Farmers first drilled into the High Plains Aquifer in the 1920s, and then rapidly increased their withdrawals after World War II, as irrigated crop yields soared 600 to 800 percent above traditional dryland methods. Today, approximately 175,000 wells draw water from the aquifer—an average of one well for every square mile of subsurface area it covers. The reserve now irrigates $20 billion worth of food and fiber production annually.

Farmers can take some solace in the fact that carbon dioxide enrichment may partially offset the effects of warmer weather on plants' demand for water. But this singular improvement probably is not enough to overcome the other drying effects on plants.[93] The multidisciplinary study of American agriculture discussed earlier concluded that a doubling of atmospheric CO_2 and associated warming would lead to a 3.5 million acre increase in irrigated cropland under the GISS general circulation model—and a 9.5 million acre increase under the warmer/drier GFDL model.[94] That would represent a net increase in U.S.

All of the water mined from the High Plains Aquifer has taken a toll, however. For every 10 gallons of water that irrigation draws out, rainfall puts back only one gallon into the aquifer. The result has been an 11 percent depletion in its groundwater supply over the last 50 years. By 2020, one-quarter of the aquifer's vast reserves (and the most accessible supplies) may be gone.

Water tables already are falling dramatically along the periphery of the High Plains Aquifer. In West Texas, the drop in the water table measures 200 feet and most farmers there have abandoned their wells. In parts of Colorado, where groundwater mining continues, the water table is falling at a rate of five feet a year. Meanwhile, Kansas has discovered much to its chagrin that it has consumed a disproportionate 40 percent of its portion of the reserve.

Neighboring Nebraska is in the fortunate position of having four-fifths of the aquifer's remaining groundwater supplies stored beneath its borders. The Cornhusker State's reserves could last another 400 years and make a big difference for farmers who want to continue growing water-intensive crops despite the potential for hotter, drier weather. Interlopers from Texas, Colorado and Kansas already are purchasing large tracts of Nebraska real estate in what is perhaps just the beginning of a mad scramble to secure long-term water supplies.

As the High Plains Aquifer continues to draw down, farmers in the Great Plains may eventually try to tap into one of the five other Great Lakes. One grand scheme to divert water from Lake Michigan already was proposed during the "greenhouse summer" of 1988, as farmers' crops withered in the fields. Unlike the other four Great Lakes, the boundaries of Lake Michigan lie entirely within the United States. In addition, the 1909 Boundary Waters Treaty (which regulates the shared use of Great Lakes water between the United States and Canada) contains no explicit provisions concerning Lake Michigan. "The status of diversions from Lake Michigan continues to be unclear," concluded one review of the treaty by a Great Lakes governors' task force. "Many view it as the most significant 'wild card' in the current system of laws, regulations and treaty obligations which could be invoked to protect the lakes."[2e] If global warming comes to pass, Great Plains farmers might press their governors to test the treaty, especially if they become desperate for water.

irrigated acreage of 8 percent to 20 percent, respectively, not to mention billions of dollars of added expenses for pumping equipment and power generation.

The study also forecast that farmers in the water-rich East would be likely to irrigate more of their crops in an effort to maintain yields as temperatures rise. Farmers in the northern Great Plains would be expected to increase irrigation requirements even more dramatically, because a rise in consumer food prices would afford them the luxury—yet only to the extent that the Ogallala aquifer and surface supplies would make the water available. (Some midwestern

Figure 11

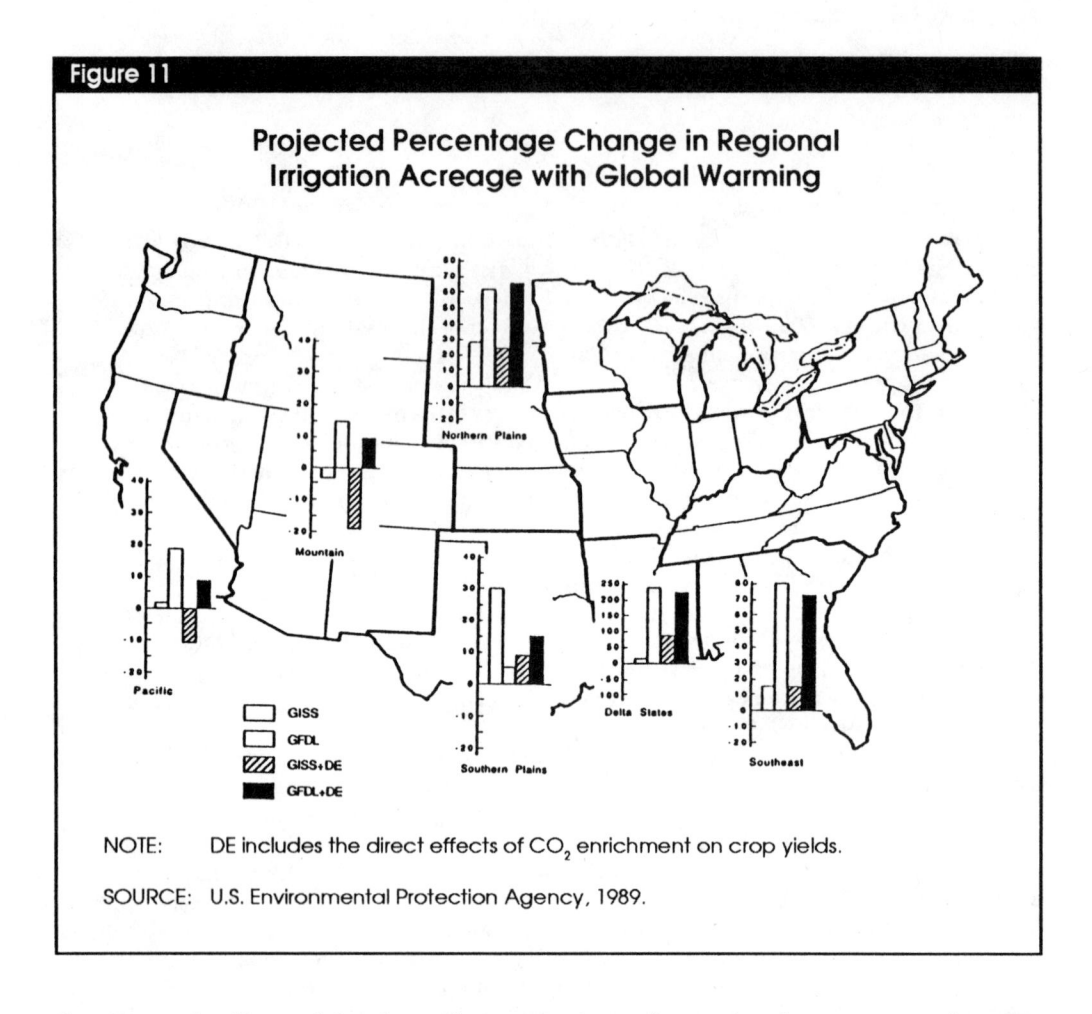

Projected Percentage Change in Regional Irrigation Acreage with Global Warming

Legend:
- ☐ GISS
- ☐ GFDL
- ▨ GISS+DE
- ■ GFDL+DE

NOTE: DE includes the direct effects of CO_2 enrichment on crop yields.

SOURCE: U.S. Environmental Protection Agency, 1989.

farmers actually might reduce their irrigation of crops such as corn and spring wheat, however, because of earlier planting dates and a shorter summer growing season.)[95] Farther to the west—where growing conditions already are marginal and may get drier still—a day of reckoning may come for growers of some especially water-intensive crops. While no farmer would be expected to give up his livelihood without a fight, the specter of global warming stacks the odds against those seeking a greater share of a shrinking water supply.

The Coming Water Shortage

Although a water supply shortage may be on the horizon, it is not for a lack of water, per se. With 1.4 trillion gallons of water running through American rivers and streams daily, only 100 billion gallons is siphoned off for societal purposes; the rest flows out to sea.[96] The real problem with the U.S. water supply is that it is not distributed evenly across the land. Some regions have plenty of water to go around, while others must go to great lengths to secure an adequate amount.

The West, in particular, could not have been settled without epic feats to overcome the region's water supply problems. In southern California, natural rainfall averages only 14 inches a year—putting it on a par with arid cities like Tripoli, Libya, and Kabul, Afghanistan. And yet, thanks to huge water projects such as the Hoover Dam, the All-American Canal, the Central Valley Project and more, the California South Coast now is the most populous desert metropolis on Earth—with nearly 20 million residents. Moreover, the desert states of Arizona and Nevada are now the fastest growing states in all of America, with population increases of 35 and 50 percent, respectively, in the 1980s alone.

California clearly has been the biggest winner in the West's water sweepstakes. More people live in the state today than in the 21 least populated American states combined. Ranked separately, California's economy would be the seventh largest in the world. And despite its semiarid, Mediterranean-like climate, California has built an agricultural economy that is twice as large as that of any other American state—with $18 billion in annual farm income. California also has the nation's most diversified farm economy. It grows nearly half of the nation's fruits and vegetables; it is the largest producer of cotton, hay, eggs and greenhouse and nursery products; and it is among the largest producing states for dozens of other crops.[97]

California has not completely overcome its water problems, however. In many respects, they are worse than ever. As it entered the 1991-92 rainy season, the Golden State's drinking water wells were at record low levels, depleted by five consecutive years of below-normal precipitation.[98] The 1991-1992 rainy season did bring above-normal precipitation to southern California, helping to replenish reservoirs there. Many coastal cities even declared the drought "over." But the drought continues in many areas north of Los Angeles. The Sierra Nevada snowpack, which stores much of California's water for the dry summer months, was at only 60 percent of normal as the spring of 1992 began. At the same time, the water level of Lake Tahoe was approaching its lowest level on record.[99]

To add insult to injury, the U.S. Supreme Court recently abrogated California's access to 660,000 acre-feet a day of Colorado River water, ruling that the water rightfully belongs to Arizona. (That volume of water, spread one foot deep, would cover 1,000 square miles—about the size of the greater Los Angeles area.) Arizona began diverting this court-won water across the Gila desert in 1992, using a new $3 billion aqueduct known as the Central Arizona Project. The water will be used principally to augment the municipal supplies of Phoenix, Tucson and other fast-growing Arizona cities. Some of the water will be used to irrigate Arizona's own desert farmlands.

Arizona's gain is California's loss. The diverted flow leaves only half as much water in the All-American Canal for California's Metropolitan Water District, which serves 15 million residents on the South Coast. Accordingly, California must find new ways to make up the loss. It could draw additional water from two dozen rivers that pour out of the Sierra Nevadas into the Central Valley. But to replenish what has been lost to Arizona would mean siphoning off four out of every five gallons of these rivers' renewable supply—twice as much as is already being diverted from these in-state sources.[100]

Conservationists have raised hackles about the effects of such dam diversions—and attendant reduced stream flows—on wildlife habitat. With the recent drought, they now say that California's water shortage has turned into a full-blown environmental crisis. Indeed, populations of chinook and coho salmon have plummeted in California rivers, and the numbers of striped bass and herring have fallen in the adjoining deltas and salt-water bays. (Lack of inflowing freshwater makes these estuaries too salty.) Since 1990, the U.S. Fish and Wildlife Service has listed five species of salmon and the minnow-like delta smelt as species threatened with extinction.[101] Waterfowl populations also have been dealt a severe blow, since water diversions have shrunk Central Valley marshes and bogs to an estimated 5 percent of their pre-1900 size. The number of ducks and geese that overwinter in these wetlands has dropped from 40 million in the 1940s to fewer than 5 million today.[102]

The U.S. Bureau of Reclamation announced in February 1992 that it would meet its obligations under the Endangered Species Act by cutting off water supplies to owners of 1 million acres of farmland in the Central Valley—the first time it has done so in its history. Federal water deliveries will continue to owners of 2.25 million acres of other California farmland, whose water rights predate construction of the huge Central Valley Project in the 1930s. But even their allocations of federally subsidized water will be cut back 25 to 50 percent. "The federal bureaucracy has looked at a fish in trouble and farmers in trouble," one farm lobbyist lamented, "and chosen the fish."[103]

Many farmers believe the only permanent solution to California's water problems is to stop trying to save endangered species and start building more surface water impoundments. Global climate change could make it harder to act on this suggestion, however, since two-thirds of the state's freshwater supply must pass through the Sacramento Delta before being drawn into distribution systems in the southern half of the state. Sea level rise could infuse more saltwater into the brackish delta, making it harder to keep saltwater at bay. Already, 140 acre-feet of freshwater must be pumped into the northern side of the delta for every 100 acre-feet of freshwater withdrawn on its southern side.[104]

If reduced rainfall also lowers the amount of freshwater entering the delta, it would become even harder to hold back the rising tide. And a rise in global temperatures could cause more winter precipitation over the Sierras to fall as rain instead of snow. Such a switch could exacerbate downstream winter flooding problems and reduce the snowpack of the Sierras during the critical summer months.[105] But while the greenhouse effect may figure prominently in the debate over whether to build new water projects in California, the whole matter could be academic, since no major reservoirs have been built in the state in the last 20 years and no new ones are planned.

That leaves groundwater as the only other major source of water left to tap in California. Groundwater normally makes up about 40 percent of all water used in the state, although its share has risen to 60 percent during the recent drought. Since California is the only western state with no groundwater regulations, the total number of wells drilled into underground supplies remains undetermined; estimates range from 1 million to 2 million wells. What is known is that nearly 5 trillion gallons *more* water is being withdrawn from the state's groundwater reserves annually than is being put back in. Recently, the U.S. Geological Survey warned that some of the state's aquifers are in danger of collapsing, which would reduce their storage capacity forever.[106] The land under San Jose, for instance, already has subsided more than eight feet in the last 50 years as a result of groundwater mining. Aquifers in the San Joaquin Valley also have been widely affected.

The way in which California has managed its water problems in recent years may not bode well for mobilizing against climate change. As researchers at the nonpartisan Pacific Institute for Studies in Development, Environment and Security told a congressional subcommittee in 1991, "The experience of the drought suggests that California will not address climatic changes with specific policies until long after the impacts are being felt."[107] Whether the recent drought in California is an early manifestation of climate change remains a matter of speculation. What is clear is that a continuation of recent trends would exacerbate the water problems in California—and perhaps throughout the West.

Factoring in Climate Change

The difficulties that water users—and farmers in particular—are facing in the West should come as a surprise to no one. John Wesley Powell, the one-armed explorer who mapped the Colorado River and braved the Grand Canyon, offered a dim prospect for agriculture in the West in his now-famous *Report of the Arid Region of the United States*. "About two-fifths of the entire area of the United States," including most of the land west of the Rockies, "has a climate so arid that agriculture cannot be pursued without irrigation," Powell wrote in 1878.[108] Yet Powell's warnings were mostly ignored. The federal government, in fact, encouraged widescale settlement and development of the West, based partly on the notion that "rain follows the plough." In a series of reports and popular essays, government-sponsored scientists wrote that tilling the soil would release dust and moisture to the sky that would help form clouds. Along with new vegetation and smoke from trains, rainfall in the West would increase, these scientists maintained.[109]

An even bigger miscalculation came after World War I, when Arthur Powell Davis (John Wesley Powell's nephew) was director of the federal government's Reclamation Service. The West was growing rapidly in the early 20th century, and a number of western states—particularly California—wanted to build a series of dams and canals on the Colorado River as a means of increasing irrigation, hydroelectric generation and flood control. That meant states upstream of these impoundments would have to agree—in perpetuity—to release sufficient quantities of Colorado River water to make these huge investments pay off.

To determine the allocation of water between the states, the Reclamation Service took measurements of the unobstructed flow of the Colorado River at Lee Ferry, Ariz., a point that divides the Upper and Lower Colorado River Basins near the Utah border. Davis's engineers determined that the average river flow past Lee Ferry was 16.4 million acre-feet a year. A more conservative figure of 15 million acre-feet was used when the Colorado River Compact was signed in 1922. States in the Upper Basin agreed to provide at least half that amount (7.5 million acre-feet) to those in the Lower Basin each year. What the Reclamation Service engineers failed to realize, however, was that the West was in the midst of its wettest 20-year period over the last 450 years when the flow measurements were taken. More recent analysis of tree-ring data suggests that the actual long-term flow at Lee Ferry is only 13.5 million acre-feet a year. In effect, the Upper Basin has been short-changed by 1.5 million acre-feet of water a year. This measurement error is becoming more of a problem as the Upper and Lower Basins start to demand their full allocations of water from the Colorado River.[110]

The allocation problem could get much worse with global warming. One analysis by Roger Revelle of the Scripps Oceanographic Institution and climatologist Paul Waggonner finds that a 3.5 degree F temperature increase would reduce the unobstructed flow of the Colorado River by nearly a third—or six times more than the 660,000 acre-feet that was reallocated from California to Arizona in 1992. If the temperature increase were accompanied by a 10 percent decrease in precipitation in the Southwest, the average flow of the Colorado River might be reduced by 40 percent, which is equivalent to the amount of water now available for use by the entire Upper Colorado River Basin.[111] Even if rainfall amounts increased 10 percent, runoff still might drop nearly 20 percent as a result of increased evaporation, according to Revelle and Waggonner's analysis.

Demand for Colorado River water today is so great that the mouth of the river usually is dry as it enters the Gulf of California. (The amount of consumption, in other words, equals the renewable supply.) If an even drier climate were to prevail in the decades ahead, all of the river's diminished supply could be consumed before the river reaches the All-American Canal or the aqueduct of the Central Arizona Project. Clearly, reallocations of water rights under the Colorado River Compact may become necessary in the event of global warming.

A similar predicament could prevail across the West, especially in regions where a high percentage of runoff is consumed by farmers and municipalities. The Missouri, Arkansas, Texas Gulf and California regions now consume 10 to 40 percent of their renewable water supply. The percentage rises to between 40

Figure 12

Colorado River Basin

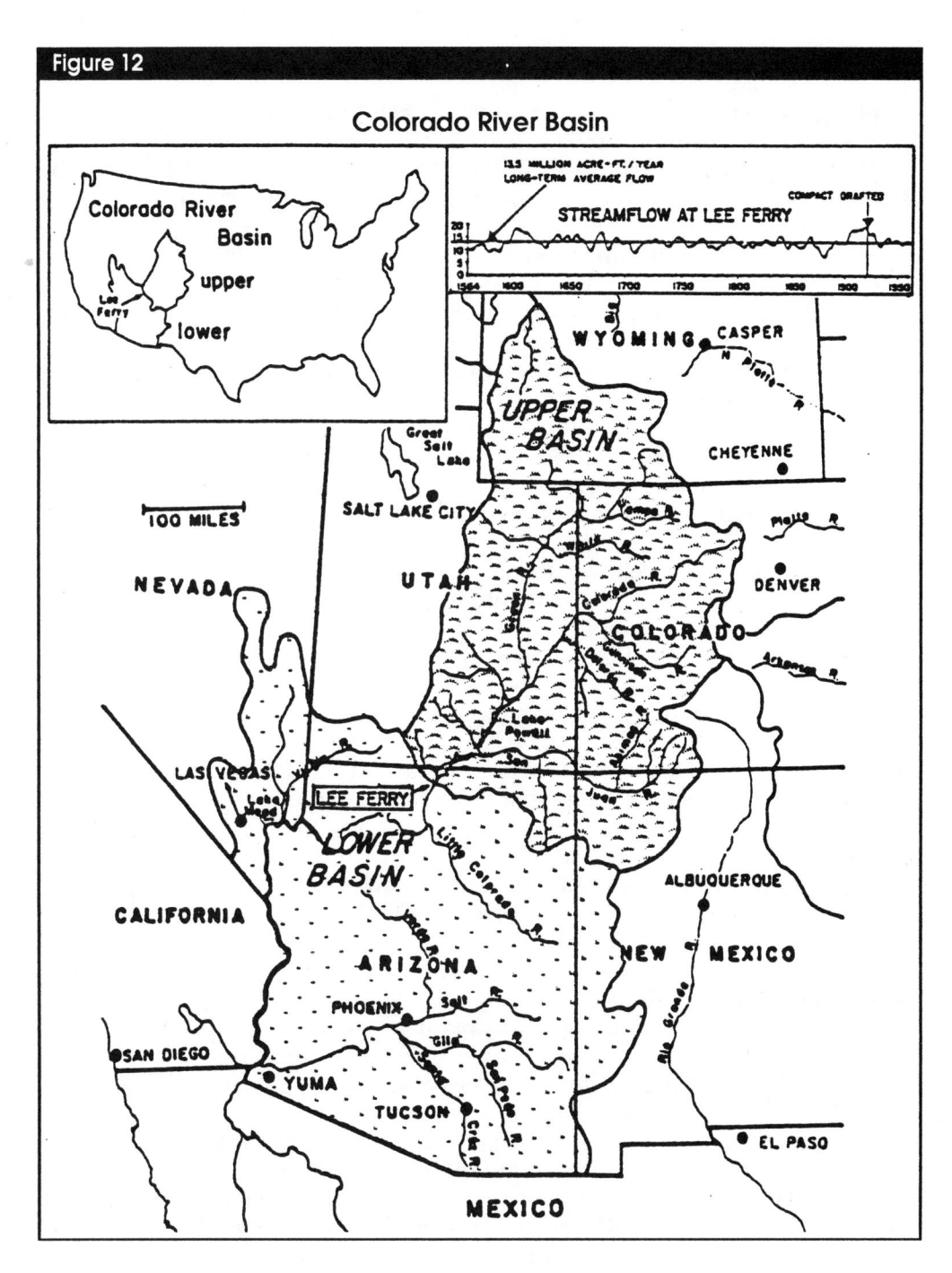

and 60 percent in the Great Basin and Rio Grande regions. By comparison, consumption in the lower Colorado River basin is 105 percent of the renewable supply. That means all of the water gets used—plus an extra 5 percent provided by groundwater withdrawals.[112]

Irrigating the Desert

Altogether, eight southwestern states—stretching from Oklahoma to California—are endowed with only 6 percent of the nation's water supply, yet they account for nearly one-third of the nation's total water consumption. The attraction of tens of millions of people to live in this region's warm, dry climate is an important factor in the lopsided supply-demand ratio, to be sure. But the main reason for the high water consumption is the extent of irrigated agriculture in the arid Southwest. With the population of the Southwest expected to grow another 40 percent by the turn of the century, a "confrontation between farm water and urban water will establish the boundaries and character of the West's [future] growth," predicts Richard Howitt, an economist at the University of California at Davis. "The clock is ticking."[113]

The protracted California drought offers an important test case of what might happen to western water consumers in a possibly warming world. California residents have long been chided for their profligate water-consuming ways. Twenty million of them live in a virtual desert climate in the southern part of the state, yet they have adorned their surroundings with swimming pools, manicured lawns and well-washed cars. Even so, the amount of water that southern California residents consume is only a few drops in the bucket compared with the state's agricultural sector. The greatest water consumers in the state are four kinds of crops: pasture and alfalfa, both used as forage for livestock, and cotton and rice, two surplus commodities principally sold overseas. Water consumption by southern Californians actually ranks fifth, behind these crops.[114]

As California's drought worsened in the latter half of the 1980s, municipal officials instituted water rationing and developed elaborate plans for cities to obtain new drinking water supplies. One effect was to raise the water bills of residents in coastal California cities by two to six times relative to pre-drought times. San Francisco's water prices jumped from $50 an acre-foot to $300 an acre-foot, for instance, and in Los Angeles, the charge rose from $230 to $545.[115] (An acre-foot of water equals about 326,000 gallons, or about what a family of four typically uses in 20 months.)

The especially drought-plagued City of Santa Barbara even decided to build a $37 million municipal desalination plant in the early 1990s. Capable of producing 10,000 acre-feet of freshwater a year, the desalination plant is the largest such facility in the nation. But the potable water is incredibly expensive to produce, costing nearly $2,000 an acre-foot. Yet it is cheaper than another option Santa Barbara considered: shipping freshwater in by tanker from British Columbia at a cost of more than $3,000 an acre-foot.[116]

While California municipalities took extraordinary measures to cope with growing water shortages during the first four years of the state's drought, California farmers were able to carry on with business as usual. Actually, their business was better than ever. The state's agricultural receipts posted records each year—rising from less than $15 billion in 1986 to $17.8 billion in 1990.[113] Since not much rain falls in California's agricultural centers anyway, the drought hardly was a factor. All that really mattered was continued access to

Agriculture

Table 2

Potential Regional Impacts of Climate Change on Water Uses: Areas of Vulnerability

	Pacific Northwest	California	Arid Western River Basins	Great Plains	Great Lakes	Mississippi	Southeast	Northeast
Irrigation	x	x	x	x			x	x
Thermoelectric Power							x	x
Industrial		x	x					x
Municipal/ Domestic		x	x					x
Water Quality		x	x	x	x	x	x	x
Navigation					x	x	x	
Flood Control	x	x		x		x	x	
Hydropower	x	x			x		x	
Recreation					x		x	

SOURCE: U.S. Environmental Protection Agency, 1989.

irrigation water, which state and federal water projects continued to provide for as little as $2.50 an acre-foot.

The principle of supply and demand gets turned on its head when the economics of irrigation water are at issue. California's agricultural sector consumes nearly six times as much water as the state's municipal and industrial sectors combined, yet the price that some California farmers pay is less than 1 percent of the amount that urban residents pay, gallon for gallon. Farmers growing water-intensive crops in the desert recognize that their livelihood can be maintained only as long as they have access to subsidized water supplies. One might think that California's protracted drought would have ended this practice. But the lesson to be drawn from California's recent experience is far from clear.

Marc Reisner, who examined California water policy in his book, *The Cadillac Desert*, observes that irrigated crops and pasture contribute comparatively little to the state's economy despite the vast quantities of water they consume. When measured against the gross state product of $575 billion in 1986, Reisner calculated that irrigated pasture contributed only $94 million to feed cows and sheep, or just one five-thousandth of California's economy. Yet this one tiny industry manages to account for fully one-seventh of the state's annual water consumption. If these pasturelands were no longer irrigated, Reisner figures, it would free up enough water to meet California's projected growth in urban water demand for the next 20 years—with half left over to maintain habitat for fish and wildlife. If California farmers stopped irrigating alfalfa as well, projected urban demand could be met for *another* 20 years without having to resort to extraordinary measures to procure new supplies.[118]

Farm interests take issue with these calculations. They point out that hay from alfalfa brought in receipts of $869 million in 1989, *and* that alfalfa is essential to maintaining the state's $2 billion dairy industry, which is second in size only to Wisconsin's giant dairy industry.[119] Cotton and rice sales, meanwhile, generate more than a billion dollars of additional state farm income, although California's rice crop alone diverts about 2.5 million acre-feet of water in non-drought years—enough to supply 25 million urban users. "People don't seem to understand," one cotton farmer complained to a reporter during the recent drought. "We'd have a helluva tough time farming out here if we didn't get that water."[120]

Critics like Reisner say that is exactly the point. Some water-intensive crops might no longer grow in the desert. But even if half of the California acreage dedicated to pastureland, alfalfa, cotton and rice were taken out of production, the state's agricultural income would decline by only 8 percent, Reisner estimates. Farmers could continue growing oranges, grapes, lemons, peaches, pears, almonds, tomatoes, beans, asparagus, avocados, artichokes, lettuce, carrots and 150 other species that consume less water than alfalfa, cotton and rice, and that have a higher value in the marketplace. At the same time, the billions of gallons of freed-up water could be used for municipal purposes and to enhance fish and waterfowl habitat.

As the California drought entered its fifth year in early 1991, California's farmers finally began to feel the pinch. California's Department of Water Resources cut off deliveries to farmers from the State Water Project, a giant network of reservoirs and aqueducts that normally provides about 1.2 million acre-feet of water to growers in the Central Valley. The federal Bureau of Reclamation also cut back by 75 percent its delivery of 7 million acre-feet of water that it normally provides to farmers in California.

Those who could pump groundwater to irrigate their crops in 1991 did so, since the state government does not not regulate those supplies. Yet the cost of pumping water from deep aquifers is expensive—ranging above $200 an acre-foot in some instances (and it may go higher still as water tables fall). As a result, farmers were forced to cut way back on alfalfa, rice and cotton production, and ranchers pared down their herds of livestock in 1991—much as Reisner and others have recommended. The result was that California's string of record farm years was broken in 1991, with revenues dropping to an estimated $17.2 billion. But it hardly constituted an economic catastrophe. In relative terms, California's farm revenues were down less than 3.5 percent from the year before, and much of the losses stemmed from a devastating freeze in December 1990 rather than from the cutbacks in irrigation water deliveries.[121]

Making a Market for Water

Water-intensive farming practices will remain under demographic pressure in the West regardless of potential changes in climate. In fact, municipalities have already turned to open market purchases to take water away from farmers. In Arizona, the cities of Phoenix, Tucson, Mesa and Scottsdale have acquired water rights to more than 50,000 acres of irrigated farmland since 1980. Some of the purchased groundwater is being pumped through the Central Arizona Project aqueduct for delivery to households as much as 100 miles away. Cities in Utah also have acquired rights to more than 100,000 acre-feet of water by buying up shares of canal and ditch companies during the 1980s. Around Reno, Nev., towns and the local electric utility are purchasing agricultural water rights for purposes of future municipal and power development. Las Vegas, meanwhile, has a standing offer to purchase water rights for $1,000 per acre-foot.[122]

Conservation groups also are getting into the act. In 1990, The Nature Conservancy purchased water rights for $300 an acre-foot from farmers in the Newlands Irrigation Project near Reno. This federal project (which marked the federal government's first large-scale attempt to make western deserts bloom) diverts water from the Truckee and Carson Rivers as they flow out of the Sierra Nevadas into the Stillwater marsh of Nevada. The Nature Conservancy's purpose is to restrict the flow of natural toxins such as arsenic and selenium, which irrigation has freed from the desert land—and deposited into the marsh—as the land was converted to agriculture.[123] The Nature Conservancy also has acquired 20,000 acre-feet of water rights from the Pittsburgh and Midway Coal Co. in Colorado in an effort to protect the humpback chub, an endangered fish in the Gunnison River.[124]

Many oil and coal companies that acquired water rights for purposes of oil shale development during the 1970s have retained their water rights, even though most of their shale development plans have since been abandoned. These companies expect the water rights to continue rising in value—perhaps making them more valuable than the oil shale itself. By one account, energy companies hold up to 400,000 acre-feet of water rights in Colorado alone.[125]

In the investor arena, Prudential-Bache Securities created two limited partnerships in the late 1980s that purchased more than $42 million worth of

> **Box 2-F**
>
> ### Farming on Water Subsidies
>
> A fundamental reason why farmers are able to grow crops on arid lands is that government agencies make irrigation water available at highly subsidized prices. At a time when some municipalities are paying more than $1,000 an acre-foot to procure new supplies, the federal Bureau of Land Reclamation continues to sell water to agricultural users for less than $20 an acre-foot—sometimes for as little as $2.50 an acre-foot. Water conservation measures rarely can compete at these low prices. And even when cost-effective water-saving practices are available, farmers are reluctant to employ them because of the "use-it-or-lose-it" principle: A farmer who does not consume all of his allotted supply from the government forfeits the conserved portion to someone else. An efficient market for buying, selling and conserving water will not emerge until these price and policy issues are reformed.
>
> Water subsidies for agriculture are deep-seated and longstanding. To encourage cultivation of the arid West, Congress decreed in 1902 that all farmers receiving irrigation water from the Bureau of Land Reclamation would be charged only enough to repay the capital cost of the project supplying them. Moreover, the farmers receive a 10-year grace period before they have to pay back any of their loans, and when they do repay, no interest is charged. The result is a 90 percent subsidy for farmers involved in federal water projects.[1f] In the case of the huge Central Valley Project in California, irrigators have paid back only 5 percent of its $931 million construction cost over the last 40 years. Altogether, federal water subsidies cost the American taxpayer about $400 million a year.[2f]
>
> Western water doctrine poses another major obstacle to reform. Traditional riparian law grants the owner of land through which water flows a right to use the water as long as his use does not unduly inconvenience other riparian owners. In the West, this doctrine has been amended in favor of prior appropriation: Those "first in time" also are "first in right." To ensure that those parties holding senior water rights do not simply hoard their supply, the water codes of most western states require riparian owners to put their water to some "beneficial use." That means a

western water rights. The partnerships plan to hold onto the rights for up to 15 years and then sell them off to the highest bidders. Another group of private investors is working on a plan to pump at least 30,000 acre-feet of water a year from the San Luis Valley of Colorado to water-thirsty suburbs around Denver. If the project succeeds, it would be the West's first major water development by a private entity—and a lucrative one at that.

American Water Development Inc., consisting of wealthy Canadian financier Samuel Belzberg, David Williams of Williams Cos., and a group of former investment bankers, figures the initial development phase of the Colorado project could cost $600 million to $900 million. But with expectations that Denver suburbs would be willing to pay more than $5,000 an acre-foot of water shipped from the company's 155,000-acre Boca Ranch, revenue from the project could reach $1.5 billion. Ultimately, American Water Development hopes to ship up to 200,000 acre-feet of water a year from its 2-billion-acre-foot reserve on the Boca Ranch to support future urban growth around the Front farmer who decides to shift to a less water-intensive crop—or institutes some other conservation practice—loses his rights to the "excess" supply.

In 1982, Congress tried to shut off subsidies to agricultural concerns that had acquired huge amounts of federal water rights over the years. It declared that subsidized water would be available only to farmers who owned 960 acres of land or less. But the big growers eluded the new restrictions by breaking up their holdings into 960-acre chunks—assigning ownership of each parcel to a different trust, partnership or corporation that was controlled by the original partners. As a result, some of the richest farmers in America—many of whom grow price-supported surplus crops like rice and cotton—continue to receive two of the government's most generous subsidies.[3f]

Farmers growing water-intensive crops in desert climes insist they have a right to irrigated water at below-market prices, and the Bureau of Land Management has been inclined to agree with them. When big growers arbitrarily carved up their land to retain access to federal water after 1982, the bureau openly encouraged the practice; and in 1987, it rejected tough new regulations proposed by its own staff that would have cut off the water supply to many of these same operations.[4f] A bill now working its way through Congress is seeking once again to prevent big landowners from dividing their holdings into small pieces.[5f]

A truly free market for water will not evolve until the "use-it-or-lose-it" principle is abandoned and water rates are raised so as to reflect their true costs—giving irrigators a financial incentive to conserve. All western states do permit water transfers under certain conditions, so long as the rights of third parties are not abridged. But satisfying these third-party provisions tends to inhibit the creation of efficient markets. To date, only Oregon permits riparian owners to retain rights to their conserved supply, although they must allocate up to 25 percent of their savings for environmental purposes approved by the state.[6f] Free market proponents believe the system cries out for reform. They note that thousands of long-term water contracts will come up for renewal in the next 15 years—just as climate change-induced water shortages may become readily apparent.

Range cities of Denver and Colorado Springs. (That is enough water to meet the municipal requirements of 800,000 residents.)[126]

Farmers are understandably wary of this and other water development schemes. In the case of the Boca Ranch, they are concerned that the groundwater pumping could lower the water table in the San Luis Valley, increasing neighboring farmers' pumping costs and requiring that new wells be drilled. Farmers also have long memories. They recall what happened to California's agricultural Owens Valley after the City of Los Angeles bought up most of its water rights in the early 20th century to foster the growth of the L.A. basin. The number of irrigated farm acres in the valley fell from 140,000 in 1920 to only 30,000 in 1950—and then the huge Owens Lake dried up. Today, only a handful of commercial farms remain in the Owens Valley. Los Angeles has acquired three-quarters of the agricultural lands there and owns most of the urban real estate as well.[127] Meanwhile, concerns are mounting that Mono Lake near Yosemite National Park may some day meet the same fate as Owens Lake.

Its water levels also have dropped considerably since southern California municipalities began to siphon water from streams that flow into it.

If fears of impending climate change hasten the scramble for water, many owners of irrigated farmland could face a fateful question: Do they continue to battle the elements and urban water interests—against increasing odds—or do they decide it is time to cash in their chips? Several bills now working their way through Congress may leave western irrigation farmers with little choice. The bills seek to raise the price of water sold from federal projects and make more of it available for use by municipalities and for environmental conservation purposes. Particular attention is being paid to "double dippers"—wealthy corporate farm operators who use subsidized water to grow crops that are in surplus and also are subsidized. (Final legislation is expected by the end of 1992.)[128]

The recent California drought also sends an important signal about the evolving struggle between farmers and others over water. Municipalities may have the financial muscle to take water away from farmers, but farmers are not likely to give up their water without exacting a high price—especially given concerns about a diminishing long-term supply. Consider what happened when California Governor Gov. Pete Wilson (R) set up a "water bank" in February 1991. The idea was for the state to purchase up to 500,000 acre-feet of water from farmers in the Central Valley, who could live without it for a year, deposit the proceeds in the "bank" and then re-sell the water to those in greatest need of additional supplies. For each forgone acre of irrigation, farmers would be paid $250 to $350.[129]

At first, very few farmers came forward. Some attorneys for local water districts recommended that irrigators hold out for prices of as much as $1,300 an acre. Rice farmers, who grow the most water-intensive of all crops in the state, were particularly concerned that selling their water rights for 1991 might come back to haunt them as they renegotiated their water supply contracts with the federal government in the years to come. As a result, only 6 percent (30,000 acre-feet) was contracted toward the state's 500,000-acre-foot goal in the first two weeks after the plan's announcement.

Then came the "Miracle March"—a momentary let-up in California's drought in 1991, during which time rain and snow accumulated at almost twice the normal rate. Infused with new hope that natural precipitation, replenished groundwater supplies and a restored Sierra snowpack would see them through the 1991 growing season, farmers rushed to sell some of their allotted irrigation water into the bank (save the rice farmers). By the end of the month, the water bank exceeded its 500,000-acre-foot goal.[130]

California's experiment with the water bank suggests that a market-based approach for allocating water can work—but perceptions about the climate are critical. If in the future farmers of irrigated cropland suspect that global warming might sustain drought-like conditions indefinitely, they may not be willing to sell their water at any reasonable price. Without water, after all, their livelihood would evaporate beneath the hot desert sun. For them, it is truly "liquid gold."

Responses of Farmers and Food Producers

As daunting as the agricultural challenges posed by the greenhouse effect may be, they appear less formidable when viewed in the context of the remarkable agricultural achievements of the last 50 years—and the potential for even more revolutionary gains in the years ahead. Whether or not global warming becomes a reality, farmers will continue to adapt their management practices much as they always have. Their decisions about crops and cultivation techniques will be based not on speculation about long-term climate trends but on the actual condition of their fields, the outlook for commodities prices and the advent of new technology. Over time, farmers may find themselves instituting more sophisticated water management practices, planting new kinds of crops and switching to new cultivation methods. But, invariably, farmers will choose the least-cost solutions available to them in an effort to produce food and fiber at returns in excess of their costs. This fundamental economic tenet will not change.

Nevertheless, if the world must come to terms with global warming, some ground rules for agriculture could change. Water agencies might discourage water consumption by raising prices for irrigation water. Commodities brokers might choose to trade for certain foods, rather than buying domestic varieties, because climate change has afforded other nations a comparative advantage. And in an effort to reduce greenhouse gas emissions, the government could enact programs to winnow herds of ruminant animals, restrict rice production in flooded paddies and limit applications of nitrogenous fertilizers.

Finally, there are steps that farmers and food producers could take on their own to harness new technologies for climate change purposes. The cellulose from perennial plant species could be converted into basic foodstuffs using high-technology methods. Genetic scientists could alter the very structure of plants to make them more nutritious and possibly more amenable to climate change. Some farmers might choose "low technology" options instead, using low-input, sustainable agricultural production methods that reduce greenhouse gas emissions and help sustain the natural environment for the long term. This section explores some of these options. They constitute greenhouse gambits for agriculture.

Tapping Water Conservation

No one can yet say for certain whether global warming would lead to chronic regional water shortages. Some places could well become wetter as a result of global climate change, making flooding a more frequent problem than drought. Nevertheless, the combination of higher temperatures and more evaporative losses increases the likelihood that some regions may dry out. In those places, everyone would have a responsibility to use water wisely, not just farmers.

Homeowners, for example, could install low-flow devices on showerheads and faucets, replace five-gallon-per-flush toilets with one-gallon toilets and cut back on watering their lawns. Residents of Santa Barbara, Calif., grew so accustomed to saving water during California's recent drought, in fact, that households continued to use 40 percent less water even after the "Miracle March" of 1991—which filled local reservoirs with rainwater and allowed the city to lift severe water restrictions.[131] Conserving water had become part of their routine.

Industrial water users also would have a vital role to play in conserving water. Companies could enact water recycling programs, control evaporative losses in cooling water and switch to less water-intensive manufacturing processes. One study of 15 major companies in the San Francisco Bay area found that implementing cost-effective water conservation measures would reduce their collective water use by a billion gallons a year.[132]

Yet as impressive as these potential savings sound, a billion gallons of water amounts to just 0.6 percent of the annual water consumption by the single largest cotton plantation in California alone. (The Boswell Trust—with 23,000 acres of irrigated cotton fields in the San Joaquin Valley—goes through a phenomenal 160 billion gallons of water a year.)[133] This giant disparity in water consumption also holds up nationally. Water consumption by agricultural users dwarfs consumption by industrial users by more than an order of magnitude.[134] Accordingly, the farm sector is by far the area where the greatest water conservation gains are to be realized. In California, farmers could free up 3 million acre-feet of water—a trillion gallons a year—if they managed to reduce their water consumption by only 10 percent.[135] And throughout the West, the amount of water available for municipal and industrial purposes would *double* if agricultural interests saved just 15 percent of the water they now consume.[136]

The high costs of irrigation can provide an economic incentive for farmers to conserve. A center pivot sprinkler system that waters a half-section of land (320 acres) costs up to $100,000 to purchase. The construction of reservoirs, wells and ditches may raise its total installation cost to between $3,500 and $12,500 per acre.[137] Then there are the costs of running and maintaining the system—and purchasing the water if it comes from off-site. In a few places, the cost of purchased irrigation water ranges above $100 an acre-foot—sometimes higher.[138] Yet for all of this expense, an unimproved, gravity-feed irrigation system may provide only half of the water to the crop.[139] The rest evaporates or runs off the fields.

As irrigation costs continue to rise, farmers are recognizing the value of water conservation. To improve the distribution of water among their crops, some farmers are leveling their fields with laser precision, lining irrigation ditches with concrete and building run-off pits to capture and recycle irrigation tailwater. Such measures can reduce run-off losses by more than 30 percent.[140] (At the same time, more pesticides and fertilizers stay in the fields where they belong, rather than running off into rivers and streams.)

Some farmers are experimenting with solar-powered surge valves, which automatically open and close irrigation spigots at strategic intervals around the clock. Today's most sophisticated irrigation systems have built-in weather stations to sense rain, wind, temperature, humidity and sunlight. Irrigation schedules are adjusted according to the weather data collected over the previous 24 hours. These systems can even shut off the flow of water to leaky or broken sprinkler valves and alert the owner to the need for repairs.[141]

Some farmers are moving away from conventional pivot irrigation systems altogether, since so much of the artificial rain they create evaporates before it hits the ground. One alternative is to use low-energy systems that apply water directly on the soil adjacent to the plants rather than in the air above them. Such trickle irrigation systems, pioneered in Israel during the 1970s, now are gaining favor in the United States and elsewhere. While they eliminate run-off losses almost entirely, these trickle systems cost up to three times as much to install as conventional sprinkler systems.[142] But in addition to applying irrigation water more efficiently, they reduce salinization—long the bane of irrigated agriculture. Since plant roots are continuously supplied with water in trickle systems, salts do not build up to the point where water uptake is impaired.

Farmers who have access to abundant water supplies, or whose purchases are subsidized by government projects, have little incentive to save water at present. Yet the need for water conservation is extremely important in places like the West, where municipal and industrial water users compete for a scarce supply. Global warming could make this competition more intense and spur the creation of an open market for water purchases. In this event, farmers could not expect to out-compete other users, simply because their own volume requirements are so high in relation to crop values. On the other hand, global warming could lead to rising crop prices and provide an economic rationale for farmers to irrigate more of their land. The unresolved question is whether additional water supplies would indeed be available at prices that farmers can afford. Making better use of water already available is therefore one of the best insurance policies that farmers can buy when it comes to global warming. Conservation measures bolster the water supply, put downward pressure on market prices and provide a margin of comfort in case farmers' water allotment is shorn by climate change and competing uses.

Growing Drought-Tolerant Crops

While there are strong incentives for farmers to develop water-conserving irrigation practices, farmers always have the option of practicing dryland farming where crop irrigation is not feasible. Unfortunately, climate change could render some rainfed acreage unsuitable for growing any traditional cash crops. Rather than abandoning this land completely, farmers might consider raising some particularly drought-hardy plants. Three planting options are triticale as an alternative to wheat, a group of plants called halophytes that are resistant to salt, and a variety of traditional perennial grasses that could grow in place of annual crops in the Great Plains. A fourth option would be to plant non-herbaceous crops and convert their cellulose into simple sugars.

Triticale: Triticale is a wheat-rye hybrid devised by plant breeders more than 100 years ago. The grain went unused until the 1960s, largely because the first hybrids produced a flour that did not rise well. Refinements made in the 1980s have improved the grain considerably; new lines produce doughs that rise as well as those made from wheat flour. What is more, the yield of triticale can match that of wheat on good soil, and on marginal soils it can out-produce wheat by up to 30 percent. Different varieties of triticale mature over a range of growing seasons and day lengths. In general, triticale thrives in places where other cereals founder—including soils considered too dry, cold, infertile, saline, sandy, acidic, alkaline, mineral deficient or toxic with boron. Triticale also appears to be notably more resistant than wheat to leaf blotch, powdery mildew, smuts, bunts and other fungal infections (although its resistance to disease could break down if it were grown in the large quantities that wheat is today). In short, triticale appears to be a crop well suited for the northern Great Plains of the United States and Canada in the event of warmer, drier growing conditions. Although most of the triticale grown at present goes for livestock feed or testing, the crop ultimately "could become a major staple" in kitchens throughout the world, the National Research Council reports.[143]

Halophytes: A diverse group of salt-tolerant plants known as halophytes could be the answer for millions of acres worldwide considered too salty for normal crop growth, or where irrigation water itself is too saline. Such lands include former farmland where poor irrigation practices have salinized the soil, coastal deserts where only seawater is available for irrigation and arid inland areas situated atop salt lakes and underground brackish water. Halophyte plants thrive in water containing up to 7,000 parts per million of salt. Scientists hope that halophytes some day may be able to grow using irrigation water with salinity levels up to 10,000 parts per million. (By comparison, sea water contains between 35,000 and 38,000 parts per million of salt.)

Agricultural scientists estimate that excess salt already has reduced productivity on about 50 million acres of irrigated land worldwide—more than all the irrigated land of the United States. The agricultural economies of arid countries such as Egypt, Iraq and Pakistan are especially threatened, since they are almost completely dependent on irrigated cropland. Although the only commercial use of halophytes at present is as forage for livestock, further refinements could lead to cultivation of these salt-tolerant plants for human consumption in such things as vegetable oils.[144]

Alternatively, halophytes could be grown on desert and saline lands to sequester carbon. One recent study estimates that 15 percent of such lands around the world—some 300 million acres—would be suitable for growing halophytes. With annual biomass yields of 3.5 to 8 tons per acre, the net carbon accumulation in halophytes could amount to 600 million tons per year, which is equivalent to about 10 percent of all carbon releases from fossil fuels each year.[145] Halophytes also could be grown as a biomass fuel to substitute for fossil fuels.

Perennial grasses: In some semi-arid inland areas, such as the Great Plains, farmers may find that they would still be able to harvest grains if they let their fields revert back to grasslands. Wes Jackson, director of the Land Institute in Salina, Kan., believes many grasses native to the Great Plains have commercial development potential but would not require the same amount of chemical- and energy-intensive inputs now applied to traditional cash crops.

The Land Institute is testing small plots of selected native perennial grasses that offer nutritional value, if not much in the way of taste.[146]

Eastern gama grass, for example, is a perennial relative of corn with a grain protein content of 27 percent—three times that of corn's and twice as high as wheat's. Another native grass, Illinois bundleflower, is a wild legume that enriches the soil with its nitrogen-fixing abilities and whose bountiful seeds traditionally have been eaten by animals. Properly cultivated, Illinois bundleflower may be able to yield more than 3,000 pounds of seeds per acre, Jackson estimates, compared with 1,800 pounds of grain per acre of Kansas winter wheat. Similarly, buffalo grass can produce up to 1,700 pounds of seed-containing burs per acre, if fertilized and irrigated. Finally, Maximillian's sunflowers produce seeds as protein-rich as soybeans and exude a chemical from their roots that kills competing vegetation, The Land Institute has found. To date, farmers in the Great Plains have expressed no real interest in growing these native grasses, however, since there is no established market for them among food producers.

A number of issues must be resolved before these native plants become commercially viable. With its test plots, The Land Institute is trying to determine whether cultivated seeds can reach their estimated yield potential and whether they would shatter if subjected to mechanical harvesting. Another unknown is whether the crops could be grown on a schedule that permits a simultaneous harvest of a polyculture of these grasses. Without pesticides, some predict that insects would become a major problem among such intermingled species. But Jackson thinks that insects would spend more time and energy locating food than in a monoculture stand, thereby limiting the number of infestations. Finally, there is a question as to whether these perennial grasses would still be able to thrive in the event of a severe warming. If the alternative were to abandon croplands altogether, however, farmers at least might be willing to give these perennial grasses a try.

Cellulose conversion: An even more radical food-growing idea has been proposed by Stephen Rawlins of the Agricultural Research Service of the U.S. Department of Agriculture.[147] He suggests that cellulose from plant species such as tulip poplar trees and kenaf could be converted into basic foodstuffs using high-technology methods. The cellulose stored in such perennial plants constitutes a living reserve of feedstock for the food system that costs nothing to store. Accordingly, it could be harvested as market demand warrants. Practically all of the biomass that these plants produce—including the lignin—could be converted to food or be used as fuel and fertilizer. But the cellulose conversion process at present remains prohibitively expensive.

Nevertheless, with advancements in biotechnological or chemical conversion processes, simple sugars could be derived from the lignocellulose in these crops—and syrups would be the main commodity. The syrups would be transported via pipeline to food production sites, allowing food storage to be based on year-round operations, rather than from harvest to harvest. That would reduce capital investments in storage facilities and losses due to pests (not to mention the risks of food shortages rising out of poor back-to-back harvests). The syrups could be made into foodstuffs for livestock, poultry and fish, or become components of edible products for humans. No technological barriers exist for developing the syrups into products such as flour for bread and pasta or mashed potato substitutes, although a massive infusion of capital would be required to develop the process. After the initial capital investment, however, the monetary difference between farm value and the market price of foods would shrink so that no price increases would be expected at the retail level, Rawlins maintains.

Engineering the Foods of the Future

The prospect of genetically engineered foods offers an even more high-tech vision of the future of agriculture. The genetic composition of plants could be manipulated to express traits that make them more commercially valuable and possibly better suited for climate stresses such as heat and drought. Already the biotechnology industry is figuring out how to "design" plants that would produce more useful oils and "create" cows and pigs that would have a higher ratio of lean meat to fat. In Britain, one genetically engineered food has been approved for market use: a yeast that produces carbon dioxide quickly, enabling bread to rise faster. In the United States, a genetically derived enzyme from calves' stomachs, chymosin, now is being used to coagulate and age up to 35 percent of the nation's cheese.[148] One company is test-marketing celery sticks that are crunchier and longer lasting than traditional varieties—and do not have strings that get caught between people's teeth.[149] Genetic engineers also have managed to block a gene that causes spoiling of some fruits and vegetables, making it possible to extend their shelf life at the grocery store. Calgene of Davis, Calif., plans to put spoil-resistant tomatoes on the market in 1993.[150]

These products are early entries in a worldwide market for agricultural and food processing biotechnology that could grow to $10 billion to $50 billion in annual sales by the end of the decade.[151] Such improvements, born through genetic engineering, could increase the dominance of major corporate players in the food sector. Agribusiness at present still consists of a chain of separate players linked together: the seed, fertilizer and pesticide suppliers sell to the crop grower who in turn sells to the food processor. Biotechnology's proprietary technology could lead to the creation of integrated firms that see the product all the way through from the farm to the supermarket.[152]

As such, genetic engineering offers a vision of a Brave New World for crop growth and food processing. Yet in promotional literature entitled "Farming: A Picture of the Future," Monsanto, a leader in genetic engineering research, assures readers that "it won't change the way things look. The products of biotechnology will be based on nature's own methods, making farming more efficient, more reliable, more environmentally friendly and—important to the farmer—more profitable." Monsanto sees a future in which "Plants will be given the built-in ability to fend off insects and disease, and to resist stress. Animals will be born vaccinated. Pigs will grow faster and produce leaner meat. Cows will produce milk more economically. And food crops will be more nutritious and easier to process."[153]

Table 3

Genetically Engineered Species

Alfalfa	Cranberry	Papaya	Spruce
Apple	Cucumber	Pea	Strawberry
Asparagus	Eggplant	Pepper	Sugarbeet
Broccoli	Flax	Plum	Sugarcane
Cabbage	Grape	Poplar	Sunflower
Carrot	Horseradish	Potato	Sweet Potato
Cauliflower	Kiwi	Raspberry	Tobacco
Celery	Lettuce	Rice	Tomato
Corn	Muskmelon	Rye	Walnut
Cotton	Oilseed Rape	Soybean	Wheat

SOURCE: Gasser and Fraley, Monsanto Co., 1992.

Familiar plant life in which genetic engineering has been demonstrated successfully.

Through the ages, plant and animal breeders have tried to pass preferred traits from one generation on to the next. Genetic engineering offers these breeders a new—indeed revolutionary—means of doing so. One promising technique employs a .22-caliber blank cartridge to shoot small metal particles coated with DNA, one or two microns in diameter, through a plant cell's wall and into its cytoplasm. To demonstrate the efficacy of this technique, Monsanto used such a biolistic "gene gun" to insert a marker gene from a firefly into corn cells, causing the cells to glow faintly.[154] Monsanto and others now are seeking to inject plants with more commercially desirable traits, so that the latest agricultural technology will be available in a traditional package—the seed.

Critical cereal grains, such as corn and wheat, have tended to respond poorly to DNA injection, however. Usually, the manipulated plants were infertile, so that the genetically desired traits could not be passed on from one generation to the next. Recently, Monsanto, Ciba-Geigy, DeKalb Genetics, BioTechnica International and Pioneer Hi-Bred International have inserted marker genes successfully in corn that have gone on to produce fertile plants, whose seeds possess the new genetic traits. In the years ahead, these companies plan to market the salable seeds through their own seed divisions. Seed corn is a $1.5-billion-a-year business in the United States alone.[155]

A similar breakthrough with wheat came about in the spring of 1992. Researchers at Monsanto and the University of Florida in Gainesville used a gene gun to insert a gene that makes fertile wheat resistant to the herbicide phosphinothicin, sold under the brand name Basta.[156] Normally, herbicides kill any plants they touch—weed or crop. Therefore, creating plants that are tolerant of certain herbicides has become a major thrust of genetic engineering among agrichemical companies. At least 27 companies, including all eight major transnational pesticide manufacturers—Bayer, Ciba-Geigy, Dow/Elanco,

Du Pont, Hoechst, ICI, Monsanto and Rhone-Poulenc—are engaged in such research.[157]

An unresolved question with herbicide-tolerant crops is whether farmers would end up using greater or lesser quantities of the manufacturers' herbicides. On the one hand, farmers would no longer have to apply large amounts of herbicides early in the growing season to keep weeds from sprouting. On the other hand, they could spray whenever weeds did appear and not have to worry about adverse effects on their crops. The manufacturers say that the net effect will be to allow farmers to use new, less toxic herbicides that are effective in much smaller doses than traditional herbicides. (Monsanto, for example, is developing row crops such as tomatoes, soybeans, cotton and tobacco that would be resistant to its popular herbicide, Roundup.)[158] But some environmental groups fear that farmers will end up applying such herbicides indiscriminately, resulting in higher toxic residues in the food and water supply.

Another thrust of genetic engineering involves the injection of natural insect pathogens into plant cells—or spraying the pathogens on to plants in the field—as an alternative to using conventional insecticides. Several companies also are field-testing crops that have been injected with the soil bacterium *Bacillus thuringiensis*, or B.t., which is a natural enemy of caterpillars and moths. The genetic manipulation of the plants' cells has given new meaning to the phrase "killer tomato." Monsanto and others have expressed the B.t. gene in the cells of tomatoes, which ruptures the alkaline stomachs of larvae that feed on the plants.[159] Since humans have acid stomachs, they are not affected by eating tomatoes containing the B.t. toxin. In the next few years, Monsanto plans to introduce cotton and potato plants that also make their own B.t. toxins. Eventually, corn, soybeans, sugar beets, rice and alfalfa could be similarly endowed.

Pesticides made from B.t. toxins already are on the market, with sales of $107 million in 1991. Mycogen of San Diego, Calif., announced recently that it had isolated new strains of the B.t. toxin that can be used against a much wider array of insects than caterpillars and moths, raising the possibility of explosive market growth in B.t. pesticides over the coming decade.[160] Pesticide manufacturers are hopeful that these natural insecticides, integrated with other pest management techniques, will reduce crop losses and lead to fewer applications of synthetic chemicals. Thus, increasing plants' genetic defenses against various environmental pests would mean that fewer insecticides would enter the environment and fewer residues would persist in the food and water supply.

Designing Plants for Climatic Stress

Recent successes in bolstering the genetic defenses of plants against pests also raise the possibility that agricultural biotechnology could be used to make plants more tolerant of heat and drought—climatic stresses that may occur more often in the event of global warming. Here again, plant breeders and seed suppliers have been selecting for such traits for many generations. But it will be much more difficult in this case for genetic engineers to replicate the success of conventional breeding techniques.

Qualitative characteristics that promote viral and insect resistance are comparatively easy to isolate genetically because they are controlled by one or a few genes and are seldom affected by the environment. Quantitative characteristics, on the other hand, are controlled by many genes and react strongly to environmental conditions. Molecular scientists therefore find it extremely difficult to isolate these multigenic characteristics—even though they control traits that are critical to determining plants' ability to cope with climatic stress.[161]

It has become common practice to cross-breed plants grown in drought-prone environments with higher-yielding relatives to produce hybrids that are both drought-hardy and relatively productive. As recently as 1991, farmers in the Midwest reported better-than-expected corn and soybean harvests, despite scattered drought conditions, largely because they had planted new varieties of drought-tolerant hybrids.[162] But while important progress continues to be made, plant breeding and genetic engineering programs have not yielded the kinds of gains that would allow farmers to put aside their worries about drought. As researchers at the Natural Resources Defense Council have observed:

> Even in this age of advanced crop technology, there are limits to what plant breeding can accomplish....The fundamental constraints are illustrated by scientists' inability to breed a corn variety that will produce economically with less than 20 inches of annual precipitation or irrigation equivalent. That the crop losses in 1988 were of comparable magnitude to the losses in the 1930s confirms that plant breeding over the last half century has not "solved" the drought problem.[163]

Since plants require some amount of water to survive—and presumably always will—it is a false expectation that genetic engineering will ever actually "solve" the drought problem. Nevertheless, genetic screening programs are an aid to farmers who are trying to raise crops under a wide range of environmentally stressed conditions. Yet farmers never will escape some of the risks inherent in their profession. Drought-tolerant crops may outperform other crops in dry years, for example; but the opposite is usually true in normal and wet years. Similarly, a crop bred to withstand early-season dryness may not be well-suited for years in which a dry period sets in closer to harvest time.[164]

In terms of making plants less susceptible to temperature-related stress, plant geneticists have focused mainly on improving plants' resistance to cold. Genetically altered bacteria have been used to make strawberries less prone to damage from frost, for example. While protection from the cold would seem destined to become less of an issue in the event of global warming, it may continue to pose a threat to a climate in transition—and it certainly remains a threat today. (California's strawberry crop and citrus groves were devastated by frost as recently as 1990.)

Some fear that altering the genetic make-up of plants could cause unintended side-effects and that there would be no way to put the "genie" back in the bottle once these genes have entered the environment. In the case of the frost-resistant strawberries, for example, protesters uprooted a test patch in Brentwood, Calif., in 1987. Part of their concern was that the genetically altered bacteria used to lower the freezing point of the strawberries also could serve as nuclei for cloud condensation. Conceivably, then, such strawberries could alter weather patterns even though they would be less influenced by the weather themselves.

Some informed observers have other reservations about genetic engineering. One concern is that a plant like rapeseed—when engineered to tolerate a herbicide—might pass the altered gene on to a weedy relative like wild mustard, making the weed more difficult to control with herbicides in agricultural settings. Precautions also must be taken to ensure that plants engineered to produce a toxin against pests do not replicate the same toxin in their fruits. Moreover, genetic alterations in food must be stable and incapable of transferring extra genetic material to humans who consume it.[165]

Because biotechnology offers tremendous promise for the food sector, however, research and development efforts will continue at a brisk pace. Recently, the Food and Drug Administration cleared the way for rapid commercialization of genetically engineered foods by announcing that it will regulate such foods no differently from foods produced by conventional means. Companies will not need government approval to proceed with commercial development of genetically engineered foods unless the changes greatly disrupt the plant's natural state.[166] This regulatory green light means that spoil-resistant tomatoes and stringless celery soon will be on supermarket shelves. But it will be a while longer before genetic engineering emerges an especially useful tool to adapt crops and crop-production methods in a climate-changed world. Genetic engineering never is likely to become a panacea for global warming, moreover. Nature still imposes limits on agriculture that biotechnology cannot overcome.

The Return of Sustainable Agriculture

Despite the success of the Green Revolution in raising crop yields multifold—and the potential for genetic engineering to tack on further gains—there is a countermovement toward so-called low-input, sustainable agriculture as well. "Sustainable" farming methods feature less high technology and more sweat of the brow, and they are gaining favor for two reasons. First, environmental problems persist with modern-day intensive farming practices—insect immunity to pesticides, fertilizer and pesticide runoff, soil erosion, falling water tables and the like. Second, consumers claim to be more health and safety conscious than ever before, preferring to eat foods that are "all natural" and "organically grown" over those that are not.

Food producers are in a quandary over how to exploit this market for natural foods—now a $1-billion-a-year industry. While touting the virtues of organically grown products, they do not want to suggest that foods grown in a conventional manner—with synthetic fertilizers, pesticides and antibiotics—are any less safe to eat. Moreover, organic foods cost 20 percent more to produce on average, and they tend to spoil more quickly because they are not treated with preservatives. This combination of factors has made food producers and consumers less enthusiastic about natural foods than polling data would otherwise suggest. Two major supermarket chains recently withdrew organically grown produce from their shelves, in fact, and some believe the organic market has reached a plateau.[167] While concerns about global warming cannot be expected to turn this situation around single-handedly, they may lend impetus to the drive to implement more organic farming practices.

There is no pat definition of low-input, sustainable agriculture. The general idea is to make soils so naturally rich in nutrients, moisture and pest inhibitors that residual demand for pesticides and synthetic fertilizers is greatly reduced—not to mention the use of energy-instensive equipment to make and apply these chemicals. Greenhouse gas emissions consequently decline. In addition, improved soil conservation practices enable the topsoil layer to stay deeper in the fields, so less carbon is liberated to the atmosphere through erosion. At the same time, the soil's enhanced ability to absorb and hold water reduces irrigation pumping requirements as well as the field's susceptibility to drought. Finally, multi-cropping and integrated pest management techniques commonly associated with sustainable farming not only reduce pesticide spraying but also make it harder for migrating pests to establish themselves on new turf. In the event of climate change, migrating pests could pose a major problem for farmers.

Sustainable agriculture also encourages frequent crop rotations and discourages monocultural farming practices. Where two or more annual crops are grown simultaneously in a single field, researchers have found that such double-cropping can provide mutual benefits. In one irrigated field in western Nebraska, for example, two-row corn windbreaks were spaced every 15 rows throughout a field of sugar beets. The wind shelter provided by the corn increased the sugar yield of the beets by 11 percent, while the greater penetration of sunlight and more rapid replenishment of carbon dioxide to leaves increased the yield of the corn by 150 percent.[168]

The most contentious issue concerning low-input, sustainable farming is whether it is economically viable. In a landmark 1989 study entitled *Alternative Agriculture*, the National Academy of Sciences reported:

> Well-managed alternative farming systems nearly always use less synthetic chemical pesticides, fertilizers, and antibiotics per unit of production than comparable conventional farms. Reduced use of these inputs lowers production costs and lessens agriculture's potential for adverse environmental and health effects without necessarily decreasing—and in some cases increasing—per acre crop yields and the productivity of livestock management systems.[169]

The academy's conclusion that farmers could reduce their production costs *and* maintain yields was a shot across the bow of companies that have made chemical-intensive farming a multi-billion-dollar enterprise. The president of the Fertilizer Institute, Gary Myers, called the report "an insult to American agriculture and the American farmer."[170] Myers and others point out that the report relied heavily on 11 favorable case studies of farms employing sustainable agriculture practices. Such farms could never be duplicated on the massive scale necessary to maintain the food supply at current prices, they maintain. Indeed, even proponents of sustainable agriculture, such as Lester Brown of the Worldwatch Institute, concede that the world's food supply would plummet by 40 percent if farmers suddenly stopped applying synthetic fertilizer on their land.[171]

Most experts agree that it would not be practical to switch to sustainable agricultural methods all at once. The challenge is in selecting the right chemical inputs and being sparing with applications. Especially in the early years of building up the soil, farmers must be prepared to maintain use of selected pesticides and synthetic fertilizers in order to preserve crop yields. Over time, however, sustainable agriculture methods are intended to wean farmers of their reliance on chemicals to fertilize the soils and protect against unwanted pests.

At present, only a tiny fraction of America's 2 million farms employ the full range of low-input, sustainable agricultural methods on any significant scale. To many skeptics, this low percentage demonstrates that farmers have failed to find an economic rationale to alter their management practices. Proponents of sustainable farming, on the other hand, say that many farmers have grown accustomed to following the instructions of seed suppliers and agrichemical manufacturers—treating entire fields on a pre-programmed schedule. With low-input agriculture, farmers must pay especially close attention to the condition of their land: scouting out pests and beneficial insects, implementing biological controls, spot-spraying insecticides and herbicides where necessary, and rotating crops frequently to maintain the vitality of the soil over time.

In the view of Wes Jackson of The Land Institute, sustainable agriculture requires greater management intensity because the approach taken is highly scientific. It also requires more farmhands because labor is being substituted for capital. "You've got to have what I call a 'high-eyes-to-acres ratio,'" Jackson says; "a lot of people watching the landscape."[172] Given this greater labor intensity, the economic question boils down to how one values time spent in the field versus off-farm employment and greater leisure time. Ultimately, for sustainable farming to catch on, the population exodus from farming communities must be reversed. An even broader demographic question is whether sustainable agricultural practices employed on a massive scale would be able to feed a burgeoning world population. And if some countries adopt these sustainable methods while others do not, how would it affect their relative ability to compete and trade in the world market?

While there are no easy answers to these questions, a recent major survey of soil conditions around the world suggests that agricultural practices are in need of reform in developed and developing countries alike. About 10 percent of the world's most productive soils—an area the size of China and India combined—have been seriously damaged by overgrazing, unsuitable farming practices and deforestation since World War II, the survey found.[173] Two-thirds of the seriously eroded land is in Asia and Africa. But even in the United States, which has the most advanced soil conservation practices in the world, one-fourth of the cropland is eroding at a rate faster than is considered sustainable

> **Box 2-G**
>
> ## Federal Policy and Alternative Agriculture
>
> Agricultural policies set at the federal level have an important bearing on the future of sustainable agriculture. Critics of federal policy say that commodity support programs cost U.S. taxpayers billions of dollars a year and lock farmers into rigid planting practices that promote dependence on chemical inputs and lead to topsoil losses. Federal price- and income-support benefits are available to corn, wheat, barley, sorghum, oats, cotton and rice farmers. Those enrolled in such programs must establish crop acreage bases, only a portion of which can be planted as a condition for program participation. The effect is to encourage farmers to plant as much of one crop as possible. If, say, a barley farmer decides that next year he wants to grow oats instead, none of those acres planted with oats will be counted in his base acreage total the year after that. Accordingly, he is compelled to continue planting barley, even if his fields would benefit from a switch to oats.
>
> Economic opportunities can be missed as well. In one recent year, a shortage of soybeans on the world market and a surplus of corn offered an opportunity for American farmers to profit from the imbalance. Because of their enrollment in price-support programs, however, most corn farmers planted corn again, despite depressed world prices. As a result, U.S. taxpayers doled out even more in price supports to American corn growers, while farmers in Argentina and Brazil profited handsomely by growing soybeans and selling them in the world market.[1g]
>
> The 1990 Farm Bill has taken the first step in addressing this problem. Now farmers are allowed to plant any crop (excluding fruits and vegetables) on up to 25 percent of their base acreage, without losing any portion of their commodity program benefits in the following year. The next steps will depend on whether concerns about the adverse environmental effects of chemical-intensive, monoculture agriculture manifest themselves in new government regulations that encourage a switch to sustainable methods. Significantly, the 1990 Farm Bill does support increases in federal funding for an existing low-input, sustainable agriculture research and education program. Congress appropriated $60 million for the program in fiscal year 1991, up from $4.5 million in appropriations in previous years.[2g]

by the U.S. Soil and Conservation Service.[174] Separately, the United Nations Food and Agricultural Organization estimates that 10 million acres of cropland are exhausted each year as a result of unsustainable farming practices, and 7 percent of the land now under cultivation could be out of production by 2000. Representatives of 124 nations attending an April 1991 meeting convened by the FAO agreed to work toward implementation of more sustainable agricultural practices worldwide.[175]

These findings do not suggest that chemical- and energy-intensive agricultural must be abandoned. On the contrary, use of high-yielding crops, more irrigation, fertilizer and pesticides has enabled twice as many people to be fed on the same amount of farmland as in 1960.[176] Considering that land clearing is the single largest source of greenhouse gas emissions within the world's agricultural sector, it makes environmental as well as economic sense for farmers to maximize yields on each acre of crops they grow.

The fundamental issue is developing means of food production that sustain the land as well as the population over the long term. The specter of global warming compounds this challenge. At the same time, it compels the consideration of policies to reduce the agricultural sector's greenhouse gas emissions while making the land base more resilient to potential climate-related stresses. Low-input, sustainable agriculture offers such a coupling of benefits. Yet it represents no more of a final solution to farmers' problems than do advances in genetic engineering. Decisions about what crops to plant and how to raise them will continue to have elements of risk, much as they always have.

Conclusions

Planet Earth is experiencing more population growth in the 1990s than ever before. Each year there are nearly 100 million additional human mouths to feed, and 6.3 billion people are expected to inhabit the planet by the turn of the century. Three billion people—mostly in the developing world—soon will be in their reproductive years. Accordingly, the population bomb will continue to explode.

Feeding 12 Billion

Demographers at the United Nations now estimate that the human race will peak at more than 11 billion people—up a full billion from its 1989 projection.[177] World Bank demographers put the ceiling at 12.5 billion. Some believe that without considerably more support for family planning, the human population will climb to 14 billion before it reaches a climax.[178]

Whatever the increase, population growth imposes a tremendous responsibility on farmers and agribusinesses. World food production must grow 18 percent in the 1990s just to keep per-capita food supplies at today's level. Given the energy- and chemical-intensive nature of modern farming methods, this rise in production will result in a concomitant increase in the farm sector's greenhouse gas emissions.

And that is only half of the picture. The prospects for economic growth compound the overall challenge. Income-driven diet improvements in developing countries could raise the demands on world agricultural resources an additional 20 to 30 percent in the 1990s. Total farm resource demand, in other words, could rise 40 to 50 percent, since more people hope to be eating better.[179]

A rising standard of living has profound implications for industrial emissions of greenhouse gases in developing countries as well. Such emissions increased almost sixteen-fold from 1950 to 1985, but they are expected to multiply at an even faster rate in the decades to come. Even if wealthier nations manage to stabilize their own CO_2 emissions at prevailing levels, the combination of population growth, improved diets and industrial expansion in developing countries could result in a *tripling* of worldwide CO_2 emissions by 2025.[180] (And if the livestock population grows apace, methane emissions from ruminants also could increase by two-thirds.)

As more and more greenhouse gases accumulate in the atmosphere, the likelihood increases that the global climate will change inexorably. While the Green Revolution has dampened greenhouse gas emissions associated with land clearing for agriculture, it has added its own set of emissions into the global warming equation through greater use of chemical- and energy-intensive inputs. Now more than ever, the food growing sector must have cooperation from the elements. Where groundwater tables are falling, the heavens must compensate for the loss of irrigation to sustain high-yielding, water-intensive crops. Where soil has eroded from the land, fields must lie fallow so that nature can begin a slow restoration process. Where pesticides have failed to protect crops (crop losses to insects have nearly doubled since 1945, despite a 1,000 percent increase in spraying), natural enemies and cold weather must keep the insects from spreading.[181] Unfortunately, the prospect of global warming imperils this cooperation. Thus, while farmers have always regarded their profession as a bit of a crap-shoot, it now may become even more of a gamble.

Of course, the greenhouse effect may not be all bad for farmers. A warmer climate would make plants (as well as bugs) less susceptible to killing frost, for example. In combination with carbon dioxide enrichment, warmer temperatures should enable plants to grow in more places—perhaps faster and better than before. The CO_2-enriched air—which is a certainty despite other doubts about climate change—may well serve as free fertilizer for most plants. This fertilization effect may enable most plants not only to grow larger and faster, but also to become more efficient users of water.

One economic study forecasting crops' response to a hypothetical change in climate found, in fact, that the nation's annual food bill might be pared by more than $9 billion—and farmers' profits might rise by more than $1 billion—because the beneficial effects of CO_2 enrichment would outweigh the adverse effects of climate change.[182] But this represents only one scenario for the future. The same economic study—using a different climate model—found that an annual income *loss* of more than $10 billion is possible if the assumed effects of global warming are more severe. Should the growing-season climate in the American heartland become hotter *and* drier, in other words, CO_2 enrichment might not make up for the heat and lack of rain.

Moreover, there are indications that plants growing outside in a CO_2-enriched atmosphere may not respond as well as test plants in controlled, growth-chamber studies.[183] And some crops like corn are not expected to benefit much from CO_2 enrichment in any case because of their so-called "C4" pathway. This limitation is unfortunate because corn accounts for two-thirds of the U.S.'s grain harvest and one-eighth of the world's. In the world commodities market, corn accounts for about three-quarters of all traded grain, and it is the primary grain used to make up food deficits in famine-prone regions.[184] Accordingly, farmers and grains merchants cannot expect CO_2 enrichment to serve as a built-in antidote to global warming.

Global Warming and Food Security

Perhaps the toughest ethical question raised by global warming is how it might affect the trade and distribution of food around the world. The United States and Canada are in the fortunate position of possessing roughly one-sixth of the world's arable land yet accounting for only one-twentieth of the world's population. Analyses performed by EPA for the U.S. Congress have come to the comforting conclusion that any domestic loss of agricultural productivity could be mitigated by exporting fewer grains abroad—as much as 70 percent fewer.[185]

Table 4

Major Net Cereal Exporters 1988 (million tons)		Major Net Cereal Importers 1988 (million tons)	
USA	98	USSR	34
France	27	Japan	28
Canada	23	China	17
Australia	15	Korea, Rep.	9
Argentina	10	Egypt	8
Thailand	6	Mexico	6
Denmark	2	Iran	5
United Kingdom	1	Italy	5
South Africa	1	Iraq	4
New Zealand	—	Saudi Arabia	3

SOURCE: U.N. Food and Agricultural Organization, 1988.

Yet the number of countries that create the world grain surplus is relatively small. Only 21 out of 172 countries, in fact, normally are net cereal exporters. The United States, Canada and the European Community alone account for three-quarters of the world's traded cereals.[186]

In a good crop year, world food production exceeds demand by about 20 percent. But when the weather turns bad in one or more of the world's grain belts—the U.S. Great Plains and Canadian prairies, the North European lowlands, the Ukraine and adjacent regions, the Australian wheat belt or the Argentine pampas—excess supplies in the world food market dissipate quickly. A 20 percent or greater decline in world crop production would wipe out the normal margin of supply over demand. Without sufficient carryover stocks, such a loss would lead to a potentially catastrophic interruption in world food trade.

In the "greenhouse summer" of 1988, the drought-ravaged U.S. grain harvest fell below domestic consumption levels for perhaps the first time in history.[187] Combined with reduced harvests in Canada and elsewhere, the world's grain reserves plunged to their lowest point in more than 20 years. The United States was fortunate, however, that a bumper crop the previous year had filled storage silos to the brim. After the 1988 drought, Lester Brown of the Worldwatch Institute warned, "The shock to the world economy of a severely reduced North American harvest in 1989 could dwarf the economic effects of the oil price hikes of the seventies. Even with a good harvest next year, global warming could easily recreate the same risky scenario time and again....Food security could replace military security as the dominant issue of the nineties and beyond."[188]

Some accuse Brown of crying wolf on looming food shortages.[189] Indeed, relatively good crop-growing weather in 1989 and 1990 replenished the world's grain reserves. But then regional drought, military conflicts and other supply disruptions caused world agricultural production to drop 1.4 percent in 1991— reportedly the largest one-year drop in the world grain harvest ever.[190]

If world crop yields eventually were to decline 10 percent as a result of adverse climate change and other factors, average world food prices would rise 7 percent (assuming constant levels of per-capita consumption), figures Pierre Crosson, a senior fellow with Resources for the Future. While this hardly constitutes a shock to the American pocketbook, it "could seriously influence the ability of food-deficit countries to pay for food imports, eroding the amount of foreign currency available for promoting development of their non-agricultural sectors," Crosson warns.[191] For those with grain available to sell, an ethical (as well as economic) question arises: Should the surplus grain be sold in the world market—driving up domestic prices—or should it be reserved for the domestic market—possibly at the expense of tens of millions of additional people going hungry abroad?

Grain sales traditionally have been a substantial source of export income for the United States. Between 1970 and 1981, the value of U.S. agricultural commodities sold abroad mushroomed from $7.3 billion to a record $43.3 billion, with corn, cotton, rice, soybeans and wheat exports leading the way. In 1981, the United States alone controlled 39 percent of total world agricultural trade and more than 70 percent of world trade in coarse grains, greater than 10 times the share of its nearest competitor—Argentina.[192]

During the 1980s, however, other nations (such as Canada, Australia and the European Community) beefed up their own farm sectors, and U.S. agricultural exports since 1987 have dipped below $30 billion. Nevertheless, the United States has remained competitive in the world market by continuing to overproduce crops relative to domestic demand (albeit with larger "set-aside" programs to control the surplus), enabling bulk commodities to be sold abroad at low market prices. This strategy provides the nation with a much needed source of exports and jobs, while assuring the world of a steady supply of affordably priced grains. Yet global climate change could put this basic agricultural strategy in doubt.

General circulation models suggest that the productive capacity of American farmland could decline (especially in the Great Plains) at the same time that high-latitude regions such as Scandinavia, northern Europe, the Ukraine, Russia, Canada and Japan benefit from longer crop-growing seasons. Such climatological shifts would create new winners and losers in the production and distribution of food, especially if policies to promote domestic agriculture are maintained in these nations.

Under these circumstances, the United States could find itself in an awkward position of recommending that those nations favored by climate change take cropland out of production while minimizing its own set-asides in an effort to make up for the domestic production shortfall.[193] Accordingly, "If protectionist sentiment rises, it most likely would be among those mid-latitude countries facing the loss of comparative advantage," Crosson believes. "Agricultural interests in those countries would pressure the governments to raise trade barriers against imports from countries favored by climate change."[194]

"There is an irony here," Crosson adds. "Under the conditions depicted, mid-latitude countries losing comparative advantage would be able to avoid rising economic and environmental costs of agricultural production only if they

Figure 13

U.S. Agricultural Export Trends
1970-87

(Chart showing Agricultural exports, Agricultural trade balance, Total trade balance, and Nonagricultural trade balance in billions of dollars from 1970 to 1987.)

SOURCE: U.S. Department of Agriculture.

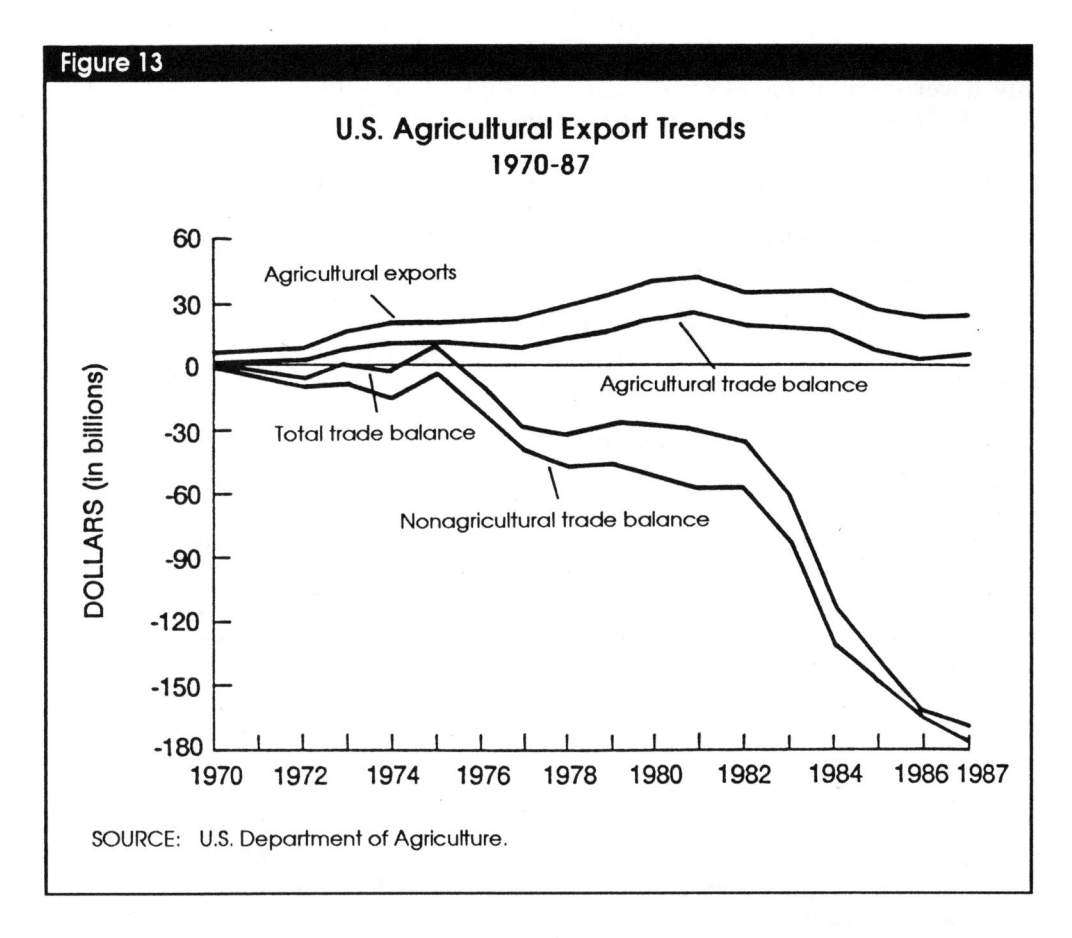

could import freely from emerging lower cost producers. But it is precisely in these disadvantaged countries that agricultural protectionism would likely strengthen." As the world's largest agricultural producer and exporter at present, the United States is the nation most vulnerable to a reduction in agricultural capacity that would compromise its geopolitical position in the world. Yet in order to maintain low food prices at home and a free-trade policy abroad, the nation would have to resist a protectionist response in the event of global warming. Such a policy, in combination with adverse climatological trends, would not bode well for domestic agricultural interests.

Farming in a Climate of Uncertainty

No one really knows, of course, what the climate may have in store. Global warming would appear more likely to alter the geography of agricultural activities around the globe than it would to change net production. But even this is sheer conjecture, dependent on the severity of climate change, the response of plants to CO_2 enrichment and the accommodations made by farmers and their governments. Scientist Martin Parry—after coordinating an exhaustive international assessment and submitting a report to the Intergovernmental Panel on Climate Change—had this to say about global warming's potential effects on agriculture:

> To date, less than a dozen detailed regional studies have been completed that serve to assess the potential impact of climatic changes on agriculture. It should be a cause for concern that we do not, at present, know whether changes of climate are likely to increase the overall productive potential for global agriculture, or to decrease it. There is therefore presently no adequate basis for *predicting* likely effects on food production at the regional or world scale. All that is possible at present is informed speculation. The risks that stem from such levels of ignorance are great.[195]

Indeed, as of 1990 only six comprehensive regional case studies had been completed—for the United States, the Soviet Union, Japan, Finland, Iceland and Saskatchewan. Substantive agricultural impact assessments have since been completed for developing nations in Asia and Africa—where food security issues assume life-and-death proportions—but the studies are still awaiting peer review.[196]

Even those studies that have been completed are circumspect because of questionable assumptions built into their analyses. Many of them take the short cut of doubling the carbon dioxide content of the atmosphere all at once, rather than letting it build up gradually, as is happening in the real world. As a result, these studies may miss some of the important transient effects of global warming. Moreover, the climate models do not forecast any changes in the frequency of extreme events that may have a tremendous bearing on farmers' fortunes. Farmers know that an increase in the frequency of drought, heat waves, torrential rains, hail, pest infestations and the like would portend much more severe losses than is suggested by a one-time rise in temperature and a one-time change in precipitation. By leaving these factors out, the models may provide an unduly optimistic portrayal of the effects of global warming.

On the other hand, most studies to date have not considered how farmers would respond to climate change by altering cultivation methods and introducing better-suited crops—maybe even genetically engineered ones. In this respect, current analyses may be overstating the negative societal effects of climate change on agriculture. "One cannot and should not attempt to predict what a climate change, either abrupt or gradual, will do to an agriculture system 50 or 75 years from now by superimposing these changed climatic conditions upon agriculture as it is today," cautions Norman J. Rosenberg, an agricultural specialist with Resources for the Future. "This, however, is what analyses, for lack of the appropriate techniques, have tended to do."[197]

Given these uncertainties, few signs point to agricultural interests factoring climate change into their strategic plans—even though the stakes in charting a proper future course for agriculture may be higher than ever. At a 1989 conference in London concerning global warming and agriculture, for example, a number of speakers remarked that food-related industry was neither capable of nor interested in funding the sort of basic research that now seems increasingly urgent. (The conference was sponsored by the United Kingdom

Figure 14

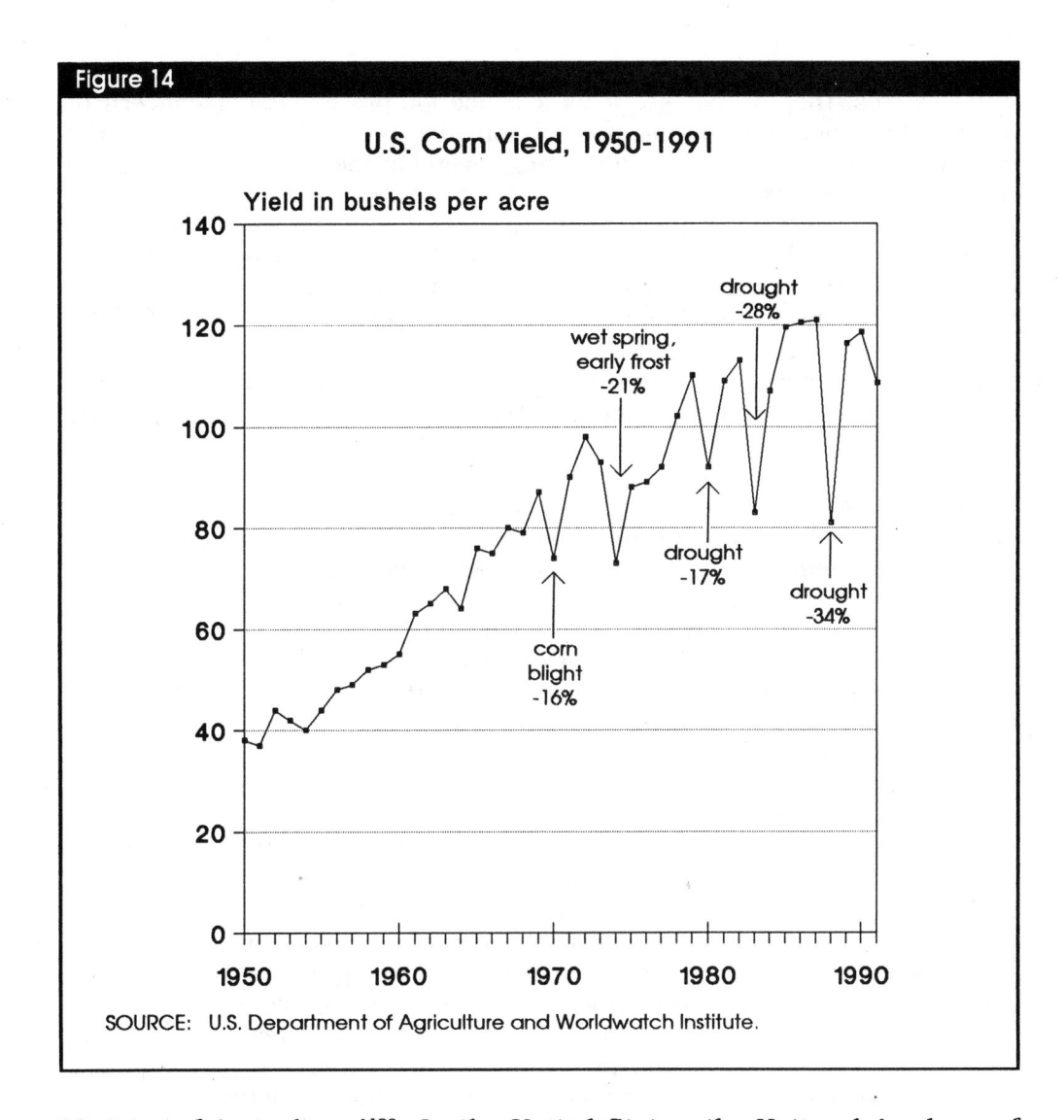

U.S. Corn Yield, 1950-1991

Yield in bushels per acre

- corn blight -16%
- wet spring, early frost -21%
- drought -17%
- drought -28%
- drought -34%

SOURCE: U.S. Department of Agriculture and Worldwatch Institute.

Ministry of Agriculture.)[198] In the United States, the National Academy of Sciences has also lamented a shortage of scientists pursuing careers in interdisciplinary or agricultural systems research—an approach that is vital to understanding the complex and convoluted potential effects of global warming on agriculture.[199]

A similar predicament reigns among water resource planners. In a survey by the Natural Hazards Research and Applications Center at the University of Colorado at Boulder, many water planners professed their belief that global warming would occur in the near future, but few thought they should be adapting their management tactics now or planning ahead for that possibility.[200] A survey by the American Association for the Advancement of Science came to the same conclusions. According to the authors of the association's 1988 report on climate change and water, "The most significant finding...was:

although managers of water supply acknowledge that climate change can harm their systems, they do not see it as a cause for present major concern or action....If responses to a problem are classified as recognition, avoidance and adaptation, the response of climate change is still in the recognition phase for managers of urban water."[201]

If the worst fears about global warming are realized, a greater portion of the world's burgeoning population seems destined to live in hunger and die of famine in the years and decades ahead. Ultimately, the world's inhabitants may have only three options: Reduce population growth, eat lower off the food chain, or pursue both of these options at once.

But most do not believe the situation has reached a point of desperation. Farmers have defied the odds before and they may well again. Farmers already are benefiting from the development of drought-tolerant hybrid crops bred by seed suppliers. In the future, they may successfully harvest triticale, salt-resistant halophytes and native perennial cereals that require fewer inputs yet may rival the yields of some traditional cash crops. Those who irrigate their fields also may realize savings by instituting water conservation measures and selling the surplus to other users.

A new revolution in agriculture also is possible. Cellulose-derived syrups could become important commodities—harvested year-round and shipped through pipelines to food production sites. Genetic engineering offers an even wider array of possibilities to develop crops that resist drought and heat stress and are inherently resistant to pests. Yet while many are counting on biotechnology to shape a better future for agriculture, unfortunately it will not provide a cure-all for every farmer's problems.

If climate change exacerbates the risks associated with conventional, energy-intensive agricultural methods, ultimately there may be a switch to farming practices that are more sustainable over the long term. Such a move would serve two important objectives: reducing the farm sector's emissions of greenhouse gases *and* increasing the land's ability to cope with climate change. Some evidence suggests that farmers practicing sustainable agriculture are able to curb their use of chemicals, energy and water—and thereby lower their greenhouse gas emissions and production costs—without suffering appreciable losses in yield. But there is no assurance that these practices, instituted on a massive scale, are economic and can sustain a growing world population.

In the final analysis, the basic greenhouse gambit facing farmers is the same one that confronts decisionmakers in the other industries examined in this book: Whether to pursue a strategy that emphasizes long-term ecological priorities over short-term production and profit maximization goals. As for the farmers' greenhouse gambit, 11, 12, perhaps 14 billion people will be vitally interested in how they move.

Notes

1. Richard D. Hylton, "Wall Street's Latest Diversification Strategy: Down on the Farm," *The New York Times*, Sept. 23, 1990.
2. U.S. Department of Commerce, *Survey of Current Business*, Washington, D.C., 1989.
3. Cynthia Rosenzweig, "How It Might Be: Agriculture," *EPA Journal*, January/February 1989. For the EPA study, see note 4.
4. U.S. Environmental Protection Agency, *The Potential Effects of Global Climate Change on the United States*, report to Congress, October 1988.
5. William Booth, "Global Warming Could Benefit U.S. Farmers," *The Washington Post*, May 17, 1990.
6. Lester R. Brown, *The Changing World Food Prospect: The Nineties and Beyond*, Worldwatch Paper #85, Worldwatch Institute, Washington, D.C., October 1988.
7. Jon R. Luoma, "Prophet of the Prairie," *Audubon*, November 1989. The energy figure cited in this article—13 barrels of oil—is too high. Average daily consumption of 2,400 calories of food per person requires the energy equivalent of 93,327 British thermal units to grow and process the food. On a yearly basis, the energy consumed per capita in U.S. food production equals 5.87 barrels of oil, assuming each barrel contains 5.8 million Btus of energy.
8. Pierre Crosson, Resources for the Future, "Climate Change and Mid-Latitude Agriculture," *Climatic Change*, October 1989.
9. Leon Hartwell Allen Jr. et al., U.S. Department of Agriculture, "Likely Effects of Climate Change Scenarios on Agriculture of the USA," in *Coping With Climate Change: Proceedings of the Second Annual North American Conference on Preparing for Climate Change*, The Climate Institute, Washington, D.C., June 1989.
10. *Ibid.* These researchers maintain that the seed yield of soy beans has increased 13 percent since 1800 as a result of carbon dioxide enrichment. They estimate that a doubling in ambient CO_2 levels to 630 parts per million by 2050 could increase the seed yield of soy beans by an additional 32 percent. Others are skeptical of these claims, however.
11. Fred Pearce, "High and Dry in the Global Greenhouse," *New Scientist*, Nov. 10, 1990. Some crops are more water-dependent than others, of course. Soy beans and green beans would perish more quickly than a drought-tolerant crop such as sorghum, for example.
12. S. Manabe and R.T. Wetherald, "Reduction in Summer Soil Wetness Induced by an Increase in Atmospheric Carbon Dioxide," *Science*, Vol. 232:626-628, 1986.
13. See note 11.
14. Martin Parry, *Climate Change and World Agriculture*, Earthscan Publications Ltd., London, England, 1990. This book summarizes the results of the study by the International Institute for Applied Systems Analysis.
15. *Ibid.*, and B. Smit, L. Ludlow, and M. Brklacich, "Implications of a Global Climatic Warming for Agriculture: A Review and Appraisal," *Journal of Environmental Quality*, October-December 1988.
16. Cynthia Rosenzweig and Martin Parry, "Implications of Climate Change for International Agriculture: Global Food Trade and Vulnerable Regions," statement prepared for the U.S. Senate Committee on Commerce, Science and Transportation, April 9, 1992.

17. See note 14.
18. Jodi L. Jacobson, "Holding Back the Sea," *State of the World: 1990*, Worldwatch Institute, W.W. Norton & Co., New York, 1990.
19. Werner Fornos, "Population Politics," *Technology Review*, February/March 1991.
20. Pierre Crosson, Resources for the Future, personal communication, March 10, 1992. Crosson's estimate considers the likelihood that most people in developing nations would spend a significant portion of any extra income they receive over the period on greater food consumption.
21. *Ibid.*
22. National Research Council, *Alternative Agriculture*, National Academy Press, Washington, D.C., 1989.
23. *Ibid.*
24. *Ibid.*
25. See note 8.
26. Lester R. Brown, "Feeding Six Billion," *World Watch*, September/October 1989.
27. U.S. Department of Agriculture, *World Grain Situation and Outlook*, Washington, D.C., December 1991; U.S. Department of Agriculture, *World Population by Country and Region*, Washington, D.C., December 1990; and Peter Weber, Worldwatch Institute, personal communication, March 7, 1992.
28. See note 16.
29. See note 8.
30. Intergovernmental Panel on Climate Change, *1992 IPCC Supplement*, World Meteorological Organization and United Nations Environment Programme, Geneva, Switzerland, February 1992.
31. Lester R. Brown, "Fall in 1991 World Grain Harvest Sets Record," Worldwatch Institute press release, Washington, D.C., Dec. 18, 1991; citing statistics of the U.N. Food and Agricultural Organization, the Fertilizer Institute and the International Fertilizer Industry Association.
32. As reported in the article by Pierre R. Crosson and Norman J. Rosenberg, "Strategies for Agriculture," *Scientific American*, September 1989.
33. See note 22.
34. *Ibid.* Since 1982, idled cropland and the introduction of new compounds that are applied more sparingly reduced the application of pesticides to 430 million pounds by 1987.
35. See note 22.
36. U.S. Environmental Protection Agency, *Pesticide Industry Sales and Usage: 1989 Market Estimate*, Office of Pesticides and Toxic Substances, Washington, D.C., July 1991.
37. See note 22. Domestic fertilizer sales peaked at $8.6 billion in 1981, which represented a fourfold increase in sales since 1970.
38. *Ibid.*
39. U.S. Congress, Office of Technology Assessment, *Changing by Degrees: Steps to Reduce Greenhouse Gases*, OTA-O-482, U.S. Government Printing Office, Washington, D.C., February 1991.
40. *Ibid.*
41. *Ibid.*
42. See note 16, and Emily Plishner and David Hunter, "Fertilizers: New World Order Resows the Market," *Chemical Week*, Sept. 25, 1991; and Lester R. Brown, "Fertilizer Engine Losing Steam," *World Watch*, September/October 1991. Future growth in fertilizer use is more likely to occur in developing than developed countries. According to the Worldwatch Institute, American farmers are applying less synthetic fertilizer than they did a decade ago, and Japanese and Western

European farmers are applying no more fertilizer than they did a decade ago. Even in agriculturally advanced Third World countries, crops are no longer responding much to added doses of fertilizer, Brown reports.

43. See note 39. Although termites are counted as a "natural" source of methane emissions, termite populations have increased because their food supply—dead wood—has grown tremendously as a result of deforestation.
44. U.S. Environmental Protection Agency, Office of Air and Radiation, Reducing Methane Emissions from *Livestock: Opportunities and Issues*, EPA 400/1-89/002, Washington, D.C., August 1989.
45. See notes 9 and 14.
46. *Ibid.*
47. See note 14.
48. Fakhri A. Bazzaz and Eric D. Fajer, "Plant Life in a CO_2-Rich World," *Scientific American*, January 1992.
49. B.J. Spalding, "A Silver Lining for the Greenhouse?" *Chemical Week*, Aug. 3, 1988.
50. See note 14.
51. *Ibid.* For historical examples of drops in herbicide and fertilizer sales because of drought, see the articles by Shelina Shariff: "Monsanto's Third Quarter Was Dry and Lean, *Chemical Week*, Nov. 7, 1990; "Fertilizer Sales Dry Up in California Drought," *Chemical Week*, April 3, 1991.
52. See note 48.
53. *Ibid.*
54. See note 30.
55. See note 48, and William K. Stevens, "Carbon Dioxide Rise May Alter Plant Life, Researchers Say," *The New York Times*, Sept. 18, 1990.
56. See note 4.
57. See note 3.
58. According to the U.S. Department of Agriculture, the 12 major agricultural states in the Midwest are the Corn Belt states of Illinois, Indiana, Iowa, Missouri and Ohio; the Central and Northern Plains states of Kansas, Nebraska, North Dakota and South Dakota; and the Lake States of Michigan, Minnesota and Wisconsin.
59. See note 4.
60. *Ibid.*
61. *Ibid.*
62. See note 20. Reference is made to the "Mink" study by Resources for the Future.
63. See note 3.
64. See, for example, Dennis Farney, "On the Great Plains, Life Becomes a Fight for Water and Survival," *The Wall Street Journal*, Aug. 18, 1989; Isabel Wilkerson, "With Rural Towns Vanishing, States Choose Which to Save," *The New York Times*, Jan. 3, 1990; and Arthur S. Brisbane, "A Farm Family Reflects North Dakota's Sense That the Best Is Over," *The Washington Post*, Jan. 7, 1990.
65. See note 4.
66. See note 14.
67. William Robbins, "Winter Wheat Farmers Fear Second Year's Crop Failure," *The New York Times*, Jan. 2, 1990.
68. Scott Kilman, "Weather Extremes Hurt Soft Red Winter Wheat With U.S. Stockpiles Projected to Become Tight," *The Wall Street Journal*, Jan. 17, 1992.
69. John Urquhart, "Canada Is Expected to Have Huge Crop of Wheat This Year," *The Wall Street Journal*, July 1, 1991.
70. See note 14.
71. *Ibid.*
72. *Ibid.*

73. *Ibid.*
74. *Ibid.*, and Barry Smit, University of Guelph, "Likely Impact of Climate Change on Canadian Agriculture," in *Coping With Climate Change: Proceedings of the Second Annual North American Conference on Preparing for Climate Change*, The Climate Institute, Washington, D.C., June 1989.
75. Richard M. Adams et al., "Global Climate Change and U.S. Agriculture," *Nature*, May 17, 1990. All information provided in this section is drawn from this article.
76. *Ibid.*
77. *Ibid.*
78. M.L. Parry and T.R. Carter, "An Assessment of Effects of Climate Change on Agriculture," *Climatic Change*, October 1989.
79. William K. Stevens, "Summertime Harm To Shield Of Ozone Detected Over U.S.," *The New York Times*, Oct. 23, 1991.
80. Janet Raloff, "Not All Plants Thrive in a 'Greenhouse,'" *Science News*, Aug. 26, 1989; and Ian Anderson, "Crops Threatened by Increases in Ultraviolet," *New Scientist*, Oct. 6, 1990.
81. Justin R. Ward, Richard A. Hardt and Thomas E. Kuhule, *Farming in the Global Greenhouse*, Natural Resources Defense Council, March 1989.
82. See note 16.
83. See notes 14 and 75.
84. For an interesting historical perspective on agriculture's impacts on the natural environment, see the article by Clive Ponting, "Historical Perspectives on Sustainable Development," *Environment*, November 1990.
85. Alan B. Durning, "Fat of the Land," *World Watch*, May/June 1991.
86. See note 22.
87. Kenneth D. Frederick, Resources for the Future, and Peter H. Gleick, University of California, Berkeley, "Water Resources and Climate Change," presented at the Workshop on Controlling and Adapting to Greenhouse Warming, sponsored by Resources for the Future, June 14-15, 1988; and "Taps Run Wild, As Water Use Doubles," *Environment Week*, July 5, 1990.
88. *Ibid.*
89. See note 22.
90. Jack Lewis, "The Ogallala Aquifer: An Underground Sea," *EPA Journal*, November/December 1990.
91. See note 32.
92. Paul E. Waggonner and Roger R. Revelle, "Climate and Water: AAAS Panel on Climatic Variability, Climate Change and the Planning and Management of U.S. Water Resources," American Association for the Advancement of Science, Washington, D.C., Sept. 27, 1988. This is a simplified explanation of an extremely complicated hydrological process. The amount of dry matter a crop produces is roughly proportional to the amount of water it transpires. The amount of water it transpires, in turn, depends on how much rain falls, how much is retained by the soils, how much is lost through evaporation from the soil surface and how much remains in the soil that the crop cannot extract. Air humidity also is a factor, as less dry matter usually is produced in less humid areas.
93. *Ibid.* One analysis coupled a hypothetical 5.5 degree F temperature increase with no net change in precipitation. Despite a doubling of ambient carbon dioxide, plant evapotranspiration still rose relative to present-day conditions because of higher temperatures and increased solar radiation. This suggests that plants' added thirst for water must be quenched by more irrigation, since rainfall amounts (in this scenario at least) did not increase.
94. See note 72.

95. See notes 4 and 75.
96. John Schefter, U.S. Geological Survey, presentation on water quality issues at the Second North American Conference on Preparing for Climate Change, sponsored by the Climate Institute, Dec. 6, 1988.
97. Marc Reisner, "The Emerald Desert," *Greenpeace*, July/August 1989. Reisner also is the author of a 1986 book on California water policy, entitled *Cadillac Desert*, published by Viking Press.
98. Lisa Lapin, "Subterranean Ocean Could Be Drying Up," *San Jose Mercury News*, April 7, 1991.
99. K. Redmond, "Summary of the 1991-1992 Rainy Season in the Far West," *Weekly Climate Bulletin*, National Weather Service, Washington, D.C., April 11, 1992.
100. Roger R. Revelle and Paul E. Waggonner, "Effects of Carbon Dioxide-Induced Climatic Change on Water Supplies in the Western United States," in *Changing Climate: Report of the Carbon Dioxide Assessment Panel of the National Academy of Sciences*, Washington, D.C., 1983.
101. Jane Gross, "A Dying Fish May Force California to Break Its Water Habits," *The New York Times*, Oct. 27, 1992.
102. Marc Reisner, "Can Anyone Win This Water War?" *National Wildlife*, October 1990.
103. Charles McCoy, "U.S. to Tighten Farmers' Water Supply in California in Effort to Save Salmon," *The Wall Street Journal*, Feb. 14, 1992. Quoted person is Steve Hall, executive director of the Farm Water Coalition.
104. See note 4 and first citation of note 87.
105. See note 11.
106. See note 98.
107. For the quote, see the article by Robert Reinhold, "Drought Taking Its Toll on California Environment," *The New York Times*, Aug. 4, 1991.
108. For the quote, see the article by Charles N. Herrick, "Science and Climate Policy: A History Lesson," *Issues in Science and Technology*, Winter 1991-92.
109. *Ibid.*
110. Barbara G. Brown, "Climate Variability and the Colorado River Compact: Implications for Responding to Climate Change," *Societal Responses to Regional Climate Change: Forecasting by Analogy*, Michael H. Glantz, editor, Westview Press, Boulder, Colo., 1988.
111. See note 100.
112. See first citation of note 87, and Tim Beardsley, "Parched Policy," *Scientific American*, May 1991.
113. Charles McCoy, "Big Farmers in West Get Subsidized Water Despite Drought Crisis," *The Wall Street Journal*, May 31, 1991.
114. Marc Reisner, "The Big Thirst," *The New York Times Sunday Magazine*, Oct. 28, 1990.
115. *Ibid.*
116. Jane Kay, "Thirsty Cities Look at Desalination," *San Francisco Examiner*, Feb. 3, 1991.
117. Nancy Rivera Brooks, "Farmers Finding Hard Row to Hoe," *The Los Angeles Times*, July 12, 1991. California's farm revenue estimates are by the Bank of America.
118. See note 114.
119. Robert Reinhold, "Drought Brings Quiet to Farms at Normally Busy Time," *The New York Times*, Feb. 6, 1991.
120. See note 113.
121. See note 117.
122. Peter Steinhart, "The Water Profiteers," *Audubon*, April 1990.

123. John Lancaster, "Buying Peace in Western Water War," *The Washington Post*, June 10, 1990.
124. Terry L. Anderson, "Water Enough for Farmer and Fish," *The Wall Street Journal*, May 15, 1991.
125. See note 122.
126. Marj Charlier, "Investors Seek to Develop Colorado Water But Age-Old Animosities Imperil Venture," *The Wall Street Journal*, April 16, 1991.
127. See note 122.
128. Lisa Lapin, "Wilson Calls for Conservation," *San Jose Mercury News*, Feb. 16, 1991.
129. Charles McCoy, "Congress Tilts to West's Cities, Wildlife, Away From Farms in Water-Policy Bills," *The Wall Street Journal*, March 17, 1992.
130. Lou Cannon and Jay Mathews, "California Welcomes Storms With Open Arms," *The Washington Post*, March 22, 1991; and Jane Gross, "California's Rice Growers Become Enemy in Drought," *The New York Times*, April 7, 1991.
131. "Conservation Technique Draws Too Much Water," *The Washington Post*, Oct. 12, 1991.
132. Kathleen M. Berry, "What If This Is a 50-Year Drought?" *The New York Times*, April 7, 1991.
133. See note 113, citing figures from the U.S. General Accounting Office.
134. See note 4.
135. See note 114.
136. See note 22.
137. See note 4.
138. See note 90.
139. See note 87. According to Marc Reisner, some farmers around San Diego reportedly pay up to $300 an acre-foot for water.
140. Daniel J. Dudek, "The Ecology of Agriculture, Environment and Economics," Environmental Defense Fund, New York, N.Y., September 1987.
141. Leslie Berkman, "Computers Tuned to Weather—and Shut-Off Valves for Sprinkler Systems," *The Los Angeles Times*, July 14, 1991.
142. See note 90.
143. "A Grain to Weather Climate Change," *Science News*, July 15, 1989.
144. "Advances Raise Hope For Crops That Grow In Salty Conditions," *The New York Times*, June 19, 1990.
145. Edward P. Glenn et al., "Growing Halophytes to Remove Carbon from the Atmosphere," *Environment*, April 1992.
146. The information on perennial grasses is drawn from the articles by Jon R. Luoma, "Prophet of the Prairie," *Audubon*, November 1989; and Chris Seitz, "Breaking New Ground," *Environmental Action*, November/December 1990.
147. Stephen L. Rawlins, "Strategies for Adapting Agriculture to Adapt to Climate Change," in proceedings of the First North American Conference on Preparing for Climate Change, sponsored by the Climate Institute, October 1987.
148. Susan Watts, "Have We the Stomach for Engineered Food?" *New Scientist*, Nov. 3, 1990; and Rebecca J. Goldburg, "Why the U.S. Should Regulate Gene-Altered Foods," *EDF Letter*, November 1991.
149. Ken Yamada, "Toward Leaner Meat and Celery Sticks Without Strings," *The Wall Street Journal*, Feb. 24, 1991.
150. *Ibid.*
151. Bob Davis and Rose Gutfeld, "Bush to Urge Rules to Boost Biotechnology," *The Wall Street Journal*, Feb. 19, 1992.

Agriculture

152. "Agriculture's Future: Seeds for Thought," *The Scientist*, March 27, 1989. In the article, John M. Hardinger, Du Pont's director of biotechnology, raises this possibility.
153. Monsanto promotional material appearing in *The Scientist*, Nov. 27, 1989.
154. Abigail Grissom, "Bioengineering Increases Yield and Research Opportunities," *The Scientist*, Jan. 21, 1991.
155. Gregory DiMorris, "Ciba-Geigy Enters the $1.5-billion/year Corn Biotech Race," *Chemical Week*, Sept. 12, 1990; and Jerry E. Bishop, "Progress Disclosed in Genetically Modifying Corn," *The Wall Street Journal*, Jan. 24, 1990.
156. Charles S. Gasser and Robert T. Fraley, "Transgenic Crops," *Scientific American*, June 1992.
157. Jane Rissler, "Biotechnology and Pest Control: Quick Fix vs. Sustainable Control," *Global Pesticide Campaigner*, January 1991.
158. Keith Schneider, "Building a Better Tomato: New Era in Biotechnology," *The New York Times*, Sept. 18, 1989.
159. *Ibid.*
160. Amal Kumar Naj, "Mycogen Scientists Find New Strains of Soil Bacterium," *The Wall Street Journal*, March 6, 1992.
161. See note 154.
162. Scott Kilman and Elyse Tanouye, "U.S. Crop Estimates Send Prices of Soybeans Plunging and Orange Juice Contracts Soaring," *The Wall Street Journal*, Oct. 14, 1991.
163. See note 81.
164. *Ibid.*
165. See note 148.
166. Malcolm Gladwell, "Biotech Food Products Won't Require Special Rules, FDA Decides," *The Washington Post*, May 26, 1992.
167. Thomas Hebert, Senate Agriculture Committee, personal communication, April 7, 1992; and Scott Kilman, "Major Companies in the Food Industry Have Little Taste for Organic Products," *The Wall Street Journal*, Jan. 10, 1992.
168. See note 32.
169. See note 22.
170. Arthur S. Brisbane, "Panel Backs 'Benign' Farming to Save Soil, Cut Chemicals," *The Washington Post*, Sept. 8, 1989.
171. See note 26.
172. See second citation of note 146.
173. *World Resources: 1992-93*, World Resources Institute, Oxford University Press, New York, N.Y., 1992. The survey was conducted by 250 soil scientists over a three-year period under the auspices of the United Nations Environment Program.
174. *Ibid.*, and Larry B. Stammer, "The Good Earth Is Being Damaged At Alarming Pace," *Los Angeles Times*, March 25, 1992.
175. Bette Hileman, "UN Group Backs Sustainable Farming Methods," *Chemical & Engineering News*, May 20, 1991.
176. Dennis T. Avery, "Mother Earth Can Feed Billions More," *The Wall Street Journal*, Sept. 19, 1991. Avery is the principal author of *Global Food Progress 1991*, published by the Hudson Institute, Indianapolis, Ind.
177. *Long-range Population Projections: Two Centuries of Population Growth, 1950-2150*, United Nations Population Fund, January 1992.
178. See note 19.
179. See note 176.
180. Susan Okie, "Developing World's Role in Global Warming Grows," *The Washington Post*, May 15, 1990. Projections are by the United Nations Population Fund.

181. See note 22.
182. See note 75.
183. See note 48.
184. See note 14.
185. See note 4.
186. See note 14.
187. Lester R. Brown and John E. Young, "Growing Food in a Warmer World," *World Watch*, November/December 1988.
188. *Ibid.*
189. See note 176 and James A. Miller, letter to the editor, "Worldwatch Wolf Keeps Crying," *The Wall Street Journal*, Jan. 21, 1992. Miller is the director of research for the Population Research Institute in Gaithersburg, Md.
190. Bruce Ingersoll, "Severe Famine in East Africa and Iraq, Food Shortages Elsewhere Seen in 1992," *The Wall Street Journal*, Dec. 19, 1991.
191. See note 8.
192. See note 22.
193. M.L. Parry and T.R. Carter, "Impact of Warming on Agriculture," *Forum for Applied Research and Public Policy*, Winter 1989.
194. See note 8.
195. See note 14.
196. See note 16.
197. Norman J. Rosenberg, *U.S. Agriculture in a Global Setting: An Agenda for the Future*, National Center for Food and Agricultural Policy, Resources for the Future, 1988.
198. Bridget Bloom, "Mixed Greenhouse Effect Predictions for British Farming's Future Climate," *Financial Times*, July 20, 1989.
199. See note 22.
200. "Little Planning for the Greenhouse Effect," *The Futurist*, November-December 1989. The article cites William Riebsame, director of the Natural Hazards Research and Applications Center at the University of Colorado at Boulder.
201. See note 88.

Box 2-A

1a. Richard D. Hylton, "Wall Street's Latest Diversification Strategy: Down on the Farm," *The New York Times*, Sept. 23, 1990; citing U.S. Department of Agriculture and Federal Reserve Bank of Kansas City statistics.
2a. *Ibid.*
3a. Scott Kilman, "Farm Failures Are Expected to Rise in '91," *The Wall Street Journal*, July 1, 1991.

Box 2-B

1b. Jim Hightower, "Sustainable Family Farming," *Issues in Science and Technology*, Fall 1989.
2b. Wes Jackson, "Third World Agricultural Tragedies," *Chemical & Engineering News*, Dec. 17, 1990.
3b. National Research Council, *Alternative Agriculture*, National Academy Press, Washington, D.C., 1989.

4b. *Ibid.*
5b. Chemical Manufacturers Association, "Appearance of Produce Versus Pesticide Use," *Chemecology,* May/June 1991.
6b. See, for example, Scott McMurray, "New Herbicides Help Farms Cut Toxic Chemicals," *The Wall Street Journal,* July 3, 1991.
7b. See, for example, Frank Edward Allen, "W.R. Grace Going 'Natural' With Bug Killer," *The Wall Street Journal,* Oct. 30, 1991.
8b. Ward Worthy, "Outlook Mixed for Pesticides Market," *Chemical & Engineering News,* Sept. 9, 1991. Statistics cited are by N. Bhushnan Mandava of SRS International, a regulatory consulting firm based in Washington, D.C.
9b. Ward Worthy, "Pesticides, Nitrates Found in U.S.. Wells," *Chemical & Engineering News,* May 6, 1991.
10b. Sandra Postel, "Halting Land Degradation," *State of the World: 1989,* Worldwatch Institute, W.W. Norton & Co., New York, 1989.
11b. Daniel J. Dudek, "The Ecology of Agriculture, Environment and Economics," Environmental Defense Fund, New York, N.Y., September 1987.

Box 2-C

1c. National Research Council, *Alternative Agriculture,* National Academy Press, Washington, D.C., 1989; and Lester R. Brown, *The Changing World Food Prospect: The Nineties and Beyond,* Worldwatch Paper #85, Worldwatch Institute, Washington, D.C., October 1988.
2c. U.S. Environmental Protection Agency, *The Potential Effects of Global Climate Change on the United States,* report to Congress, October 1988.
3c. Justin R. Ward, Richard A. Hardt and Thomas E. Kuhule, *Farming in the Global Greenhouse,* Natural Resources Defense Council, March 1989.
4c. Martin Parry, *Climate Change and World Agriculture,* Earthscan Publications Ltd., London, England, 1990.
5c. See note 2c.
6c. *Ibid.*
7c. *Ibid.*
8c. *Ibid.*

Box 2-D

1d. Stanley Changnon, Illinois Water Survey, personal communication, Feb. 14, 1992.
2d. Sue Shellenbarger, "Weather Keeps Grain, Soybean Traders on Edge As Mixed Outlook Emerges for This Year's Crops," *The Wall Street Journal,* May 22, 1989.
3d. Jeffrey Taylor, "Corn Rises as Hot Conditions in Midwest Prompt Feats of Crop Damage at Key Development Stage," *The Wall Street Journal,* July 25, 1991.
4d. Scott Kilman, "Agriculture Agency Proposes Requiring Aflatoxin Testing for All Corn Exports," *The Wall Street Journal,* Feb. 14, 1990.

Box 2-E

1e. Most of the information in this box has been excerpted from the article by Jack Lewis, "The Ogallala Aquifer: An Underground Sea," *EPA Journal*, November/December 1990.
2e. For the quotation, see the article, "Canada Sends U.S. Note About Stance On Water Diversion," *The Wall Street Journal*, July 25, 1988.

Box 2-F

1f. Terry L. Anderson, "Water Enough for Farmer and Fish," *The Wall Street Journal*, May 15, 1991.
2f. Charles McCoy, "Congress Tilts to West's Cities, Wildlife, Away From Farms in Water-Policy Bills," *The Wall Street Journal*, March 17, 1992.
3f. For example, the California cotton-farming trust controlled by J.G. Boswell, who ranks among the "Forbes 400" richest Americans, obtains subsidized water at an estimated taxpayer expense of $2 million a year, according to an analysis by the U.S. General Accounting Office. The cotton grown by the Boswell trust also receives commodity support payments ranging up to $8 million annually, because cotton is a surplus crop. See the article by Charles McCoy, "Big Farmers in West Get Subsidized Water Despite Drought Crisis," *The Wall Street Journal*, May 31, 1991.
4f. See the citation at the end of 3f.
5f. See note 2f.
6f. Kenneth D. Frederick and Peter H. Gleick, "Water Resources and Climate Change," presented at the Workshop on Controlling and Adapting to Greenhouse Warming, sponsored by Resources for the Future, June 14-15, 1988; and Marc Reisner, "The Emerald Desert," *Greenpeace*, July/August 1989.

Box 2-G

1g. U.S. General Accounting Office, *1990 Farm Bill: Opportunities for Change*, GAO/RCED-90-12, U.S. Government Printing Office, Washington, D.C., April 1990.
2g. Janet Edmond, "Congress Passes 1990 Farm Bill With Key Groundwater Protection Provisions," Environmental and Energy Study Institute, Special Report, Nov. 2, 1990.

Chapter 3
The Forest Products Industry

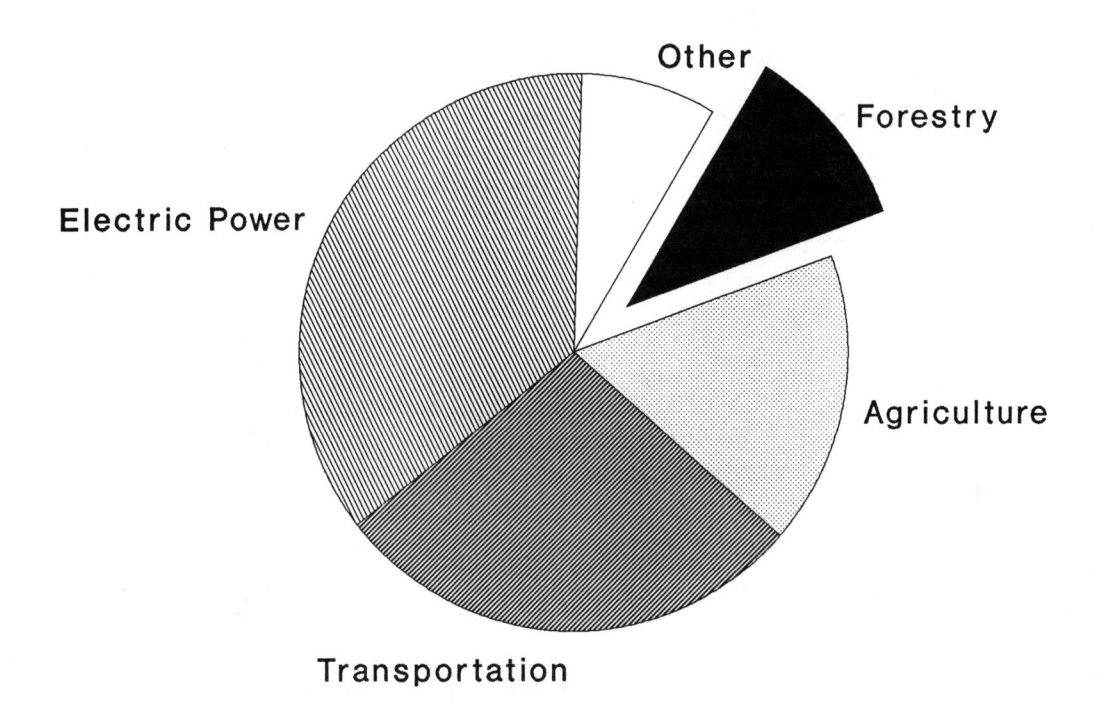

Relative contribution to global warming potential.
(Forestry's contribution reflects tropical deforestation.)

Contents of Chapter 3

Introduction 157

Effects on Forests
       Southern Roots 161
       Looking into the Crystal Ball 163
       Trees on the Move 168
       Getting from Here to There 170
       Fortified Air: CO_2 Enrichment 173
       Pests, Fire, Wind and Chaos 176
       Climate Change and Timber Management 179

Responses of the Forest Products Industry
       Addressing Pollution Problems 183
       Industry Views on the Greenhouse Effect 186
       Engineering Drought-Tolerant Trees 189
       Growing New Crops 190
       Acquiring New Land 193
       Looking for Places to Grow 199

Carbon Sequestration Potential
       Minding the Carbon Store 205
       Offsetting Utility Emissions 209
       Power to Burn? 211
       A New Pitch: Eco-friendly Products 212

Conclusions
       Forestry in the 21st Century 217

Boxes
       Box 3-A: U.S. Forest Use and Ownership Trends 164
       Box 3-B: Trees, Climate and the Global Carbon Cycle 174
       Box 3-C: Bridging the 'Generation Gap'
                 in U.S. Forest Resources 184
       Box 3-D: Old-Growth Forests 196
       Box 3-E: 'New Forestry' and the U.S. Forest Service 202
       Box 3-F: Conservation Reserve Program 208
       Box 3-G: Greenways: Paths to the Future 214

Notes 222

Introduction

To be a forester requires patience and a long-term planning perspective. Unlike farmers, who plant and reap crops in a single growing season, foresters must wait 25, 50, even 75 years for harvest time. As their resource grows and matures, foresters have a number of silvicultural options to choose from in an effort to maximize timber yield. But each time they plant a new seedling, they must take on faith that climate will remain their long-term benevolent ally. As one forest products company remarked in a recent annual report, "While modern management methods improve fiber yield and assure a renewable supply of timber, forestry remains a trust between man and nature."[1]

The prospect of global warming places this vital trust in doubt. Such uncertainty, in turn, raises the stakes for a number of strategic questions already facing the forest products industry:

- Will there be enough available timber to meet future demand?
- Is the industry sufficiently diversified in its land holdings and growing stock?
- Are commercial plantations adequately protected against pests, disease and adverse weather conditions?
- Will "new forestry" management practices reconcile competing demands for forest resources without pricing wood products out of the market?

How forest products companies respond to these and other questions in an era of possibly changing climate is likely to have a profound influence on their future.

At the moment, it is too early to say whether a carbon-dioxide-enriched environment will help or hurt the forest products industry. Trees (like most other plants) should benefit from the CO_2 fertilization effect, even if the climate itself becomes somewhat hotter and drier. Forests' ability to migrate possibly hundreds of miles to keep pace with a changing climate is of greater concern, however, as is their ability to stand up against other adversities such as possibly increased frequency of droughts, wildfires and insect attacks. With the ultimate effects of global warming on America's forests still undetermined, it behooves foresters to employ strategies that are flexible and highly adaptive to changing climatic conditions.

Within the forest products industry, the future is likely to be far different from the past for reasons beyond the weather. As the availability of virgin timber diminishes, dependence on second- and third-generation forests is increasing. Growing trees on rotation—rather than harvesting mature forests that grow on their own—entails a much greater stake in the land and the climate.

The region where the forest products industry is concentrating most of its timber-growing assets is the South. The warm, wet climate there favors quick-rotation species grown on vast tree plantations. In the event of severe global warming, however, southern forests could dry out and revert to grasslands—jeopardizing the industry's primary resource base. On the other hand, global warming could lead to an even warmer and wetter climate in the South, allowing the region's forests to become more productive than ever. Climate models have yet to sort out this vital research question.

On another matter, prospects are more certain. Since trees and most forest products offer a "sink" for carbon dioxide—the primary contributor to the greenhouse effect—the forest products industry is in a unique position to ameliorate global warming by planting more trees—and to aggravate it by cutting too many down for short-lived purposes. Therefore, consumption of wood products in a greenhouse world would be dictated by whatever sustainable forestry practices can provide and whatever the global carbon budget itself can afford.

Fortunately for the forest products industry, wood is a renewable resource that distinguishes itself from most other materials with which it competes. Processes for manufacturing wood into finished products tend to be much less energy-intensive than those of alternatives. As a result, policies to address global warming could lead to an overall expansion of wood-products markets. It is conceivable that other industries in need of carbon offsets might even pay forest products companies to expand their tree-growing base.

In a policy sense, forests already have gained stature because they represent a partial antidote to global warming. As a global warehouse for carbon, trees use photosynthesis to sequester about 2 to 3 billion tons of carbon a year. Each additional acre of growing timber removes 1.25 tons of carbon (on average) from the atmosphere—offsetting about one-quarter as much carbon as the "average" American citizen produces in a year.[2]

Partly for this reason, tree planting has become a much-vaunted civic activity. None other than President George Bush is the nation's most conspicuous arborist. With his "America the Beautiful" plan, Bush seeks to boost the nation's tree-planting rate by an extra billion trees a year during the 1990s—thereby offsetting 2 to 5 percent of U.S. carbon dioxide emissions into the atmosphere after the turn of the century.[3]

Counteracting the greenhouse effect is not the only reason trees are in demand, however. In urban areas, trees shade buildings from the hot summer sun, break the cold winter winds, muffle noise and beautify their surroundings at all times. In rural areas, trees keep soils from eroding, hold critical watersheds in place and serve as a natural filter for man-made pollutants. In the energy sector, fuels from wood generate heat and power affordably without necessarily altering the global carbon budget. And in the consumer marketplace, forest products enjoy a relatively "eco-friendly" reputation, despite their impacts on natural forest ecosystems. For the most part, products from the forest are renewable, recyclable and biodegradable. These attributes will continue to favor the forest products industry regardless of whether policies are enacted to combat the greenhouse effect.

To date, no forest products company has chosen to act in the face of uncertainty about the greenhouse effect. The "crystal balls" of climate research apparently still provide too cloudy a picture for them to move decisively. Unless

Figure 1

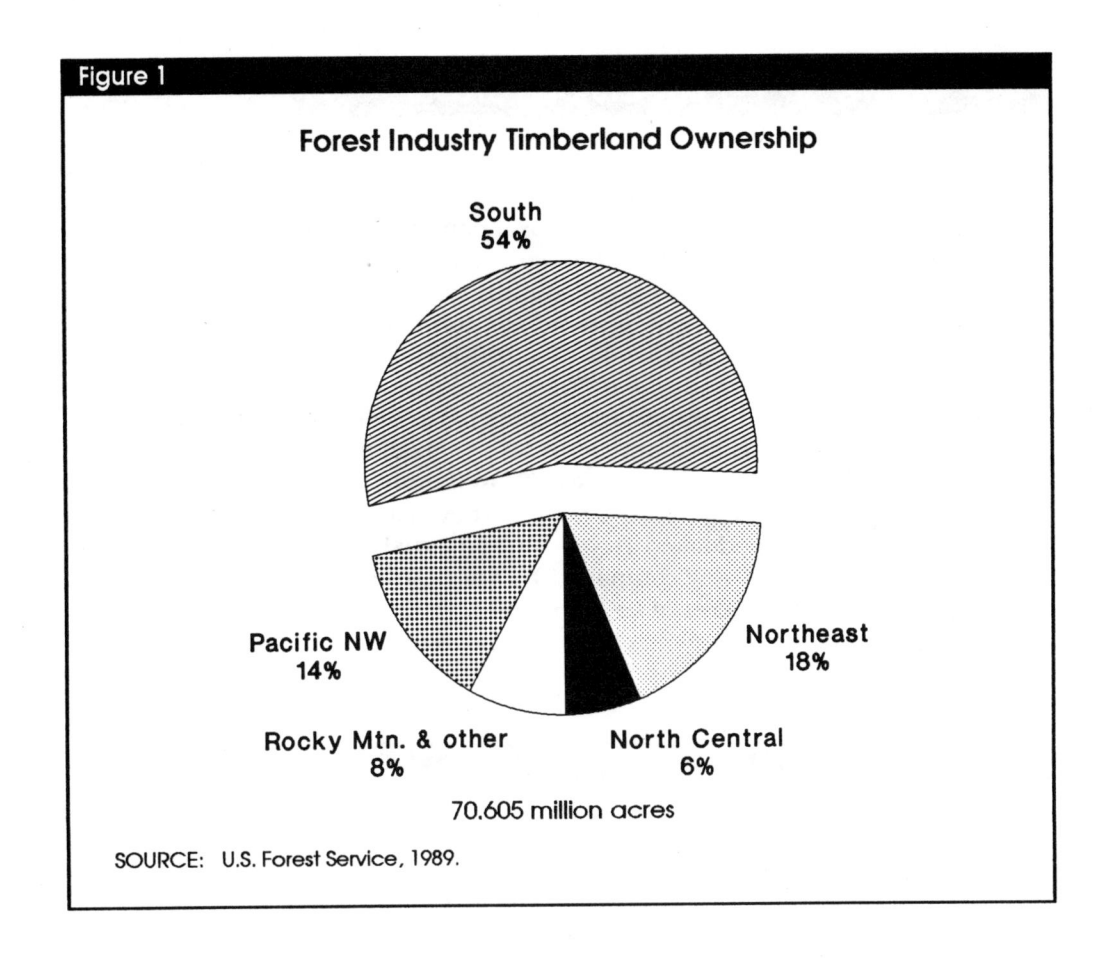

Forest Industry Timberland Ownership

- South 54%
- Northeast 18%
- Pacific NW 14%
- Rocky Mtn. & other 8%
- North Central 6%

70.605 million acres

SOURCE: U.S. Forest Service, 1989.

evidence of a threat becomes more convincing, industry executives will remain understandably reluctant to publicize the prospect of a diminution of timber-growing assets. As Paul Kramer, professor emeritus of botany at Duke University, surmises: "Until meteorologists develop the methods and models to more accurately predict how the climate will change globally and regionally, it would appear more productive for plant physiologists and forest scientists to concentrate on improving their understanding of how current weather extremes affect the physiological processes that control tree growth. Since drought is the most common cause of reduced tree growth," Kramer adds, "it is especially important to see if tolerance to drought can be improved."[4]

One thing is known for certain: As long as forest products companies stake their reputation as "tree growing" companies, planting and nurturing new seedlings to replenish the harvest of mature stock, they will be playing an important role in keeping global warming in check. Whether they are able to turn the greenhouse effect more to their advantage remains an open question. Perhaps more than any other industry, the forest products industry has the opportunity to profit from climate change—or lose because of it.

Accordingly, its greenhouse gambit must be chosen wisely.

Effects on Forests

Wood is the highest-valued agricultural commodity in the United States.[5] On a tonnage basis, the nation's wood consumption exceeds that of all other structural materials combined—steel, aluminum, plastics, brick and concrete.[6] The average American's yearly demand for lumber, paper and other forest products equals the volume of wood in a 100-foot-tall tree.[7]

The American forest products industry ranks among the top 10 employers in 40 American states. It employs more than 1.6 million individuals, boasts an annual payroll of $43.5 billion and generates sales of more than $100 billion a year.[8] Forest products companies own 70 million of the 483 million acres of commercial timberland in the United States. They also serve as the nation's chief tree growers, a fact not overlooked in corporate advertising slogans. Weyerhaeuser bills itself as the "tree growing company," having planted more than 2 billion tree seedlings on 5.6 million acres of its own land over the last 20 years.[9]

But Weyerhaeuser hardly has a monopoly on the tree-growing claim. More than a dozen major forest products companies planted 800 million tree

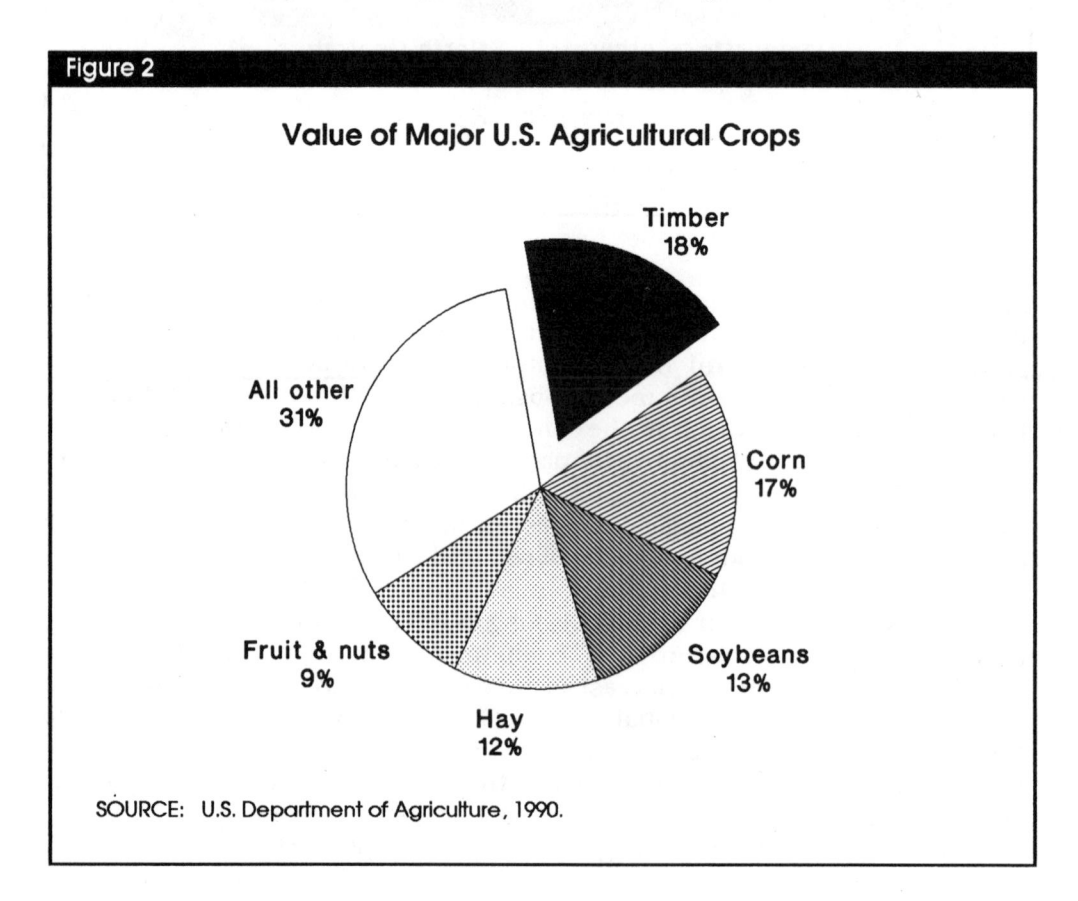

Figure 2

Value of Major U.S. Agricultural Crops

- Timber 18%
- Corn 17%
- Soybeans 13%
- Hay 12%
- Fruit & nuts 9%
- All other 31%

SOURCE: U.S. Department of Agriculture, 1990.

seedlings on nearly 500,000 acres of industry-owned land in 1990 alone.[10] By comparison, the 136-million-acre network of government-owned commercial timberlands accounted for only 18 percent of the 1.9 billion tree seedlings planted over 3 million acres in 1990. Non-industrial private forest owners, who own 57 percent of the nation's timberlands, planted the remaining 42 percent of the seedlings.[11]

Southern Roots

Of every 10 seedlings planted in 1990, seven are rooted in southern soil. The ratio is closer to eight of 10 for the forest products industry.[12] The warm, hospitable climate of the South—not to mention its flat, accessible terrain—makes it the ideal region for managing plantations of fast-growing trees. Loblolly pine is the most popular of all the commercial species. As a softwood, it is in the category that constitutes four-fifths of all timber consumption in the United States. Compared with northern conifers, which rest for long winter periods, loblolly grows more uniformly throughout the year, producing a coarser grain with more even spacing between the annual rings. More importantly from an investment standpoint, loblolly matures naturally in about 35 years, which is twice as fast as western Douglas-fir or northern white pine. With the aid of intensive management practices, loblolly pine plantations can be harvested in rotations of 18 to 25 years, accelerating forest products companies' return on investment.

Each of the South's four major forest types—loblolly, oak-gum-cypress, oak-pine and oak-hickory—are in fact relatively fast-growing and highly productive species, yielding at least 85 cubic feet of industrial roundwood per acre a year. The South's most productive timberlands can produce 120 cubic feet or more per acre a year. Since the South accounts for 53 percent of the nation's highly productive timberlands, yielding 80 to 120 cubic feet of industrial roundwood per acre a year, and 45 percent of the nation's most productive timberlands, yielding more than 120 cubic feet per acre a year, it is not surprising that the forest products industry has chosen to concentrate more than half of its timberland acreage in the region.[13]

Many forest products companies have longstanding ties to the South, including James River, The Mead Corp., Stone Container, Temple-Inland and Union Camp. But the industry took a symbolic turn toward the South in the late 1970s, when Georgia-Pacific moved its headquarters from Portland, Ore., to Atlanta. According to the Wilderness Society, the nation's seven largest producers of plywood—Boise Cascade, Weyerhaeuser, Champion International, Georgia-Pacific, Louisiana-Pacific, International Paper and Willamette Industries—have reduced their production capacity in the Pacific Northwest (west of the Cascade Mountains) by two-thirds since 1978, while increasing their plywood capacity in the South by a tremendous 380 percent. At the same time, their lumber production capacity west of the Cascades has fallen by 45 percent, while rising 64 percent in the South.[14] (Industry sources point out that actual lumber production has remained fairly constant on the Pacific Coast,

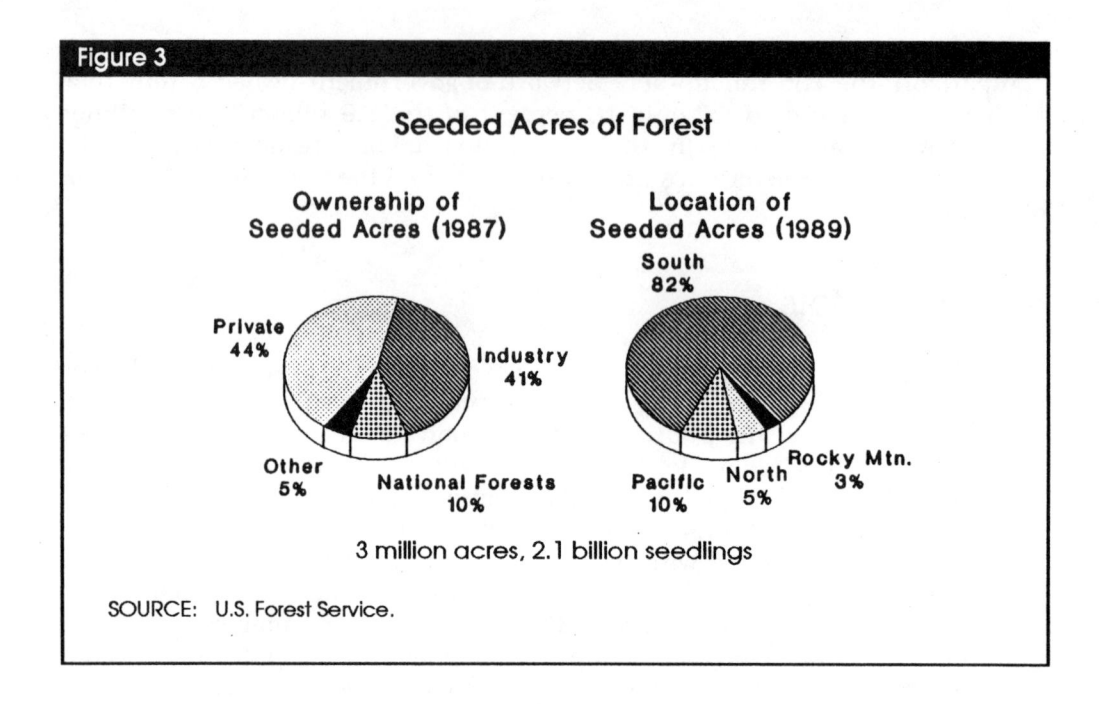

> **Figure 3**
>
> **Seeded Acres of Forest**
>
> Ownership of Seeded Acres (1987)
> - Private 44%
> - Industry 41%
> - National Forests 10%
> - Other 5%
>
> Location of Seeded Acres (1989)
> - South 82%
> - Pacific 10%
> - North 5%
> - Rocky Mtn. 3%
>
> 3 million acres, 2.1 billion seedlings
>
> SOURCE: U.S. Forest Service.

while increasing 40 percent in the South since 1978.)[15] Today, five of the top 10 and nine of the top 15 lumber-producing states are in the South.[16] And considering the harvesting restrictions in the Pacific Northwest to protect old-growth forests and endangered species such as the northern spotted owl and Sockeye salmon, the industry's southern momentum seems likely to continue.

Timber already is the single largest agricultural product of the South, constituting a $60-billion-a-year business. The forest products industry is also the region's leading employer—ahead of textiles and apparel—providing 630,000 jobs in manufacturing lumber and wood, pulp and paper products. Although Washington, Oregon and California remain the largest timber-producing states in the country, they employ fewer than half as many forest industry workers.[17] The South has in fact become the nation's largest regional timber supplier, accounting for 53 percent of the volume harvested in the United States each year.[18]

Intensively managed pine forests cover 22 million acres of the South—an area half the size of Georgia. By 2030, these plantations are expected to nearly double again in size, led by expansion of forest industry activities in the region.[19] Already the industry has acquired most of the land it needs to carry out its strategic plans. With 38 million acres of owned timberland in the South as of 1987, the industry is expected to purchase only 1 million additional acres through 2040, mainly in the South Central region.[20] By comparison, the industry acquired more than 6 million acres in the South over the last 35 years. Thus, the groundwork has been laid for major expansion of forest industry activities in the South in the decades ahead.

Looking into the Crystal Ball

What if the climate were to change so that forests of the South became more or less productive than they are now? Would the forest products industry continue to consolidate its holdings there? Where else might it look to acquire new growing stock? Admittedly, these speculative questions are not likely to be addressed until the industry's top executives are convinced that global warming is in the offing—a determination that may be more than a decade away. In the interim, they are likely to carry forward with existing plans, not factoring in the possibility of climate change, even though global warming may well occur during the life span of their forests and mills.

This is not to suggest that forest industry executives shun a long-term view of their business. On the contrary, they know they must stagger their growing stock over many age classes in order to avoid periodic supply disruptions. Moreover, the industry's pulp and paper mills must be built to last, processing two or three rotations of forest growth before their own retirement. The concern is that global warming may hasten the demise of these long-lived assets. Accordingly, the industry may find it prudent to develop a contingency plan. As a preliminary step, some companies are consulting two "crystal balls" of climate research. One sees past climate changes with a high degree of clarity. The other forecasts future changes, but gives a blurrier image.

Paleo-climatic records: Paleo-climatic records offer a clear window into the past. Following the trail of fossilized pollen left behind by ancient forests, scientists can track their movement about the continent in response to changing climatic conditions. Eighteen thousand years ago, when the Earth was in the grips of its last major Ice Age, spruce forests plunged southward (a step ahead of the advancing glaciers), reaching all the way to the northern edge of the Coastal Plain, from Georgia to east Texas.[21] At the end of the Ice Age some 12,000 years ago, the borders between forests and prairies, conifers and hardwoods changed once again. This time, the spruce forests chased the retreating glaciers back to the north, moving all the way across the border that now separates the United States and Canada.

At the peak of the Altithermal period 8,000 years ago, temperatures climbed a few degrees Fahrenheit higher than they are now, and most tree species reached locations 60 to 250 miles north of their present distributions.[21] Others headed up the sides of mountains as the climate warmed. Hemlock and white pine grew at elevations 1,000 feet higher than where they are found today. At the same time, the treeless midwestern prairies bulged to the south and east— covering an area larger than today's Great Plains—before the climate began to moderate once again about 6,000 years ago.

That the forests moved great distances in response to past temperature changes is not as significant as the rates at which they traveled. Researchers have determined that spruce forests moved at the slowest pace across the country—migrating only six miles a century. Maple, chestnut and balsam fir traveled faster, at rates up to 12 miles a century. Beech trees migrated up to 15 miles a century—extending their range from Georgia all the way to southern

(continued on p. 166)

Box 3-A

U.S. Forest Use and Ownership Trends

In the early 1600s, before European settlers came to America, forest covered about half of the U.S. land area—some 1.1 billion acres (including Alaska). In theory, a squirrel could travel from Maine all the way to the Mississippi River without ever setting foot on the ground. For the new settlers, however, the endless forest posed an obstacle to preferred agricultural uses of the land. By one account, it became "almost a moral duty of citizens to 'subdue' the forest wilderness and 'improve' its use for the benefit of society."[1a]

So began a centuries-long crusade to clear the forest for crop production, with the felled timber used to build shelters and provide energy. In 1850, wood provided 95 percent of the nation's energy requirements.[2a] Yet the clearing of the forest had just begun in earnest. Over the next 60 years, 190 million acres of forest land was cleared, more than in the previous 250 years of settlement. Per-capita consumption of timber peaked in 1905 at 530 board-feet—nearly three times today's level.[3a]

By the start of World War I, 60 percent of America's virgin forests, representing 1 million square miles of land, had been cleared for farming, logging and other activities. Forests since then have managed to reclaim one-third of this denuded land, including retired hill farms in New England, worn-out cotton fields in the South and marginal croplands in the Midwest. Altogether, this reforested area equals the size of Texas.

Today, the nation's forests encompass nearly 750 million acres in all 50 states—roughly one-third of the total U.S. land area—and forest cover is holding relatively stable. The U.S. Forest Service estimates that forest land declined about 6.5 percent from 1953 to 1987 (approximately 52 million acres) and will decline another 2.2 percent (or 17 million acres) by 2010.[4a] With agricultural expansion grinding to a halt since the 1930s, today's development pressures come in many other forms: urban development, highway and airport construction, creation of reservoirs, expansion of surface mines and more.

American forests now represent less than 10 percent of the planet's forest coverage, yet they produce more than 20 percent of the world's industrial forest products. The amount of commercial timberland in the United States totals 483 million acres. (Commercial timberland, by definition, is capable of growing more than 20 cubic feet of timber per acre a year). Governments own 28 percent of this timberland, including 18 percent in the national forests and 5 percent in state forests. The forest products industry owns about 15 percent of the total, or slightly more than 70 million acres. However, the lion's share—some 276 million acres—is owned by non-industrial private forest landowners (NIPFs).[5a]

Half of the nation's annual timber harvest of 80 billion board-feet comes from land owned by NIPFs. Another 30 percent of the harvest is on timberlands owned by industry, and about 20 percent of the harvest occurs on government lands.[6a] Competing uses are taking some of the timberlands out of production, however. In the Pacific Northwest, efforts to maintain habitat for the northern spotted owl are expected to put more than 5 million acres of old growth timber off-limits to loggers.[7a] As this land is set aside, it will increase by more than 50 percent the acreage of old growth trees protected within national parks, wilderness areas and by other administrative or legislative edicts.[8a]

A greater threat to foresters over the long term may be that many of the nation's

Decline in Virgin Forests

1620   1850   1992

SOURCE: The Wilderness Society.

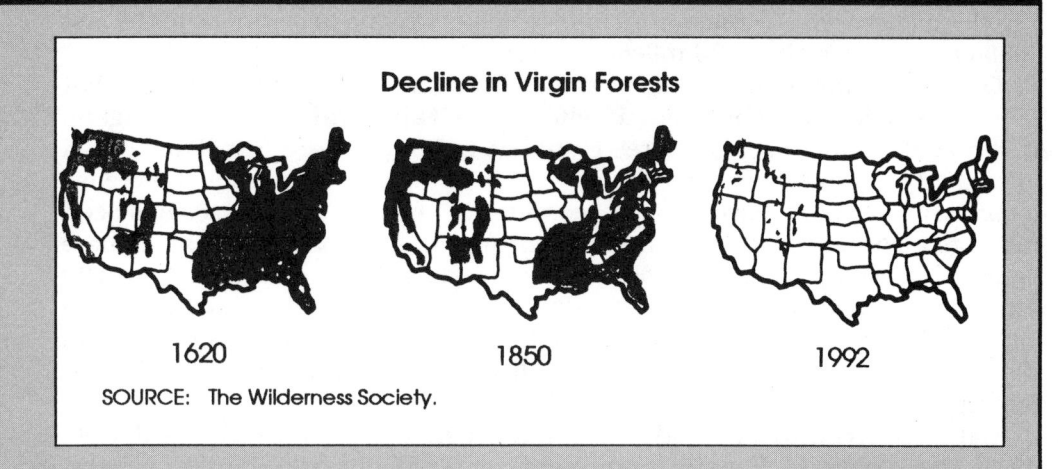

7 million NIPFs are converting their property for other uses. In Washington state, NIPFs own 22 percent of the commercial timber base—much of it prime, low-elevation, second-generation growth. But state officials estimate that more than 500,000 acres of this forest land—3 percent of the state's current timber base—soon will be lost to urban sprawl.[9a] Similar real estate development pressures have cropped up in northern California and the Northeast. In some instances, public funds have been expended to purchase environmentally sensitive, privately held tracts to preserve them as forests.

The greatest potential shortage of private timberland may be in the South, where more than half of the nation's timber is logged. According to the U.S. Forest Service, a majority of that region's NIPFs are not replanting their forests after harvest. By one estimate, as much as 18.8 million acres of southern private timberland could be converted to other uses by 1995.[10a] Already, private softwood timber inventories in the region are decreasing because of an imbalance between harvests and new growing stock. After the year 2000, private southern hardwood timber inventories also could decline in the South. (Revised Forest Service estimates suggest that the NIPF situation in the region is not quite as serious as was believed a few years ago, however.)

To make up for some of the shortfall in replanting by NIPFs, the forest products industry could step up its own tree-planting programs. The American Forestry Association estimates that industry-owned nurseries could expand seedlings production by 50 percent if given the proper incentives and a few years' lead time.[11a] Over the long term, such a commitment to replanting by the forest products industry might overcome the deficit created by NIPFs. Already, the net annual growth of trees exceeds harvests in every region of the country—even in the West and South. The nation's timber growth averages 22 billion cubic feet a year, while harvests and natural mortality average 16 billion cubic feet a year, a surplus of 27 percent.[12a]

Maintaining the surplus in growing stock will be critical to the expansion of the forest products industry in the 21st century. Per-capita consumption of wood products in the United States—which had been falling for more than 60 years—began to rise again around 1970, and it continues to rise today. By 2040, demand for U.S. wood products could rise another 50 percent.[13a]

Canada.[22] Oaks and pines were among the swiftest travelers, migrating 19 to 25 miles a century. The maximum known dispersal rate in response to the last glacial retreat was about 30 miles a century.[23]

One disturbing conclusion to be drawn from this paleo-climatic data is that forests plod at a slower pace than the climate itself appears capable of changing. As a rule of thumb, each 1 degree F change in temperature translates into a latitudinal shift in climate of 40 miles.[24] Accordingly, even a modest increase in temperature could disrupt forest ecosystems, since no tree species has been known to migrate more than 30 miles a century. Making matters worse, climate models suggest that ambient temperatures could rise as much over the next 100 years as they did in six millennia—from 12,000 to 6,000 years ago. It is conceivable that North Carolina's climate, say, could reside in New England after the year 2050. Such a climate shift would exceed forest communities' demonstrated rate of migration by 10, 30, even 50 times.

General circulation models: If paleo-climatic evidence raises doubts about the ability of forests to respond to rapid climate change, general circulation models—the other crystal ball of climate research—serve only to compound this concern. While general circulation models remain highly unreliable predictors of future climate, especially at a regional level, they do suggest that a CO_2-enriched environment may bring about temperature and precipitation changes that stress forest ecosystems, particularly in the South.

Since the climate models have different built-in assumptions, they come to varying conclusions about how much and how fast the atmosphere will warm, and which regions of the globe will become wetter or drier. In the Southeast, for example, the models' predictions of increases in average annual temperature range from 4 degrees F to more than 8 degrees F, after doubling the atmospheric content of carbon dioxide. (The Southeast region consists of Virginia, North Carolina, South Carolina, Georgia, Florida and Alabama.) Moreover, two of the models suggest that the Southeast will experience an increase in rainfall, while the other two models forecast drier times.[25]

Two models offer a relatively hopeful vision for the Southeast. One developed at the National Center for Atmospheric Research projects an average summer temperature increase of 4 degrees F and a slight increase in annual precipitation. Another developed at Oregon State University projects a summer temperature increase of about 6 degrees F with a similar increase in annual precipitation.[26] Under these conditions, the timber yields of southern forests might actually increase. Areas abandoned by loblolly pine could be occupied by highly productive longleaf and slash pine, extending the range of these most southerly pine forests from the Gulf Coast to the Piedmont District. In the extreme South, subtropical evergreen trees might spread northward from their present bases in southern Florida and southern Texas (although their expansion would be limited to those areas free of severe summer drought and freezing winter temperatures).[27]

The other two climate models offer a more pessimistic outlook for the sustainability of commercial forestry in the South. The Goddard Institute for Space Studies model forecasts a temperature rise of 6.5 degrees F in the Southeast—with annual precipitation staying about even—so that the region's

Table 1

Factor	Certainty of Change	Consequence for Forests
Climate		
Temperature	Moderate	High
Temperature load		
Evaporative demand		
Heat sums		
Precipitation	Low	High
Quantity		
Spatial & temporal distribution		
Climatic Extremes	Low/moderate	High
Late/early frosts		
Droughts		
Floods		
Rain or snow		
Wind		
Atmospheric Variables		
Carbon Dioxide	High	Variable-unknown
Ozone		
Tropospheric	High	High
Stratospheric-UV(B)	Moderate	Low-unknown

SOURCE: T. Hinckley, T. Wolf, University of Washington.

soils tend to dry out. The other model, developed at the Geophysical Fluid Dynamics Laboratory, projects an even hotter and drier climate for the region. Some of this model's early forecasts were dubbed the "doomsday scenario" because they suggested the South might inherit an arid, steppe-like climate—not unlike what now exists in the Mediterranean and subtropical Africa and Asia. If that were the case, 18 important tree species would no longer flourish in the southern half of the Southeast; scrub, savanna or otherwise sparse forest conditions would become the norm. Even under the milder climate scenarios that this model and others have projected, tree-growing conditions in much of the Southeast could become marginal. Forests in South Carolina, for example, might support only half as much biomass eventually as existing stands.[28]

Trees on the Move

If global warming reaches the proportions projected by some general circulation models, entire forest communities may seek to migrate hundreds of miles north of their present locations, a trek that would take 200 to 500 years to complete, based on the paleo-climatic record. One team of researchers led by Jonathan Overpeck has adapted the climate-pollen record of the last 18,000 years to gauge future changes in forest composition according to three of the climate models.[29] (The National Center for Atmospheric Research model was not considered in this study.) The results, following a doubled-CO_2 simulation, indicate that loblolly and slash pine would move up the Eastern seaboard, infiltrating the hardwood forests of New Jersey and possibly southern New England. At the same time, spruce that now grows from Georgia to Maine along the Appalachian Mountain chain, and throughout New England, would virtually disappear from the United States—shifting its range northward by as much as 600 miles.

This may actually be good news for commercial timber interests of the North. The productivity of most northern forests would be expected to rise significantly as the mix of species came to feature fewer evergreens and more fast-growing hardwoods, especially maple and oak species migrating from the central-eastern United States. Sugar maples, on the other hand, could die out throughout their entire range in the United States (save extreme northern New England) under the climate scenarios considered here.[30] Such a loss would deal a blow to the fall tourism industry and sound a death knell for the maple syrup industry, which already has suffered a series of poor sap runs since 1980.[31]

Farther to the north, the climate suited for spruce, white pine and northern hardwood trees could shift by more than 400 miles as a result of global warming, reaching the Hudson Bay region of the Canadian boreal forest.[32] The boreal forests in turn could move 320 to 640 miles closer to the North Pole, perhaps converting portions of the tundra to a tree-studded landscape. The circumpolar coverage of the boreal forests, however, might decline by more than a third eventually. Canada, for instance, could gain approximately 175 million acres of boreal forests in its northern reaches but lose more than 400 million acres in its southern flank.[33] A mixed conifer-hardwood forest, featuring sugar maples and white pines, eventually might replace the southern boreal forests (now dominated by white and black spruce) in a transition period spanning four centuries.[34]

For the American West, the climate models suggest that forests may come to favor drought-tolerant hard-pine species, such as lodgepole pine and ponderosa pine, at the expense of fir, hemlock, larch and spruce species. In the northern Rocky Mountains, upslope shifts of several thousand vertical feet would be possible for ponderosa pine, Douglas-fir and western hemlock. And along the coastal mountains of California and Oregon, Douglas-fir could give way to other western pine species if the climate warms to a persistent temperature above 48 degrees F. (Douglas-fir requires a "winter chill" to ensure proper timing of seed germination and establishment of seedlings before the

The Forest Products Industry

Figure 4

Potential Shifts in Range of Hemlock and Sugar Maple

Hemlock

A B C

Sugar Maple

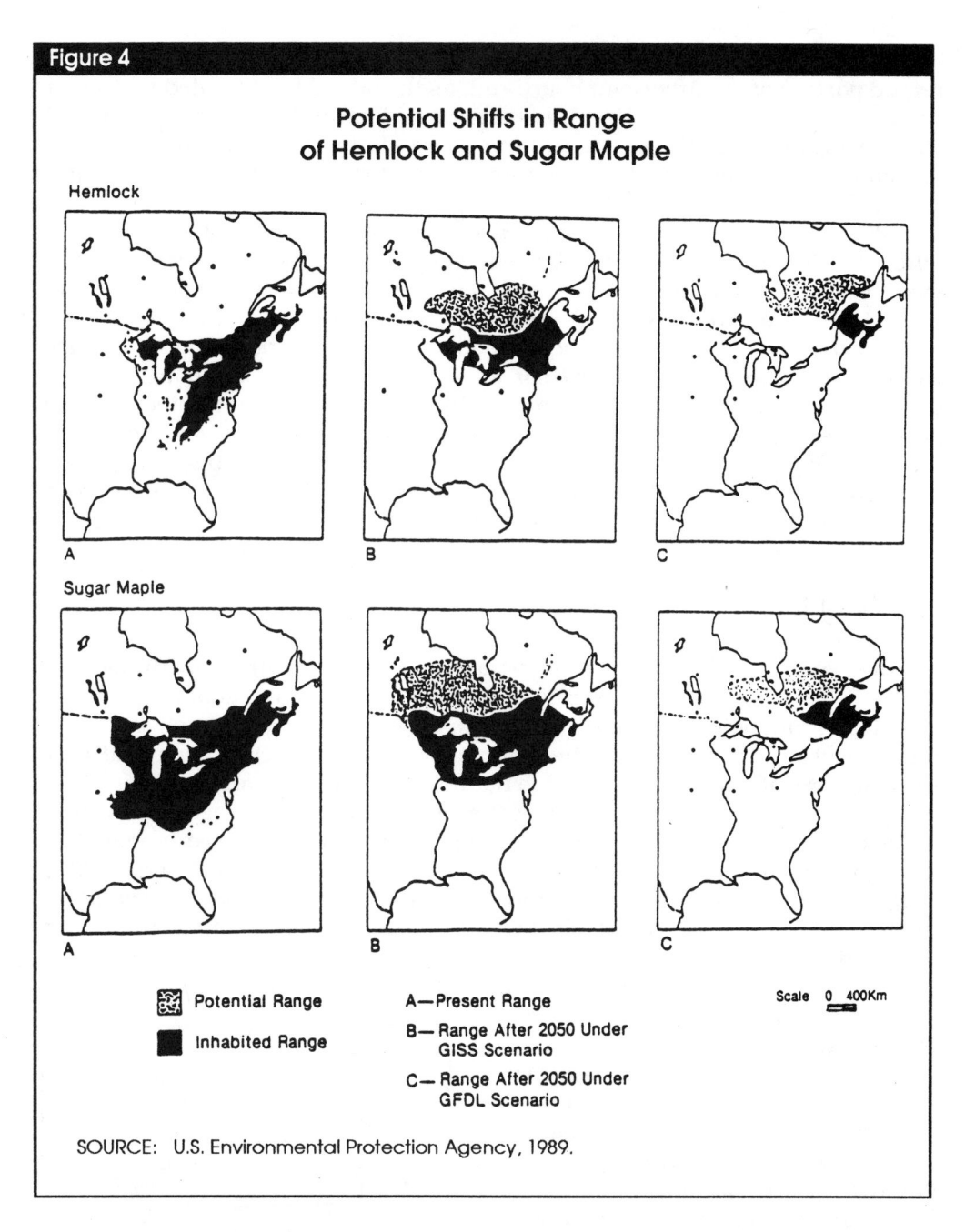

A B C

- Potential Range
- Inhabited Range

A—Present Range
B—Range After 2050 Under GISS Scenario
C—Range After 2050 Under GFDL Scenario

Scale 0 400Km

SOURCE: U.S. Environmental Protection Agency, 1989.

onset of summer drought.)[35] In the event of drastic climate change, standing biomass in the West could be reduced to 60 percent of current levels, with the species composition of the west-side Sierra Nevada forests coming to resemble that of the drier forests on the east side of the Sierra Nevada.[36]

Perhaps the most disturbing prospect for foresters is that global warming could expand the region of the United States where trees do not grow at all.

During the Altithermal Period 8,000 years ago, the Great Plains covered a record portion of the American heartland, as the added heat tended to dry out the soil. Based on this paleo-climatic data and several of the CO_2-enriched climate simulations, the prairies of the Great Plains could sweep as far east as Pennsylvania and New York, despite a modest increase in regional precipitation.[37]

Of course, future climate changes need not resemble those of the past. The climate of North America could become generally warmer *and* wetter in the event of global warming. In this case, fast-growing southern species could spread to the north, covering more of the continent with highly productive families of trees. At the same time, carbon dioxide enrichment might enhance the growth of trees and bolster their tolerance of drought, enabling them to encroach on the prairies. Some foresters believe they may have detected early signs of an enhanced greenhouse effect in industry plantations, some of which are maturing faster than expected. With positive signals such as this, some plantation managers believe the onus remains on climate modelers to prove that rising CO_2 levels will do more to harm the forests than to benefit them.[38]

Getting from Here to There

Ultimately, what matters most is the way climate change may unfold. At first, CO_2 enrichment may give trees a photosynthetic boost, enabling them to grow faster and better until thermal lags in the climate system catch up. Then temperature and precipitation changes may begin to stress the forests and spur their movement toward new suitable habitats. Such a migration process may take 500 years or more—but only if the rate of climatic change is slow and methodical. In the event of rapid warming, trees would be forced to march at an unprecedented rate. The vegetation changes simulated *over the next 200 to 500 years* in the Oregon State and Goddard Institute general circulation models, for example, are comparable to the maximum vegetation changes that occurred *over 1,000 years* at the end of the last Ice Age. The Geophysical Laboratory's doomsday scenario compresses *3,000 years* of vegetative change into a transition period of only 200 to 500 years.[39]

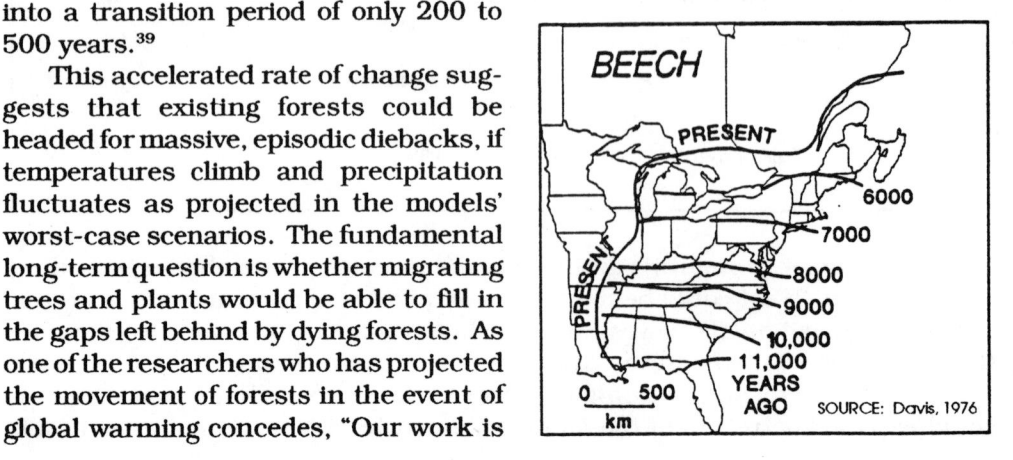

This accelerated rate of change suggests that existing forests could be headed for massive, episodic diebacks, if temperatures climb and precipitation fluctuates as projected in the models' worst-case scenarios. The fundamental long-term question is whether migrating trees and plants would be able to fill in the gaps left behind by dying forests. As one of the researchers who has projected the movement of forests in the event of global warming concedes, "Our work is simply showing you where habitat for birch or oak or spruce will be, not whether the trees will make it there."[40]

A brief example may help to illustrate this important point. The beech tree is a commercially important hardwood species that now grows across the entire eastern United States north of Florida. Using the Goddard Institute general circulation model, foresters project that climate change would force beech trees out of the United States entirely, save New England, or using the gloomier Geophysical Laboratory model, save northern Maine. The models also assume that new growing habitat would emerge in Quebec and Ontario, albeit a smaller area than is lost in the United States.[41]

In the real world, beech trees are relatively slow movers. Their seeds are dispersed mainly by jays, and they migrate at a rate of fewer than 15 miles a century. If the climate models prove correct, they would have to retreat 350 to 650 miles from their southern borders as the globe warms, which is up to 40 times their known dispersal rate.[42] In a worst-case scenario, the canopy of a beech forest would offer an illusion that all is well—but only for a time. As the warming intensifies, mature trees would begin producing fewer seeds and reproductive rates would start to decline. Weakened trees would become more susceptible to disease, insects and fire, and less able to withstand the wind. Eventually, the surviving trees would fail to flower and fruit, and no young beeches would rise from the forest floor to take their place. In the end, beeches would go extinct because their natural migration capabilities failed to keep up with the rapid pace of climate change.

Foresters could try to avoid this fate by planting southern ecotypes in northern locations, hoping that soil and climatic conditions accommodate the move. But the soils of boreal forests are vastly different from those under southern deciduous forests, raising questions as to whether northern regions can support the southerly species. Beech trees, for one, appear to be highly specialized; three distinct subspecies grow in different regions of the country. It is uncertain how well the southern subspecies would fare in the North.[43]

The migratory tendencies of animals must be considered as well, since animals exert a profound influence on forest ecosystems through selective browsing of seedlings, wood infestation (by insects) and dispersal of seeds (mainly by birds). In the event of climate change, birds, insects and microorganisms would be able to react much more quickly than the trees themselves, whose response times would lag decades behind. Accordingly, forest communities would not migrate as a unit, but in distinct parts. How this disassociated response to climate change would affect the redevelopment of healthy forests in new locations is an important yet unresolved research question.[44]

Even if forests are able to make the extended journey, they would be likely to encounter significant changes in growing conditions once they arrived. The planet's day length regime, for example, becomes increasingly lopsided as one moves away from the equator. Closer to the poles, days are longer in summer and shorter in winter. In addition, severe depletion of the ozone layer in northern latitudes could expose forests to more damaging ultraviolet rays that stunt their growth. (The same holds true for trees that migrate to higher elevations, since the intensity of ultraviolet light also increases with altitude.) All of these conditions could reduce forest productivity and hinder their reproduction and survival capabilities.[45]

There remains a possibility, however, that trees may not have to migrate nearly as far as some models suggest in order to acclimate to global warming. "The concern that climate change will outpace the ability of species to migrate may be somewhat overstated," contends Michael S. Coffman, manager of planning for the northeastern operations of Champion International. "Species do not necessarily have to advance their range along an 'edge' or 'front' as is often hypothesized," he told a group of foresters at a 1990 conference on climate change. "In fact, the range of most species does not end in an 'edge' but rather occurs as a declining mosaic of 'islands' whose habitats (soils and macro/micro-climate) are still favorable to a particular species," he said. Adjacent habitats sometimes depict environmental shifts equal to several hundred miles of climate gradient, Coffman points out. As temperatures change and begin to match the other characteristics favorable to these species, the islands may grow and consolidate. Thus, it is conceivable that not only will migration occur faster than expected, but "many species will only need to migrate several kilometers rather than several hundred to remain healthy and viable," Coffman concludes.[46]

Where these islands of habitat are found, it may behoove foresters to use them as "stepping stones" to aid in the tree migration process. "The prudent course for program managers is to establish forests containing tree species at the northern edges of their ranges," researchers from the Natural Resources Defense Council wrote in a 1989 report, *Farming in the Greenhouse*.[47] Foresters could plant "stands of white oak in Minnesota, bald cypress in southern Illinois, longleaf pine in North Carolina and black mangrove in Florida. These biotic reserves may serve as 'advance' seed sources to facilitate the migration of species as the climate changes," the researchers maintained.

If such an effort is undertaken, foresters must be mindful of the manmade roadblocks now crisscrossing the landscape. Houses, roads, urban centers and millions of acres of cropland now cover some formerly suitable islands of forest habitat. In the event of climate change, a valuable commercial species such as loblolly pine may have sought to use these islands as stepping stones to increase its potential range by 40 million acres, according to some climate models. But portions of this path are now blocked by "concrete jungles" and by recalcitrant farmers who do not want their farms to revert to forests. That may confine the expansion routes for loblolly pine and other species to steeply sloping highlands, where there are fewer competing uses for the land. Such difficult terrain would increase the costs of stand management, at the same time that shallow, coarse-textured soils would reduce the forests' productivity.[48] Even if the added annual costs of stand management totaled only a dollar an acre (a conservative estimate to be sure), the extra expense would total tens of millions of dollars a year for the forest products industry's relocated forest plantations.

Fortified Air: CO_2 Enrichment

While most climate-forest analyses performed to date have failed to account for all of the factors hindering forests' migration capabilities, they also have tended to leave out one attribute of the greenhouse effect that affects forests favorably: carbon dioxide enrichment. Since CO_2 enrichment is positively correlated with photosynthesis and water use efficiency by plants and trees, climate change may not affect forests as severely as some studies suggest.

Numerous laboratory tests show that most trees respond favorably when the amount of carbon dioxide in the atmosphere rises. One laboratory study of sweetgum (a southern hardwood commonly used in furniture and veneer) detected a 75 percent jump in productivity when the trees were exposed to air containing 600 parts per million of carbon dioxide—nearly double today's ambient level.[49] A recent outdoor test of white pine and big-tooth aspen yielded a 20 to 30 percent increase in biomass accumulation over one summer growing season, relative to untreated counterparts, when they were exposed to 700 parts per million of CO_2 in open-top chambers.[50]

"In other words, the level of CO_2 now available in the atmosphere may be stunting growth," concluded a group of scientists who reviewed several CO_2 enrichment studies. "More CO_2," they said, "may help trees grow faster."[51] Perhaps more importantly, trees grown in a CO_2-enriched environment are likely to perform better in an otherwise stressed environment. The aspens and white pines, for example, grew well despite nutrient-poor soils in Michigan. The study of sweetgum, moreover, found that the seedlings continued to grow well even when water-stressed, with productivity rates increasing 60 percent above nominal levels. Yet some studies suggest that low nutrient levels ultimately limit the boost offered by the CO_2 fertilization effect, even though CO_2-enriched plants do better in nutrient-poor soils than non-CO_2-enriched plants.[52]

Other studies have found that tree species generally are able to use water more efficiently as CO_2 levels rise.[53] This improvement is caused by constriction of the tiny openings in the leaves of trees, known as stomata, when more CO_2 is present. With evapotranspiration reduced, leaf water potential and cell turgor are maintained at adequate levels for longer periods, postponing dehydration in the event of drought.

Another favorable result of CO_2 enrichment is that the optimal temperature range for photosynthesis increases by as much as 9 degrees F among many plants in laboratory tests. (This upward shift is fostered by the increased rate at which carbon is broken down into organic enzymes and the decreased rate at which photorespiration occurs.) Thus, global warming could allow for faster accumulation of biomass because of higher temperatures and CO_2 enrichment, and reduced vulnerability to drought because of lower stomatal conductance. These biological responses to climate change could enable biomass to grow in more hostile environments. "If carbon dioxide levels rise as predicted, we could see a shifting of the boundaries, with the forest moving toward the prairie and the prairie edging out the desert," reports Dr. Hyrum Johnson, a U.S. Department of Agriculture scientist who has studied the effects of CO_2 enrichment on

> **Box 3-B**
>
> ### Trees, Climate and the Global Carbon Cycle
>
> Trees have been in existence for some 350 million years, since before the age of the dinosaurs. Large ferns and other vascular plants were the world's first tree-like structures on the super-continent known as Pangea. They were followed by seed plants that spread huge swaths of evergreen—primitive ancestors of today's coniferous forests. After Pangea split into the seven continents as we know them today, the forests grew exposed to a much wider range of climatic conditions. Lush evergreen forests remained clustered around the equator, where tropical rainforests are now. Temperate forests—featuring a mixture of broad-leaved deciduous trees and coniferous softwoods—evolved in mid-latitude regions. And closer to the poles, spruce-dominated boreal forests spread from Alaska to Labrador in North America and across northern Eurasia.[1b]
>
> Eventually, forests grew to cover two-fifths of the Earth's land surface. The associated rise in photosynthetic activity greatly altered the composition of gases in the atmosphere—and the global carbon cycle. Trees filter carbon dioxide gas through chloroplasts in their leaves, converting carbon to starches and sugars, returning the leftover oxygen to the air. In this way, trees came to serve as a lung for the planet, sustaining oxygen in the atmosphere while converting CO_2 into biomass with the aid of photosynthesis and carboxylation. (Ocean phytoplankton are the other Earth's other lung, taking up about half as much CO_2 as terrestrial plants and trees.)[2b]
>
> The Earth's breathing pattern is easily revealed by plotting the difference between summertime and wintertime levels of airborne carbon dioxide. During summer in the Northern Hemisphere, when photosynthetic activity is at its peak, some of the carbon dioxide leaves the atmosphere as the world's forests and other biomass draw a collective breath. In the winter, as this photosynthetic activity declines, the atmospheric level rises again. At the equator, the annual flux amounts to 5 parts per million out of 356 parts per million as a yearly average. Closer to the poles, the difference between winter and summer concentrations is as much as 15 parts per million, because the seasonal changes are more pronounced.[3b]
>
> Despite these seasonal variances, the yearly balance of carbon dioxide in the

woody plants.[54] This would be contrary to what happened during the Altithermal period, when the prairies encroached on the forests.

A number of caveats are in order, however. First, test simulations are rarely conducted over long periods. Some studies suggest that the photosynthetic rate of herbaceous plants may slow down once the plants have adjusted to a CO_2-enriched environment, making any boosts in productivity short-lived.[55] Moreover, laboratory tests offer only a crude simulation of real-world conditions. More outdoor tests using open-top chambers will be necessary before scientists can draw any firm conclusions about the long-term response of trees and plants to CO_2 enrichment.

A second caveat is that not all trees respond favorably to CO_2 enrichment. Studies of loblolly pine and Sitka spruce seedlings find no significant differences in water retention under normal and CO_2-enriched conditions.[56] In the event of drought stress, other species sharing the forest ecosystem, such as atmosphere remained fairly constant for more than 10,000 years after the last Ice Age. But the planet's breathing pattern recently has failed to maintain a balance in the global carbon cycle—now that humanity releases 6 billion tons of carbon from fossil fuels directly into the atmosphere each year.

A decline in forest coverage also is partly responsible for the carbon upset. More than 2 billion acres of the 12 billion acres of forests that covered the Earth before the Neolithic revolution have disappeared.[4b] Virgin tropical rainforests (and some old-growth stands in temperate regions) are being felled at rapid rates, often for short-lived purposes—fuelwood burning, providing pulp for disposable paper and simply clearing space for an expanding human population. The result is that humanity's treatment of the forests on a global level is exacerbating the buildup of carbon dioxide in the atmosphere rather than offsetting the increase.

According to the United Nations, the Earth is losing 42 million acres of tropical forests each year to development—an area greater than the size of Florida. (There are roughly 5 billion acres of tropical forest altogether.) Even more alarming, the rate of deforestation is 50 percent greater than only a decade ago. From a global warming standpoint, tropical forests are the worst kind to lose, since they are able to fix five to 10 times as much carbon per acre as temperate forests—owing to their higher rate of photosynthesis. (And depending on their soils, tropical forests can sequester 20 to 100 times as much carbon per acre as the cropland to which they are often converted.)[5b]

Estimates of the enormous amount of carbon dioxide and other greenhouse gas emissions that result from slash-and-burn clearing and decay of felled trees range from 1 billion tons to 2.5 billion tons of carbon annually, or about 17 to 42 percent as much as that produced by combustion of fossil fuels. While reforestation in theory could absorb all of the excess carbon from fossil fuel burning and deforestation, it would require a huge land mass of 1.15 billion acres, nearly equal to the size of the western United States.[6b] Although a billion acres of degraded and abandoned land exists worldwide, no one expects that amount of reforestation to occur. Moreover, the storage of carbon in trees is only temporary. After trees die, they eventually give back the carbon they had sequestered. Accordingly, tree planting is a helpful—but by no means lasting—solution to global warming.

sweetgum, may actually outcompete these species.[57] Moreover, competition among tree species may diminish their *collective* response to CO_2 enrichment. Recent experiments in two deciduous temperate forests and one Mexican rain forest found that tree seedling communities were not more productive in a CO_2-enriched environment when different species were grown together.[58] (Of course, this particular finding may not be of concern to industrial forest plantation managers who raise monoculture stands. But owners of coniferous plantations might have to defend their crops against encroachment by broad-leafed species benefiting from CO_2 enrichment.)

A third limitation of most CO_2-enrichment studies is that they usually test only one variable at a time. Researchers may deprive a CO_2-enriched plant of water, for example, to gauge its water use efficiency under drought stress. In the real world, climatic stresses such as drought and heat usually initiate a productivity decline, followed by biotic stresses such as defoliating insects and root-infecting fungi that ultimately causes plant mortality.[59] In the case of global warming, several important variables are likely to change at once: atmospheric CO_2 content, precipitation amounts, temperature variability, pest attacks and infiltration by competing species, to name a few. Such forces working together will determine the fate of the forests in the event of climate change, not the sum total of such forces acting on their own. "Thus it is unwise to attempt to extrapolate from short-term experiments to long-term effects of exposure to high concentrations of CO_2," warns Duke University's Paul Kramer.[60]

Pests, Fire, Wind and Chaos

When one talks to foresters about the ways in which climate affects their livelihood, much less concern is focused on the possible long-term consequences of global warming than on extreme weather-related events that may decimate their forest groves at any time. The fact is, drought-induced pest outbreaks and wildfires kill more American trees each year than loggers fell with chain saws and axes.[61] Occasionally, a killer hurricane fells more trees in a 72-hour period than the forest products industry harvests in an entire year.[62] Climate-related events such as these pose the clearest threat to the successful establishment of new tree plantations—and to their survival through harvest time decades later.

Chaos: Suppose a climate modeler were to issue a seemingly reassuring climate forecast that no change in average precipitation was expected in a particular timber-growing region over a given decade. If the decade began with three years of severe drought, followed by three years of extreme wetness, and then three years of normal precipitation, "average" precipitation over the period would show no change. Yet the initial drought period would kill many seedlings, and the subsequent wet period would spread root damage among surviving saplings; this is why the frequency of extreme events matters more to foresters and farmers than averages. Consequently, all weather predictions must be taken with a grain of salt, since chaotic weather patterns are an inherent part of natural climate variability.[63]

Chaotic weather changes could be especially damaging to trees grown in regions subject to freezing temperatures. Unseasonable warming during the winter months might induce trees to leave their dormant phase and "break bud" as if it were springtime. If freezing weather returned after the buds were out, the foliage could be decimated. Abnormally warm and wet weather during late autumn also could fail to send the signal for trees to "harden off" for winter, similarly exposing them to frost.[64] Ozone pollution and excessive nitrogen may exacerbate some trees' vulnerability to the cold (such as red spruce), since these pollutants also inhibit the tendency to harden off. (Nitric acid is a component of acid rain.)[65]

Pests: Perhaps the best way to understand the impact of climate anomalies on forests is to consider their relationships to pests, fire and wind. Longer, warmer growing seasons may permit one to three additional generations of pests to ravage the forests each year—while enabling them to reach places they previously avoided because of the cold.[66] In the Pacific Northwest, the balsam wooly aphid might be able to extend its range upward into stands of subalpine firs, which have a low resistance to this juice-sucking insect.[67]

Forests suffering from drought are prone to attack by boring insects. Severe spruce budworm outbreaks occurred in 1912 and 1949, for example, when fall weather was abnormally warm and dry. More recently, generally dry weather conditions have contributed to a spruce budworm epidemic in 2.7 million acres of forest in the Blue Mountains of Oregon.[68] Meanwhile, back on the East Coast, researchers have found southern pine beetles spreading from the Coastal Plain and Piedmont regions of the South into the surrounding mountains—also because of extended drought during the latter half of the 1980s.[69] If this climate pattern persists, infestations of southern pine beetles could increase throughout the South, particularly among even-aged pine stands. At present, southern pine beetles cause average losses of $30 million a year.

Fire: Another effect of drought on forests was vividly displayed in the rash of fires that plagued the American West during the sizzling summer of 1988. Extremely hot and dry conditions that persisted most of that summer may have

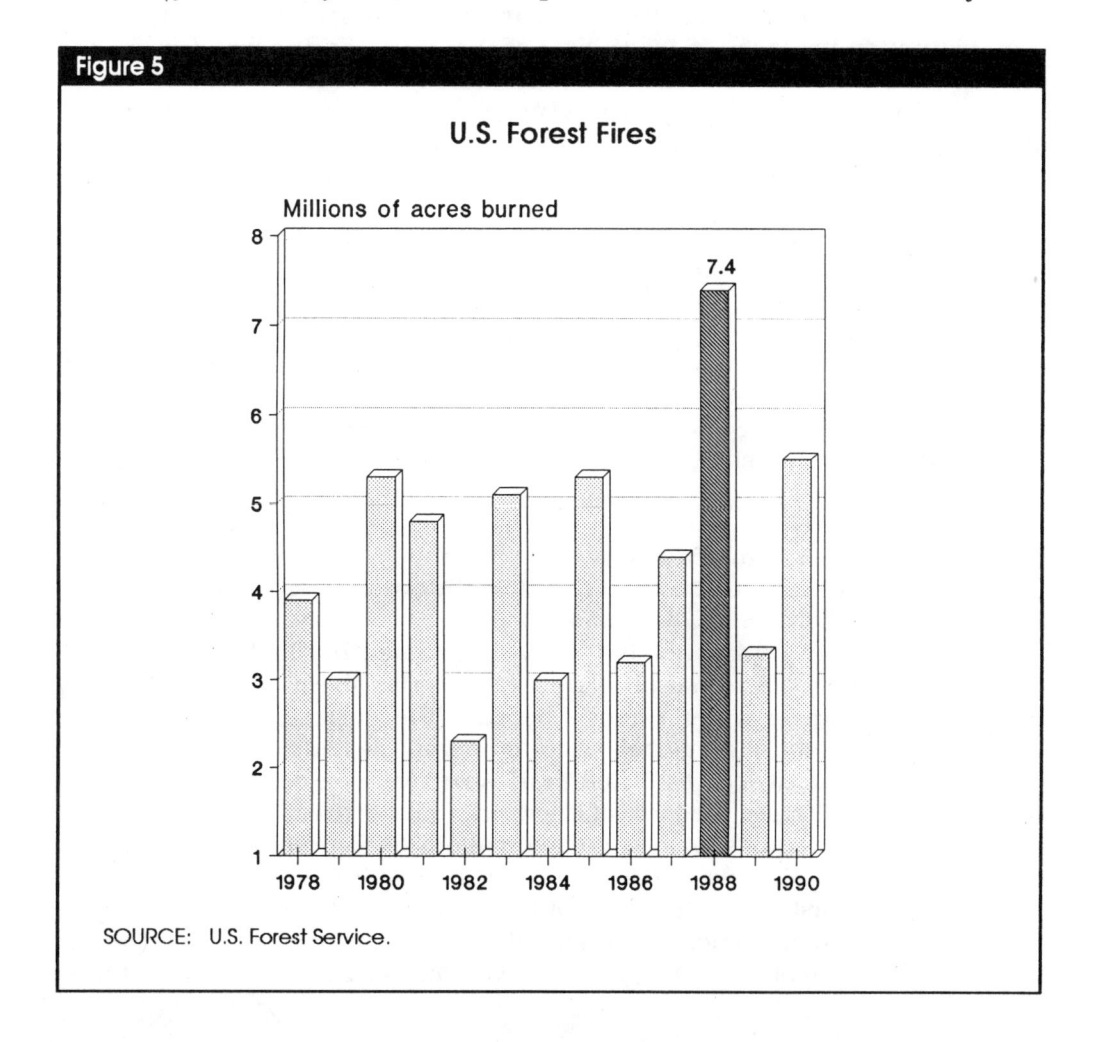

Figure 5

U.S. Forest Fires

Millions of acres burned

SOURCE: U.S. Forest Service.

been a harbinger of global warming. Nearly 7.5 million acres of forest were swept by fire that year—including one-quarter of Yellowstone National Park—the most in 60 years.[70] Even more devastating fires struck drought-plagued western Canada the following summer. A record 15 million acres burned in Canada in 1989—25 percent more than in the previous record year of 1980. After the 1980 fire season the Canadian government reported that "an upward trend in burned area is linked both to the increased use of the land by people, *and to worsening climatic trends in the past two to three decades.*"[71]

One study by researchers at the University of California at Berkeley concludes that a two-fold increase in fires of several hundred acres would be likely in the event of climate change and that a three-fold increase would be expected in major fires of more than 25,000 acres.[72] A separate study cited by the U.S. Environmental Protection Agency estimates that total fire occurrence might increase 8 percent as a result of global warming, with fire-suppression costs rising 20 percent.[73]

Forestry officials point out that the incidence of fire is lower on industry plantations because such trees are comparatively young and vigorous. Moreover, plantation managers remove most of the "litter" on the forest floor. By contrast, the litter in Yellowstone National Park had accumulated for decades, serving as kindling when the 1988 fires broke out. Not all impacts of forest fires are negative, however. They tend to clear out the litter and rejuvenate life on the forest floor, while leaving many standing trees alive and well. In Yellowstone, only 20,000 acres of the nearly 1 million touched by flame in 1988 were so scorched that plant life could not renew itself in short order. Moreover, important timber species such as lodgepole pine and western larch regenerate especially well after fire. Were the incidence of wildfire to increase as a result of global warming, their populations, at least, would be likely to grow.[74]

Wind: Perhaps the most extreme weather change that may result from global warming is an increase in the frequency and intensity of killer hurricanes. In September 1989, Hurricane Hugo slammed into the Marion National Forest, north of Charleston, S.C., packing winds of 175 miles an hour. Fifty percent of the trees in this 250,000-acre preserve came down—containing enough wood to circle the globe seven times if cut into one-inch-by-twelve-inch boards.[75] Altogether, Hurricane Hugo toppled 5 million acres of woodlands—more than hurricanes Camille and Frederick, the eruption of Mount St. Helens and the 1988 Yellowstone fires combined.[76]

The strength of hurricanes is largely determined by the temperature of ocean waters; the more sea surface temperatures rise above 80 degrees F, and the deeper the warm water extends beneath the surface, the greater the hurricanes' strength. With global warming average sea surface temperatures could increase several degrees F, and the destructive force of hurricanes could rise by 40 percent, since the force of the wind is proportional to the cube of its speed.[77] If hurricanes of similar or greater magnitude than Hugo make landfall in the years ahead, they may pose the single greatest threat to forests growing adjacent to the East and Gulf coasts of the United States.

At the same time, hurricanes may serve one constructive purpose: blowing the seeds of trees to places where they have never grown before. As such, these powerful storms could aid in the forest migration process that global warming might compel. Meanwhile, trees in the blowdown area would provide a rich carbon base on which new forests could sprout—introducing, perhaps, new species whose seeds originated from some earlier point along the hurricane's track.

Climate Change and Timber Management

Spring planting: While pests, fire and wind raise serious concerns for foresters as their stands mature, the greatest threats of all from global warming are likely to come at the beginning and end of the harvest cycle.[78] In order for seedlings to survive, they must have suitable air and soil temperatures and sufficient moisture. When, for instance, a severe drought struck Douglas-fir forests of southern Oregon in 1988, newly planted seedlings in their first field season suffered many casualties, even though better-established groves were barely affected. Should adverse climate change increase drought, wind and temperature stress, seedling survival rates could drop from 80 to 90 percent on average to 35 to 50 percent or less, especially in exposed clear-cut areas.[79]

Problems also could develop if spring rains gave way more quickly to summer dryness, in effect, allowing the onset of summer conditions in the latter half of spring (traditionally a critical period for the establishment of seedlings). To compensate for this seasonal change, forest nurseries might consider planting their seedlings earlier in the spring—keeping their fingers crossed that frost would not return. Alternatively, they could provide more containerized seedlings to reduce transplant shock and permit more rapid growth during a compressed spring-planting season. And if all else fails, foresters have the option of pulling newly planted seedlings out of the ground and returning them to cold storage until more favorable planting conditions prevail.

Drier soil conditions also would increase the importance of proper soil preparation and herbaceous weed control, so that seedlings in their early growth phases would establish an adequate root system and maintain a sufficient water supply. At the other end of the harvesting cycle, winter logging operations could be extended by good weather—or hampered by mild, rainy weather that thaws the ground and mires logging equipment in mud. Overall, costs of plantation management would be likely to rise as a result of climate change.

Earlywood/latewood: Another consideration for foresters is that initiation of trees' growth phase up to eight weeks earlier in the spring could increase the proportion of so-called "earlywood" relative to "latewood."[80] These two kinds of wood are revealed when a tree is cut and one examines the annual growth rings. The light, thin-walled earlywood cells are produced during wet spring weather; the dense, thick-walled latewood cells are produced during dry summer months. Latewood contains up to four times as much cellulosic material (and hence carbon) as earlywood. Lumber yields generally decline as the percentage of earlywood in the total harvest increases, since lumber containing mostly earlywood often warps. In addition, a shorter period of latewood production creates wood of lower density and, thus, of poorer quality and strength.

Lower-density wood also increases the overall volume of wood required by paper mills and decreases pulp yields (all other things being equal), causing the plants to run below their rated capacity.[81] Expanding plant capacity on a retrofit basis to compensate for this deficiency would entail substantial redesign of the mills and a large investment of capital. Building new pulp and paper mills also is an expensive proposition, averaging $500 million per plant.[82]

Forest products companies believe they may be able to overcome the earlywood problem by selecting families of trees that produce denser wood (i.e., have a higher specific gravity). Some companies, in fact, already consider specific gravity in their genetic screening programs.[83] Researchers also point out that seedlings planted earlier in the spring could make a faster transition from earlywood to latewood production. One study of Douglas-fir appears to confirm this hypothesis.[84]

Water use and sea level rise: The pulp and paper industry is highly dependent on waterways as a means of transporting raw logs, manufacturing pulp and disposing of effluent. Diminished stream flow caused by episodic drought could cause some companies to curtail or suspend their mill operations so as not to exceed permissible water quality standards. In the South, most of the pulp and paper industry is concentrated in coastal areas with good access to water supplies and international ports. This does not eliminate water concerns entirely, however. If sea level rises significantly as a result of global warming, some low-lying pulp mills might have to relocate or invest considerable sums to protect their operations from invading waters.

One analysis of the pulp and paper industry in the South concludes that 16 mills constituting nearly 15 percent of the region's operating capacity could be inundated if sea level were to rise by more than nine feet.[85] In the Pacific Northwest, the output of only a single mill, representing barely 2 percent of that region's considerably smaller pulp-and-paper capacity, would be jeopardized by a nine-foot rise in sea level. The majority of plants in the Pacific Northwest are sawmills or other solidwood facilities that could relocate with greater ease and less capital investment in case of sea level rise. (The cost of a new sawmill usually ranges from $10 million to $20 million.)

These projected impacts in this analysis are probably overstated, since the rise in sea level during the 21st century is not expected to be more than three feet. Moreover, the lead time with sea level rise is such that most existing mills should be retired before water encroachment develops into a serious problem. The data do suggest, however, that sea level rise warrants consideration in the siting of new mills.

Timber harvesting: One of the greatest threats to the forest products industry in the long run is that climate change may lower the yield curves of growing stock. Normally, trees are harvested about midway through their growth cycle, as the rate of biomass accumulation reaches its peak. (That way, more timber can be harvested over a series of rotations than if the trees are allowed to reach old age, by which time additional biomass accumulation has essentially stopped.) "For longer-rotation trees, the problem may be more difficult in that the environmental changes over their expected lifetime may be so great as to restrict their growth and perhaps force their demise before maturity," cautions Norman Rosenberg of Resources for the Future, an environmental think tank in Washington, D.C. "This condition could force a premature harvest," he says, "thereby compromising the economic returns."[86]

Studies by Daniel Botkin of the University of California at Santa Barbara add to this concern. Botkin, considered the "father of forest stand models," has performed a number of hypothetical studies of the effects of climate change on forest productivity and timber yields. One of his case studies examined what might happen to jack pine growing in central Michigan. Normally, these trees would be harvested at their mid-point of succession—about 50 years. With the onset of global warming, however, timber yields of jack pine might begin to decline rapidly, with no prospect for regeneration. Accordingly, the longer one waits to harvest the stands after climate change sets in, the lower the timber yields will be. This creates an incentive to harvest growing stands as soon as possible, Botkin concludes, even though the woodlands would be decimated by the extreme harvest and prices would be depressed by the flood of timber entering the market.[87]

On a more macro-economic level, Clark Binkley of Yale University's School of Forestry has considered what might happen to timber income in the United States and elsewhere as a result of major shifts in the locations of temperate and boreal forests. His conclusion, based on an analysis of the Goddard Institute climate model, is that expansion of forest-growing habitat in northerly latitudes would benefit countries such as Canada, Finland and the Soviet Union, since greater timber inventories would lead to an expansion of production capacity. In turn, harvest rates would increase in these countries and timber prices would

Table 2

World's Top 15 Timber Producers, 1988*

Country	Volume (million cubic meters)	Share of Total (percent)
USA	417	25
USSR	305	18
Canada	173	10
China	98	6
Brazil	67	4
Sweden	48	3
Finland	46	3
Indonesia	40	2
Malaysia	36	2
France	32	2
W. Germany	31	2
Japan	28	2
India	24	2
Poland	20	1
Australia	18	1
Others	281	17
World Total	1,664	100

* Includes all wood products except fuelwood and charcoal.

SOURCE: U.N. Food and Agricultural Organization, 1990.

fall. Countries lacking boreal forests would be on the flip-side of these developments, however. Binckley estimates that income from timber sales in the eastern United States might fall as much as 20 percent through the year 2030 as a result of projected global warming. Timber income in the western United States might fall as much as 26 percent.[88]

On this basis, global climate change does not appear to be a welcome prospect for investors in the American forest products industry. Yet one must recall that future climate may bear little resemblance to that portrayed in the Goddard climate model (or any of the other general circulation models). Nevertheless, the certainty of CO_2 enrichment combined with the possibility of migrating forests, greater pest infiltration, more forest fires, stronger wind storms and overall chaos in the weather poses a major challenge for the forest products industry. Industry officials point out that they have already dealt with year-to-year variability in the climate that resembles some of the scenarios for global warming and still met their planting and harvesting requirements. The question is not so much whether they will be able to continue to do so in the future, but at what cost.

Responses of the Forest Products Industry

Industry foresters are wary of predictions of calamity caused by climate change. The geographic resolution of general circulation models is much too coarse, they say, to allow any firm predictions about the fate of the forests. Yet ecological response models tend to be placed on top of this shaky foundation, and economic analyses are conducted on top of that. Accordingly, they fear that a propagation of errors throughout the review process may lead to unfounded conclusions about climate change and its effects on the forest products industry.

Since actual evidence of climate change remains nearly as inconclusive as the models projecting it, the forest products industry has been largely in a wait-and-see mode with respect to the issue. "If current practices are any indication," James Woodman, director of the atmospheric impact research program at North Carolina State University, observed in 1987, "significant changes in seedling survival rates, growth rates of mature trees, or incidence of disease or insects would have to be detected first before most managers would let the changes influence their decisions. A manager's willingness to act on this new information," Woodman said, "will depend on the financial risks the organization is willing to accept."[89] Even now, forest industry executives show few signs that they are acting on new information about the greenhouse effect, although their level of interest continues to rise.

Addressing Pollution Problems

The forest products industry took its first organized look at the global warming issue in 1984. The National Forest Products Association in cooperation with the Conservation Foundation and the Society of American Foresters sponsored a conference in Boulder, Colo., to "address the effects of the 'greenhouse phenomenon' on climatic conditions and the risks and opportunities for future forest management."[90] A likely impetus for the conference was the release of two major studies on climate change by the National Academy of Sciences and the U.S. Environmental Protection Agency in late 1983.

Five years later—following the greenhouse summer of 1988—representatives of three trade associations formed the "Forest Industry/Global Climate Change Task Force." This task force consists of representatives of the American Paper Institute and the American Forest Council as well as the National Forest Products Association. It provides the industry with a collective voice on policy questions related to climate change. In public statements, the task force has recommended that priority be given to improving the resolution of general circulation models, ferreting out the possible regional effects of climate change and calculating the associated impacts on forest system reproduction, migration and productivity. Before the nation launches a major program to

> **Box 3-C**
>
> ### Bridging the 'Generation Gap' in U.S. Forest Resources
>
> While most forest products companies regard themselves as tree-growing companies, the great timber barons of the early 20th century—and some of their modern-day successors—have garnered reputations as plunderers of the forest. The rapid harvest of ancient redwood groves by takeover artist Charles Hurwitz, who purchased Pacific Lumber Co. in 1986, is but a recent example in which timber interests have cashed in on a natural inheritance passed down through the generations. At the turn of the century, it was the Paul Bunyon timberlands in Minnesota that fell before the axe—with timber interests pulling up stakes and heading West soon after the land was cleared. Virtually all logging before World War II, in fact, took place on privately owned tracts, and very little acreage was replanted.
>
> President Theodore Roosevelt was among the first American statesmen to recognize the value of preserving forest resources for future generations. In 1905, he established a national forest system, mostly on federal lands west of the Mississippi River, declaring: "A people without children would face a hopeless future; a country without trees is almost as hopeless." Roosevelt went on to appoint Gifford Pinchot, a close friend and a member of his administration, to serve as the U.S. Forest Service's first director.
>
> It was Pinchot who laid out a conservation paradigm that remains official Forest Service policy today: to manage the multiple, competing uses of forests in a manner that raises current living standards yet affords these same opportunities to future generations. Still, Pinchot thought of forests mainly as a commodity, often referring to them as "tree factories" and "wood farms."[1c] Nevertheless, the national forest system he helped create was left relatively untouched by commercial loggers for four decades.
>
> The clearcutting and lack of replanting on private land in the first half of the century created what some foresters now refer to as the "The Gap"—a decades-long

ameliorate the greenhouse effect, the task force emphasizes, policymakers should consider whether it might be more beneficial to address other problems with a more tangible environmental impact.[91]

As for science-related research, forest products companies are relying mainly on the National Council of the Paper Industry for Air and Stream Improvement, an industry-supported research group, to address global warming as part of a broad array of environmental issues that affect forest health. NCASI in turn cooperates with government agencies such as the U.S. Forest Service and the Environmental Protection Agency as well as universities and member companies on research projects.

NCASI began formal study of the global climate change issue in 1990, spending $250,000 as part of its research program on air quality and forest health. (NCASI has a total of eight such research programs.) In 1991, NCASI budgeted $400,000 specifically for climate change research, marking the first time it had earmarked funds explicitly for that purpose. It is spending about $400,000 on climate change research in 1992 as well.[92] NCASI's current research priorities include: "(1) evaluating the potential effects of global change period in which few second-generation trees on private land would be of merchantable size. To bridge The Gap, forestry interests turned to the nation's last great untouched preserve—the national forest system created by President Roosevelt. Harvesting the old growth in these forests helped avert a timber supply crisis during the post-World War II housing boom. While concerns about overharvesting began to mount again in the 1960s—prompting Weyerhaeuser and other forest products companies to step up their own reforestation efforts—the size of timber cuts in the national forests continued to grow. In the 1970s and 1980s, the ancient timber in national forests again played a crucial role in alleviating a potential lumber shortage—this time for post-war baby boomers purchasing new, wood-framed homes of their own.

Today, the harvest of old-growth timber in the continental United States is almost complete. Only 5 percent of all the virgin stands in existence in 1600 remain. Those still around are almost exclusively in the Northwest, nestled near the Pacific coast or draped along the Cascade Range.

Efforts to preserve the remaining old-growth stands will make it harder to bridge The Gap in the years to come. In 1982, the U.S. Forest Service issued a major econometric study that forecast a 45 percent jump in global demand for U.S. wood and wood-fiber products by the turn of the century, on the heels of a 90 percent increase since the early 1950s.[2c] A more recent projection forecasts a 50 percent increase in demand for U.S. forest products—to 29 billion board-feet—by the year 2040. Over this same 50-year period, however, the total acreage of commercial timberland is expected to shrink by 4 percent, to approximately 460 million acres.[3c] Such projections of rising consumer demand and diminishing timberland acreage prompted the American Forest Council, an industry trade group, to declare in 1989 that "the forest products industry faces a timber supply crisis of unprecedented proportions."[4c]

For now, The Gap lives on.

on forests; (2) determining the role of forests and forest products in the global carbon cycle; and (3) evaluating the validity of current models used to simulate the response of forests to climate change scenarios."[93] Climate change research now is one of NCASI's largest areas of forest-related expenditure. Wildlife studies garner a larger share of NCASI's current budget (which is not surprising considering the controversy over the spotted owl and other endangered species). Acid rain, on the other hand, is receiving only half as much funding—$220,000—in 1992.

In many respects, climate-related research picks up where the forest products industry's studies on acid rain and ozone pollution left off. Acid rain is an environmental problem that the industry has been wrangling with for more than two decades. The industry's handling of that problem may be indicative of the way it approaches the global warming issue as well: Research will be paramount. Confronted with alarming evidence of forest damage in Europe and a decline of high-elevation red spruce in the United States, the forest products industry and the government-funded National Acid Prepicitation Assessment Program embarked on a 10-year, multi-million-dollar research effort in 1981.

A decade later, NAPAP reported that it had found no conclusive evidence that acid deposition was causing widespread decline of the nation's forest resources, despite some localized damage.[94]

While the NAPAP report is considered the definitive assessment of American forest health, not all forestry groups agree with its conclusions. The American Forestry Association, for one, claimed in 1987 that a variety of pollutants, including ozone, airborne sulfur, excessive nitrogen, toxic aluminum and other metals—along with acid rain—are combining with natural stresses and contributing to a decline of American forests, especially in the East. While the forest products industry fully acknowledges ozone pollution as a serious concern, it dismissed the association's broader findings as a ploy by conservation-minded foresters who favored amendments to the Clean Air Act. "We see no convincing evidence that further controlling air pollution on a national scale will improve the health of the forests," declared John Thorner, environmental counsel for the National Forest Products Association, in 1987. "We do see risks in acting," he added, such as raising the price for electricity, which pulp and paper companies consume in great quantities.[95]

Such tension may surface again in the greenhouse debate. Just as power plants and automobiles are leading causes of ozone pollution and acid rain, they also are chief contributors to the greenhouse effect. Forest industry leaders insist that they would favor tougher controls on these emissions sources if they believed that the health of the forests were at stake. Yet forest industry executives are reluctant to advocate greenhouse gas controls for at least three reasons. First and foremost, they see no compelling evidence that climate change is indeed jeopardizing the health of their forests. (And even if a decline in forest productivity were underway, it would be difficult to blame on climate change, since many other environmental factors are at work and because it is not yet clear whether global warming has begun.) Second, policies to address global warming may have broad economic impacts that adversely affect the highly sensitive forest products business. Third, industry executives do not want to alarm shareholders unnecessarily about the condition of their greatest asset—the forests—unless they are convinced that a change in government policies would be likely to improve their prospects.

Industry Views on the Greenhouse Effect

A prudent first step for forest products companies would be to gather more information on these outstanding scientific and policy-related questions. As a 1986 report prepared for EPA by the ICF consulting firm concluded, "In the case of CO_2-related projects, simply raising the issue and incorporating it into the (forest products) firm's ongoing planning process may be a good start."[96] Many companies have indeed taken this first step, or at least contributed money to the industry's National Council to study the issue on their behalf.

But a 1990 survey conducted by the Investor Responsibility Research Center finds that the climate change issue has yet to become a major factor in the strategic planning of the forest products industry. Ten companies possess-

Table 3

Commercial Uses of U.S. Forests

Type	Description	Total size (thousands of acres)	Management intensity
Managed natural forest	Commercial forest land, recreation and wilderness areas	470,000	Medium
Suburban forest	Specimen, shade, ornamental trees	200,000	High
Urban forest	Park, street trees	100,000	High
Plantation forest	Row pattern planting with regular spacing	50,000	High
Christmas tree plantation	Short rotation, holiday market	500	High
Forest tree nursery	Seedling production areas	15	Very high
Forest tree seed orchard	Seed production areas	10	Very high
Arboretums	Trees established for study, enjoyment	10	Very high

SOURCE: William H. Smith, Yale University, 1991.

ing nearly half of the industry's owned timberlands consistently ranked global warming as presenting the least threat to their timberlands, when compared with eight other environmental stresses over the next five and 50 years. In addition, more than 70 percent of the survey respondents strongly agreed with the statement, "Greater scientific certainty is required before our company includes global warming in its strategic planning."

Public comments made by major forest products companies are consistent with results of the IRRC survey. Georgia-Pacific remarked in a environmental report to shareholders issued after the drought summer of 1988: "Since forests could be affected by global warming, we must be aware of environmental changes and, specifically, the possibility of changing weather patterns. Given many scientific uncertainties, Georgia-Pacific sees no valid strategy for action on our forests at this time."[97] On the contrary, adds Richard Good, Georgia-Pacific's group director of corporate and investor communications, "We would argue that a far greater—and real—risk to the environment is ill-conceived public policy that harms the economic environment. Nothing will subordinate environmental concerns to bread and butter issues more quickly than a nasty recession, which could be triggered or aggravated by costly regulatory measures at the federal and/or state level."[98]

Besides, there may be benefits to global warming under certain conditions. As Georgia-Pacific's environmental report noted, "Some increase in CO_2 levels actually increases photosynthesis activity, thereby causing additional tree growth; warming also increases the ability of the atmosphere to hold water; rainfall in coastal areas actually could increase; forest growth may be enhanced by longer growing cycles due to global warming."

Few other forest products companies have issued written statements concerning the greenhouse effect. But oral comments of industry spokesmen make similar points. David Frankil, Champion International's director of government affairs and a member of the Forest Industry/Global Climate Change Task Force, told attendees at a May 1990 conference in Washington, D.C., on forests and climate change: "If global warming occurs, there is sure to be major economic and environmental change. But there is massive uncertainty about where and when it might occur. Before we act, we need as complete a knowledge base as possible. We do not advocate rushing to implement policies that are not cost-effective." Frankil went on to say that "maintaining a healthy tree base is the best insurance policy against stress caused by climate change," adding that carbon sequestration through "active forest management is the best way to work against global warming."[99]

At Champion International, managers of its forest operations brief the company's vice president of timberlands quarterly on any important developments in climate change research. In addition to tracking studies conducted by NCASI and other outside groups, Champion has conducted its own research on the development of drought-resistant loblolly pine in Texas and drought-resistant slash pine and sand pine in Florida. The company is not planning any specific actions on the greenhouse effect until more is known about it, however.[100]

Perhaps the company that has done the most to incorporate climate change research into its strategic program is Weyerhaeuser. Stirred by findings of the government reports issued in the early 1980s, Weyerhaeuser became a charter member of the Corporate Affiliates Program at the National Center for Atmospheric Research in 1985 to facilitate scientific exchanges on the subject. Since then, Weyerhaeuser has committed staff time of an in-house librarian and a researcher to stay abreast of the latest scientific developments. More significantly, in 1986 it adjusted its ongoing research priorities to become "technologically prepared" for the greenhouse effect.[101]

Peter Farnum, Weyerhaeuser's manager of strategic biology research, told attendees at the 1990 conference on forests and climate change that while others "have stressed that the rate of temperature change will be unprecedented in terms of natural history, I want to stress that during the next 40 years the rate of technological change will be unprecedented with respect to history. Thus the real challenge is for us to direct that process of technological change so we are prepared for the greenhouse effect when it occurs."[102] Weyerhaeuser's aim, in short, is to develop trees that can stand up to global warming.

Engineering Drought-Tolerant Trees

Genetic selection of trees is hardly a new idea, of course. Pines in Europe and America have long been bred for good form, reduced forking and pest resistance. But selecting for characteristics that may become critical in the event of climate change is a relatively new concept. Weyerhaeuser's genetic research program points to a greenhouse gambit that many forest products companies may in fact come to employ in the years ahead.

In Weyerhaeuser's case, a decision was made in 1980 to expand the company's research on trees' genetic tolerance of drought. That summer, Weyerhaeuser's tree plantations in Arkansas and Oklahoma—spread over 1.5 million acres—endured some of the hottest and driest weather in that region's recent history. Average summertime temperatures ran 6 degrees F above normal, and only two inches of rain fell from practically the beginning of the summer to the end. Loblolly pine seedlings that had been planted that year or during the previous year experienced significant mortality; many of the plantations had to be replanted.[103]

The loblolly seedlings that fared the worst during the drought had been planted in shallow, sandy soils. It was also determined that seedlings of loblolly genetic families whose parentage was from North Carolina (where Weyerhaeuser owns 600,000 acres of timberland) succumbed more to the drought than local loblolly seedlings. This finding was particularly unfortunate, because the North Carolina provenance had a superior growth rate in non-drought years. Based on its 1980 experience, Weyerhaeuser decided it would no longer plant the fast-growing North Carolina stock on 600,000 acres of land in Arkansas and Oklahoma that had experienced soil moisture deficits of 12 inches or more during the drought.[104]

In 1986, Weyerhaeuser accelerated its research on drought-resistant trees, based on new information concerning the greenhouse effect. Some general circulation models developed in the mid-1980s painted a particularly grim picture of the future for loblolly pine growing west of the Mississippi River. They inferred that average summertime temperatures in the region could rise 9 degrees F by 2080—3 degrees higher than the average for the summer of 1980. Making matters worse, the June through October soil moisture deficit could reach nearly 21 inches, practically twice the limit set by Weyerhaeuser for North Carolina loblolly stock planted in Arkansas and Oklahoma. If climate change occurs, "The future for loblolly looks grim indeed on shallow or sandy soils in eastern Oklahoma and western Arkansas," concluded one study by a team of university forestry professors.[105]

Even if this drastic scenario were to be realized, Weyerhaeuser remains hopeful that advances in genetic engineering would be able to meet the challenge. "Considering the tremendous advances in forest biological technology over the last 30 years," Farnum says, "it is not unrealistic to think that an ambitious genetic engineering program would be able to overcome potentially severe soil moisture deficits during the next 90 years of technology development."[106] Indeed, Weyerhaeuser and other companies already are exploring laboratory techniques to yield huge numbers of genetically identical tree embryos that could make seed orchards obsolete.[107] Someday, it may even be possible to implant tree embryos with especially drought-tolerant genes. Whether such an ambitious genetic engineering program would be able to overcome a soil moisture deficit as great as 21 inches, however, remains to be seen.

Drought is not the only threat posed by global warming, moreover. Pest infiltration and availability of soil nutrients are other considerations. Fortunately, genetic selection can be used for purposes of insect resistance, nutrient uptake and other growth characteristics of trees. Care must be taken not to develop a completely homogenous growing stock, however, lest the trees fall prey to other pests or disease that could wipe out an entire crop. While genetically selected plantations are potentially more susceptible to unforeseen environmental circumstances than genetically diverse forests, many researchers insist that the risks associated with gene-screening programs are limited.

Researchers R.C. Kellison and R.J. Weir of North Carolina University say it "borders on the ludicrous" to think that the genome of loblolly pine or Douglas-fir could be changed drastically in one generation, for example. "Forest trees, which have occupied a position on Earth for thousands of years and which have been without direct selection until about 35 years ago, are extremely heterozygous and therefore genetically diverse," Kellison and Weir maintain. "In comparison, the genetic diversity of corn, which has been selectively bred annually for hundreds of years, continues to exhibit tremendous variability....It seems equally farfetched to suggest that the offspring from one generation [of trees] will be significantly more prone to pest attack than are unimproved offspring." Nevertheless, "The best insurance policy is to maintain a broad genetic base from the start," Kellison and Weir aver.[108]

Growing New Crops

Someday, genetic engineering programs may create new families of trees that thrive under harsh weather conditions. Today, the emphasis is on understanding the specific climatic tolerances of existing commercial species, since this is the first step toward identifying genetic traits that may become more desirable in a greenhouse world. The next step will be for foresters to obtain more detailed information about the potential rate of growth of species at different locations within their range, species' prospects for territorial expansion and their means of migration, their resistance to new pests and diseases, and their susceptibility to fires and blowdowns.

A chart plotting the existing range and potential range of loblolly pine, for instance, finds relatively little overlap between the current growing range and the one projected under several climate scenarios. In areas of possible abandonment, serious consideration could be given to introducing new families of loblolly—or even entirely new species—once the existing rotations are harvested. The "lost pines" of Texas, for example, are an isolated family of loblolly known to be especially drought-hardy. If the climate of, say, Arkansas and Oklahoma were to become hotter and drier as a result of global warming, foresters could plant lost pine seedlings there in the soils with the poorest moisture retention. For that matter, Ponderosa pine from eastern Oregon—which get by on even less annual rainfall than the lost pines of Texas—could become a future planting option outside of their native territory.

Perhaps the most remarkable story of transplantation involves the Monterey pine of California. American foresters regard this conifer as an ornamental shrub with little commercial value. Yet Monterey pine has been bred to grow fast, straight and tall in Chile, Australia and New Zealand. In fact, these countries now are the lowest-cost producers of wood fiber in the world, all because a few Monterey pine left California decades ago—much by accident—in the ballast of ships headed Down Under.[109]

Transplantation of trees would not be required in all regions, however, since there are places where an overlap of current and projected growing ranges would be expected. One such place of overlap is east-central Tennessee.[110] Many species growing there are more or less near the midpoint of their north-to-south growing ranges. Thus, the initial phases of warming would not be likely to produce such extreme changes in temperature or precipitation that their growth would be stunted.[111] And to the extent that changes in climate did have an impact, less commercially valuable species such as shortleaf pine would be likely to be affected first.[112]

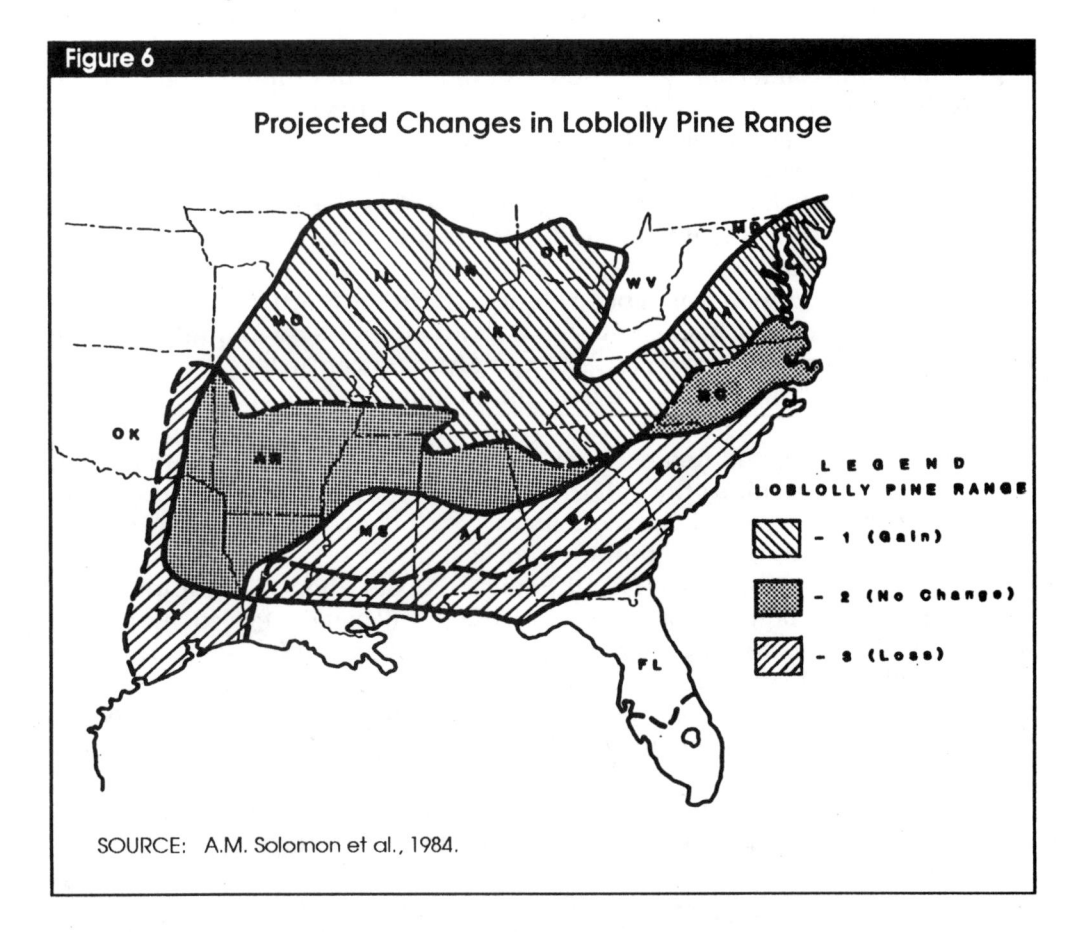

Figure 6

Projected Changes in Loblolly Pine Range

LEGEND
LOBLOLLY PINE RANGE
- 1 (Gain)
- 2 (No Change)
- 3 (Loss)

SOURCE: A.M. Solomon et al., 1984.

Farther to the north, the initial phases of global warming could promote climatic conditions favorable to the growth and expansion of both spruce-fir and boreal forests. As the warming continues, however, optimal climate conditions for spruce-fir might disappear faster than for the boreal forests. As a result, a forester's strategy could be to allow hardwoods to grow into the overstory of spruce-fir forests and harvest the dying softwood. Once the transition to hardwoods is complete, productivity on some northern sites actually could triple, because hardwoods accumulate biomass so much faster than softwoods. Eventually, however, the climate might become so warm that much of the boreal forests would die back as well as the remaining spruce-fir forests. At that time, it might become necessary to plant entirely new species, rather than let natural migration run its course.

The place where foresters would be likely to do their first experimenting with new crops is not in the North, however, but along the southern Coastal Plain, where the impacts of climate change may be the most immediate. Slash pine would be expected to expand northward, naturally filling a void left by retreating loblolly stands. Because slash pine generally grows straighter than loblolly, it could prove to be a superior replacement. If a subtropical climate were to develop in the South, it might make even more sense to plant fast-growing Caribbean pine or Eucalyptus, which are not native to North America.

Eucalyptus from Australia has been transplanted to many parts of the world, including California. It is known for its excellent pulping qualities and extremely productive plantations. Moreover, Eucalyptus matures in only seven years, so it is less vulnerable to long-term climatic changes wherever it grows. Eucalyptus is prone to cell damage from rapid drops in temperature, however (not just from freezing temperatures alone), so foresters have to weigh the consequences of exposing it to a variable temperate climate before they decide to plant the species.

A more radical option would be to plant kenaf in the South as an alternative to wood pulp for making high-quality newsprint. Kenaf is an annual fiber crop that has been grown as cordage in Asia and Africa for centuries. Seeds mature into 14-foot-tall plants in just five months, yielding about 12 tons of dry plant per acre. As such, kenaf produces nine times more biomass than wood on a per-acre/per-year basis, and its fiber cost is 12 to 20 percent below that of wood. Moreover, kenaf contains only half the lignin of wood, so energy requirements for pulping are reduced by 20 percent as well.[113] (Lignin must be removed from the cellulose fiber during the pulping process).

A joint venture company—Kenaf Paper of Texas—hopes to complete the nation's first kenaf pulp mill by the end of 1992. It will blend recycled newspaper fibers in a 50-50 mix with kenaf fiber grown in the Rio Grande Valley. The $47-million mill should be capable of producing about 27,500 metric tons of newsprint annually, equal to about 0.3 percent of U.S. newsprint demand. Equity partners in the project include Bechtel Enterprises, Sequa Investments and Kenaf International, which will provide the fiber for the mill.[114] CIP Paper Products of Montreal, Canada—North America's second largest newsprint manufacturer—also maintains a peripheral interest in the project.[115]

At present, the United States obtains two-thirds of its newsprint from Canada, at a cost of $3 billion a year. Large-scale kenaf production—grown in warm, well-irrigated fields south of the Mason-Dixon line—could reduce this deficit. At the same time, it could help newsprint manufacturers to scale up recycled newsprint capacity—now in short supply.

The forest products industry has barely scratched the surface when it comes to programs to introduce new species that may be of commercial value in a greenhouse world. Altogether, more than 50,000 forest tree species are known to exist. But only about 400 of these species are involved in testing and breeding programs. Of these, 60 tree species—all with commercial value—have been bred to the point that seed orchards or conservation stands now exist for them. The paucity of spending on genetic research for forest trees—estimated at only $5 million annually worldwide—is of great concern to the National Research Council, which advises the federal government on scientific and technical matters. Its Committee on Managing Global Genetic Resources concluded in 1991 that "no adequate global strategy exists for systematically identifying, sampling, testing and breeding trees with potential use."[114] The committee recommended a tenfold increase in the number of species included in study programs, with special attention given to development of improved varieties of trees for use in industry, agroforestry and rehabilitation of degraded lands.

Acquiring New Land

Perhaps the greatest greenhouse gambit that forest products companies could take would be to change their strategies on resource acquisition. At first blush, the question might seem as basic as whether to reverse the trend toward consolidation of timberland holdings in the South and diversify into other regions where global warming appears more likely to enhance forest productivity. Within the forest products industry, however, there is an even more fundamental question to address: whether to own timberlands at all.

From a business's perspective, investing in commercial tree plantations is a remarkably long-term proposition. Prized softwoods such as Douglas-fir and white fir grow on rotations of 50, 60, even 70 years; jack pine also requires up to 50 years to reach merchantable size. Northern hardwoods like maples, beeches and birches also require a half-century or more to reach the top of their growth curve. Some forest products companies would rather acquire mature timber on the open market than commit themselves to this painstaking development process. Yet for this alternative approach to work, the government and private timberland owners must make sufficient quantities of wood available at prices these companies can afford.

As long as timber can be acquired at reasonable prices on the open market, some forest products companies believe there is little point in acquiring an equity stake in the forests for themselves. Executives from Pope & Talbot, an integrated wood products firm based in Portland, Ore., stressed this point in a presentation before investment analysts in 1988:

> Tree farming looks to us to be a low return-on-investment business, particularly when something is factored in for risk....[I]f that's not bad enough, when you get ready to sell the trees you've been growing—and that may take 40 to 50 years or longer—who will be your competition in the log market? If you are going to be competing with other "farmers," you probably have a reasonable chance, because they too will need to have some return on their capital. But when you begin trying to sell the trees, you're going to be competing with governments—state and federal, or even foreign governments. Governments do not need to earn a return on their timberland investment, and so may not price their logs accordingly.[117]

Many forest industry executives take exception to this point of view, however. Since the government is an oligopoly supplier of timber, they say, the net effect is to drive log prices up—not down—relative to what "economic efficiency" dictates.[118] Moreover, the U.S. Forest Service, which supplies about one-eighth of the timber harvested within the United States each year, has expenses that go beyond building logging roads in the national forests and presiding over the sale of timber at public auctions. It is required to reforest cutover areas, protect watersheds, prevent forest fires and offer recreational opportunities. These other factors tend to drive up the price of timber on public lands.

In any event, the Forest Service still manages to make timber available at prices companies are willing to pay—and often at prices below its own costs of managing the national forests. In fiscal year 1988, for example, timber-sales receipts failed to cover the Forest Service's expenditures in 74 of 120 national forests, including the majority of forests in seven of nine Forest Service regions around the country.[119] In that sense, the Forest Service has in fact provided a safety net for companies with little timber inventory of their own, which might otherwise be priced out of the open market for timber.

While below-cost timber sales are less common now than they were a few years ago—partly because harvesting restrictions have driven up auction prices—timber sales from the national forests soared during the 1980s. The low-cost government sales enabled companies with small timberland holdings to achieve generally higher returns on equity than timber companies with large timberland holdings. The disparity in stock prices in turn made large timberland holders more vulnerable to takeovers: Witness the buyouts of Diamond International in 1982, St. Regis Paper in 1985, Crown Zellerbach and Pacific Lumber in 1986, and Great Northern Nekoosa in 1990—all major landowners.

Now the specter of global warming looms as yet another factor that may place large timberland owners at risk. Commercial species that grew well on company-owned plantations in the last rotation might not fare as well the next time around—if the climate were to take a decided turn for the worse in the meantime. Barring the successful introduction of new species (which remains a distinct possibility), tree-farming companies might become saddled with huge tracts of timberland that are no longer suited for commercial operations. Meanwhile, other forest products companies not burdened with these unproductive assets would be free to go wherever the wood was available—and be in a stronger financial position to buy it.

Figure 7

Major Forest Regions of the United States

WESTERN REGIONS
Pacific Northwest
- Douglas fir/hemlock/fir

California
- Pine/fir/redwood

Northern Rockies
- Pine/fir/birch

Southern Rockies
- Pinyon/juniper/pine

EASTERN REGIONS
Northeast
- Spruce fir
- Maple/beech/birch

Central
- Maple/beech/birch
- Oak/hickory

Southeast
- Loblolly, shortleaf, slash pine

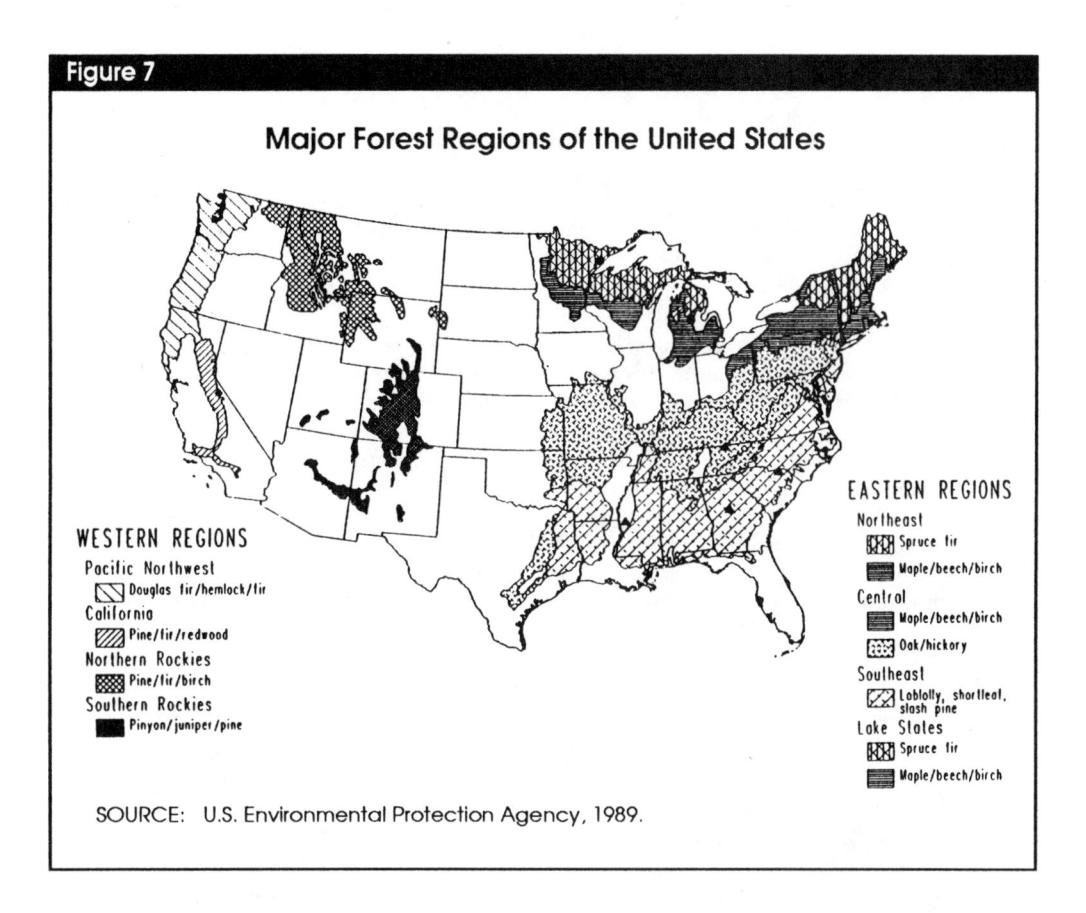

Lake States
- Spruce fir
- Maple/beech/birch

SOURCE: U.S. Environmental Protection Agency, 1989.

Some forest products companies are simply too large to subject their entire operations to the risks of acquiring timber on the open market, however. More importantly, they are able to manage their industry plantations in ways that result in harvests of relatively more trees in less space and time. By building economies of scale and shortening rotation cycles, tree-farming companies can earn a faster return on their investment and reduce their exposure to possible future supply shortages.

Indeed, planned cutbacks in government timber sales—prompted by the record harvests of the 1980s and protection of the northern spotted owl in 1991—have pushed lumber prices to record levels. Profits have surged as well for large timberland holders like Weyerhaeuser, Louisiana-Pacific and Boise-Cascade. "For the immediate term, this is good for us," a Boise-Cascade executive observed recently. "But the long-term outlook for timber is depressing....People are starting to see there's going to be no relief from the timber crisis. Availability is never going to be what it once was."[120]

Global warming has the potential to exacerbate this "crisis." For companies committed to tree farming, the question may become whether to sell certain timberlands and acquire others to assure a steady future supply. Presumably, areas considered especially vulnerable to drought would be on the chopping

(continued on p. 198)

Box 3-D

Old-Growth Forests

Along the Pacific Coast from San Francisco to southeastern Alaska, ancient groves of Sitka spruce, Douglas-fir, giant redwoods and other species tower more than 200 feet into the air. Yet this thin strip of coastal rainforests may disappear far more quickly than the tropical rainforests being felled around the world. Compared with pre-colonial times, only 10 to 30 percent of these U.S. coastal stands remain—and they are highly fragmented—whereas 93 percent of the Amazon rainforest remains intact. The rest of America's temperate rainforests could be wiped out in 15 to 60 years, except for those stands within protected national parks and wilderness areas.[1d]

While about 95 percent of America's virgin forests have vanished since the 1600s, the U.S. government saw no reason to spare ancient parcels from the axe until recently. Forest Service researchers reported back in 1952, for example, that old-growth forests were nothing more than "biological deserts." Such "decadent" mature trees should make way for new stands of rapidly growing trees, they reasoned, whose development would increase the growing stock of timber.[2d] Many foresters continue to hold to this point of view today.

Commercial interests covet old-growth stands for their tremendous harvest volumes, tight grain, lack of knots and durability. But foresters now are discovering a number of ecological reasons for preserving old-growth stands. Far from being biological deserts, these groves are home to hundreds of plant and animal species, many of which are threatened with extinction. About 150 endangered species reside in the national forests, as well as 1,000 to 1,300 other species that are candidates for such designation.[3d]

Much attention has been lavished recently on the northern spotted owl, a reclusive bird that lives and forages in the ancient forests of the Cascade Range and Olympic Peninsula of Washington. To protect 2,800 remaining pairs of these owls, the U.S. Fish and Wildlife Service has designated some 5.4 million acres of these woods as prime owl habitat—an area greater than the size of Massachusetts.[4d] Should these areas remain off-limits to loggers, the federal timber harvest in Oregon, Washington and northern California could fall below 2 billion board feet, nearly two-thirds under the harvest levels of the mid-1980s. The timber industry fears an economic catastrophe, with 50,000 to 100,000 jobs at stake because of the proposed logging restrictions. The government estimates that 32,000 logging jobs could be lost in the region.[5d]

While there are potentially high economic costs associated with preserving old-growth forests, there also are compelling ecological reasons to save them—besides providing habitat for the spotted owl, Sockeye salmon and other endangered species. The recent discovery of a promising anti-cancer drug, taxol, made from the bark of the Pacific yew tree, is one such indicator of the value of old-growth forests.[6d] (The yew is a slow-growing evergreen species found in "climax" forests of the Pacific Northwest.)

From a greenhouse perspective, old-growth forests also store considerable amounts of carbon that might otherwise enter the atmosphere. While timber interests argue that it would be better to cut down these "decadent" trees and plant new forests that accumulate carbon faster, it would take at least 250 years and eight successful rotations of new trees to sequester as much carbon as would be lost in the initial harvest of the old-growth trees, according to a study by forestry researchers at the University of Washington and Oregon State University.[7d] (This study assumes that 55 percent of the wood would release its carbon to the atmosphere soon after harvest; the remaining 45 percent would be converted to long-term storage in buildings and other structures, with a 2 percent annual loss rate. Some forestry researchers believe that the amount of of carbon stored in wood products is higher than the percentage assumed in this study.)

These university researchers also estimated that logging operations in the Pacific Northwest have released 1.5 billion to 1.8 billion tons of carbon into the atmosphere over the last 100 years with the conversion of more than three-quarters of the region's old-growth forests into second- and third-generation stands. "Given the small area we are considering, a mere 0.017 percent of the Earth's land surface, old-growth forest conversion [in the Pacific Northwest] appears to account for a noteworthy 2 percent of the total carbon released [worldwide] because of land use changes in the last 100 years," they noted in their study.[8d]

Nevertheless, the possibility remains that an even greater amount of carbon would have been released to the atmosphere if more energy-intensive materials had been used in place of the wood. One projection by the Wood Science Laboratory in Corvallis, Ore., considers the potential future effects of a 1.45-billion-board-feet reduction in the annual timber harvest on federal lands (compared with the 1983 to 1987 average). If nonrenewable structural materials are used to make up for the expected decline in availability of wood products, it could lead to net increased carbon emissions of approximately 3.5 million tons a year.[9d] Even if the harvest reduction were made up by logging elsewhere around the world, carbon emissions would be likely to increase if the felled timber came from tropical rainforests, which typically are not replanted, or from the huge Siberian taiga, which regenerates only very slowly.[10d] In other words, harvesting old-growth forests in the United States may be the best of the available resource options from a greenhouse standpoint.

While there are many other reasons to preserve old-growth forests, one set of circumstances in which most foresters agree that it makes sense to harvest them is when they are in a dieback situation already. Better to chop down most of the trees and make use of the wood—even if it results in the eventual release of carbon dioxide to the atmosphere—than to let the them decay naturally and emit methane, a far more potent greenhouse gas.[11d]

Areas with Old-Growth Remnants

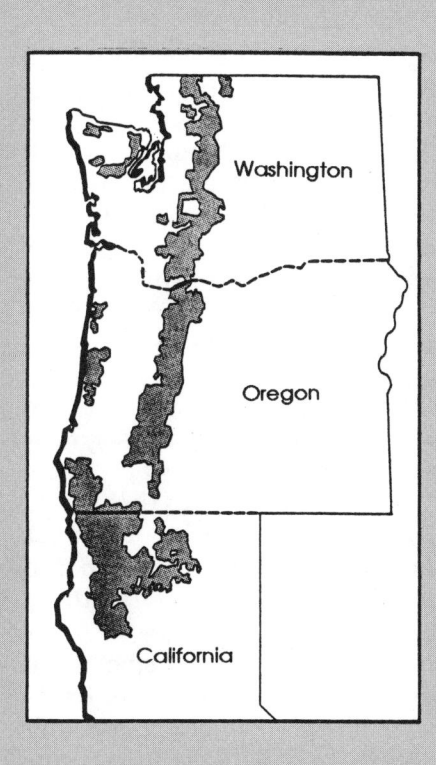

block first, while new acquisitions would emphasize geographical diversification, so as to protect against any calamitous change in climate that may befall one region or another. If the betting was that the South would become hotter and drier, resource holdings and growing stock would likely increase in the North—and might eventually surpass those in the South as the regional disparities of global warming grew readily apparent.

At the moment, however, the outlook remains decidedly muddled. One can imagine a scenario in which speculation about the greenhouse effect drives up northern land values relative to southern timberland prices, creating a distinct cost advantage for southern land acquisitions. If forest products companies then introduced new drought- and pest-resistant trees species in the South, they might be able to maintain timberland productivity in the region even if the climate turned hotter and drier. It is also possible that the South could become warmer and wetter as a result of climate change, raising the productivity of timberlands there without any extraordinary efforts by forest products companies.

Until companies are confident that they can gain a comparative advantage by moving their base of operations or introducing a new species on existing timberlands, they are likely to shy away from such bold actions. Since no company has "inside knowledge" about the greenhouse effect, no company can be sure what its regional effects might be. By the same token, if one company actually feels the effects of global warming, the thinking goes, so will all of its competitors.

"A reduction in regional timber supply—regardless of cause—would affect regional competitors somewhat equally," explains one conceptual analysis of the likely impacts of climate change on the forest products industry.[121] With local markets absorbing the impacts through higher prices, "Competitors with a higher degree of self-sufficiency will increase their competitiveness and force competitors out of the region," the assessment continues. "If the effect on supply reaches a point that the region as a whole loses its competitiveness with other regions, even the best regional competitors will close plants," this analysis concludes. Accordingly, the best strategy at the moment may be one of "no regrets": acquiring assets for reasons other than climate change but also with the prospect they may rise in value if warming does occur.

On that basis, Georgia-Pacific's $3.2 billion hostile takeover of Great Northern Nekoosa in 1990 could have been a brilliant—if unwitting—greenhouse gambit. Georgia-Pacific bought the New England-based forest products company mainly to obtain pulp, paper and containerboard facilities that would complement its own regional production mix. In the process, "GP" acquired 3.2 million acres of timberland, including 2.1 million acres in Maine and 436,000 acres in Wisconsin, making it the largest private landowner in each of those states—but only for a short while.[122]

Great Northern Nekoosa's northern timberlands (and many of its northern mills) were considered to have little strategic value to Georgia-Pacific, so GP sold most of them off in 1991. As GP communications director Richard Good later explained, when it comes to producing timber, "We're basically a southern pine plantation company."[123] Beyond that, GP had assumed a huge debt load when it bought Great Northern Nekoosa and wanted to pare it down. The sale of most of its newly acquired assets in Maine to Bowater Inc. was thought to be such good news, in fact, that GP took out full-page newspaper advertisements to publicize the sale.[124]

Now Bowater, a producer of market pulp, newsprint and groundwood publication papers, is the repositor of most of GNN's assets in Maine. With the $300 million purchase from GP, Bowater has acquired several mills and 2.1 million acres of timberland not far from its headquarters in Darien, Conn.—effectively doubling its timberland holdings. (Bowater has the option to purchase GP's outstanding 20 percent interest in the Maine properties after June 30, 1992.)[125] Georgia-Pacific, meanwhile, has retained nearly 1.1 million acres of GNN's southern land holdings in Arkansas, Georgia, Mississippi and Virginia. On that basis, the deal appears to suit each party's regional strengths. But if Georgia's climate someday resides in Maine, only Bowater will realize the advantage of a greenhouse gambit. GP had the option, but chose not to take it.

Looking for Places to Grow

Northeast: Major land deals have had a relatively short history in the Northeast. Before 1980, much of the region's timberlands had been under the relatively benign stewardship of large corporate forest land holders for nearly a century. The sale of Diamond International to British real estate tycoon Sir David Goldsmith marked a turning point in 1982. Goldsmith bought the company—with 1.5 million acres of timberland holdings—strictly for the purpose of selling off its assets. About 186,000 acres of Diamond International timberlands eventually made their way into the hands of real estate developers.

The legacy of the Diamond International sale was to point out the premium that nontimber interests will pay for choice parcels of rustic land in New England. The 1980s were in fact a boom decade for real estate in which development pressures encroached on the region's forests. Measures have since been taken to preserve more of the remaining northern timberlands for logging and recreational purposes.[126] The downturn in New England real estate prices now presents a buying opportunity for foresters and conservationists. But the size of any future sales is not likely to be nearly as large as the recent three-way deal brokered between Great Northern Nekoosa, Georgia-Pacific and Bowater.

Ultimately, large-scale land-buying opportunities in the Northeast are limited by the region's comparatively small size. Forest industry holdings are concentrated in a 10-million-acre swath that cuts across the Adirondacks of New York, northern Vermont and New Hampshire into northern Maine, encompassing an area about four times the size of Yellowstone National Park. Several land sales have taken place quietly in New England in recent years: Oxford Paper to Boise Cascade (600,000 acres), St. Regis Paper to Champion International (760,000 acres), and Hudson Pulp & Paper to Georgia-Pacific (520,000 acres).[127] At the beginning of 1992, nine tracts totaling 400,000 acres were either on the market or expected to be offered soon.[128] Given climate model indications that the productivity of Northeastern forests could increase, these tracts may draw the interest of timberland speculators.

North Central: The North Central portion of United States is another well forested region, but it is less densely populated than the Northeast and less subject to real estate development pressures. The aspen-birch forests surrounding the Great Lakes are a major source of fiber for the pulpwood industry of the North. White ash from elm-ash-cottonwood forests are used in a number of specialty wood products, and red pine, jack pine and spruce-fir plantations also are valuable sources of pulpwood and other softwood materials. For these reasons, forest products companies already have a major presence in the North Central region and could expand their operations there.

Since these forests are well away from the coastline and closest to the Great Plains, however, climate models tend to pick them as forests likely to feel ill-effects of global warming. Some forest response models suggest that prairie forbs and grasses could replace mixed conifer-hardwood forests in the states bordering the Great Lakes.[129] One evaluation of the Minnesota Superior National Forest has forecast such a rapid decline in the productivity of spruce-fir trees that timber yields would start to decline by the year 2000, with most of the trees wiped out by 2020—and all of them gone by 2040.[130]

Another climate-related assessment of the boreal/mixed hardwood forest in northwest Michigan rendered this dire outlook: "An immediate dieback of most species dominates the initial 75 years of warming, including local extinction of boreal spruce and paper birch populations....The forecast for industry land-holdings on such sites appears quite dim."[131] Rather than recommending any further land purchases in that area, this analysis concluded, "Quick divestiture with the onset of warming would seem warranted, whether the holding industry is concerned with pulp or lumber. Early cessation of reproduction, large mortality losses in sawtimber, and reduced growth in poletimber would severely reduce commercial yields after the initial mortality was harvested."

Some industry foresters do not believe the changes would be as severe as portrayed here. One-third of the mixed boreal/mixed hardwood forest in Michigan consists of hardwoods that should flourish during the initial phases of the warming. Another quarter consists of non-boreal aspen and spruce trees that should continue to grow because they are rooted in deep, moisture-retaining soils. Red, white and jack pine that inhabit shallower, sandy soils would have greater difficulty surviving. But even they are known to grow in places that have warmer climates than Michigan, such as California, Iowa and Indiana.[132]

Many trees native to the Great Lakes states also grow across the border in Canada. Presumably, the effects of climate change would be slower to take a toll on these forests farther to the north. Even here, however, the prospects may not be especially good in the event of a severe warming, according to several studies. Climate scenarios raise concerns about the health of spruce within the first 75 years of projected warming "and good reason for considering divestiture thereafter."[133] Commercial dominance of merchantable boles of sugar maple and white pine—migrating north from the United States—would not develop until 150 years after the demise of the spruce forests, this study estimated. The boreal forests of western Ontario are a major resource for the Canadian pulp and paper industry.

Figure 8

Lumber Production by Region

Billions of board feet

[Bar chart showing lumber production by region (South, Pacific NW, Rockies & SW, Northeast, North Central) for years 1952, 62, 70, 76, 86 (actual) and 2000, 10, 20, 30, 40 (projected)]

SOURCE: U.S. Forest Service, 1989.

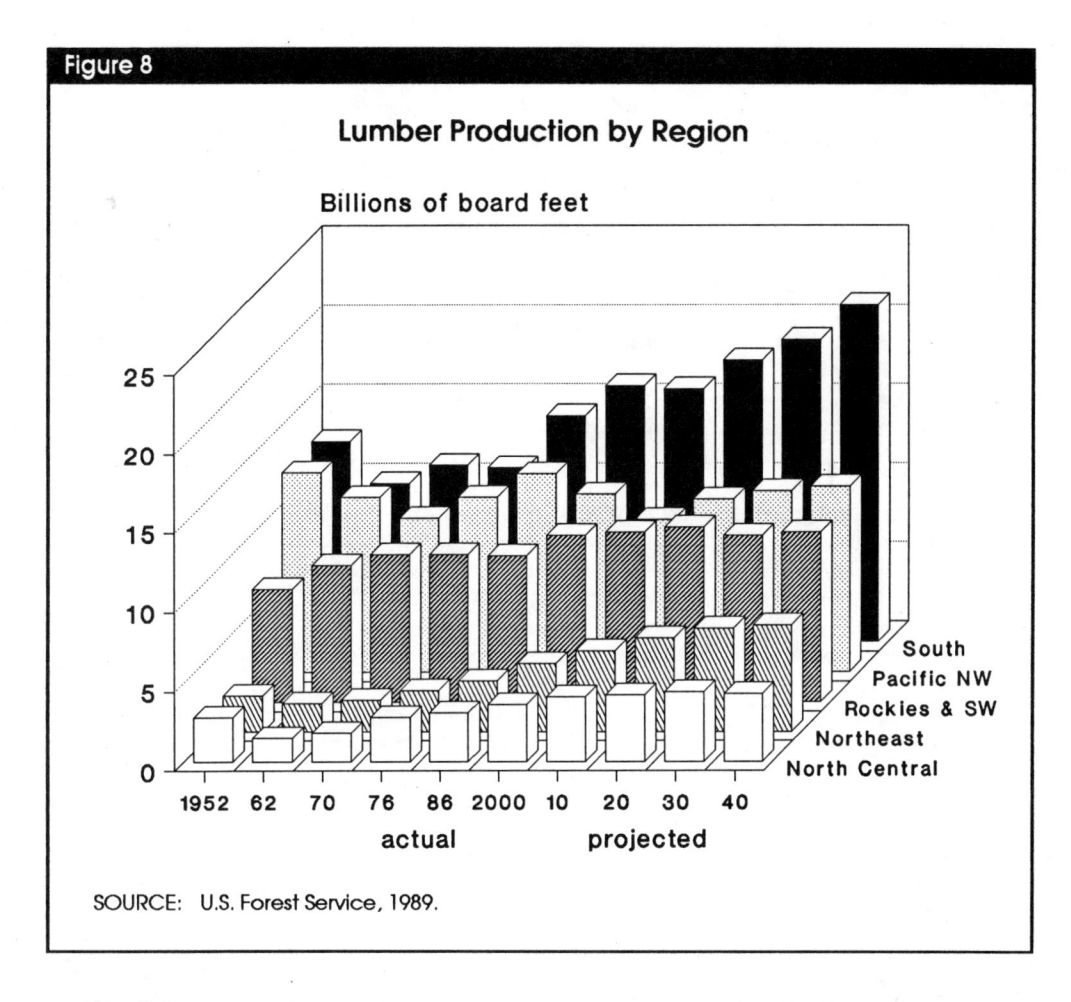

Pacific Northwest: One other place within the United States where timber interests could seek refuge against the greenhouse effect is the same place they turned after the harvest of North Central timberlands earlier this century—the Pacific Northwest. Buffered by the Pacific Ocean and blessed with mountain ranges that support many climate ecosytems, the Northwest is sure to continue as an important timber-growing region of the United States, regardless of climate change. New management regimes already are emerging for these species as younger second-growth stands far outnumber old-growth ones.

One review of general circulation models suggests that global warming could have a favorable effect on some of the most important commercial species of the region. "The overall outlook for [Douglas-fir and western hemlock] is for an increase in their areas of natural range, increased yields, shorter rotations, and increased opportunities to manage these species profitably," according to one assessment.[134] On the other hand, an earlier onset of summer drought or the lack of a cool-down period in the winter could adversely affect the growth of Douglas-fir. Accordingly, the internal dynamics

(continued on p. 204)

> **Box 3-E**
>
> ## 'New Forestry' and the U.S. Forest Service
>
> With the ascendance of environmental issues in managing the nation's forests, the U.S. Forest Service has begun to redefine its mission. Managers of the national forests are placing greater emphasis on preserving soils, watersheds and wildlife habitat and less emphasis on "getting out the cut." While much remains to be learned about this evolving biological science, it may enhance the forests' ability to weather the stresses and strains of climate change.
>
> "New Forestry" is the generic term used to describe this new conservation ethic. Within the Forest Service, the program is known as "New Perspectives." "We are looking at multiple-use management from a new perspective," proclaimed Secretary of Agriculture Clayton Yeutter in June 1990, as the Forest Service announced plans to expand wilderness areas and protect other special places. "Where timber and mineral production and livestock grazing cannot be accomplished in an environmentally acceptable manner," Yeutter vowed, "production levels will be reduced."[1e]
>
> The latest Forest Service plan (issued in June 1992) calls for a 10 percent reduction in harvest volumes on national forest lands, to about 10 billion board-feet a year. By 1995, clearcutting on national forest land will be reduced by about 70 percent from fiscal year 1988 levels, when four-fifths of all timber harvested in the national forests was clear-cut.[2e] On timberlands where harvests are planned, yields per acre are expected to fall 10 or 20 percent because of the more selective harvesting methods used—meaning that more acreage will have to be cut in order to reach the production targets. But the land that is logged presumably will be treated with greater care than those harvested using traditional clearcut methods.
>
> In essence, New Forestry is kinder, gentler forestry. In Oregon's Siskiyou National Forest, where the New Perspectives program is in full swing, the Forest Service has banned clearcutting of trees and allowed harvests of two or three age classes simultaneously to mimic the effects of natural losses resulting from wildfires and pest outbreaks. Selected old-growth Douglas-fir and dead snags are being spared harvest to maintain habitat for nesting birds and boring insects, while some downed logs and woody debris remain on the forest floor to provide dens for small mammals and replenish nutrients in the soil.[3e]
>
> In other national forests where clearcuts still occur, Forest Service managers are doing more to protect riparian habitat—maintaining wide buffer zones along rivers and streams. Another recent trend has been to cluster clearcuts in close proximity, rather than dispersing them throughout the forest, in order to minimize fragmentation of biological corridors and old-growth stands. Some Forest Service managers hope to extend harvest cycles in selected areas beyond 100 years to allow "new" old-growth groves to mature.
>
> New Forestry is not without its dissenters from the old school of forestry, however. William Atkinson, a forestry professor at Oregon State University, asked Oregon foresters convening for their 1990 annual meeting: "Have you been out in the woods recently to see the new cuttings on national forest lands? You have to look pretty far anymore to find a good old-fashioned clearcut. Most New Forestry units look like the logger got half way through and walked away in disgust."[4e]
>
> It is not just that New Forestry methods leave merchantable timber in the woods. Harvesting methods also tend to be more expensive, since selective logging often is coupled with fewer logging roads and more helicopter lifts, cable systems and ground

skidtrails. Inevitably, residual stands are damaged as selected trees are felled. Moreover, isolated standing wood becomes more vulnerable to windfall and insect attack. At the same time, understory vegetation exposed to sunlight grows rapidly and becomes tinder for fire. Stands of Douglas-fir often fail to regenerate; instead, they are crowded out by cedar, hemlock and true firs.[5c]

Some foresters dispute another basic tenet of New Forestry—that the new management technique resembles natural patterns of growth, decay and regeneration of the forests. If 10 percent of a particular hillside used to burn every 50 years, say, an equivalent amount should be logged over the next 50 years, according to New Forestry thinking. The size of a previously burned tract, moreover, should dictate how large an area to clear during the next harvest. But some believe New Forestry is misreading the signs of nature. Silviculturalist David Smith of Yale University argued as long ago as 1970 that "essentially even-aged stand structure generated by lethal disturbances over sizable areas is more nearly the norm of nature."[6c] In natural stands featuring many kinds of tree species, he said, "the very large difference in tree diameter is not the result of significant differences in age but of markedly different rates at which various species grow in association with each other." On this basis, some foresters insist that clearcutting remains the superior management option.

Ultimately, decisions on how to manage the forests will be left to those who own them. Some are urging the Forest Service to take further steps to designate more lands as wilderness areas, making them off-limits to development of any kind. George Frampton of the Wilderness Society, for one, believes that the United States should double its 90 million acres of wilderness lands over the next quarter-century. Brock Evans of the National Audubon Society has suggested that the Forest Service divide its lands in two. One set of lands would be administered by a "U.S. Logging Agency," for purposes similar to those performed by today's Forest Service managers. The rest of the land would be given to an agency whose job is to preserve the forests in their natural state, such as the National Park Service.[7c]

Jeff DeBonis, executive director of the Association of Forest Service Employees for Environmental Ethics, believes that even more radical steps should be taken. He urges that "all uncut Forest Service land" be preserved as "gene pools and wildlife corridors." A former timber sales planner in Oregon's Willamette National Forest, DeBonis has enlisted more than 1,500 of the Forest Service's 39,000 employees in his crusade for timber management reform.[8c]

Randall O'Toole, author of *Reforming the Forest Service*, advocates a more libertarian approach. He believes that market value should be charged for all amenities that national forests provide—recreation, mining, fishing and animal grazing. That way, Forest Service managers would have an incentive to manage each tract for its highest-valued purpose, even if it is for something other than timber harvesting.[9c] Still others believe the national forests should be sold off altogether on the assumption that private landowners would do a more responsible job of caring for the nation's timberlands.[10c]

For now, anyway, management of America's 191 million acres of national forest land remains in the hands of the U.S. Forest Service. While some question the depth of its commitment to New Forestry—and others question the benefits of New Forestry altogether—the days of wide-swath clearcutting do appear to be nearing an end. With the specter of global warming looming on the horizon, such an emphasis on conservation biology appears to have arrived at an opportune time.

of climate change may determine the outcome rather than the ultimate degree of warming.

In other respects, the fate of the Pacific Northwest forests is as much in the hands of political forces as climatological ones. The region is home to more than a third of the nation's most productive timberlands. But with harvests there declining, its contribution to the nation's timber supply is expected to fall six percentage points over the next 50 years, from 25 to 19 percent, and this projection does not include the harvesting restrictions resulting from protection of the spotted owl and other endangered species.[135]

Increased availability of northern hardwoods is expected to make up for some of the shortfall out West. By the turn of the century, in fact, Northeastern and North Central forests may once again exceed Northwestern forests in total yield, completing a harvest cycle that began after World War II. But the lion's share of production is expected to continue to come from the South. In 2040, that region's highly productive forests are expected to provide fully half of the nation's timber, according to the Forest Service's most recent planning assessment.[136]

The rest of the supply will have to come from abroad. The United States already is a net importer of lumber, mainly from Canada. A handful of companies own pulp and paper mills and veneer plants in Central and South America. Others maintain licensing agreements to import tropical hardwoods.[137] With the breakup of the Soviet Union, several companies are eyeing the vast Siberian taiga, an uncut forest that is larger than the Brazilian rainforest and twice the size of the continental United States.[138]

Indeed, plenty of forests are still standing around the world (albeit in increasingly remote places). The Intergovernmental Panel on Climate Change estimates that 1.9 billion acres of forests remain in temperate regions, another 2.3 billion acres are in boreal zones and 4.7 billion acres remain in tropical regions.[139] In recent years, however, the international community has come to appreciate these forests as more than undeveloped "wood factories." They now represent a vast storehouse for carbon and genetic diversity as well. Therefore, in a greenhouse world, the emphasis may have to shift from harvesting trees in places where there are many to planting trees in places where there are few.

Carbon Sequestration Potential

Regardless of how forest products companies prepare for an uncertain future, one trend seems certain: They will have to rely more on their own land and that owned by non-industrial private forest landowners to stave off possible timber shortages in the 21st century. The U.S. Forest Service estimates that it will be 50 years before harvests in the national forests once again rival the record levels posted in the mid-1980s, when timber extractions averaged 12.2 billion board-feet a year. These planned harvest reductions come despite a Forest Service projection that the nation's demand for hardwoods will rise by 79 percent over the next half-century and that demand for softwoods will grow by 35 percent over the period.[140]

"The bottom line is this," says Con Schallau, chief economist for the American Forest Resource Alliance: "Unless the resolution of the growing debates over the management of forest land can accommodate timber harvesting, forests in the South and [imports from] Canada cannot compensate for the planned reduction in harvesting from the national forests. In fact, contrary to the Forest Service forecast, consumption of wood products will decrease, and consumer prices for wood products will rise."[141]

To make up for the potential shortfall of timber, the greenhouse effect may emerge the industry's greatest ally for two reasons. First, carbon dioxide enrichment may stimulate the growth of trees, allowing them to mature more quickly. Second, trees' ability to sequester carbon from the atmosphere promotes the concept that planting trees is good for the planet as well as for commerce—and that tree-planting efforts should be encouraged at every turn.

Minding the Carbon Store

The forest products industry has long encouraged private landowners to grow and harvest successive forest crops. The American Tree Farm System, established in 1940, has enrolled 71,000 privately owned tree farms in all 50 states. The tree farms cover 95 million acres, an area larger than Japan.[142] Today, many forest products companies have stepped up their efforts to promote tree planting among the general public. Westvaco Corp., for one, has distributed 200,000 seedlings among its employees, shareholders and clients. International Paper, for another, is making 10 million seedlings available for use in state and federal tree-planting programs.[143]

Meanwhile, the American Forestry Association is in the midst of a four-year program, known as "Global ReLeaf," to spur the planting of 100 million trees in U.S. metropolitan areas. Besides fixing carbon directly, these urban and suburban trees provide shade for homes and buildings, thereby lowering air-conditioning bills. The association estimates that the direct effect of carbon

Figure 9

Commercial Timberland Ownership

- NIPFs 57%
- Federal 20%
- State & County 7%
- Indian 1%
- Forest Industry 15%

483.3 million acres

SOURCE: U.S. Forest Service, 1989.

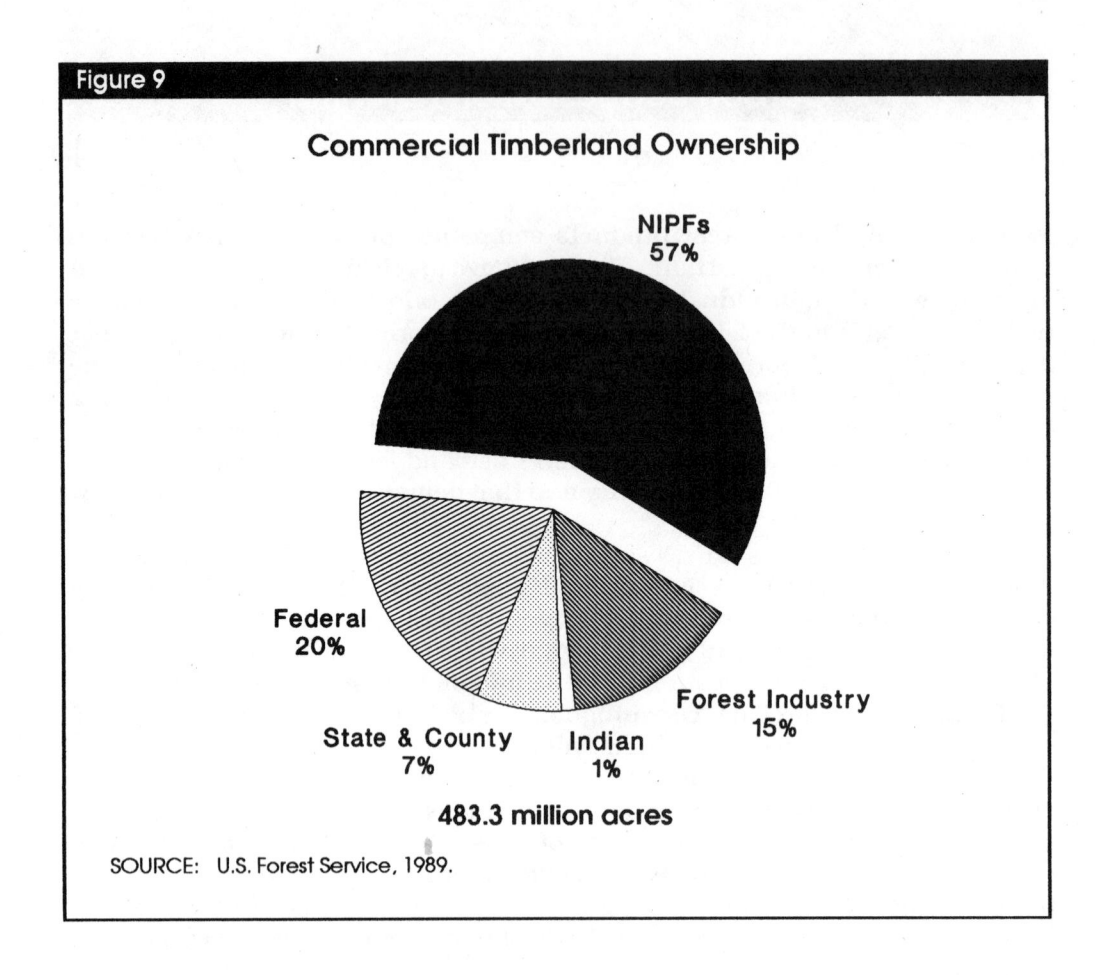

sequestration and the indirect effect of reduced electricity demand will keep 5 million tons of carbon out of the atmosphere each year.[144]

Non-industrial private forest landowners, known as NIPFs, figure prominently in these reforestation efforts, since they own more productive timberland than the forest products industry, the federal government and the 50 state governments combined. Altogether, roughly 7 million NIPFs possess more than 275 million acres of productive U.S. timberland—an area about the size of France, Spain and Portugal.[145] Half of the nation's annual timber harvest, in fact, comes from land owned by NIPFs; industry lands provide only about 30 percent of the harvest, and government lands, the remaining 20 percent.[146]

The future contribution of NIPFs is in doubt, however. Woodlot owners throughout the country are feeling economic pressure to sell off portions of their land to real estate developers or to convert it into agricultural land. Others have simply neglected their land after logging it, allowing natural regeneration to run its course. These developments blunt the momentum of public and private reforestation efforts and place a vital portion of the nation's timber supply at risk.

The specter of the greenhouse effect could revitalize NIPFs' tree-planting efforts by highlighting the ability of trees to sequester carbon from the

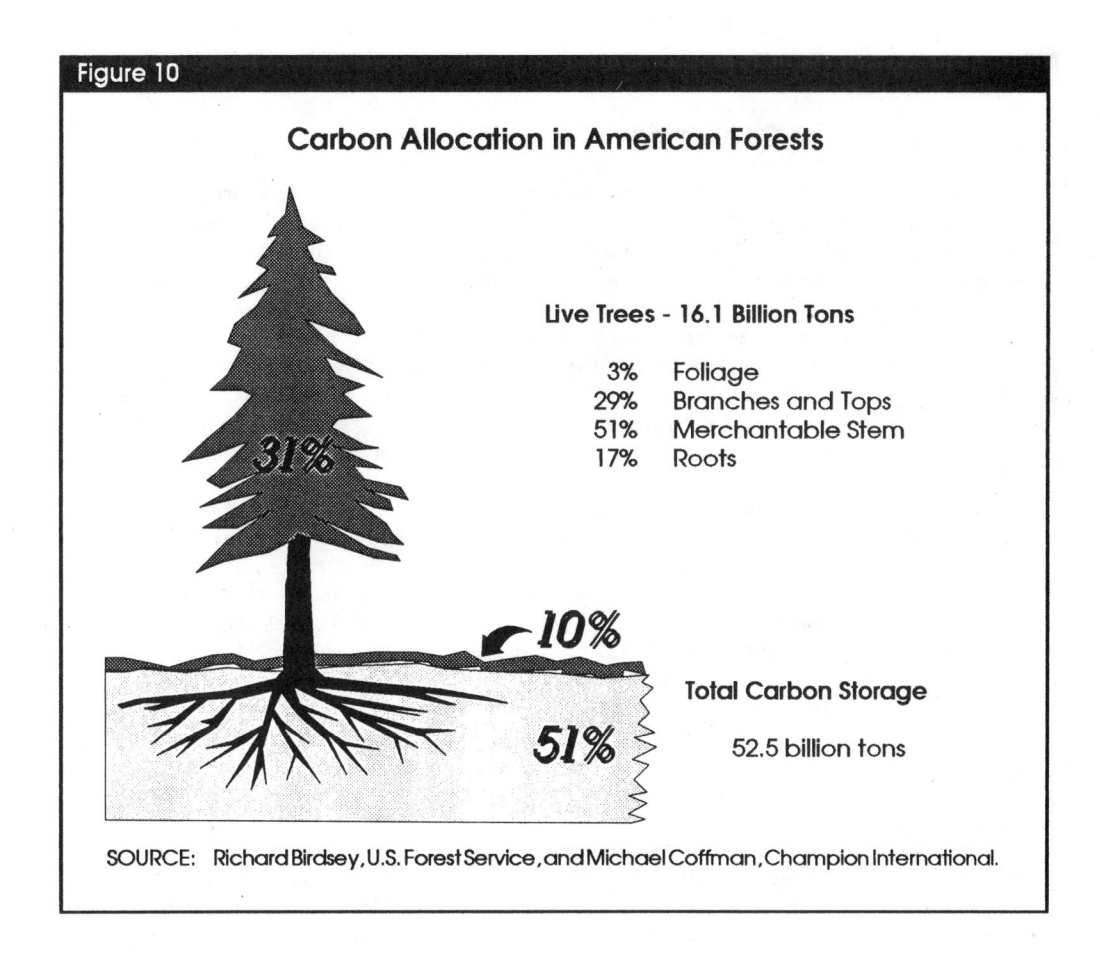

> **Figure 10**
>
> **Carbon Allocation in American Forests**
>
> **Live Trees - 16.1 Billion Tons**
>
> | | |
> |---|---|
> | 3% | Foliage |
> | 29% | Branches and Tops |
> | 51% | Merchantable Stem |
> | 17% | Roots |
>
> 31%
>
> 10%
>
> 51%
>
> **Total Carbon Storage**
>
> 52.5 billion tons
>
> SOURCE: Richard Birdsey, U.S. Forest Service, and Michael Coffman, Champion International.

atmosphere. Acre for acre, temperate American forests store 20 times more carbon than croplands do. Yet less than a third of the carbon is harbored in the trees themselves; about three-fifths is bound to the roots and adjacent soils.[147] Thus, by planting trees instead of crops, NIPFs are able to make a far greater contribution to sequestering carbon from the air. By the time a replanted southern pine forest reaches its 50th birthday, for example, its soil *alone* will contain nine times more carbon than if it were converted to cropland or other purposes. Soils of the North fix an even greater amount of carbon, since they are generally richer and moister than the soils of the South.

What would be the benefit of converting all marginal cropland and pastureland in the nation to forests? One theoretical study by Peter Parks of the Center for Resource and Environmental Policy Analysis at Duke University examined 62 million acres of marginal crop and pasture lands for which reliable economic data exist. Of this amount, more than 23 million acres could be turned into forests profitably, Parks determined, at an establishment cost of $1.5 billion. Such an investment would not only increase the global sequestration of carbon by nearly 1 percent a year, it would raise the nation's merchantable timber volume by 1.5 billion cubic feet annually, helping to alleviate a potential future

> **Box 3-F**
>
> ## Conservation Reserve Program
>
> Since the 1930s, the federal government has instituted many programs to reward private foresters to maintain their timberlands in active rotations. These include the Agricultural Conservation Program and the Forestry Incentive Program, both of which provide incentives to plant trees on logged land. By far the largest government-sponsored reforestation program began in 1985. The Conservation Reserve Program pays farmers to take highly erodible cropland out of production and permits trees to be grown in their stead. Over the first five years of the program, however, farmers planted trees on only 2 million acres of 34 million acres of cropland taken out of production.[1f] Most program participation to date has been in the Great Plains, where climate and soil conditions do not favor the growth of forests.
>
> To the extent that the Conservation Reserve Program has led to the establishment of new forests, they have been mainly even-aged pine stands planted in the South. Some conservationists believe the program should direct more attention to restoring diverse hardwood forests farther north, such as in the eastern half of the Corn Belt. Before their conversion to agriculture, states like Ohio, Indiana and Illinois featured abundant mixed stands of oak, hickory, maple and beech trees. Now these three states alone have 8 million acres of cropland eroding at more than twice the rate required to be eligible for the Conservation Reserve Program. While these croplands are prime candidates for the establishment of new hardwood forests, farmers remain reluctant to convert their croplands unless government payments are high enough to cover revenue losses that result from taking cropland out of production.[2f]
>
> As part of the 1990 Farm Bill, Congress did vote to expand the Conservation Reserve Program from 40 million acres to 45 million acres. Also in 1990, the U.S. Department of Agriculture relaxed the soil erosion criteria for eligibility in the program as a way of increasing the number of trees planted.[3f] This has been a mixed blessing as far as carbon storage is concerned, however. While more acreage is now eligible for tree planting, there is still no means of ensuring that the trees will be placed on soils experiencing the worst erosion. (Only about 30 percent of the most erodible soil in the nation is currently enrolled in the Conservation Reserve Program; farmers continue to till the rest.) The failure to take this highly erodible cropland out of production is responsible not only for tremendous topsoil losses; it also is freeing carbon from the soil, whereas planting trees instead would increase the carbon deposits multifold. Further efforts to get farmers to plant trees on highly erodible soil would be one of the most effective ways to sequester carbon from the atmosphere.

timber shortage. More than 80 percent of the increase in carbon uptake, Parks calculates, would result from softwood trees being planted in the southern United States.[148] Another study of cropland conversion potential by Ralph Alig of the U.S. Forest Service's Forestry Science Lab in Research Triangle Park, N.C., has reached similar conclusions. Alig calculates that the South contains more than two-thirds of the nation's marginal lands that could be converted profitably to forests—61 million acres out of the 84 million acres identified in his study.[149]

One caveat should be added to these studies, however. While they use the atmospheric buildup of carbon dioxide as a rationale for converting cropland to forests, the studies do not consider that global warming could force the opposite to happen: that is, forests could convert to grasslands as a result of climate change, especially in the South, where most of the studies' land-conversion potential is foreseen. Thus, the mammoth tree-planting efforts envisioned in these studies could be for naught if the weather fails to cooperate.

Despite the limitations of these studies, the general point remains that carbon sequestration to combat the greenhouse effect could become a larger, ennobling mission for foresters throughout the nation. The purpose of the "America the Beautiful" program, in which President Bush seeks the planting of 10 billion extra trees over a decade, as well as the more modest "Global ReLeaf" program sponsored by the American Forestry Association, is to instill such a sense of duty and honor among foresters. The Forest Service estimates that if all non-industrial private forest owners rallied to this cause—taking cost-effective steps to maximize the productivity of their lands—the nation's annual timber growth would increase nearly 4 billion cubic feet a year, or more than 15 percent. At the same time, the additional 28 million tons of carbon sequestered each year would mitigate 2 percent of annual net U.S. carbon emissions.[150]

Offsetting Utility Emissions

The forest products industry is not the only industry with a vested interest in reforestation programs. The electric power industry, as the nation's largest emitter of carbon dioxide, also is looking at reforestation programs as a means to offset power plant emissions. Applied Energy Services, an independent power producer based in Arlington, Va., already has acted on this principle through its support of reforestation programs abroad. In 1989, AES donated $2 million toward the planting of 52 million trees in Guatemala to offset 15 million tons of carbon emissions expected during the 40-year operating life of a coal-fired power plant it has built in Connecticut. (The plant is rated at 180 megawatts.) AES also donated $2 million to The Nature Conservancy in 1991 to offset emissions from another coal plant it is building in Hawaii. The Nature Conservancy will use the money to buy and preserve a 225-square-mile tract of endangered rainforest in Paraguay.[151]

So far, no other American utility or independent power producer has followed AES's example—but that may soon change. As part of their programs to reduce CO_2 emissions 20 percent by the year 2010, Southern California Edison, New England Electric System and the Los Angeles Department of Water and Power are considering funding reforestation projects around the globe.[152] In addition, The Netherlands Electricity Generating Board has committed funds to replant more than 1.7 million acres of tropical forest in Bolivia, Ecuador and Indonesia. Its purpose is to offset emissions from two new thermal power plants under construction in Rotterdam—and to avoid paying a carbon tax on the plants' CO_2 emissions.[153]

Meanwhile, in the U.S. Congress, Reps. Jim Cooper (D-Tenn.) and Mike

Synar (D-Okla.) have drafted a bill that would mandate forestry or conservation offsets for new utility or industrial facilities that emit more than 100,000 tons of CO_2 annually.[154] With a framework for "emissions trading" already created by the Clean Air Act, it is possible that more power companies will turn to timber growers as a means of offsetting incremental CO_2 emissions—as AES has done voluntarily—in the event that policies are enacted to control global warming.

If the market for CO_2 offsets develops, it is likely to be a boon for the forest products industry (not to mention NIPFs). One study by Gregg Marland of the Oak Ridge National Laboratory determined that a plantation of fast-growing American sycamore trees totaling 30.6 miles in diameter could offset the emissions of a new 1,000-megawatt coal-fired plant.[155] Extrapolating from Marland's research, Tufts University's William Moomaw has estimated that 30 million acres of trees—or an area about the size of the state of Mississippi—could offset 45 million tons of carbon emitted annually from 25,000 megawatts of new U.S. electric generating capacity that is expected to be on-line by 1996.[156] To sequester all of the excess carbon that enters the atmosphere each year—about 3 billion metric tons worldwide—an area about the size of the western United States would be required—some 1.15 billion acres.[157]

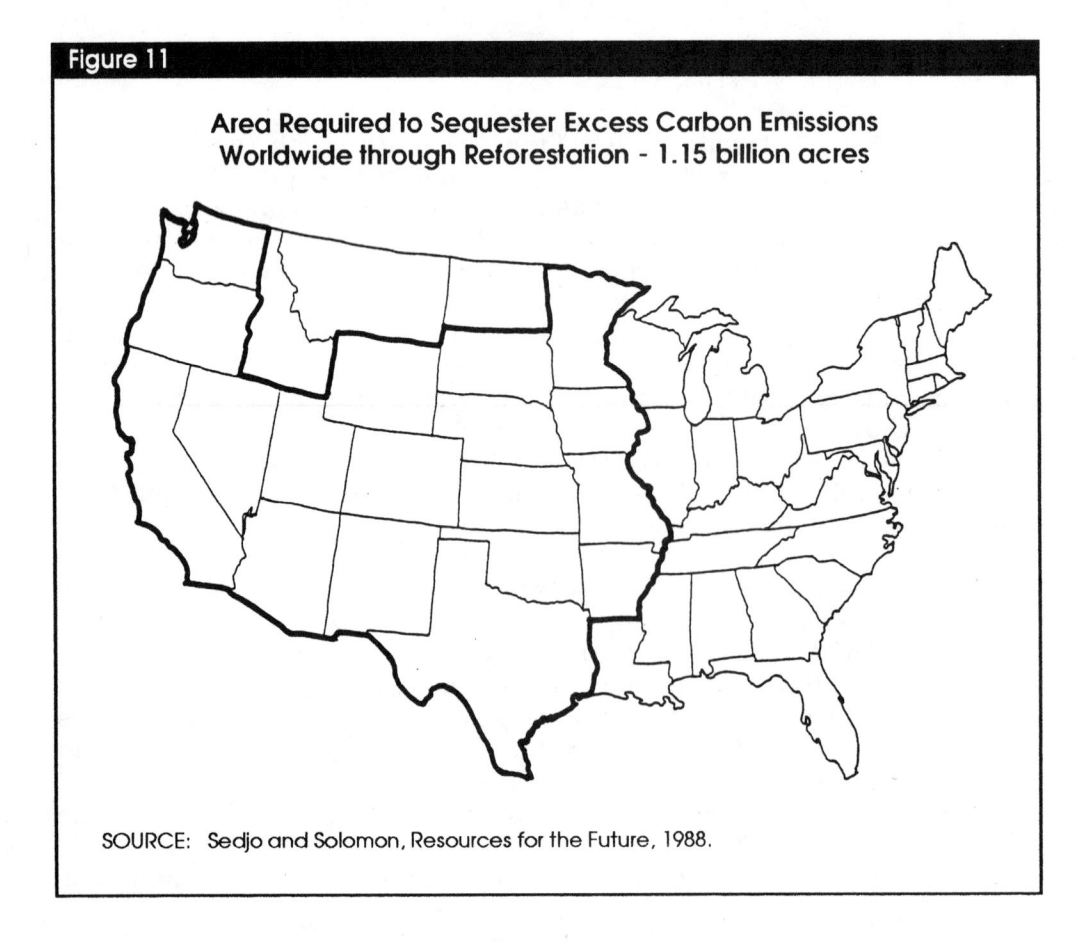

Figure 11

Area Required to Sequester Excess Carbon Emissions Worldwide through Reforestation - 1.15 billion acres

SOURCE: Sedjo and Solomon, Resources for the Future, 1988.

In theory, enough land is available through reforestation and afforestation to offset all of the world's excess carbon emissions. The Intergovernmental Panel on Climate Change has identified between 1.25 billion and 2.5 billion acres of such available land worldwide: 125 million to 310 million acres in temperate zones, 125 million to 370 million acres in boreal zones and 1 billion to 1.85 billion acres in tropical zones. The Intergovernmental Panel's preliminary assessment is that afforestation in boreal and temperate zones might range from $30 to $60 per ton of carbon sequestered, and from $10 to $30 a ton in tropical zones.[158] Land use decisions are based on many factors, however, so the ultimate costs of reforestation remain uncertain. Moreover, the rate of carbon accumulation decreases as trees mature, and the amount of carbon that remains stored in forests and forest products depends on how they are managed and used. Ultimately, tree-planting efforts must be regarded as a stop-gap means to sequester carbon from the atmosphere until non-fossil energy sources emerge as a more permanent solution to the atmospheric CO_2 problem.

Power to Burn?

There are other ways in which trees can be used to ameliorate global warming besides sequestering carbon directly. One option is to burn wood in its own right as an alternative to fossil fuels. Already, wood and wood waste provide about 2.7 quadrillion British thermal units (quads) of energy in the United States, contributing roughly 4 percent of the nation's annual energy production.[159] In fact, more than half the wood already harvested from U.S. forests each year is used as firewood, mainly in residential fireplaces and wood furnaces.[160]

With the steep rise in energy prices since the early 1970s, forest products companies have made greater use of bark, sawdust and other wood waste to power their own factories. Currently, about three-quarters of the process energy required by solid-wood products mills and more than one-half that used in pulp and paper mills comes from burning wood-waste residues. Over the same period, the pulp-and-paper industry's use of conventional fuels and electricity has fallen by more than 30 percent.[161]

If timber companies were to turn to biomass energy production in a big way, they could look beyond their own factories and the residential sector to supply potentially huge energy markets in transportation and electric power industries. The National Research Council estimates that conventional energy production from wood and wood waste has the potential to nearly triple to 7.5 quads, while the development of new wood and herbaceous "energy crops" could supply another 4 quads of power for these markets.[162] (At present, energy consumption of all forms in the United States amounts to about 81 quads annually.)

Lynn Wright, director of the short rotation wood crops program at Oak Ridge National Laboratory, estimates that American sycamore trees grown in intensive rotation cycles of four to 12 years could provide fuel at a cost of $36 per dry ton. Expressed in British thermal units, this would amount to a delivery cost of $2.12 per Btu in a best-case scenario. As such, wood fuel would still cost more than coal, which Wright estimates to be $1.56 per Btu, but slightly less than natural gas, which Wright pegs at $2.26 per Btu. On that basis, Wright concludes that wood-derived ethanol fuel for transportation has the potential to displace 3 percent of the nation's fuel-carbon emissions annually, or about 34 million tons of carbon. Alternatively, wood used in an electricity mode has the potential to displace 5 percent of such emissions, or 59 million tons.[163]

A similar analysis by the World Resources Institute finds a more favorable outlook for ethanol fuel derived from fast-growing tree plantations, with fuel-carbon displacement reaching 5 or 6 percent.[164] In addition, EPA has reported to Congress that all forms of biomass energy production (including that from agricultural crops) could reduce the nation's carbon emissions by as much as 10 percent.[165] Adding in the sequestration of 2 to 5 percent of the nation's CO_2 emissions expected as a result of the America the Beautiful tree-planting program during the 1990s, it is clear that wood has a potentially important potential role to play in policies to address global warming.

Nevertheless, the forest products industry is not likely to encourage the development of wood as a major fuel source as long as it remains concerned about a shortage of timber for use in conventional wood products. A major tree-planting program, if successful, could alleviate fears of such a shortage, however, and launch the industry into a gambit for marketing biomass fuels in the 21st century. In the meantime, the industry understands that the greater the effort made to reforest the landscape, the more bountiful the future raw material supply will become for conventional wood products (and the more carbon that will be sequestered from the atmosphere). A greater long-term supply should keep costs down in turn and help maintain the viability of wood products in the marketplace.

A New Pitch: Eco-friendly Products

If concerns about global warming intensify, the forest products industry may reap yet another important benefit in the marketplace. It could promote its products as environmentally preferable alternatives to those manufactured by competing industries. In a fossil-fuel dependent economy such as ours, the forest products industry enjoys two distinct marketing advantages. First, it consumes far less oil, coal and natural gas than other industries because of its in-house use of wood and wood waste (not to mention renewable hydropower). Second, and perhaps more important, most of its products require far less energy to manufacture—and, hence, emit far less carbon dioxide—than competing materials.

Studies of the energy-intensity of wood products versus alternative building materials date back to 1976, when the National Academy of Sciences issued a report from the Committee on Renewable Resources for Industrial Materials.[166] This groundbreaking study determined that structures built with aluminum or brick typically require five times as much energy to erect as those made from wood. If oil were the energy source used to manufacture

Figure 12

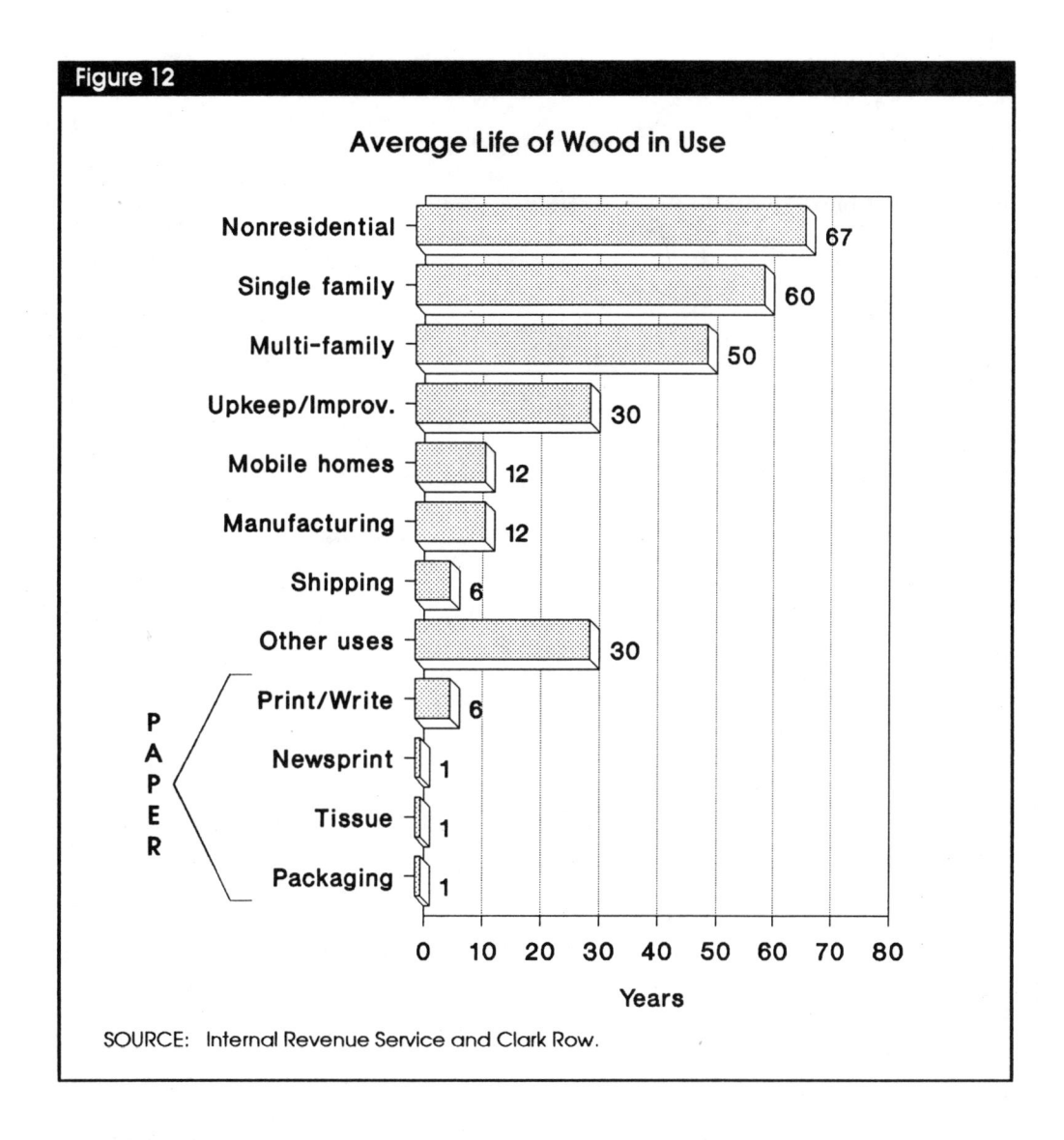

Average Life of Wood in Use

Category	Years
Nonresidential	67
Single family	60
Multi-family	50
Upkeep/Improv.	30
Mobile homes	12
Manufacturing	12
Shipping	6
Other uses	30
PAPER — Print/Write	6
PAPER — Newsprint	1
PAPER — Tissue	1
PAPER — Packaging	1

SOURCE: Internal Revenue Service and Clark Row.

the aluminum, 25 pounds of CO_2 would enter the atmosphere for five pounds of wood it displaces.

The forest products industry is in the process of updating this analysis.[167] One analysis by industry consultant Clark Row estimates that lumber emits less than 300 pounds of carbon per metric ton of product manufactured. By comparison, energy-intensive cement emits more than 4,000 pounds of carbon per ton—nearly 14 times that of lumber. The numbers go up from there: Steel emits 26 times more carbon per ton of product than lumber; aluminum, 34 times more; and petroleum-based plastic materials, 170 times more.[168] With numbers like these, the forest products industry is trying to draw a sharp distinction between the CO_2-intensity of its products and those of its major competitors.

> **Box 3-G**
>
> ### Greenways: Path to the Future
>
> The old-growth debate now raging in the Pacific Northwest juxtaposes the desire to maximize timber harvests with the quest to preserve forest diversity and integrity. This conflict is likely to grow more intense with the debate over global warming. While virtually all of America's virgin timberlands have been cut down, U.S. forest ecosystems remain relatively diverse—supporting about 850 tree species—compared with the more intensively managed temperate forests of Europe. More species of trees exist in just the Appalachian Mountains, in fact, than in all of Europe. (To put this comparison in its proper context, however, three acres of a Malaysian rain forest contain more species of trees than all of the United States.)[1g]
>
> In the event of global warming, wilderness areas that have eluded the chain saws and logging roads may become primary paths of migration for thousands of plant and animal species in search of new homes. Man-altered environments create roadblocks for such migration, however, and not all obstructions are as visible as roads and buildings. In the case of the northern spotted owl, this shy bird has a natural reluctance to cross any large open spaces, such as heavily logged or clearcut land (or for that matter even large natural barriers such as the Columbia River Gorge). To avoid isolating the remaining population of spotted owls, the U.S. Forest Service is attempting to implement a timber harvesting policy that builds a network of preserved areas into an integrated whole—a greenway—allowing the population to travel back and forth.
>
> The same approach may become necessary for other species if climate change compels them to move. "Changing species ranges will not neatly accommodate current patterns of resource ownership," warned William Moomaw of the World Resources Institute in testimony before Congress in 1989. "Those plant and animal communities around which our national forests, national parks and wildlife refuges are built will suddenly be forced to move north" in the event of climate change, he predicted. "Will we attempt to establish a whole new set of mobile species refuges and national forests to track their migration? And if global warming is proceeding rapidly, would such a strategy make sense even if it were physically possible?"[2g]
>
> Such questions have spawned a lively debate as to whether intensively managed forests would stand up to the strain of climate change as well as old-growth forests. Environmental Protection Agency analyst Steve Young has observed:
>
>> Trees in the old forest are survivors that have made the grade over millions of years of evolution and hundreds of years of growth as individuals. They have stood up to a wide range of threats, including weather extremes, diseases, insect infestations, fires and pollution. Thus each old-growth

Since wood does not last forever, though, the CO_2 benefit of using lumber and related wood products declines as the life of the product shortens. Fortunately, wood has many long-term uses. Amortization schedules developed by the Internal Revenue Service indicate that wood in single-family homes is expected to last for 60 years; in home improvement projects, 30 years; and in mobile structures and manufacturing, about six years.[169] Of course, in many instances, wood lasts far longer than the averages suggest.

forest contains trees with strong survival potential bred into their genes. In contrast, the 'supertrees' developed by humans have been genetically engineered over a period of mere decades. Moreover, uniform stands of trees of the same species, genetic heritage and age are highly vulnerable to attacks from diseases and insect pests.[5g]

The sheltered areas under old-growth stands also create a protected microclimate, insulating the understory from harsher conditions that may develop above the foresty canopy. In some areas of the Pacific Northwest, needles of the forest's conifers actually capture moisture from fog and clouds, accounting for up to one-quarter of the local rainfall.[3g] If a drier climate were to prevail in the Northwest because of global warming, this "rainmaking" capability could be crucial. There are also preliminary indications that fungi and lichen, which anchor the forest food chain, are produced primarily in old-growth areas and spread subsequently to second-growth areas. If that is the case, cutting old growth could reduce the long-term vitality of surrounding second-growth forests and impede the migration of species.[4g]

Some forest industry officials dismiss these claims as so much "hooey." They say that old-growth stands nearing the end of their natural lives are much more likely to succumb to the stresses of climate change than young, vigorous stands. Moreover, they emphasize that a tremendous gene pool remains in genetically improved growing stock. What matters, they say, is that foresters maintain a diversity of age classes that includes middle-aged and young stands as well as old growth.[6g]

If global warming begins in earnest, however, triage may come to dictate timber management decisions more than anything else. Already, President Bush has convened a Cabinet-level Endangered Species Committee, known as the "God Squad," to determine whether timber-related jobs in the Northwest are more worth preserving than the spotted owl. In May 1992, by a vote of five to two, the God Squad approved harvesting old-growth on about 1,700 acres of federal land in Oregon in an effort to salvage about 1,000 logging jobs in the region.[7g]

The God Squad had met only twice before, both times in 1979; once to approve building the Grayrocks Dam in Wyoming despite a threat posed to whooping cranes on the Platte River in Nebraska, and another time to recommend against construction of the Tellico Dam in Tennessee to preserve a small fish known as the snail darter. (In the case of the snail darter, Congress later approved a law allowing the dam to be built despite the threat to the fish.) In the event of climate change, the God Squad (or its equivalent) may have to meet much more often. Its mission: to decide whether the preservation of local jobs is more important than the migration and preservation of species.

Paper products are one use of wood that does not last long, however. Virtually all paper (save that in books and archives) is thrown away less than a year after manufacture. Moreover, paper manufacturing is far more energy-intensive than ordinary lumber production. A ton of paper produces nearly 3,500 pounds of carbon during production—nearly as much as concrete.[170] Paper recycling partially addresses the carbon emissions problem by leaving carbon stored in trees that otherwise would be processed into pulp.[171] Whether paper recycling saves energy compared with a virgin fiber operation depends on the fuel source used for each. If a recycling plant purchases fossil-fueled electricity while a virgin paper mill burns in-house wood wastes, the recycling plant actually creates more net CO_2 emissions than the virgin paper mill.[172]

Accordingly, the energy used to turn wood into products—and the uses of those products—matter most in terms of wood's role in the global carbon cycle. If *all* of America's remaining 5 million acres of old-growth forests were cut down and replaced with young stands grown on 60-year rotations, for example, carbon emissions would *increase* because the old growth will store more carbon than these managed stands over the long term. If the harvested wood were used for short-lived purposes, moreover, the net increase in carbon emissions would amount to three years' worth of domestic fossil energy consumption, according to one recent analysis. If the harvested wood were used as a durable material to replace aluminum, plastic or brick, however, the net effect of chopping down even the old-growth would be to *reduce* carbon emissions by seven years' worth of U.S. fossil fuel consumption.[173] In other words, the global carbon cycle would be more in balance if the world used more durable products made from wood and fewer competing products made with nonrenewable fossil fuels.

The carbon cycle would benefit most, of course, if carbon were retained in old-growth stands *and* young trees were planted for carbon sequestration besides. But then the forest products industry would have less wood available to turn into products. Accordingly, conservation of materials and efficiency in manufacturing is a must for the forest products industry. Fortunately, the industry has made significant strides in this area already. It is now common practice for companies to derive several types of roundwood products from the same merchantable bole, for example, aided by laser-guided milling. That way, less wood waste is left to burn or throw away. In addition, more uses have been found for coarse and fine wood byproducts and residues at each step of the manufacturing process. Debarked slabs of sawmill chips and edgings are made into pulp chips and planer shavings for particle board. Residues not suitable for these purposes increasingly find their way into boilers as a replacement for fossil fuels.

According to industry consultant Clark Row, product recovery of plywood and lumber could rise by more than 20 percent over the next 50 years, while 15 percent efficiency gains are possible for pulpwood.[174] The aggregate effect of such efficiency improvements and recycling gains would be to limit demand for virgin timber. When combined with reforestation efforts, the forest products industry could overcome its fears of a pending supply shortage. Affordable pricing and an "eco-friendly" reputation could in fact make wood products ever more appealing to consumers. Such a prospect bodes well for the industry and for the atmosphere.

Conclusions

The old adage about "seeing the forest through the trees" seems well suited to conclude a discussion of the forest products industry and climate change. At present, the industry is mired in a thicket of environmental questions concerning timber management practices, protection of endangered species, diminution of air and water quality and the like—not to mention the impact of other industries' pollution on the health of the forests themselves. Now the specter of climate change hangs over these issues, much as a forest canopy shrouds a stand of trees. For the forest products industry to see its way through the mire, it must focus on a single, overarching objective: to preserve the long-term value of forestry resources—and promote its own business in the process.

Forestry in the 21st Century

The industry's long-term challenge—and its greatest worry—is to assure a steady supply of commercial timber, its basic raw material. With the harvest of virgin U.S. timber almost complete, and second-generation forests years away from filling "The Gap" in merchantable supply, forest products companies are concerned that they may not be able to meet the projected rising demand for wood products in the 21st century. "New Forestry" management techniques and protection of endangered species serve only to compound the industry's fears of a looming supply shortage. Despite a protracted slump in the housing market, lumber prices touched record levels in early 1992, indicating that the era of cheap wood may indeed be over.

Global warming could add to the industry's woes—or prove to be its salvation. In a direct sense, carbon dioxide enrichment should accelerate the growth of forests and bolster trees' water-use efficiency. Forests also could become more productive if the climate gets warmer and wetter. Yet some general circulation models raise a disturbing possibility that portions of the country might dry out and become less hospitable to forests in the event of global warming. One such region is the South, where the industry has placed half of its assets and 80 percent of its new tree seedlings.

By the time these trees mature a quarter-century from now, it should be clear what the climate has in store. Before then, no one may know for sure whether the enhanced greenhouse effect will benefit or harm the forests. The greenhouse gambit falls during this interim period. It compels forest industry executives to think anew about their long-term plans and ways to amend those plans in light of the emerging potential risks.

There is little evidence to suggest that forest products companies have amended their basic strategies to date. Seven of the ten major forest products companies surveyed by the Investor Responsibility Research Center in 1990 agreed with the statement, "Greater scientific certainty is required before our company includes global warming in its strategic planning." The crystal balls of climate research apparently are still too fuzzy to compel them to act on premonitions about global change.

As a result, forest products companies continue to buy, sell and trade timberland much as before, basing their holdings on the geographic mix of their mills and customers—not on speculation about the greenhouse effect. They continue to plant the same mix of species that have always served them well—loblolly pine in the South, Douglas-fir in the West and mixed hardwoods in the North—rather than planting southern genotypes at their extreme northern limits, while switching to new crops like Eucalyptus and kenaf fiber in the far South.

This is not to suggest that forest products companies are doing things that make no sense from a greenhouse perspective. On the contrary, they are sowing nearly a billion seedlings on their own timberlands each year and encouraging non-industrial private forest landowners to do the same; these vigorous young stands will sequester carbon as they grow. At the same time, industry researchers are breeding trees to resist drought and pests better, while expanding the horizons of genetic engineering. Forest products companies also are increasing per-acre yields on harvested land, producing less waste in milling processes and extending the useful life of disposable products through recycling. Recently, the industry has even begun to publicize these efforts to show how they can reduce the amount of carbon entering the atmosphere.

Despite these moves, the evolution in thinking within the forest products industry has not changed much from many years ago. The industry's purpose remains to sell as many wood products as the market demands and to harvest trees accordingly. This is a sound objective in a strict business sense. Even environmental arguments have been conjured to defend it. One recent study notes how projected harvest limitations in old-growth forests could result in about "117 cargoes of tankers the size of the *Exxon Valdez*" if more energy-intensive materials are called upon to make up the shortfall in building materials.[175] But while it is important to distinguish the lower energy-intensity of wood products, this analysis sidesteps the environmental controversy surrounding the harvest of old-growth itself.

Most observers do not believe the forest products industry is formulating long-term strategy on the basis of environmental priorities. On the contrary, there is a "mandate for change" in forestry research objectives, according to one analysis by the National Academy of Sciences. "Past approaches to forestry research employing conservation and preservation paradigms [have] proven inadequate," the academy reported in 1990. Foresters have "inherited [Gifford] Pinchot's 'tree-farm' view of the forest and his belief in scientific forestry as the road to wise use of forestry resources. Research priorities have long been dominated by commodity production goals."[176]

"To help overcome a deficiency in knowledge," the academy continued, "a new research paradigm will need to be adopted—an environmental paradigm." Among the new research objectives identified by the academy are understanding how forests and climate affect each other, curbing the loss of biological diversity, preserving pristine forest areas, and sustaining growing demand for

Figure 13

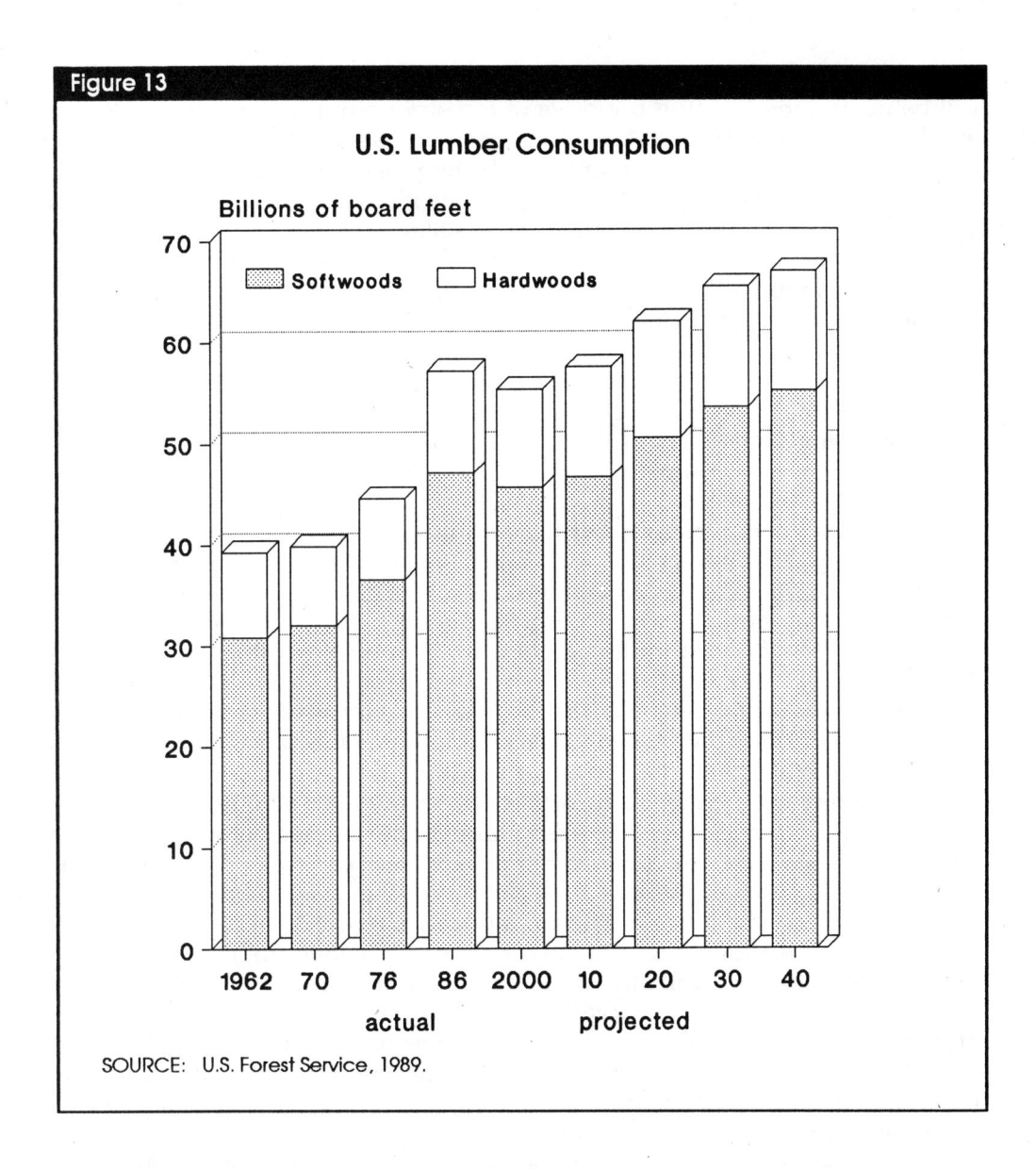

U.S. Lumber Consumption

Billions of board feet

[Bar chart with Softwoods and Hardwoods categories, showing values from 1962 to 2040. Years 1962, 70, 76, 86 are actual; 2000, 10, 20, 30, 40 are projected.]

SOURCE: U.S. Forest Service, 1989.

wood and wood products with less available timber. Looking ahead to the next century, the academy concluded:

> The twenty-first century forester, farmer, government warden, and park manager will need to recognize the interdependence of each of their forms of natural resource management on the others. To succeed, resource management must be considered in the context of an ecosystem, where resource development, conservation, and protection are considered simultaneously. Competition for resources will give way to cooperative management strategies, where conservation and resource managment are linked in sustainable resource systems.[177]

In other words, forests will not be managed solely for timber anymore. There will be many cross-cutting priorities of forestry management, and some will work at cross-purposes. At least the specter of global climate change offers common ground on one key point: Both the planet and the forest products industry would benefit from efforts to promote reforestation and the judicious use of timber to maintain a healthy carbon balance on the land and in the air.

Should the forest products industry seek to maximize the effect of a greenhouse gambit, there is much it could do. It could set ambitious new tree-planting goals for the American Tree Farm System, a program launched 50 years ago by the industry to teach private landowners how to manage and maintain their forest property. It could seek further amendments to the Conservation Reserve Program and other government-funded programs to expand tree-planting around the nation. It could even advocate a carbon tax on fossil fuels and support the establishment of an emissions trading program that includes carbon offsets.

A few energy companies already have volunteered to pay others to grow trees as a means of offsetting power plant emissions. Forest products companies could share in this bounty. Such a program would sequester more carbon in the short term, buy time for energy companies to switch to non-carbon alternatives over the medium term and help alleviate a possible timber supply shortage over the long term. Once these carbon-storing trees have served their initial purpose, they could be chopped down and converted to wood products that continue to store the carbon for decades longer. Depending on the success of such reforestation efforts, the forest products industry could even make a push into biomass energy production, which would have a far greater effect in offsetting carbon emissions than tree-planting efforts themselves.

One final thing the American forest products industry could do is international in scope. It could send a message around the world to preserve tropical rainforests by ending its own determination to harvest dwindling old-growth timber at home (including that in America's own coastal rainforests). By some projections, America's remaining old-growth forests could disappear faster than virgin tropical rainforests, with the exception of tracts afforded wilderness protection. By supporting efforts to set aside more old-growth as greenways and biological preserves, the industry could blunt the argument of developing nations that the United States is more interested in preserving their natural resources than its own.

The loss of tropical forests around the world is contributing perhaps 10 percent of the global buildup of greenhouse gases. President Bush expressed his concern for the plight of the rainforests in June 1992 by pledging $150 million in additional assistance to help developing nations preserve their forests.[178] If the preservation of American old-growth stands were to set an example for those who possess the real terrestrial lungs and gene pool of the planet—the tropical-forested nations—then the American forest products industry would make a contribution that is far greater than the president's contribution and its own comparatively limited ability to sequester carbon.

There are many possible reasons why the American forest products industry has not stepped forward on the greenhouse issue. Perhaps because of onerous environmental regulations affecting its own timber-harvesting and manufacturing operations, it has not called for tougher controls on other industries' pollution. (This acquiescence comes despite some localized evidence that ozone pollution and acid rain are impairing the health of U.S. forests.) If the American forest products industry were to announce strong support of policies to address climate change, however, it would further distinguish itself from competing, energy-intensive industries that are responsible for the lion's share of greenhouse gas emissions.

It may also be that forest industry executives are inclined to go along with others who oppose climate change legislation because they see no evidence of a decline in the productivity of their forests and do not wish to alarm investors about the prospect. Yet no one could impugn the industry's motives if it moved to protect its timberlands on the basis that other industries' pollution was putting these assets at risk. Should the perception grow that the forest products industry is in fact coming to terms with this potentially serious environmental problem faster and more effectively than other industries, its image among consumers and investors would seem likely to improve. Forest products companies—and the people they employ—would be counted among the Earth's caretakers, nurturing a natural resource that is vital to the nation's economic health and the world's ecological well-being.

The basic question is whether timber industry executives should take a leadership position now—before any definitive signs of climate change emerge—or keep their fingers crossed that dire scenarios of global change do not unfold. Considering the very long-term nature of this industry's investments, the stakes are extremely high in making the right decision. With thousands of people at work in the forests and hundreds more carrying on research in laboratories, the forest products industry has an effective early warning system in place to detect any adverse signs of climate change. But the risk in waiting for the alarm bell to sound is that it may become too late to do much of anything constructive about the problem. The greater risk for this industry, therefore, may be to pass on the greenhouse gambit rather than to put it into play.

Notes

1. Potlatch Corp. annual report, San Francisco, Calif., 1989.
2. R. Neil Sampson and Thomas E. Hamilton, "Forestry Opportunities in the United States to Mitigate Effects of Global Warming," *Forests and Global Warming*, proceedings of a conference sponsored by the American Forestry Association, R. Neil Sampson and Dwight Hair, editors, Washington, D.C., May 1991. According to Sampson, average tree growth ranges from 47 to 100 cubic feet per acre per year in the United States. To determine average annual carbon sequestration per acre, the growth rate can be multiplied by a constant, 33.4. Accordingly, a northern hardwood forest sequesters approximately 1,600 pounds of carbon per acre per year, whereas a well-sited southern pine plantation accumulates more than 3,300 pounds.
3. U.S. Department of Agriculture, "America the Beautiful" brochure, Washington, D.C., 1989.
4. Paul J. Kramer and J. Charles Lee, "Forestry Research Needs and Strategies," *The Greenhouse Effect, Climate Change and U.S. Forests*, William E. Shands and John S. Hoffman, editors, The Conservation Foundation, Washington, D.C., 1987.
5. U.S. Forest Service, *An Analysis of the Timber Situation in the United States*, Resource Planning Assessment Technical Document supporting RPA Assessment, RM-199, U.S. Department of Agriculture, Washington, D.C., December 1990.
6. National Research Council, Committee on Forestry Research, *Forestry Research: A Mandate for Change*, National Academy Press, Washington, D.C., 1990.
7. Roger Rosenblatt, "Trees," *Life* Magazine, May 1990.
8. "Facts and Figures 1991," American Forest Council, Washington, D.C., July 1991.
9. Weyerhaeuser Co. annual report, Tacoma, Wash., 1989.
10. See note 5.
11. *Ibid.*
12. U.S. Forest Service, *FY 1990 U.S. Forest Planting Report; Forest Planting Seeding and Silvicultural Treatments in the United States*, U.S. Department of Agriculture, Washington, D.C., 1989. For purposes of this report, the South is defined as Alabama, Arkansas, Florida, Georgia, Kentucky, Louisiana, Mississippi, Missouri, North Carolina, Oklahoma, South Carolina, Tennessee, Texas, Virginia and West Virginia.
13. See note 5.
14. Jeffrey T. Olson and H. Michael Anderson, "Owl Delivers Message in Forest Management," *Forum for Applied Research and Public Policy*, Fall 1991.
15. "Lumber Production Comparison: Coast and Inland Region to United States 1978 - 1990," statistical chart compiled by the National Forest Products Association, Washington, D.C., 1991.
16. "Lumber Production in Leading States: 1982 - 1989," statistics compiled by the Western Wood Products Association and the Bureau of Census, Current Industrial Reports, Lumber Production and Mill Stocks, Series MA-24T.
17. See note 8.
18. U.S. Forest Service, *Forest Statistics of the United States*, PNW-RB-168, U.S. Department of Agriculture, Washington, D.C., September 1989.
19. See note 5.
20. *Ibid.*
21. "The Potential Impact of Rapid Climatic Change on Forests in the United States,"

Policy Options for Stabilizing Global Climate - Report to Congress, Volume II, U.S. Environmental Protection Agency, February 1989; and Cooperative Holocene Mapping Project, "Climatic Changes of the Last 18,000 Years: Observations and Model Simulations," *Science*, Aug. 26, 1988
22. Leslie Roberts, "How Fast Can Trees Migrate?" *Science*, Feb. 10, 1989.
23. See first citation of note 21.
24. *Ibid.*
25. William W. Kellogg and Zong-Ci Zhao, "Sensitivity of Soil Mositure to Doubling of Carbon Dioxide in Climate Model Experiments. Part I: North America," *Journal of Climate*, April 1988.
26. *Ibid.*
27. Ian Woodward, "Plants in the Greenhouse World," *New Scientist*, May 6, 1990.
28. See first citation of note 21.
29. Jonathan T. Overpeck, Patrick J. Bartlein and Thompson Webb III, "Potential Magnitude of Future Vegetation Change in Eastern North America: Comparisons with the Past," *Science*, Nov. 1, 1991.
30. See note 22.
31. Andrea Heil, "Sugarers Have Seen Bad Get Worse: Changes in Region's Weather Have Industry Wondering and Worrying," *Valley News*, Feb. 25, 1991.
32. See first citation of note 21.
33. Peter N. Duinker, "Climate Change and Forest Management, Policy and Land Use," *Land Use Policy*, April 1990.
34. Allen M. Solomon and Darrel C. West, "Simulating Forest Ecosystem Responses to Expected Climate Change in Eastern North America: Applications to Decision Making in the Forest Industry," *The Greenhouse Effect, Climate Change and U.S. Forests*, William E. Shands and John S. Hoffman, editors, The Conservation Foundation, Washington, D.C., 1987.
35. Jerry W. Leverenz and Deborah J. Lev, "Effects of Carbon-Dioxide Induced Climate Changes on the Natural Ranges of Six Major Commercial Tree Species in the Western United States," *The Greenhouse Effect, Climate Change and U.S. Forests*, William E. Shands and John S. Hoffman, editors, The Conservation Foundation, Washington, D.C., 1987. Some studies suggest that winter chilling of Douglas-fir requires a temperature below 45 degrees Fahrenheit, rather than 48 degrees. In that case, global warming could eliminate the species from lower elevation sites along the entire West Coast of the United States, along the Willamette Valley of Oregon, and in the Sierra Range of California.
36. See first citation of note 21.
37. See note 29.
38. Personal communication with forest industry executives, Washington, D.C., Jan. 14, 1992.
39. See note 29.
40. William Booth, "Vast Changes Predicted for America's Forests," *The Washington Post*, Nov. 1, 1991.
41. See note 22.
42. See note 21.
43. See note 22.
44. See first citation of note 21.
45. *Ibid.*
46. Michael S. Coffman, Champion International Corp., "Changes in Regional Forest Management Planning Due to Global Climate Change," manuscript presented at North American Conference on Forestry Responses to Climate Change, sponsored by the Climate Institute, Washington, D.C., May 17, 1990.

47. Justin R. Ward, Richard A. Hardt and Thomas E. Kuhule, *Farming in the Greenhouse: What Global Warming Means for American Agriculture*, Natural Resources Defense Council, Washington, D.C., March 1989.
48. W. Frank Miller, Philip M. Dougherty, and George L. Switzer, "Effect of Rising Carbon Dioxide and Potential Climate Change on Loblolly Pine Distribution, Growth, Survival and Productivity," *The Greenhouse Effect, Climate Change and U.S. Forests*, William E. Shands and John S. Hoffman, editors, The Conservation Foundation, Washington, D.C., 1987.
49. Paul J. Kramer and Nasser Sionit, "Effects of Increasing Carbon Dioxide Concentration on the Physiology and Growth of Forest Trees," *The Greenhouse Effect, Climate Change and U.S. Forests*, William E. Shands and John S. Hoffman, editors, The Conservation Foundation, Washington, D.C., 1987.
50. "Trees Thrive Under Elevated CO_2 Levels," *Chemical & Engineering News*, Oct. 21, 1991.
51. Robin Sandenburgh, Carol Taylor and John S. Hoffman, "Rising Carbon Dioxide, Climate Change, and Forest Management: An Overview," *The Greenhouse Effect, Climate Change and U.S. Forests*, William E. Shands and John S. Hoffman, editors, The Conservation Foundation, Washington, D.C., 1987.
52. Fakhri A. Bazzaz and Eric D. Fajer, "Plant Life in a CO_2-Rich World," *Scientific American*, January 1992.
53. See note 49.
54. "CO_2 Rise May Favor Trees Over Grasslands," *The New York Times*, Jan. 15, 1991.
55. See note 52.
56. See note 48.
57. William K. Stevens, "Carbon Dioxide Rise May Alter Plant Life, Researchers Say," *The New York Times*, Sept. 18, 1990. A recent study at Duke University has affirmed this hypothesis.
58. See note 55.
59. William H. Smith, "Air Pollution and Forest Damage," *Chemical & Engineering News*, Nov. 11, 1991. Smith is the Clifton R. Musser Professor of Forest Biology in the School of Forestry and Environmental Studies at Yale University.
60. See note 49.
61. See note 2.
62. See note 7.
63. Such chaos in the weather may be partly responsible for a decline in sugar maples and hardwoods in the Adirondack Mountains of New York and in Quebec, according to the U.S. Forest Service, as relatively warm weather and severe drought in that region during the 1960s was followed by unusually cold and fierce winters in the 1970s and early 1980s.
64. See note 49.
65. James J. MacKenzie and Mohamed T. El-Ashry, *Ill Winds: Airborne Pollution's Toll on Trees and Crops*, World Resources Institute, Washington, D.C., September 1988.
66. See first citation of note 21.
67. World Resources Institute, "Forests and Rangelands," *World Resources: 1990-91*, Oxford University Press, New York, 1990.
68. See note 51, and Tom Kenworthy, "'Unraveling' of Ecosystem Looms in Oregon Forests," *The Washington Post*, May 15, 1992.
69. National Acid Precipitation Assessment Program, *Acidic Deposition: State of Science and Technology*, "Changes in Forest Health and Productivity in the United States and Canada," Report 16, U.S. Government Printing Office, Washington, D.C., December 1990.

70. David S. Wilson, "Worst Forest Fire Year Appears to Be at an End," *The New York Times*, Nov. 10, 1988.
71. See note 51. The quote appears in a 1981 U.S. Department of Agriculture report, which paraphrased the Canadian government's report. Emphasis added.
72. Michael Fosberg, U.S. Forest Service, *Proceedings of the Second North American Conference on Preparing for Climate Change*, The Climate Institute, Washington, D.C., 1989
73. Joel Smith, U.S. Environmental Protection Agency, *Proceedings of the Second North American Conference on Preparing for Climate Change*, The Climate Institute, Washington, D.C., 1989.
74. See note 35.
75. Frank Graham Jr., "Matchsticks!", *Audubon*, January 1990.
76. See note 7.
77. Donald G. Friedman, The Travelers Corp., *Proceedings of the Second North American Conference on Preparing for Climate Change*, The Climate Institute, Washington, D.C., 1989.
78. James L. Regens, Frederick W. Cubbage and Donald G. Hodges, "Greenhouse Gases, Climate Change, and U.S. Forest Markets," *Environment*, May 1989.
79. See note 35.
80. Norman J. Rosenberg et al., *Policy Options for Adaptation to Climate Change*, Resources for the Future, Washington, D.C., March 1989.
81. See note 78.
82. *Ibid.*
83. See note 38. Weyerhaeuser is one such company.
84. R.W. Kennedy, "Specific Gravity of Early- and Late-Flushing Douglas-fir Trees," *Tappi*, 1970 (Vol. 53, No. 8).
85. See note 78.
86. See note 80.
87. Daniel Botkin, University of California, presentation at the North American Conference on Forestry Responses to Climate Change, sponsored by the Climate Institute, Washington, D.C., May 16, 1990.
88. Clark S. Binkley, "A Case Study of the Effects of CO_2-Induced Climatic Warming on Forest Growth and the Forest Sector: Economic Effects on the World's Forest Sector," *The Impact of Climatic Variation on Agriculture*, M.L. Parry, T.R. Carter, and N.T. Konjin, editors, D. Riedel Publishing Co., Dordrecht, The Netherlands, 1987.
89. James N. Woodman, "Potential Impact of Carbon Dioxide-Induced Climate Changes on Management of Douglas-Fir and Western Hemlock," *The Greenhouse Effect, Climate Change and U.S. Forests*, William E. Shands and John S. Hoffman, editors, The Conservation Foundation, Washington, D.C., 1987.
90. "'Greenhouse Effect' on Forest Management to be Examined at June 1984 Conference," National Forest Products Association press release, Washington, D.C., May 15, 1984.
91. David Frankil, Champion International Corp., presentation at the North American Conference on Forestry Responses to Climate Change, sponsored by the Climate Institute, Washington, D.C., May 17, 1990.
92. See note 38.
93. Eric D. Vance, National Council of the Paper Industry for Air and Stream Improvement, personal communication, Dec. 20, 1991.
94. See note 69.
95. Philip Shabecoff, "Forestry Group Seeks Tougher Pollution Controls," *The New York Times*, Nov. 22, 1987.

96. See note 12. The ICF report referred to in the text is entitled, *Strategic Options for Enhancing Forest Industry Productivity in the Face of Rising Atmospheric Carbon Dioxide.*
97. "The Georgia-Pacific Forest: Environmental Issues and Answers," Georgia-Pacific Corp., Atlanta, Ga., Oct. 18, 1988.
98. Richard Good, Georgia-Pacific Corp., personal communication, March 2, 1990.
99. See note 91.
100. Michael Coffman, Champion International, personal communication, May 15, 1990, and Jan. 14, 1991.
101. Peter Farnum, Weyerhaeuser Co., "Creating New Stands: Adapting to the Greenhouse Effect Through Technological Preparedness," presentation at the North American Conference on Forestry Responses to Climate Change, sponsored by the Climate Institute, Washington, D.C., May 16, 1990.
102. *Ibid.*
103. C.C. Lambeth et al., "Large-Scale Planting of North Carolina Loblolly Pine in Arkansas and Oklahoma: A Case of Gain Versus Risk," *Journal of Forestry*, December 1984.
104. *Ibid.*
105. See note 48.
106. Peter Farnum, Weyerhaeuser Co., personal communication, March 1, 1992.
107. See note 101.
108. R.C. Kellison and R.J. Weir, "Breeding Strategies in Forest Tree Populations to Buffer Against Elevated Atmospheric Carbon Dioxide Levels," *The Greenhouse Effect, Climate Change and U.S. Forests*, William E. Shands and John S. Hoffman, editors, The Conservation Foundation, Washington, D.C., 1987.
109. See note 38.
110. See note 34.
111. *Ibid.*
112. See first citation of note 21.
113. B.J. Spalding, "Kenaf: A Cheaper Pulp for Brighter U.S. Newsprint," *Chemical Week*, Feb. 10, 1988.
114. M.S. Ward, BeCon Construction Co., personal communication, Oct. 31, 1990.
115. Douglas Wilson, CIP Paper Products Co., personal communication, Oct. 31, 1990.
116. National Research Council, *Managing Global Genetic Resources: Forest Trees*, National Academy Press, Washington, D.C., 1991.
117. Pope & Talbot Inc., "A Presentation to the Paper and Forest Products Industry Analysts Group by Pope & Talbot," Portland, Ore., April 20, 1988.
118. Perry R. Hagenstein et al., "Below-Cost Sales: Impact on Timber Prices," *Journal of Forestry*, August 1987. The "economic efficiency" model constructed in this analysis found that log prices in the Pacific Northwest would fall by one-third "if public-land timber harvests were set so as to maximize the discounted net benefits of the flow of timber."
119. U.S. Forest Service, "Timber Sale Program Annual Report, FY 1988 Test, Forest Level Information," U.S. Department of Agriculture, Washington, D.C., 1989.
120. Carrie Dolan, "Lumber Prices Soar on Worries About Supply, But Anxiety Grows Over Future Industry Effect," *The Wall Street Journal*, Feb. 19, 1992; and Chip Johnson, "Some Wood-Product Prices Hit Highs, Sparked by Spotted Owl Litigation, Low Timber Output," *The Wall Street Journal*, June 12, 1991.
121. Dietmar W. Rose, Alan R. Ek, and Keith L. Belli, "A Conceptual Framework for Assessing Impacts of Carbon Dioxide Change on Forest Industries," *The Greenhouse Effect, Climate Change and U.S. Forests*, William E. Shands and John S. Hoffman, editors, The Conservation Foundation, Washington, D.C., 1987.

122. For background on the sale, see James R. Schiffman, "Georgia-Pacific Bid Gets Cool Response But Stock Price of Target Nekoosa Soars," *The Wall Street Journal*, Nov. 1, 1989.
123. Richard Good, Georgia-Pacific Corp., personal communication, Nov. 26, 1991.
124. These advertisements appeared in several editions of *The Wall Street Journal* in January 1992.
125. James C. Hyatt, "Georgia-Pacific To Sell Bowater Interest in Mills," *The Wall Street Journal*, Oct. 11, 1991.
126. Richard Ober, "Time to Move," *Forest Notes*, New Hampshire Society for Protection of Forests, July/August 1990. This article describes measures taken by a New England governors' task force on northern forest lands.
127. Robert Anderberg, "Wall Street and the Great North Woods," *The Amicus Journal*, Winter 1989.
128. David Stipp, "Forests of the Northeast Get Breathing Space," *The Wall Street Journal*, Jan. 15, 1992.
129. See note 29.
130. See note 46. The study referred to in the text was performed by Daniel Botkin of the University of California.
131. See note 34.
132. Michael Coffman, Champion International, personal communication, Dec. 26, 1991.
133. See note 34.
134. See note 89.
135. See note 5.
136. *Ibid.*
137. Conrad B. MacKerron, Investor Responsibility Research Center, personal communication, Jan. 16, 1992. MacKerron's research is for a forthcoming book on innovative projects in tropical forested regions.
138. William K. Stevens, "Experts Say Logging of Vast Siberian Forest Could Foster Warming," *The New York Times*, Jan. 28, 1992. This article reports that Georgia-Pacific, Louisiana-Pacific and Weyerhaeuser are negotiating to cut Siberian forests.
139. Intergovernmental Panel on Climate Change, *1992 IPCC Supplement*, World Meteorological Organization and United Nations Environment Programme, Geneva, Switzerland, February 1992.
140. U.S. Forest Service, *An Analysis of the Timber Situation in the United States: 1989 - 2040, Part II: The Future Resource Situation*, U.S. Department of Agriculture, Washington, D.C., 1989.
141. Con H. Schallau, "Is Nation's Supply of Timber Threatened," *Forum for Applied Research and Public Policy*, Fall 1991.
142. See note 8.
143. Business Brief, *The Wall Street Journal*, March 8, 1990.
144. Gregory Byrne, "Let 100 Million Trees Bloom," *Science*, Oct. 21, 1988.
145. See note 8, and Dale Russakoff, "They're All Tree Huggers Now," *The Washington Post*, Oct. 5, 1989.
146. See note 8.
147. Richard Birdsey, U.S. Forest Service, "Changes in Forest Carbon Storage from Increasing Forest Area and Timber Growth," *Forests and Global Warming*, proceedings of a conference sponsored by the American Forestry Association, R. Neil Sampson and Dwight Hair, editors, Washington, D.C., May 1991.
148. Peter Parks, Duke University, "Opportunities to Increase Forest Area and Timber Growth on Marginal Crop and Pasture Land," *Forests and Global Warming*,

proceedings of a conference sponsored by the American Forestry Association, R. Neil Sampson and Dwight Hair, editors, Washington, D.C., May 1991.
149. Ralph Alig, U.S. Forest Service, presentation at the North American Conference on Forestry Responses to Climate Change, sponsored by the Climate Institute, Washington, D.C., May 16, 1990.
150. William Moomaw, World Resources Institute, prepared testimony before the Senate Agriculture Committee, Dec. 1, 1988. Professor Moomaw now is at Tufts University.
151. Barry Cassell, "AES Corp. Buys Into Forest CO_2 `Sink' to Offset Coal Plant Emissions," *The Energy Report*, June 3, 1991.
152. See note 137.
153. "Others Get Much Out of Dutch," *Environment*, November 1991.
154. Peter Passell, "Greenhouse Gamblers," *The New York Times*, June 19, 1991.
155. Gregg Marland, "Reforestation: Pulling Greenhouse Gases Out of Thin Air," *Biologue*, April/May 1989. Marland's calculation assumes that the coal plant has a 38 percent efficiency rating and is operated at 70 percent of capacity. It also assumes that the sycamore trees are spaced at five-foot intervals and are harvested and replaced with new seedlings every four years.
156. See note 150. The 25,000 megawatt capacity figure is based on a 1987 forecast of new electric power requirements by the North American Electric Reliability Council.
157. Roger A. Sedjo, "Forests: A Tool to Moderate Global Warming?" *Environment*, January/February 1989. This calculation assumes that one cubic meter of tree biomass contains 0.26 tons of carbon. Accordingly, one acre of forest would sequester about 2.52 tons of carbon.
158. See note 139.
159. Susan Williams and Kevin Porter, *Power Plays: Profiles of America's Leading Independent Renewable Electricity Developers*, Investor Responsibility Research Center, Washington, D.C., 1989.
160. U.S. Forest Service, *An Analysis of the Timber Situation in the United States: 1989 - 2040, Part I: The Current Resource and Use Situation*, U.S. Department of Agriculture, Washington, D.C., 1989.
161. U.S. Congress, Office of Technology Assessment, *Changing by Degrees: Steps to Reducing Greenhouse Gases*, OTA-O-482, U.S. Government Printing Office, Washington, D.C., February 1991.
162. National Research Council, Committee on Alternative Energy Research and Development Strategies, *Confronting Climate Change: Strategies for Energy Research and Development*, National Academy Press, Washington, D.C., 1990.
163. Lynn Wright, Oak Ridge National Laboratory, presentation at the North American Conference on Forestry Responses to Climate Change, sponsored by the Climate Institute, Washington, D.C., May 16, 1990.
164. See note 150.
165. See first citation of note 21.
166. National Research Council, *Renewable Resources for Industrial Materials*, National Academy of Sciences, National Academy Press, Washington, D.C., 1976.
167. For one cradle-to-grave analysis, see Peter Koch, "Wood Vs. Non-Wood Materials in U.S. Residential Construction: Some Energy-Related Implications," Wood Science Laboratory Inc., Corvallis, Mont., October 1991.
168. Clark Row, forest economic consultant, and Robert B. Phelps, U.S. Forest Service, "Carbon Cycle Impacts of Improving Forest Products Utilization and Recycling," manuscript prepared for North American Conference on Forestry Responses to Climate Change, sponsored by the Climate Institute, Washington, D.C., May 16, 1990.

169. *Ibid.*
170. *Ibid.*
171. See note 161. There is not a one-for-one relationship between recycled paper and saving trees, however. That is because it takes about 1.4 pounds of wastepaper to produce one pound of finished recycled paper. In addition, the content of most virgin paper is itself only half virgin fibers; the rest consists of manufacturing and forest residues and wastepaper. See, for example, the manuscript by Steve Conway, Scott Paper Co., "Recycled Paper: Balancing Hype with Reality," June 11, 1991.
172. See the citation at the end of note 171.
173. C.D. Oliver, J.A. Kershaw Jr. and T.M. Hinckley, "Effect of Harvest of Old-Growth Douglas-fir and Subsequent Management on Carbon Dioxide Levels in the Atmosphere," paper repesented at the Society of American Foresters National Conference, Washington, D.C., July 29-Aug. 1, 1990.
174. Clark Row, forest economic consultant, presentation at the North American Conference on Forestry Responses to Climate Change, sponsored by the Climate Institute, Washington, D.C., May 16, 1990.
175. See note 6.
176. *Ibid.*
177. See note 167.
178. "Bush Pledges $150 Million for Forest Preservation," *The Wall Street Journal*, June 2, 1992.

Box 3-A

1a. Douglas W. MacCleery, U.S. Forest Service, "Condition and Trends of U.S. Forests: A Brief Overview," U.S. Department of Agriculture, Washington, D.C., undated manuscript.
2a. *Ibid.*
3a. U.S. Forest Service, *An Analysis of the Timber Situation in the United States: 1989 - 2040, Part I: The Current Resource and Use Situation*, U.S. Department of Agriculture, Washington, D.C., 1989.
4a. U.S. Forest Service, *An Analysis of the Timber Situation in the United States: 1989 - 2040, Part II: The Future Resource Situation*, U.S. Department of Agriculture, Washington, D.C., 1989.
5a. See note 3a.
6a. *Ibid.*
7a. Charles McCoy, "U.S. Agency Offers Scaled-Back Plan To Protect Owls," *The Wall Street Journal*, Jan. 9, 1992.
8a. "Facts and Figures 1991," American Forest Council, Washington, D.C., July 1991.
9a. Keith Ervin, "Timber Economy Evolves," an *Audubon Activist* special report, undated.
10a. See note 4a.
11a. Neil Sampson, American Forestry Association, presentation at the North American Conference on Forestry Responses to Climate Change, sponsored by the Climate Institute, Washington, D.C., May 15, 1990.
12a. U.S. Forest Service, *Forest Statistics of the United States*, PNW-RB-168, U.S. Department of Agriculture, Washington, D.C., September 1989.
13a. See note 4a.

Box 3-B

1b. One interesting description of the evolution of life on Earth appears in the book by Anne H. Ehrlich and Paul R. Ehrlich, *Earth*, published by Franklin Watts, New York, 1987.
2b. Jonathan Scurlock and David Hall, "The Carbon Cycle," Inside Science #51, *New Scientist*, Nov. 2, 1991.
3b. Pieter P. Tans, Inez Y. Fung and Taro Takahashi, "Observational Constraints on the Global Atmospheric CO_2 Budget," *Science*, March 23, 1990.
4b. Carl Sagan, Owen B. Toon and James B. Pollack, "Anthropogenic Albedo Changes and the Earth's Climate," *Science*, Dec. 21, 1979.
5b. "The Potential Impact of Rapid Climatic Change on Forests in the United States," *Policy Options for Stabilizing Global Climate - Report to Congress*, Volume II, U.S. Environmental Protection Agency, February 1989.
6b. Roger A. Sedjo, "Forests: A Tool to Moderate Global Warming?" *Environment*, January/February 1989. This calculation assumes that one cubic meter of tree biomass contains 0.26 tons of carbon. Accordingly, one acre of forest would sequester about 2.52 tons of carbon.

Box 3-C

1c. National Research Council, Committee on Forestry Research, *Forestry Research: A Mandate for Change*, National Academy Press, Washington, D.C., 1990.
2c. U.S. Forest Service, *Analysis of the U.S. Timber Situation, 1952 - 2030*, U.S. Department of Agriculture, Washington, D.C., 1982.
3c. U.S. Forest Service, *An Analysis of the Timber Situation in the United States: 1989 - 2040, Part II: The Future Resource Situation*, U.S. Department of Agriculture, Washington, D.C., 1989.
4c. For the quote, see note 1c.

Box 3-D

1d. Leigh Dayton, "New Life for Old Forest," *New Scientist*, Oct. 13, 1990; and Timothy Egan, "With Fate of the Forests at Stake, Power Saws and Arguments Echo," *The New York Times*, March 20, 1989.
2d. National Research Council, Committee on Forestry Research, *Forestry Research: A Mandate for Change*, National Academy Press, Washington, D.C., 1990.
3d. Michael Lipske, "Who Runs America's Forests," *National Wildlife*, October/November 1990.
4d. Charles McCoy, "U.S. Agency Offers Scaled-Back Plan To Protect Owls," *The Wall Street Journal*, Jan. 9, 1992.
5d. Bruce Ingersoll, "U.S. Forest Service Plans a Cutback in Logging Allowed," *The Wall Street Journal*, June 8, 1990.
6d. Marilyn Chase, "Cancer Drug May Save Many Human Lives—At Cost of Rare Trees," *The Wall Street Journal*, April 9, 1991.
7d. Jerry F. Franklin, Mark E. Harmon and William K. Ferrell, "Effects of Carbon Storage on Conversion of Old-Growth Forests to Young Forests, *Science*, Feb. 9, 1990.
8d. *Ibid.* Of the carbon removed at harvest, about 55 percent is assumed to be lost to the atmosphere from paper production, fuel consumption and decomposition.

The remainder is converted to boards and plywood that continue to store carbon, usually for decades.

9d. Peter Koch, "Wood Vs. Non-Wood Materials in U.S. Residential Construction: Some Energy-Related Implications," Wood Science Laboratory Inc., Corvallis, Mont., October 1991.

10d. William K. Stevens, "Experts Say Logging of Vast Siberian Forest Could Foster Warming," *The New York Times*, Jan. 28, 1992.

11d. Jerry F. Franklin, University of Washington, presentation at the North American Conference on Forestry Responses to Climate Change, sponsored by the Climate Institute, Washington, D.C., May 15, 1990.

Box 3-E

1e. Bruce Ingersoll, "U.S. Forest Service Plans a Cutback in Logging Allowed," *The Wall Street Journal*, June 8, 1990.

2e. Bruce Ingersoll, "Logging Policy of U.S. Forests to Be Changed," *The Wall Street Journal*, June 4, 1992; and U.S. Forest Service, *An Analysis of the Timber Situation in the United States*, Resource Planning Assessment Technical Document supporting RPA Assessment, RM-199, U.S. Department of Agriculture, Washington, D.C., December 1990.

3e. Seth Zuckermam, "New Forestry: New Hype?" *Sierra*, March/April 1992.

4e. William Atkinson, Oregon State University, "Another View of New Forestry," paper presented at the Annual Meeting of the Oregon Society of American Foresters, Eugene, Ore., May 4, 1990.

5e. The definitive early work on this subject is by Leo A. Isaac, U.S. Forest Service, "Place of Partial Cutting in Old-growth Stands of the Douglas-fir Region," USDA Research Paper No. 16, Pacific Northwest Forest and Range Experiment Station, March 1956.

6e. David M. Smith, "Applied Ecology and the New Forests," paper presented at the Western Forestry and Conservation Associations annual meeting, 1970.

7e. Jim Stiak, "Memos to the Chief," *Sierra*, July/August 1990.

8e. *Ibid.*

9e. Randall O'Toole, "Reforming the Forest Service," *FREE Perspectives*, Foundation for Research on Economics and the Environment, January 1990.

10e. See, for example, Lawrence Solomon, "Save the Forests—Sell the Trees," *The Wall Street Journal*, Aug. 25, 1990. Solomon is executive director of Environment Probe, an environmental organization, based in Toronto, Canada.

Box 3-F

1f. U.S. General Accounting Office, "Forest Service: Timber Harvesting, Planting, Assistance Programs and Tax Provisions," GAO/RCED-90-107BR, Washington, D.C., April 1990; and Fred Cubbage, University of Georgia, presentation at the North American Conference on Forestry Responses to Climate Change, sponsored by the Climate Institute, Washington, D.C., May 17, 1990.

2f. Justin R. Ward, Richard A. Hardt and Thomas E. Kuhule, *Farming in the Greenhouse: What Global Warming Means for American Agriculture*, Natural Resources Defense Council, Washington, D.C., March 1989.

3f. U.S. General Accounting Office, *1990 Farm Bill: Opportunities for Change*, GAO/RCED-90-12, U.S. Government Printing Office, Washington, D.C., April 1990.

Box 3-G

1g. Roger Rosenblatt, "Trees," *Life* Magazine, May 1990.
2g. William Moomaw, World Resources Institute, prepared testimony before the Senate Agriculture Committee, Dec. 1, 1988. Professor Moomaw is now at Tufts University.
3g. Leigh Dayton, "New Life for Old Forest," *New Scientist*, Oct. 13, 1990.
4g. Jerry F. Franklin, University of Washington, presentation at the North American Conference on Forestry Responses to Climate Change, sponsored by the Climate Institute, Washington, D.C., May 15, 1990.
5g. Steve Young, "Tree Slaughter: Your Taxes at Work," *The Washington Post*, Aug. 13, 1989.
6g. Personal communication with members of the forest products industry, Washington, D.C., Jan. 14, 1991.
7g. Jonathan Weil, "Panel Allows Logging in Site Vital to Owl," *The Wall Street Journal*, May 15, 1992.

Chapter 4
The Automobile Industry

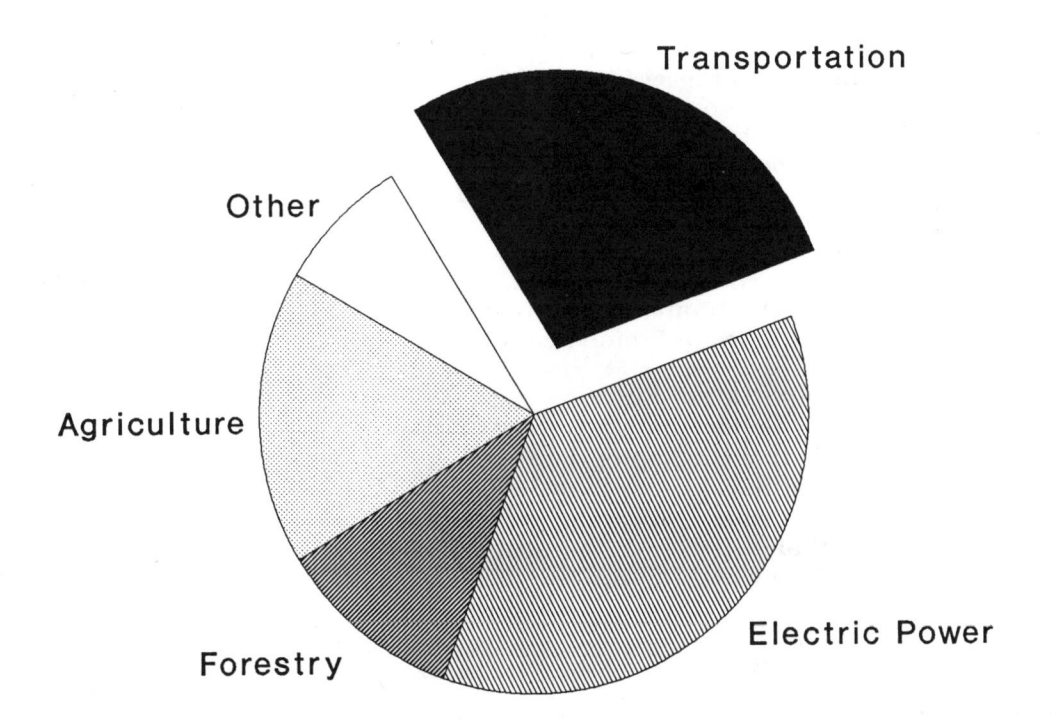

Relative contribution to global warming potential.

Contents of Chapter 4

Introduction
    Adding Up the Greenhouse Gases 236
    Many Options, No Magic Solutions 239
    Taking a Global View 242

Fuel Efficiency Potential
    Feeling the Heat 249
    CAFE: The Legacy of Round I 250
    Fuel Economy in Reverse 252
    Power Is Back in Style 256
    Gambling on Bigger Cars 258
    CAFE: Time for Round II? 260
    Potential for Fuel Economy Gains 262
        Overhead Cam and Multivalve Engines 265
        Variable-Valve and Lean-Burn Engines 267
        Two-Stroke Engines 269
        Diesel Engines 271
        Improved Transmissions 273
        Lightweight and Aerodynamic Designs 275
        Fuel-Efficient Prototype Cars 278

Alternative Transportation Fuels
    Gearing Up for Another Energy Crisis? 281
    In Search of the Appropriate Fuel 284
        Reformulated Gasoline 287
        Methanol 291
        Ethanol 296
        Compressed Natural Gas 300
        Hydrogen 304
        Electricity 307
        Batteries 313

Conclusions
    Driving into the 21st Century 317
    The First Move 320
    The Right Fuels 325

Boxes
    Box 4-A: The Rise and Fall of Mass Transit 246
    Box 4-B: EPA Mileage Ratings and the 'Mileage Gap' 254
    Box 4-C: Cars and Auto Safety 276
    Box 4-D: Cars, Fuels and the Clean Air Act 288
    Box 4-E: Assessing the 'Impact' 310
    Box 4-F: Battery Research 314

Notes 327

Introduction

In 1867, a German engineer named Nikolaus Otto patented a machine that effectively shrank the world: the four-stroke, internal combustion, reciprocating piston engine. If not the greatest invention since the wheel, Otto's motor certainly served as the ideal complement. Along with Rudolf Diesel, who patented the diesel engine in 1892, Otto revolutionized transportation, allowing people to travel faster to destinations farther over the horizon.

Since their introduction around the turn of the century, motorized vehicles have multiplied at a faster rate than the human population. Nearly 600 million cars, trucks and buses roam the globe today—one for every 10 of the planet's inhabitants. If the explosion continues, the motor vehicle population could top 1 billion before 2025.

While the love affair with the automobile is universal, in no country has the attraction been stronger than in the United States. Americans who comprise only 4 percent of the human population own 33 percent of the world's cars and trucks.[1] They travel 2.1 trillion miles a year in them—the equivalent of more than 11,000 round trips to the Sun. In the process, U.S. motor vehicles consume 135 billion gallons of carbon-based fuel annually, releasing about 350 million metric tons of gaseous carbon (not to mention other more noxious pollutants) into the atmosphere.

The advent of mass transit systems and jet aircraft has not curtailed enthusiasm for this century-old technology. Americans continue to rely on their cars, vans and pick-up trucks for four out of every five miles of intercity travel.[2] No other form of transportation has emerged to rival the automobile as the most convenient and economical means of getting around. Nor has any other machine better encapsulated the American spirit. "It's not just a car," boasts a General Motors television advertisement. "It's your freedom!"

The romance with the automobile seems likely to continue well into the future. "There is no evidence to suggest that the automobile will not be the predominant form of transportation in 2020," predicts Lester Hoel, a University of Virginia professor and former chairman of the Transportation Research Board of the National Research Council. "This hardy little beast is characterized by flexibility for the user, permitting him to go wherever he wants and whenever he wants. It has a remarkable capacity to survive, changing its color, size and shape to conform to the current environment. Although the vehicle may be technologically unrecognizable as it adapts to the computer age, its essential features will remain."[3]

While the auto reigns supreme, the days of conventional Otto and diesel engines may be numbered. Their reliance on carbon-based fuels makes them vulnerable on at least three counts. First, the world's supply of crude oil, the wellspring for these engines, is diminishing and could be exhausted in the 21st century. In the meantime, the Organization of Petroleum Exporting Countries—with 85 percent of the proven reserves—will be the world's dominant oil merchants.

Second, more than a half-billion vehicular tailpipes constitute the world's largest source of air pollution—accounting for more than half of all urban air pollution in the United States.[4] While auto makers have made tremendous strides toward building cleaner-burning cars—cutting average U.S. tailpipe emissions 90 percent since 1970—progress on the pollution front has been impeded by the growing number of cars and trucks on the road, the greater number of miles they travel and the increasingly congested conditions in which they operate.

Finally, there is an emissions source that no vehicle can avoid as long as fossil carbon remains in the fuel tank. That source is carbon dioxide, a greenhouse gas, and therein lies the third and potentially most serious threat to gasoline and diesel engines' longevity. In the past, auto engineers found ways to deal with conventional auto emissions, such as hydrocarbons and soot, by making engine and exhaust modifications. Other pollutants, such as carbon monoxide and oxides of nitrogen, have been curbed by the introduction of catalytic converters and, more recently, reformulated gasolines. But essentially all of the carbon in the fuel ends up as CO_2 (except for partially burned carbon that forms soot). As a result, carbon dioxide will remain as a byproduct of combustion in gasoline- and diesel-powered vehicles, no matter how "clean" the fuel-burning process is and no matter how advanced exhaust systems may become.

Adding Up the Greenhouse Gases

The auto industry is practically unique among major industries in that it is a major contributor of greenhouse gases yet relatively impervious to the likely weather-related effects of climate change. During the "greenhouse summer" of 1988, several non-air-conditioned auto assembly plants did shut down temporarily after temperatures on the production floor rose to 130 degrees F.[5] Auto air-conditioning also has been in demand because of the growing population in the South and West and a desire to "beat the heat" elsewhere. (More than 90 percent of new American cars now feature such factory-installed equipment.)[6] Yet in most respects, the auto industry has been able to carry on with its business plans as if climate were not a factor, even though vehicular emissions of carbon dioxide, ozone precursors and other greenhouse gases could spur a rise in the Earth's temperature over time.

Carbon dioxide: Simple arithmetic reveals the significance of motor vehicles as producers of carbon dioxide. A gallon of gasoline as it enters the tank weighs about 6.2 pounds, and 87 percent of that weight is in the form of carbon. When the fuel burns thoroughly in an engine cylinder, each carbon atom bonds with two oxygen atoms. This combustive process drives the pistons that turn the engine crankshaft. But the resulting carbon dioxide—now weighing nearly four times as much as the carbon that went in—is vented as a waste product through an exhaust valve. Hence, each gallon of gasoline entering the gas tank leaves the tailpipe as 19.7 pounds of carbon dioxide.[7] (For diesel fuel, the figure is closer to 21.6 pounds of CO_2.)

Figure 1

**CO$_2$ Emissions
Transportation Sector - 1990**

Industry 32%
Buildings 36%
Transportation 32%

Rail, marine 7%
Aircraft 14%
Heavy trucks 14%
Light trucks 20%
Automobiles 43%

Total carbon emissions
1.3 billion metric tons

Transportation
420 million metric tons

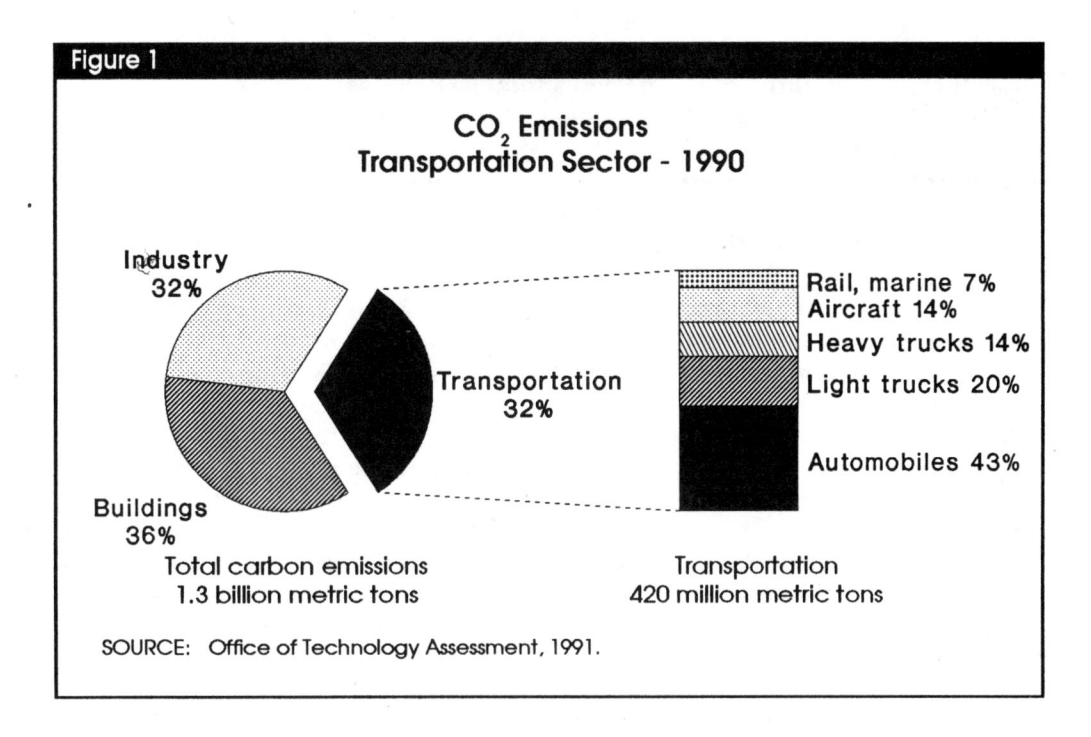

SOURCE: Office of Technology Assessment, 1991.

These pounds of CO$_2$ add up fast when one considers how much fuel a vehicle consumes. A gasoline-powered car with a 15-gallon tank emits 300 pounds of CO$_2$ between refuelings. The distance that car travels on a tank of gas depends, of course, on its on-road fuel economy. A car getting 18 miles per gallon consumes more than 5,500 gallons of gasoline and releases approximately 13.4 metric tons of carbon (in the form of carbon dioxide) over 100,000 miles of travel. A car getting 36 mpg, on the other hand, produces only 6.7 tons of carbon while traveling just as far over its 10-year average lifetime. That makes fuel economy an important factor in determining a vehicle's relative contribution of CO$_2$ to the atmosphere.

For the nation as a whole, more than 260 million tons of carbon (equal to nearly 1 billion tons of CO$_2$) are vented from cars and light-duty trucks each year; heavy trucks emit another 60 million tons annually. Other modes of transportation—airplanes, ships, pipelines and railroads—release 100 million tons of carbon in the United States each year. Accordingly, fuel combustion in the transportation sector accounts for about 32 percent of the nation's 1.3 billion tons of annual carbon emissions to the atmosphere. The light-duty fleet alone is responsible for about 20 percent of the nation's total.[8] (The light-duty fleet consists of passenger cars and small trucks weighing less than 8,500 pounds.)

These percentages represent only the end-use of motor fuels, however, not the energy required to process them. Every barrel of oil used directly in transportation consumes another quarter-barrel of oil to extract, produce, refine and distribute the fuel.[9] Moreover, manufacturing and repairing motor vehicles requires a tremendous amount of energy, as does building and maintaining transportation infrastructure. Such industrial activities boost annual transportation-related carbon emissions to more than 500 million tons altogether, or two-fifths of the nation's total CO_2 emissions.[10]

While these numbers are immense, they seem less conspicuous when viewed in a global context. The United States accounts for slightly more than one-fifth of the world's CO_2 emissions resulting from the burning of fossil fuels.[11] Since American cars and trucks are responsible for about 22 percent of such emissions domestically, the U.S. vehicular fleet contributes roughly 5 percent of CO_2 emissions from fossil fuel consumption around the globe (not including emissions related to fuel processing, vehicle manufacturing and road-building.)

Moreover, CO_2 represents only about 60 percent of the total weighted amount of greenhouse gases entering the atmosphere. (Methane, chlorofluorocarbons, nitrous oxide and other gases make up the other 40 percent.)[12] On a global basis, then, carbon dioxide released directly from U.S. motor vehicles accounts for about 3 percent of worldwide greenhouse gas emissions. Motor vehicles in all other countries emit a similar amount of CO_2 into the atmosphere.[13]

Smog-forming pollutants: An analysis of auto-related greenhouse gas emissions is not complete until several other pollutants are considered. Motor vehicles account for two-thirds of anthropogenic carbon monoxide emissions, nearly half of nitrogen oxides emissions and two-fifths of hydrocarbon emissions in the 24 industrialized nations comprising the Organization for Economic Cooperation and Development, including the United States.[14] Tailpipe emissions are far worse in countries without sophisticated emissions control devices, such as in Eastern Europe, the former Soviet Union and most developing nations. But their auto-related greenhouse gas emissions do not rival those in the West because fewer people own cars in those countries and they drive them fewer miles each year.

Even so, automotive emissions in many cities around the world contribute to "smog" pollution, which is caused by the reaction of non-methane hydrocarbons and oxides of nitrogen in the presence of sunlight and heat. Such tropospheric ozone is a noxious greenhouse gas, although its precise contribution to global warming potential has not been determined.[15]

Another greenhouse gas that comes directly from vehicle exhaust is nitrous oxide. It traps infrared heat 270 times more efficiently than does carbon dioxide and is emitted from cars equipped with catalytic converters. Nitrous oxide is not considered a major source of vehicular-related greenhouse gases, however, because N_2O emissions are comparatively small—equivalent to less than 2 percent of CO_2 emissions from automobiles.[16]

Finally, carbon monoxide is a byproduct of incomplete fuel combustion that causes headaches and heart-related stress. Although not a greenhouse gas, carbon monoxide contributes to the atmospheric buildup of methane, which has more than 20 times the heat-trapping capability of carbon dioxide. Carbon monoxide and methane react with the hydroxyl radical (OH), an oxidizing agent that effectively removes pollutants from the atmosphere. As carbon monoxide-vehicle exhausts increase, less hydroxyl is available to react with methane, allowing methane to linger in the air. Such competition for the hydroxyl radical has contributed to a rising concentration of methane in the atmosphere, which is increasing about 0.6 percent a year. The exact role of vehicle exhausts in spurring this increase (and a corresponding loss of hydroxyl) has not been quantified.[17]

Chlorofluorocarbons: Until recently, scientists believed that chlorofluorocarbons also played a major role in global warming potential. These synthetic chemicals are used in the automotive sector mainly as refrigerants in car air conditioners and also as solvents and degreasers for vehicles' electronic circuitry and as blowing agents for dashboard padding and seat cushions. (Car makers outside the United States use far fewer CFCs because of the low percentage of cars equipped with air conditioning.)

In terms of radiative forcing potential, CFC-12 is the most potent of all major greenhouse gases, with approximately 7,500 times the heat-trapping potential of CO_2.[18] On this basis, Ford Motor Co. once estimated that three charges of CFC-12 refrigerant vented from a mobile air conditioner over a car's lifetime—about 7.5 pounds by weight—had the same global warming potential as 100,000 miles worth of CO_2 vented from a car's tailpipe.[19] If that were the case, the American automobile industry's use of CFCs would add another 1 percent to global warming potential—on top of the 3 percent or so stemming from U.S. vehicular emissions of CO_2 and other smog-forming pollutants.[20]

Recent scientific evidence suggests, however, that most of the radiative forcing potential of CFCs is offset by their depletion of the ozone layer, which cools the stratosphere.[21] Because of this destruction, auto makers and their CFC suppliers have pledged to phase out use of conventional CFCs. One alternative refrigerant, HFC-134a, will be introduced in 1993 model-year cars.[22]

As such non-chlorinated replacements for CFCs enter the market, automobile manufacturers may claim that they will have taken an important step toward restoring the ozone layer and reducing global warming potential. Yet the residual impacts of many CFC substitutes will not be negligible. Some of the alternatives are thought to have up to one-sixth of the radiative forcing potential of CFC-12, so their potency as a greenhouse gas will be up to 1,000 times greater than that of CO_2 on a molecular basis.[23] Moreover, since non-chlorinated substitutes should not damage the ozone layer, they will not have a counteractive effect in terms of cooling the stratosphere.[24] As a result, they may contribute even more to net global warming potential than the fully halogenated CFCs they replace.

Many Options, No Magic Solutions

Progress by the auto industry in reducing greenhouse gas emissions overall will not come fast or easily. Car manufacturers in the United States will achieve modest near-term reductions through compliance with provisions of the 1990 Clean Air Act. This legislation mandates tighter controls on automotive emissions of CFCs and smog-forming pollutants, while encouraging the development of alternative fuels. Some of these fuels have a low carbon content—or contain no carbon at all—and may be able to curb the auto industry's CO_2 emissions dramatically. These alternative fuels are not likely to become a major source of power for internal combustion engines until well after the turn of the century, however.

In the meantime, much of the emphasis will be on improving the fuel economy of gasoline- and diesel-powered vehicles as a means of combatting global warming. The most thrifty cars on the road already produce only one-third of a pound of CO_2 per vehicle mile traveled compared with today's average of one full pound per mile. Therefore, the greatest near-term potential to curb CO_2 emissions is to shift the mix of autos toward the more fuel-economic models now available.

Fuel economy gains achieved by the American auto industry over the last 20 years have been impressive. Average mileage ratings of new cars have doubled since 1973, rising from 14 mpg to 28 mpg, as measured by the U.S. Environmental Protection Agency. At the same time, light truck fuel economy has increased 60 percent, from 13 mpg to 21 mpg. Had this improvement not occurred, the light-duty fleet in the United States would be consuming 10.5 million barrels of oil a day (instead of 6.5 million barrels), the nation's annual fuel bill would be at least $60 billion higher, and carbon emissions would be 170 million tons greater than they are now, according to an analysis by the American Council for an Energy-Efficient Economy.[25]

Conservation advocates believe that maintaining the trend toward improved vehicular fuel economy is one of the most immediate and effective ways to ameliorate global warming. According to a recent ACEEE report:

> If U.S. new light vehicle fuel economy is improved 40 percent by 2001, as is being proposed by Congress, carbon emissions will be 120 million metric tons per year lower by the year 2005 than they would be if new light vehicle fuel economy remains at today's level. Although this is only 9 percent of current U.S. carbon emissions, no other single improvement in end-use energy efficiency, or plausible switch to low-carbon, nuclear or renewable fuel, will yield reductions as large by the year 2005.[26]

Most auto makers vigorously oppose the effort to raise corporate average fuel economy (CAFE) standards, however, insisting that future gains must be incremental and cost-effective. "Contrary to the myths spread by CAFE advocates," contends Thomas Hanna, president and chief executive officer of the Motor Vehicle Manufacturers Association of the United States, "these unrealistic [mileage] standards would have no appreciable impact on oil imports or global climate change. And because there is no magic fuel economy technology, the only way to substantially raise fuel economy is to make smaller cars and trucks, and ration the availability of larger vehicles." This, Hanna says, "would raise the number of highway deaths and injuries, limit consumer choice...and place thousands of auto-related jobs at risk."[27]

While the value of raising CAFE standards to address global warming is a matter of heated debate, all agree that the American auto industry will have to map its future course wisely if it is to remain competitive in the global marketplace. About 12 million American jobs—or one in nearly every seven jobs—depend on the motor vehicle industry and related industries. Motor vehicle output alone represents 3.5 percent of U.S. gross national product.[28]

Figure 2

**Corporate Average Fuel Economy
1974-1991**

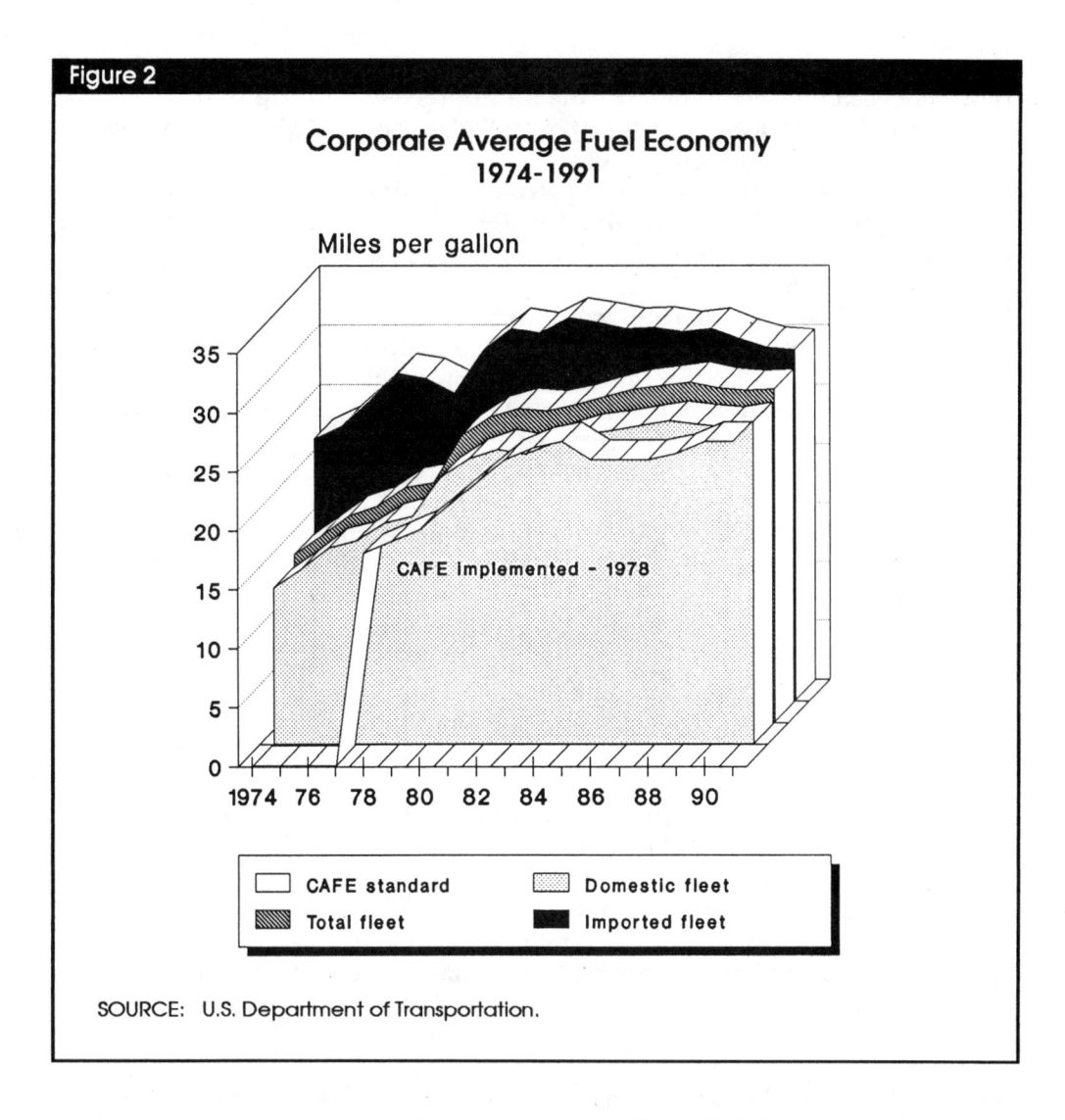

Miles per gallon

CAFE implemented - 1978

Legend:
- CAFE standard
- Domestic fleet
- Total fleet
- Imported fleet

SOURCE: U.S. Department of Transportation.

Fears about the deteriorating condition of the global environment notwithstanding, American auto makers believe they are in a fight for their own survival. The Big Three's share of the U.S. auto market has dropped more than 20 percentage points since 1970. The labor force of domestic auto makers has shrunk by more than 500,000 employees, while foreign auto makers have created 100,000 jobs in America, mainly by siting assembly plants here. The United States has ceded more of its new car market to foreign car manufacturers, in fact, than any other nation (excluding intra-European trade).[29]

The contraction of the domestic auto industry and America's growing reliance on foreign cars and imported oil has become a major election-year issue and a flashpoint in international relations. The United States imported $52 billion more motor vehicles and equipment parts than it exported in 1990.[30] Forty billion dollars of this deficit was with Japan, prompting

President Bush and the chief executives of the Big Three auto makers to seek trade concessions from Japan during an unprecedented trip to the Far East in early 1992. America's energy import bill also topped $50 billion in 1990—almost all of which went for oil and two-thirds of which was consumed by the domestic transportation sector. Petroleum shipments from Arab OPEC nations to the United States doubled in the five years preceding the Iraqi war (which put many Middle Eastern oil fields out of commission in 1991).[31] Were it not for the trade deficits in autos and oil, in fact, the United States' $108 billion trade imbalance in 1990 would have been virtually nil.

Some believe the United States now has a historic opportunity to address some of its most serious economic and environmental problems by formulating new policies for the automotive industry. If American auto makers were to embark on a strategy to raise the fuel economy of the nation's auto fleet substantially, it "would allow us to make singular gains against global warming, oil spills, dependence on imported oil, and further declines in international competitiveness," says transportation consultant Michael Walsh, who headed the U.S. Environmental Protection Agency's oversight of the auto industry during the 1970s.[32]

The auto industry is not so sure. It remains concerned about consumers' embrace of such fuel-saving measures, noting that few car buyers express much interest in high-mpg cars at present. "This is not a simple drill," cautions Albert Slechter, director of government affairs for Chrysler Corp. "If people reject these [more fuel-economical] vehicles after we've built them, we've got a major disaster on our hands."[33]

Fuel economy gains in any case would buffer the auto industry against future oil price increases and be likely to slash the nation's imported oil bill. A 12.5-miles-per-gallon increase for passenger cars, from 27.5 mpg to 40 mpg, and a 9.3-mpg increase for light-duty trucks, from 20.7 mpg to 30 mpg, would likely reduce U.S. oil demand by 1.5 million to 2 million barrels a day by 2010, according to an analysis by the congressional Office of Technology Assessment.[34] The upper bound of this estimate nearly equals the amount of oil that the United States imported from all of the Persian Gulf in 1990 and is about one-fifth as much oil as is thought to reside in Alaska's Arctic National Wildlife Refuge. Moreover, the fuel savings would represent a 100 million ton annual reduction in the nation's carbon emissions (emitted in the form of carbon dioxide), equal to 7.7 percent of the present-day total.[35]

Taking A Global View

If the opportunity for significant fuel economy gains is not realized in the American auto sector, many observers fear that the nation's reliance on foreign oil as well as its transportation-related CO_2 emissions will increase inexorably. The number of vehicles on the road and the miles they are driven each year seems destined to grow, even if fuel economy gains fail to keep pace. Making matters worse, driving conditions are likely to become more congested, causing a reduction in on-road fuel economy.

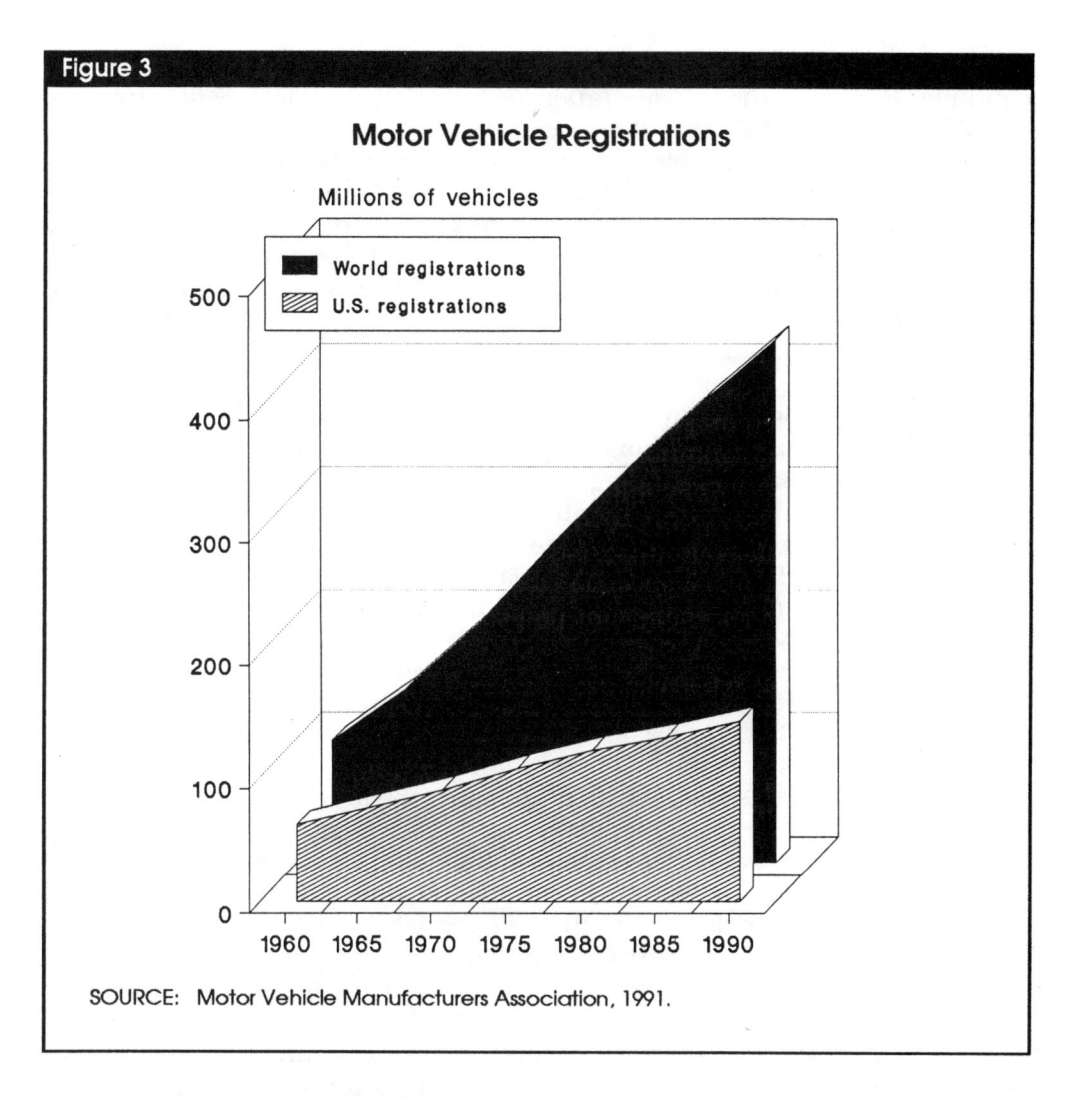

> **Figure 3**
>
> **Motor Vehicle Registrations**
>
> *[Area chart showing Millions of vehicles from 1960 to 1990, comparing World registrations and U.S. registrations. World registrations rise from about 130 million in 1960 to about 460 million in 1990; U.S. registrations rise from about 75 million to about 150 million over the same period.]*
>
> SOURCE: Motor Vehicle Manufacturers Association, 1991.

The U.S. motor vehicle population is rapidly approaching parity with the human population in America; one motor vehicle is registered for every 1.3 people.[36] Yet U.S. demand for cars and trucks continues to grow. More than 30 million vehicles were added to the domestic fleet during the 1980s alone. If the trend continues, the number of registered vehicles in the United States could top 250 million by 2010—an increase of 60 million.

With more cars on the road, bumper-to-bumper traffic has become commonplace. More than two-thirds of all urban interstate roads are now congested during peak travel times. The Federal Highway Administration estimates that such stop-and-go driving wastes 1.4 billion gallons of fuel a year, or 2 percent of the energy used for highway passenger transportation. By 2005, the figure could rise to 7.3 billion gallons of fuel wasted each year—7 percent of projected oil use for highway passenger transport.[37] Adding in the trend toward more travel by airplane, total petroleum demand in the U.S. transportation sector could rise from 11 million to 14 million barrels a day by 2000 if present trends continue. Similarly, the transportation sector's CO_2 emissions could rise from 1.7 billion tons to 2 billion tons a year.[38]

Overseas markets for motor vehicles are not nearly as saturated as they are in the United States. On a per-capita basis, Japan has only three-fifths as many registered vehicles as the United States; Western Europe, only half as many vehicles. And these are the most economically developed regions of the world, where more than 70 percent of all vehicles are found. In the former Soviet Union, only one vehicle is registered for every 13 people; in Egypt, there is just one vehicle per 80 people; and in India and China, one for every 214 people.[39]

With the demise of centrally planned economies in Eastern Europe, a huge new car market is opening up that consists of 100 million prospective buyers. Already, auto companies ranging from Volkswagen to General Motors, Fiat, Renault and Suzuki are building new factories or retooling old plants in Czechoslovakia, (formerly) East Germany, Hungary, Poland, the former Soviet republics and Yugoslavia to meet the anticipated demand.[40] Throughout Europe, the size of the automotive population fleet is expected to double over the next 20 years. Asian and Latin American vehicular fleets are expected to *more* than double over the same period. India—with 120 million middle-class citizens—represents another potentially huge new auto market. China plans to quadruple its auto output to 600,000 cars annually in the next decade alone.

It is because of these burgeoning markets—rather than in spite of them—that the auto industry is a part of the potential global warming problem and must be part of any eventual solution. Already, the industry's contribution of carbon dioxide into the atmosphere represents 14 percent of all such emissions from fossil fuels worldwide.[41] (Ozone-forming tailpipe emissions and possibly CFCs are in addition to this.) Unless market demand for motor vehicles experiences a dramatic downturn, the auto industry is on course to becoming one of the very largest collective CO_2 emissions sources by shortly after the turn of the century—rivaling emissions from fossil-fueled electric-generating plants.[42]

All told, vehicular emissions of carbon dioxide could increase nearly 50 percent worldwide through 2010, according to an analysis by the World Resources Institute, an environmental research organization in Washington, D.C.[43] The overriding factor in the increase would be a 50 percent increase in the world's motor vehicle fleet—reaching nearly 900 million by 2010. Any improvements in fuel economy over the period are likely to be overwhelmed by this larger number of cars and trucks on the road and the greater distances they travel. The World Resources Institute study concluded that vehicular carbon emissions are likely to reach 1.3 billion tons worldwide by 2010, given current trends. That is about as much as the United States produces today from all carbon emission sources.

To reverse this trend within the auto industry, the onus will fall mainly on 15 large companies that command more than 85 percent of the world's motor vehicles market. Cars and trucks throughout the world have been based on designs originated by these companies. In the future, these same companies will be counted on to offer alternative forms of transportation that do not compound the threat of global warming. This is their greenhouse gambit.

Figure 4

Motor Vehicle Production - 1989

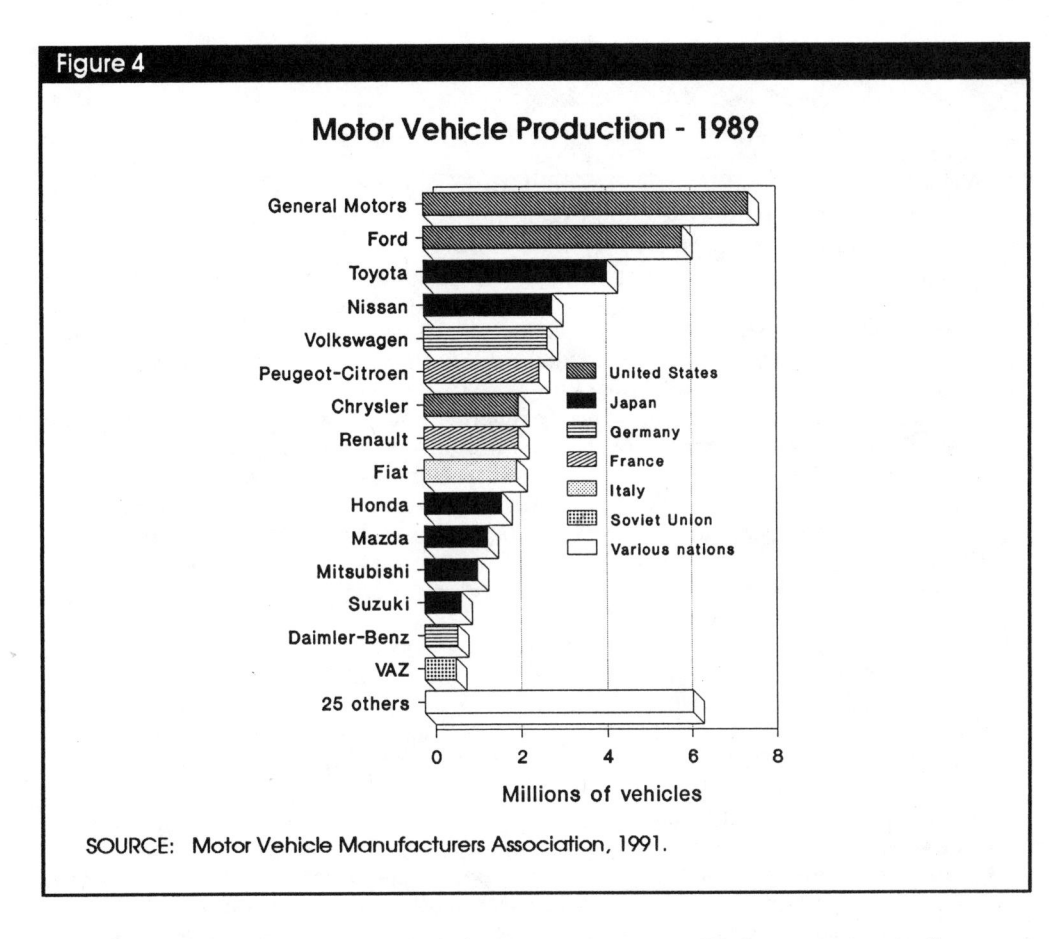

General Motors	United States
Ford	United States
Toyota	Japan
Nissan	Japan
Volkswagen	Germany
Peugeot-Citroen	France
Chrysler	United States
Renault	France
Fiat	Italy
Honda	Japan
Mazda	Japan
Mitsubishi	Japan
Suzuki	Japan
Daimler-Benz	Germany
VAZ	Soviet Union
25 others	Various nations

Millions of vehicles (0 to 8)

SOURCE: Motor Vehicle Manufacturers Association, 1991.

Essentially, the auto makers have two options. In the near term, they can design engines and car bodies that result in far more efficient use of carbon-based fuels. A bit farther down the road, they can build cars and trucks to run on non-fossil-carbon energy sources, including electricity and hydrogen derived from nuclear or renewable generating sources. That way, no CO_2 would be released to the atmosphere regardless of the amount of fuel consumed.

Transportation policymakers eventually may decide it is time to switch direction and emphasize the development of alternative forms of transportation—such as passenger buses, trolley cars and "bullet" trains. Considering the deteriorating infrastructure of roads and bridges—and the increasingly congested conditions that motorists face—pursuing this third option may make the most sense from a societal standpoint. Yet most observers put the least stock in this option. The automobile and its accoutrements have become an integral part of modern-day living and of the landscape itself. The automobile will be with us for some time to come.

The remainder of this chapter explores the first two options in detail: increasing the efficiency of Otto and diesel-powered vehicles and switching to non-fossil-carbon transportation fuels. Conclusions follow those discussions.

(continued on p. 248)

> **Box 4-A**
>
> ## The Rise and Fall of Mass Transit
>
> The heyday of U.S. mass transit came in the first half of the 20th century and peaked in the late 1940s. Once 24 billion passengers took the bus, trolley, subway or commuter train to work each year. Now the number of riders is not much higher than it was in 1900, despite a huge increase in the labor force.[1a] The reason for the decline can be found in any driveway: the automobile and all the convenience it has to offer. Seven out of eight American households now own at least one car, and more than half own two vehicles or more.[2a] Meanwhile, nine out of every 10 trolley and light-rail systems that once operated in the country have since been dismantled.
>
> America and the automobile grew up together in the 20th century, transforming the landscape forever. Two percent of the nation—an area larger than the state of New York—has been paved over. Major metropolitan areas devote half of their land, on average, to roadways, parking lots and service stations.[3a] (Los Angeles has set aside two-thirds of its land to accommodate the almighty automobile.) Since World War II, suburban developments have also sprung up on inexpensive land outside the reach of most public transit systems. As a result, modern American living almost demands the use of an automobile.
>
> Today, residential population densities are low, commuting distances are long, and origin-destination pairs are too diverse to make mass transit affordable in many locations. Consequently, 86 percent of those who commute take to the road in private vehicles—and 73 percent drive alone. Moreover, 16 percent drive to work in particularly fuel-inefficient vans, jeeps and pick-up trucks.[4a] Yet commuting accounts for less than one-third of the miles that people travel in motor vehicles each year. With access to the nation's 43,000-mile interstate highway system never far away, getting from one major city to the next by car is very simple. Private automobiles, in fact, account for 81 percent of all intercity passenger miles that Americans travel each year. Adding in the other transportation marvel of the 20th century, airplanes, Americans are more mobile than ever before—traveling more than 12,000 miles a year on average—compared with 436 miles in 1900.[5a] Buses account for only 1.2 percent of intercity travel, however; and trains, 0.6 percent.[6a]
>
> Too much of a good thing can cause problems, of course. In wasted gasoline consumption alone, motor vehicles squander some 2.2 billion gallons in stop-and-go traffic, according to the Federal Highway Administration, representing 2 percent of the energy used for highway passenger transport.[7a] The cost in lost economic productivity is far greater. In 1987, America's 110 million commuters spent 2 billion hours stuck in traffic, costing the United States $16 billion in lost productivity and wasted fuel.[8a] The U.S. Transportation Department estimates that growth in the gross national product between 1990 and 1995 could be reduced 3.2 percent (relative to unconstrained levels) if the nation's road system continues to deteriorate.[9a]
>
> There is every reason to believe that the nation's road system will continue to decline. Highways are crumbling under the strain of traffic loads growing five times faster than the rate of new road capacity. The federal highway system has 77,000 deficient bridges; one fails about every other day. Forty percent of all interstate highway pavement is in fair or poor condition, according to the Federal Highway

Administration, and conditions are not expected to improve overall even if funding levels are boosted substantially.

To bring the nation's road system up to "minimum engineering standards" will require $550 to $650 billion in spending over the next 20 years, the Federal Highway Administration estimates. If the investment is not made, highway congestion is expected to get many times worse than it already is.[10a] As a result, fuel wasted in traffic jams could rise to 7.3 billion gallons a year by 2005, or 7 percent of projected oil use for highway passenger transport.[11a]

A transportation bill passed by Congress in the fall of 1991 may alleviate some of the congestion. It authorizes spending nearly $120 billion on a variety of highway and bridge projects through the 1997 fiscal year, and $32 billion more for mass transit. Perhaps more significantly, states will have the authority to switch up to $50 billion of the federal highway funds into mass transit programs—and help even out a growing imbalance in funding levels for highways and public transportation.[12a]

Only a decade ago, the Department of Transportation spent just twice as much on highways as it did on mass transit. Today, the ratio stands at five to one—as funding levels for mass transit were cut 50 percent (in real dollars) during the 1980s. Had Congress passed the 1991 transportation bill as it was originally submitted by the Bush administration, the ratio would have dropped to 10 to one.[13a]

Congress's increased support for mass transit suggests that public transportation may be staging a comeback. California, Colorado, Florida, Maine and New Jersey are among states that have canceled highway expansion projects in recent years in favor of increased support for mass transit.[14a] Up to 30 cities—including Buffalo, Miami, Minneapolis, Portland (Ore.), Sacramento and St. Louis—have built or are in the process of building urban light-rail systems. Atlanta, Dallas, Los Angeles, San Diego, San Francisco, Seattle, northern Virginia and southern Florida are converting abandoned freight track for use as passenger-rail transit. Even Boston is renovating its century-old trolley lines.[15a]

In another realm, a few companies are beginning to open satellite offices and neighborhood work centers within walking and biking distance of the homes of their employees. The air quality improvement plan drafted for Los Angeles provides incentives to encourage such interspersion of housing and employment centers. Even full-blown telecommuting is becoming a reality, where employees work out of offices right in their homes.

But the idea of kicking the auto habit is still a pipe dream for most commuters. The vast majority of the American labor force continue to drive more than 20 miles round-trip to work each day, spending the better part of an hour in their cars, racking up 425 billion commuter miles a year—putting more than 400 billion pounds of carbon dioxide into the atmosphere.[16a] If the present trend continues—and the revival of mass transit stalls out—commuters could spend an average of four hours a day in their cars by 2005, moving at 5 miles an hour for a 20-mile-round-trip commute.[17a] Under these conditions, the vehicles' carbon dioxide emissions would grow immensely, since the engines would be running for longer periods at very low speeds, making for extremely inefficient use of fuel.[18a] Keeping traffic moving at reasonable speeds is one of the best ways to hold down CO_2 emissions—short of getting people out of their cars altogether.

Fuel Efficiency Potential

International concern over the greenhouse effect comes at a time when the American auto industry is frankly more worried about its own well-being. Detroit's auto makers have been hemorrhaging badly since the nation slipped into recession in 1990. With sales sluggish and facilities idled, General Motors, Ford and Chrysler collectively lost $7.5 billion in 1991, making it the worst year in history for the American auto industry.

General Motors, the world's largest industrial concern, lost $4.45 billion in 1991 and announced plans to close 21 factories and slash 74,000 jobs from its payroll by 1995. Only a year before, GM had taken a $2 billion charge to write off more than a half-dozen other North American assembly plants it had shuttered during the 1980s. "We've accepted [the need for] a smaller base on which to become profitable," conceded GM's chairman and chief executive officer, Robert Stempel, in announcing the latest cutbacks. "GM will become a much different corporation."[44]

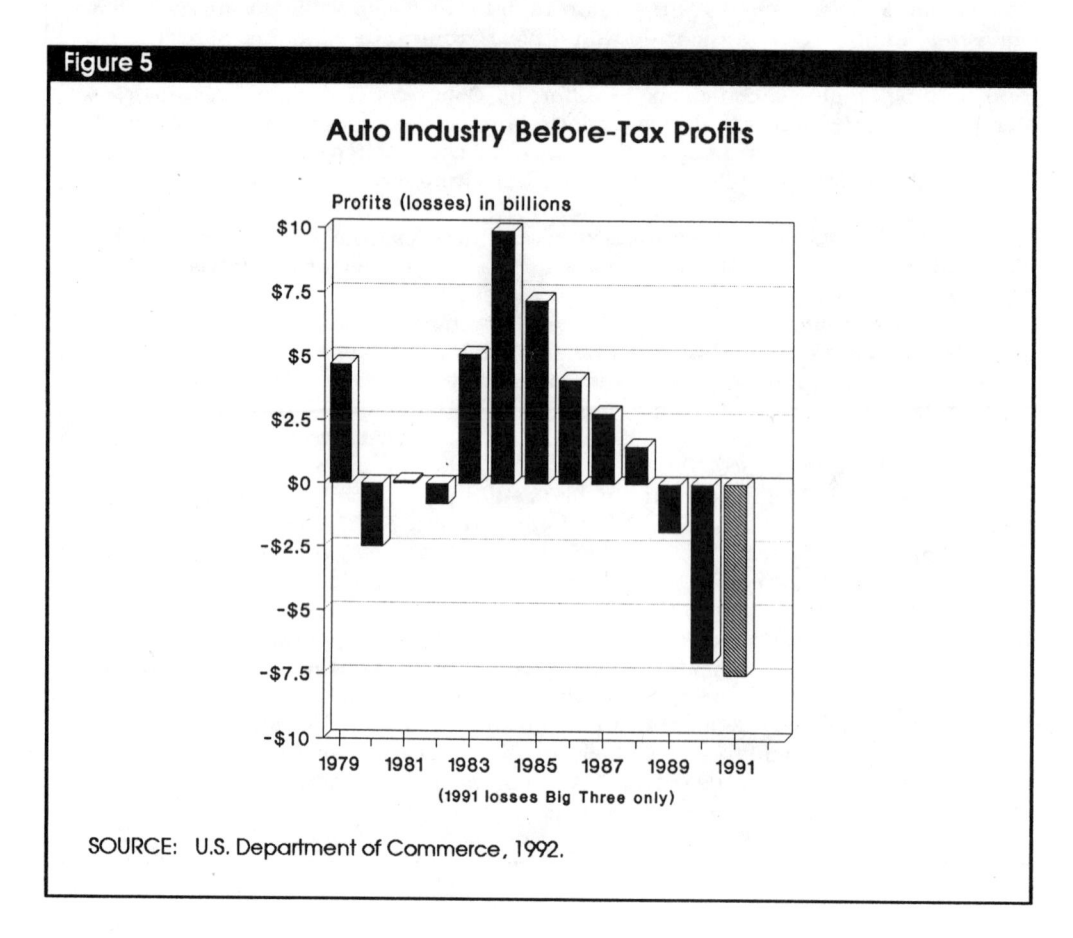

Figure 5

Auto Industry Before-Tax Profits

Profits (losses) in billions

(1991 losses Big Three only)

SOURCE: U.S. Department of Commerce, 1992.

From a global standpoint, the "Big Three" auto makers had shrunk to the "Big Two" even before the onset of the recent recession. General Motors and Ford produced a total of 13.6 million vehicles worldwide in 1989—more than the four next largest car companies combined—but Chrysler produced only 2.2 million vehicles, tying it for seventh place with Peugeot-Citroen of France.[45] In 1991, for the first time ever, Honda and Toyota exceeded Chrysler in U.S. passenger car sales (although Chrysler still sold enough minivans and pick-up trucks to qualify it for third place in total light-duty vehicle sales).[46]

Feeling the Heat

Environmentalists are urging America's auto makers to combat the greenhouse effect by raising the gas-mileage standards of their fleets. But in a sense, Detroit is still paying a price for the last two times that fuel economy emerged as a priority among American car buyers. The oil price shocks of 1973 and 1979 caused gasoline prices to soar and created long lines at service stations. Motorists suddenly wanted more fuel-economical cars to drive, but the Big Three had relatively few models to offer. That opened the door for fuel-thrifty Japanese auto makers, who were anxious to dispel the impression that "made in Japan" meant products of shabby quality.

Now that notion among American consumers has been completely reversed. In 1990, 28 of 31 best-rated models for quality in America were Japanese-made, according to a *Consumer Reports* survey; all but one of the 33 worst-rated models were American-made.[47] Moreover, an influential list of the "Top 10 Trouble-Free Cars" sold in America, compiled by the automotive marketing firm, J.D. Power & Associates, found seven Japanese brand names and two German brand names making the grade in 1990. The only American brand to make the top 10: Buick, which ranked fifth.[48] (In the 1991 J.D. Power survey, Ford placed eighth; Lincoln, ninth; and Chrysler, eleventh.)

While American car makers have made demonstrable progress in improving the quality of their cars in recent years, the damage essentially has been done. Japanese car makers have capitalized on their newfound reputation for quality to seize a greater share of the U.S. auto market—even as consumers' interest in fuel economy has faded. As recently as 1978, all 10 best selling cars in America carried a Big Three brand name. But by 1990, four of the 10 most popular cars had Japanese nameplates, and a fifth was jointly engineered by a Japanese company. In 1991, Japanese auto makers captured a record 31 percent of the U.S. passenger car market.[49] The top selling car in America—for three years running—has been the Honda Accord, making it the first foreign nameplate ever to top the American automotive sales charts.[50]

In terms of global sales, nine of the 20 largest auto companies in the world now are based in Japan. These companies account for 30 percent of world vehicle production and may soon eclipse production by the Big Three.[51] Altogether, the world's 40 major auto companies have the capacity to build five cars for every four prospective car buyers. And because the United States offers the largest and most open market, 70 percent of this excess capacity finds its way on to the North American continent.[52] The result is that the world auto market has become more competitive than ever, and the Big Three auto makers are feeling the heat.

CAFE: The Legacy of Round I

Car makers spend billions of dollars annually on advertising to gain an advantage in the giant American marketplace. Today's car commercials tout the safety, performance, comfort, styling and reliability of particular models—reflecting consumers' strong interest in these features. Occasionally, commercials also make a point about fuel economy. But with gasoline today selling for less than the price of bottled water, consumers are not particularly concerned with how far a gallon of fuel will take them. When their gas gauges are nearing empty, they know that a $20 fill-up will put them back on the road back for another 300 to 400 miles—and that is comfort enough for them.

Consumer attitudes about automotive fuel efficiency have been known to change, however. After oil prices plummeted to $10 a barrel in late 1985, only 3 percent of American car buyers listed fuel economy as their primary criterion in selecting a car, according to a 1987 survey. Yet seven years earlier—with the Iran hostage crisis and long gas lines fresh in their minds—fully one-third of prospective car buyers ranked fuel economy as their top priority.[53] Moreover, at the time of the 1980 survey, real U.S. gasoline prices were approaching their highest level in history.

Fortunately for American consumers, auto makers in 1980 were well on their way toward shrinking the costs of fuel in relation to a car's total operating budget, and they had Congress partly to thank for that.[54] Five years earlier, Congress established corporate average fuel economy (CAFE) standards for makers of light-duty vehicles sold in the United States. Between 1978 and 1985, each manufacturer of passenger cars was to raise its fleetwide average from 18 mpg to 27.5 mpg, with interim steps along the way. Auto makers that failed to meet interim goals or the final standard would be assessed a $50 penalty for each mile per gallon that its fleet average fell beneath the target, multiplied by the number of cars it produced that year.

The CAFE law initially posed little challenge to European and Japanese car importers, since their own domestic markets already were dominated by sales of compact and subcompact cars—85 percent in Europe and 95 percent in Japan. But for full-line American manufacturers, who sold mainly intermediate- and standard-size vehicles, the situation was altogether different. In testimony before Congress in 1974, Chrysler executives warned that the CAFE law would effectively "outlaw full-size sedans and station wagons," General Motors executives said "it would restrict the availability of five- and six-passenger cars regardless of consumer needs," and Ford executives maintained it would lead to "a Ford product line consisting of either all sub-Pinto-sized vehicles or some mix of vehicles ranging from a subsubcompact to, perhaps, a Maverick."[55]

While cars' interior and exterior dimensions did shrink somewhat after 1974, the dire predictions of the auto industry were not realized.[56] American-

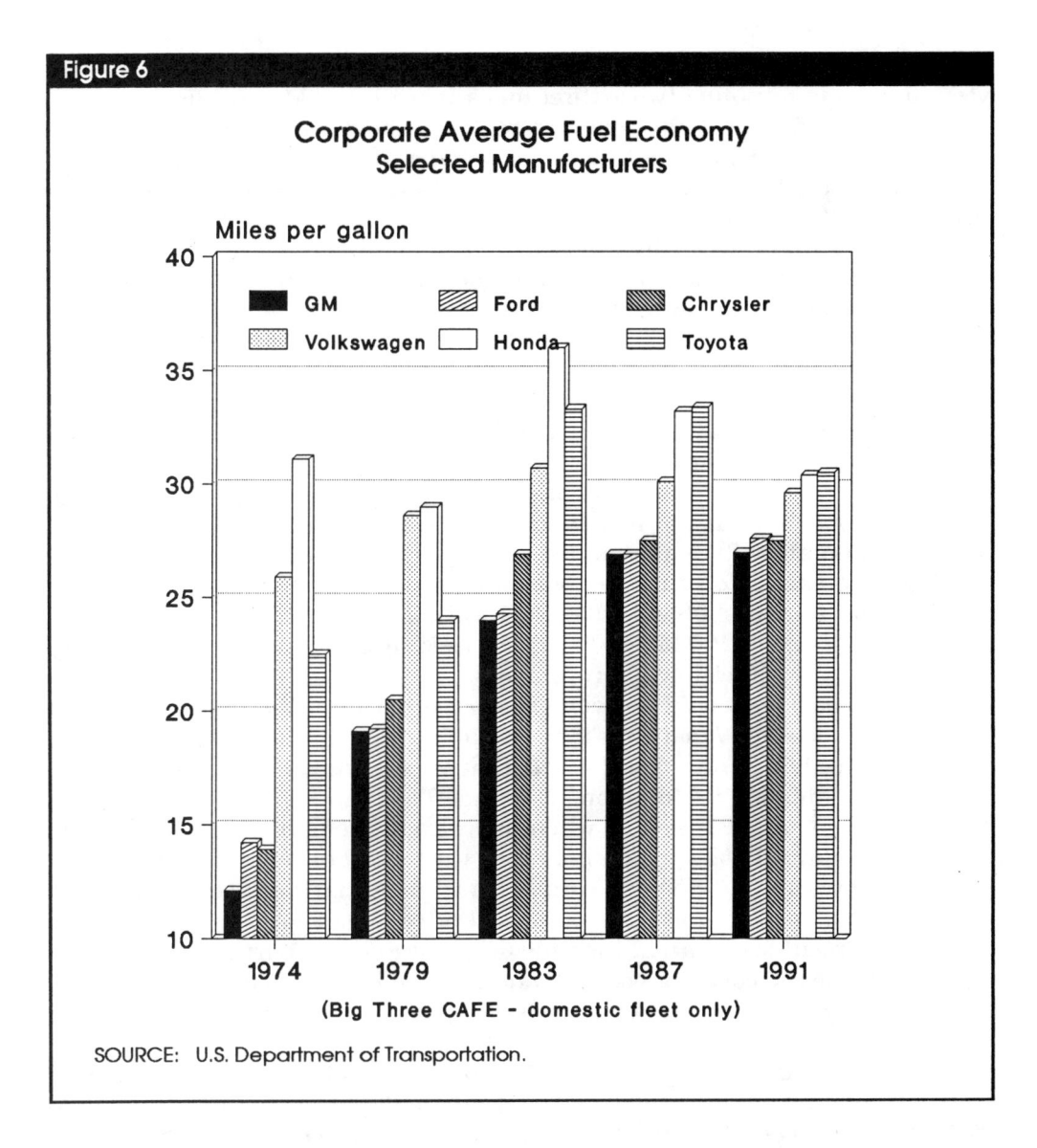

Figure 6

Corporate Average Fuel Economy
Selected Manufacturers

[Bar chart showing Miles per gallon for GM, Ford, Chrysler, Volkswagen, Honda, and Toyota in years 1974, 1979, 1983, 1987, 1991]

(Big Three CAFE - domestic fleet only)

SOURCE: U.S. Department of Transportation.

made subcompact and compact cars peaked at 37 percent of the new car market in 1981. (Imports accounted for another 27 percent of retail sales that year.) In 1990, U.S. intermediate- and standard-size vehicles still accounted for 41 percent of the new car market.[57] Meanwhile, the fuel economy of the domestic fleet soared more than 80 percent in just eight years, rising from 13.2 mpg in 1974 to 24.2 mpg in 1981. (The average fuel economy of imported cars rose 37 percent over the period, from 22.2 mpg to 29.6 mpg.)[58]

Initially, Detroit realized tremendous fuel economy savings by reducing the curb weight of its vehicles. Domestic cars that hefted 4,000 pounds on average in the early 1970s weighed only 3,100 pounds by the mid-1980s. Each 250-pound weight reduction boosted fuel economy by one mile per gallon on average. Sleeker exterior dimensions that reduce aerodynamic drag also yielded savings of another two to three miles per gallon.[59] Meanwhile, under the hood, four- and six-cylinder engines replaced most gas-guzzling V-8s, and electronic fuel injectors took over for comparatively clumsy carburetors, permitting more precise metering of fuel. Finally, conversion to front-wheel-drive transmissions maintained cars' interior space, even as their exterior dimensions shrank markedly.

All of these modifications hardly made cars less fun to drive (with the possible exception of the muscle cars built in the late 1960s and early 1970s). Today's cars offer superior roadhandling, braking and acceleration. Off the line, an average 1990 car goes from 0 to 60 miles per hour in 12.1 seconds, compared with 14.4 seconds for comparable cars built in 1982.[60] (This increase in acceleration has been at the expense of fuel economy ratings, however, which otherwise would be 2 mpg higher than they are today.)

Best of all for cost-conscious consumers, the cost of these improvements has been far less than the fuel savings gained. An analysis by Oak Ridge National Laboratory finds that added costs averaged $333 for each of the 110 million or so cars sold in the United States from 1975 through 1987. This amounts to a total cost of $40 billion in 1987 dollars. The 35 billion gallons in fuel savings totaled $260 billion, however, yielding $6 in fuel savings for every dollar invested in efficiency improvements.[61]

The fuel savings would be even greater were it not for the falling price of gasoline. While some Americans may recall that gasoline sold for less than 25 cents a gallon shortly after World War II, that price is equivalent to paying $1.50 today, after adjusting for inflation. A typical fill-up now costs only half of the record price set in 1981, in fact, when the price at the pump jumped to $2.00 a gallon in today's dollars.[62] Compounding the benefit of lower gasoline prices, fuel economy gains now enable cars to travel much farther between fill-ups. Factored on a cost-per-mile basis, the real price of gasoline today actually is about 50 percent less than at the time of the Arab oil embargo of 1973.[63]

Expressing the costs for gasoline and motor oil in terms of cents per mile, the high water mark appears to have come in 1975—after the Arab oil embargo but before the commencement of the CAFE standards. In that year, the inflation-adjusted cost for gasoline and motor oil was 11 cents a mile. By 1979, the cost had dropped to 7 cents a mile. But then it rose again to 9 cents a mile in 1980, following the second oil price shock. With the collapse of oil prices at the end of 1985, the cost fell all the way down to 5 cents a mile, and it has hovered at that level ever since.[64]

Fuel Economy in Reverse

Given how inexpensive it has become to fuel an automobile, it was perhaps inevitable that the CAFE standards would be criticized as outliving their usefulness. In 1988, President Reagan's transportation secretary, James Brumley, characterized CAFE as a "perverse law" that effectively exported jobs in the automotive industry to other countries. Brumley—along with Ford and

General Motors (but not Chrysler, which was required under its 1979 federal loan agreement to meet the CAFE standards)—called for CAFE's outright repeal shortly before Reagan left office.

Brumley's concern echoed complaints by members of the United Auto Workers union who are alarmed by the tremendous loss of jobs in the American auto industry beginning with the 1974-75 recession. The union's beef with the CAFE law is that American car makers are tempted to meet the standard by importing more high-mpg cars from abroad to make up for the low-mileage ones built in the United States. Congress tried to prevent this switch-over by creating a dual-fleet system that forbids car companies to average the fuel economies of their imported fleets with those models made domestically. Despite the provision, vehicles imported by U.S. auto makers quadrupled between 1976 and 1989.[65] Critics charge that CAFE essentially has amounted to a tax on large cars—where U.S. manufacturers have an advantage—while importers of small, fuel-efficient cars remain unconstrained.[66]

CAFE has had other unintended side effects as well. A domestic vehicle must have at least 75 percent "domestic content" for CAFE's purposes. Ford decided in 1991, however, to add about a dozen foreign parts to its big Crown Victoria so that the vehicle's domestic content would fall below 75 percent. About 400 U.S. workers lost their jobs as a result of the decision to buy car seats, fuel tanks and windshield glass from Mexico.[67] Now the V-8-powered car—with an EPA mileage rating of only 21 mpg—is considered an import for CAFE purposes. As such, Ford gets to count the Crown Victoria among other more fuel-economical imports like the Korean-made Ford Festiva, which gets 32 mpg.

Muddying the waters even more, the Reagan administration rolled back the CAFE standard to 26 mpg for three years, beginning in 1986. This further reduced pressure on Ford and General Motors—the domestic companies that were having the most trouble meeting the economy standard—to produce fuel-thrifty models in the United States. Moreover, it provided a windfall in terms of credits that auto makers earn when their fleets exceed the CAFE mileage targets. Ford, for example, racked up $103 million in credits in 1986 alone, which it could use to offset any monetary penalties that might be assessed against it for failing to achieve the CAFE target over the next three years.[68]

Despite all this maneuvering, Japanese auto makers appear to have been the biggest corporate beneficiaries of the American experience with CAFE. They, too, qualified for large CAFE credits, as the fuel economy of their fleets exceeded the mileage standard by as much as 11 mpg shortly after the law went into effect.[69] Moreover, they agreed in 1981 to set a voluntary quota on the number of cars they imported to the United States and to increase their domestic manufacturing base. (At the time, only Honda had an American assembly plant.) As of mid-1991, seven Japanese companies had a total of eight American assembly plants capable of building 2 million cars and trucks a year. Yet until Mazda announced in 1992 that its model 626 sedans assembled in Michigan had 75 percent U.S. content, none of these companies regarded its Japanese nameplate, American-assembled cars as "domestic" models for CAFE purposes.[70] Just as Ford now considers its Crown Victoria an import, Japanese auto makers have not wanted to run afoul of the CAFE standard by creating a

> **Box 4-B**
>
> ## EPA Mileage Ratings and the 'Mileage Gap'
>
> The U.S. Environmental Protection Agency evaluates the fuel economy of all cars and light-trucks sold in the United States. The ratings serve two purposes: First, to calculate vehicle manufacturers' corporate average fuel economy (CAFE), a sales-weighted measure of their domestic and imported fleets; and second, to let potential buyers know how many miles a particular model will travel on a gallon of gas—in both city and highway driving. The EPA tests use a dynamometer to maximize the engines' efficiency. The vehicles are accelerated gently, brought to gradual stops and never raised to speeds faster than 55 miles per hour. While this exercise hardly simulates real-world driving, it forms the basis of the CAFE rating.
>
> The EPA mileage ratings displayed in windows of new light-duty vehicles are adjusted downward, however, to reflect actual driving conditions. The mileage rating for highway driving is reduced 22 percent, while the rating for city driving is lowered 10 percent. EPA assumes that 45 percent of driving takes place on highways and 55 percent in cities. Accordingly, the composite EPA mileage rating for automobiles factors in a 15 percent "mileage gap." A car rated for CAFE purposes at 28 mpg, for instance, is rated at only 23.8 mpg on the showroom floor. (In this chapter, mileage ratings cited for specific models include the 15 percent mileage gap, whereas CAFE fleet averages do not.)
>
> From the time the car is driven off the dealership lot, chances are it will never attain even the downward-adjusted level of fuel economy. There are three principal reasons for this. First, the share of driving done in urban areas has risen substantially since the early 1970s, when EPA assigned the 55-percent share. The city-driving share topped 63 percent in 1987 and is projected to reach 72 percent by 2010.[1b] The stops and starts associated with city driving waste fuel. On that basis, on-road fuel economy is expected to decline 3.1 percent over the next 20 years.[2b]
>
> Second, traffic congestion is increasing in urban areas. The U.S. Department of Energy forecasts a 1.6 percent increase in vehicle miles traveled each year, outpacing any increase in urban road capacity. Vehicles get their worst mileage when they are frequently accelerating and braking in stop-and-go traffic. The projected rise in

domestic fleet that failed to meet the 27.5 mpg standard. (Some observers add that Japanese car makers' insistence on using auto parts from Japan makes it nearly impossible to call any of their models "domestics.")

Having gained a foothold in the American market with the sale of inexpensive yet reliable "econoboxes" in the 1970s, Japanese auto makers have since entered the upscale car market. Today, Japanese luxury cars sport brand names such as Acura, Infiniti and Lexus. "Three major Japanese auto companies [Honda, Nissan and Toyota] have developed upscale divisions to market bigger, heavier and more powerful vehicles," observes Robert G. Liberatore, Chrysler Corp.'s executive director of public policy and legislative affairs. "Because their base fleet is subcompacts, and with their vast backlog of accumulated credits, they can offer these vehicles with little regard to their impact on CAFE."[71]

Indeed, sensing the mood of American car buyers, Japanese car makers now traffic congestion is therefore expected to reduce on-road fuel economy an additional 9.1 percent.

Third, an increase in average highway speeds penalizes on-road fuel economy. Average highway speeds have risen from 55.8 mph in 1975 to 59.7 mph in 1987 and may increase to 66 mph by 2010.[3b] The projected 6.3-mph increase could reduce on-road fuel economy an additional 1.6 percent by 2010.

Adding together the increase in urban driving loads, traffic congestion and highway speeds, the mileage gap could reach nearly 30 percent by 2010—well above the 15 percent adjustment factor now used in EPA's composite fuel-economy rating. This suggests that a car's CAFE rating would have to increase from 28 mpg to 32.2 mpg just to maintain actual on-road fuel economy of 23.8 mpg. Despite an improvement in CAFE ratings, in other words, vehicular carbon dioxide emissions may not be abated.

On a brighter note, an analysis by the congressional Office of Technology Assessment concludes that the mileage gap may widen—but not to the level of 30 percent.[4b] Replacement of engine carburetors with fuel injectors, on-board diagnostic systems that improve engine efficiency, recovery of fuel now lost to evaporative emissions and reduced fuel enrichment during cold starts (as a means of reducing carbon monoxide emissions) will help to offset some of the mileage gap. Moreover, OTA believes that highway speeds will not increase as much as suggested by recent trends, because of deteriorating road conditions and increased traffic congestion.

It should be noted that cars generally get their best mileage when traveling at steady speeds between 35 and 45 mph. A car cruising at 65 mph consumes 20 to 30 percent more fuel per mile, on average, than a car going 55 mph, and a car traveling 75 mph consumes up to 30 to 40 percent more fuel per mile.[5b] Even many fuel-efficient models do not save much gas when the vehicle is driven over 55 mph. Accordingly, many conservation advocates stress the importance of maintaining steady speeds of 55 mph or below—regardless of the vehicle's fuel economy rating. A car's on-road fuel efficiency also is maximized by keeping its tires properly inflated and its engine properly tuned.

seem inclined to demonstrate that they can make cars just as powerful—and hence fuel-consuming—as their American counterparts. Honda, which used to make four-cylinder engines exclusively, has outfitted its upscale Acura Legend with a V-6 automatic transmission. Its 1992 model gets only 19 mpg in city driving and 24 on the highway, for a composite rating of 21.2 mpg. Nissan's Infiniti Q-45 has V-8 power and a composite mileage rating of only 18.7 mpg. And Toyota's highly successful luxury sedan, the Lexus LS 400, achieves 250 horsepower with its own V-8 engine and is rated at just 20.2 mpg. Not to be "out-muscled," Mitsubishi introduced a 300 horsepower turbocharged sports car in 1990, the 3000 GT. The 1992 model gets 20.7 mpg, still nearly 7 mpg below the CAFE standard.

At the same time that Japanese car makers are moving into the mainstays of the American market, U.S. auto makers have basically conceded their role in the small-car market. Chrysler discontinued sales of its subcompact Omni/

Horizon in early 1990. Similarly, Ford suspended production of its best-selling Escort model in the spring of 1989, introducing a new Escort—jointly engineered by Mazda—one year later.[72]

General Motors now produces nearly all of its smallest cars in partnership with Japanese companies, with the notable exception of its new Saturn division. GM's Geo Metro, the most fuel-efficient car sold in America (getting 58 mpg on the highway), features a three-cylinder Suzuki engine and is built by Cami-Canada. The sportier Geo Storm is built by Isuzu. And the four-door Geo Prizm, a virtual clone of Toyota's popular Corolla model, is built at a GM-Toyota joint venture factory in Fremont, Calif. (The Corolla is counted as an import, incidentally, while the Prizm is a domestic, even though both models roll off the same assembly line.)[73]

All of the 10 most economical cars now on the market have engines designed, engineered and built by Japanese companies. Four are variations of the Geo Metro. Two others are versions of the Suzuki Swift, and the remaining four are versions of the Honda Civic.[74]

"The response of domestic automakers [to the CAFE mandate] has been to cede their smaller car lines to the competition, the justification being that U.S. car makers have never been adept at designing and marketing smaller cars," concludes Deborah Lynn Bleviss, executive director of the International Institute for Energy Conservation in Washington, D.C., and author of a 1988 book on auto efficiency trends. "In the current market, such a decision has merit: consumers are no longer as interested in smaller cars as they once were. But should the market change, domestic auto manufacturers could find themselves without a manufacturing base to meet renewed buyer demand for smaller, fuel-efficient cars," Bleviss warns.[75]

Power Is Back in Style

Such a dramatic change in the auto marketplace could be brought about by a government program to combat the greenhouse effect in the years ahead. Presumably, further increases in the CAFE standard and/or a carbon tax on gasoline would rank high in the government's battle plan, since the light-duty fleet (including passenger cars) is responsible for one-fifth of the nation's carbon dioxide emissions. Rather than looking forward at this possibility, however, American and Japanese auto makers—as well as those buying their cars—appear to be looking back on the era of cheap gasoline prices and saying, "Gee, I could have had a V-8!"

Make no mistake, power is back in style. The automotive fleet rolling off the assembly line in 1992 (averaging both domestic and imported models) is rated at the CAFE minimum of 27.5 mpg—the poorest showing since 1984. The fuel economy of new cars sold in the United States actually peaked in 1988 at 28.8 mpg. The fuel economy of the new car fleet was down 4 percent by the 1990 model-year, with the average car gaining more than 6 percent in weight and 10 percent in horsepower. The trend has continued since then. Auto makers are building fewer four-cylinder engines and more of the powerful V-6s and V-8s, adding more accessories and weight to their vehicles and emphasizing acceleration over fuel economy at nearly every turn.

American drivers clearly have a preference for cars that accelerate quickly. When test-driving a vehicle, prospective car buyers often "floor" the gas pedal to see how fast the car can pick up speed, even though they rarely require such power in everyday driving. One study of driving patterns finds that drivers' rate of acceleration rarely exceeds 3.3 miles per hour per second, yet today's cars are capable of accelerating 5 mph per second.[76] (In other words, they can go from 0 mph to 60 mph in about 12 seconds.) The gain in acceleration is achieved by adding extra horsepower in relation to the car's weight; this horsepower-to-weight ratio has increased from 64 horsepower per ton to 80 horsepower per ton since 1982.[77] Yet the horsepower gain means that cars run at "part-load" conditions more often, reducing overall engine efficiency. According to the American Council for an Energy Efficient Economy, the 2.3-second gain in average 0-to-60 acceleration times has reduced average new car fuel economy nearly 10 percent, or more than two miles per gallon, relative to what it would have been at the 1982 performance level.

As cars have become more powerful in recent years, they also have added weight. Air conditioning and power steering became standard equipment in most vehicles during the 1980s, and airbags and anti-lock brakes seem destined to do the same in the 1990s. Such safety equipment adds 60 to 90 pounds of weight to the vehicle. Taken together, these four technologies decrease average fuel economy some 3 to 5 percent.[78] A four-wheel-drive option adds another 150 to 200 pounds to a vehicle's weight, and the fuel economy penalty can be as great as 15 percent when four-wheel drive is engaged.[79]

Another important development concerning fuel economy has been the growth in demand for compact pick-up trucks, sport utility vans and four-wheel-drive jeeps. Contrary to the image often portrayed in television advertising, these vehicles mainly haul kids to school and groceries from the supermarket—rather than hauling themselves up steep mountainsides. According to the Bureau of the Census, only one light truck in five actually hauls any freight.[80] But no matter. People value these vehicles for their versatility, durability and rugged good looks. As a result, light-truck vehicle miles traveled tripled between 1970 and 1985, compared with a 38 percent increase for autos.[81] Now they account for one-third of all light-duty sales (which combines passenger cars and trucks weighing less than 8,500 pounds), up from 15 percent in 1970.[82]

Since pick-up trucks, vans and jeeps sometimes are put to uses that require extra horsepower and maneuverability, the federal government has been comparatively lenient with CAFE requirements. The existing standard is 19.1 mpg for four-wheel-drive vehicles and 20.7 mpg for two-wheel-drive light trucks. Because of their greater fuel consumption, light-duty trucks consume about 30 percent of the motor fuel used by households even though they account for less than a quarter of household vehicle miles traveled.[83] To put this consumption in terms of the greenhouse effect, the light-duty fleet—excluding passenger cars—is responsible for nearly 7 percent of the nation's carbon dioxide emissions.

Gambling on Bigger Cars

The fuel economy trend internationally mirrors what has been happening in the United States. The most dramatic gains in fuel efficiency came in the first dozen years after the Arab oil embargo, between 1973 and 1985. New car efficiencies rose from 21 to 31 mpg in the United Kingdom; from 23 to 30 mpg in Japan; and from 23 to 31 mpg in (formerly) West Germany.[84] Improvements since 1985 have been virtually nil.

If the late 1980s were a harbinger of the 1990s, it is clear that fuel economy is on the skids. Chrysler, for example, outfitted its entire fleet with four-cylinder engines throughout the 1980s until the 1987 model year. As of 1990, however, Chrysler was building 1 million vehicles with V-6 engines, mainly for its popular line of minivans that have become the most profitable part of its business. Moreover, to show that it can still offer a true hot rod, Chrysler introduced a 10-cylinder, 400-horsepower roadster in 1992 called the Viper. As a result of these moves, Chrysler forfeited a 1- to 2-mpg domestic fleet advantage it held over Ford and Chrysler as recently as 1987. "If you tell me that gasoline will be at $2 a gallon in three years, I would say we'd all be searching for ways for better fuel efficiency, and spending money to find tenths of a mile per gallon" in fuel savings, explains Albert Slechter, Chrysler's director of federal affairs. "But that kind of mentality doesn't exist right now."[85] Now the CAFE rating for each of the Big Three is virtually identical.

At Ford and General Motors, auto executives are thinking that new V-8 engines will rev up sales as well. In the 1991 model year, Ford decided to offer a V-8 in its Thunderbird model for the first time in years. With a 5.0 liter engine, this Thunderbird gets 20.2 mpg in combined city/highway driving. Ford could have chosen to make the engine more economical, but it opted for faster acceleration instead. Consequently, the car goes from 0 to 60 miles an hour a half-second faster than it would have otherwise, but it also gets 0.4 miles less per gallon.[86] Ford introduced a modular V-8 engine in 1992, touted as the most economical V-8 it has ever built. A Lincoln Town Car equipped with this engine gets 25 mpg on the highway and 18 in the city, for a composite rating of 21.2 mpg—still well below the 27.5 mpg CAFE standard.

General Motors has pursued a similar engine-design strategy. Beginning with the 1991 model year, it put larger V-8 engines in many of its own luxury cars. As with Ford, GM's newest V-8 engine is relatively economical for its size—getting 26 mpg on the highway. It even represents a 1-mpg improvement over the smaller V-8 engine in its 1990 models. Like Ford's decision with the Thunderbird, however, GM ruled out an even greater fuel economy gain, choosing instead to boost the engine's horsepower rating from 180 to 200.[87]

To the extent that foreign competitors try to "out-muscle" the Big Three auto makers in the large-car market, the American manufacturers do hold a certain fuel economy advantage. The Lincoln Town Car gets better mileage than the Lexus LS 400, for instance, even though the Continental is larger and heavier. By the same token, GM's mid-sized Chevrolet Lumina, when equipped with a 3.1-liter V-6, gets better mileage than a similarly equipped compact Toyota Camry. In fact, when gasoline consumption is measured by the ton-mile,

Figure 7

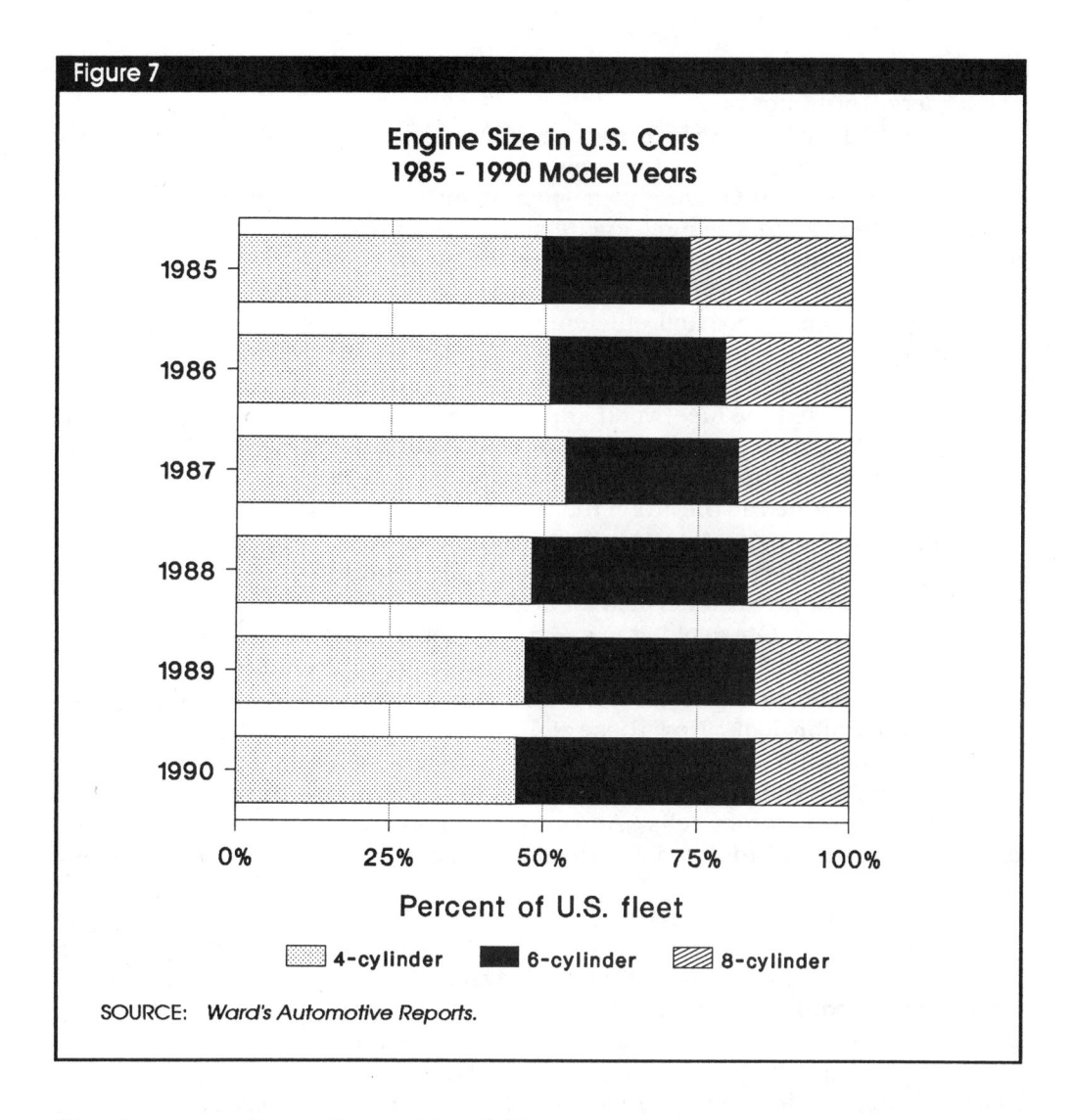

Engine Size in U.S. Cars
1985 - 1990 Model Years

Percent of U.S. fleet

☐ 4-cylinder   ■ 6-cylinder   ▨ 8-cylinder

SOURCE: *Ward's Automotive Reports.*

thereby normalizing for weight differences between cars, the overall fuel economy of American cars actually exceeds those of Japanese cars sold in the United States by a narrow margin. European vehicles sold in the United States are far less efficient than both the American and Japanese fleets when measured this way.[88] Nevertheless, the Japanese do retain an edge in small engine technology. Many of their four-cylinder cars have the pep of a typical American six-cylinder model, yet deliver superior gas mileage.

The strategic question facing auto makers today is whether the current thrust toward large cars with big engines will be maintained. As of the summer of 1990, the Big Three auto makers operated 37 auto plants in North America, 29 of which built cars that failed to meet the 27.5 mpg CAFE standard.[89] The auto makers say that retooling these plants to attain even higher mileage standards would entail considerable time and expense. The end result, they add, would lead to substantially smaller cars with higher sticker prices that consumers would not buy.

Recent history tends to bear out the auto makers. General Motors, for one, learned a costly lesson in 1986 when it downsized its line of luxury cars—Cadillacs, Buicks and Oldsmobiles—just as oil prices were plummeting. Car buyers turned down the new models in the showrooms. In subsequent marketing studies, company executives came to realize that a significant portion of new car buyers actually prefer large, heavy, rear-wheel-drive vehicles, even though newer, more fuel-efficient models are (in most respects) functionally superior.

Consequently, America's largest auto maker did an about-face. Its Cadillac Division lengthened the Fleetwood and DeVille models by nine inches in 1989, and the Buick Riviera grew by nearly a foot. "The '86-'88 [Riviera model] was designed when everyone thought gas was going to $3 a gallon," a Buick spokesman explained as the larger models rolled off the assembly line. "It didn't, and people want a more classic, substantial look."[90]

For the same reason, General Motors scrapped plans to retire its rear-wheel-drive Chevrolet Caprice model in favor of a mid-size, front-wheel-drive car. With the fall in oil prices, GM decided to spend six months and hundreds of millions of dollars to renovate two factories making the Caprice and its sibling cars—the Buick Roadmaster wagon and the Cadillac Fleetwood Brougham—giving them a new aerodynamic look. Now these eight-cylinder giants, weighing more than 4,000 pounds each, are among the least fuel-efficient full-size cars on the market—averaging about 20 mpg. But they also are among the most profitable cars that General Motors sells. Accordingly, GM has little incentive (let alone historical precedent) to downsize these cars again into smaller, front-wheel-drive models.

Ultimately, auto manufacturers have to look well into the future to make their strategic planning and marketing decisions. Domestic manufacturers redesign their entire product line only every eight years or so, and many technologies require five years of lead time. Consequently, a new technology slated for development today would not be expected to reach the market until 1997, or to reach a high level of market penetration until after the year 2000.[91] Moreover, high-sales-volume models require $1 billion in capital spending before the first car is rolled out of production. Given these constraints, the Big Three auto makers already are in critical design stages for the 1997 and 1998 model-years, and many of models are due to be outfitted with six-cylinder and V-8 engines, based on the expectation that gasoline prices will continue to be low. Therefore, consumers can expect new cars to be only marginally more fuel-efficient than the ones already on the road.

CAFE: Time for Round II?

If Congress were to raise CAFE standards now, the production plans conceived by the auto makers in the wake of falling oil prices would have to be overhauled completely. Moreover, the new strategic plan would have to reconcile the severe financial constraints under which the American auto industry is now operating with the need for accelerated research and development and substantial retooling. At present, momentum is taking the industry in the opposite direction. Faced with a severe cash crunch, the Big Three are abandoning certain new product-development programs, jettisoning marginal assembly plants and selling off other assets. In such a recessionary environment, new government policies to address the greenhouse effect are the last thing they want. As General Motors' Robert Stempel remarked to reporters during an announcement of the company's $1.1 billion loss for the third quarter of 1991, "This isn't the time for Washington to pile on more costly and ineffective regulations like the federal fuel-economy standards."[92]

Nevertheless, Sen. Richard Bryan (D-Nev.) has sponsored a bill that would raise corporate average fuel efficiency standards 40 percent over a decade's time as a means of reducing autos' carbon dioxide emissions as well as the nation's dependence on imported oil. Bryan has faced intense opposition from the auto industry ever since he drafted the bill in 1989, and its prospects for passage remain uncertain. Contrary to the original CAFE law, Bryan's bill requires each of the auto manufacturers to raise the fuel economy of its fleets by the same percentage rather than by a fixed amount. Accordingly, some companies would have to attain higher mileage standards than others.

As Bryan's bill was proposed originally, car makers would have to boost their 1988 CAFE averages 20 percent by 1996 and 40 percent by 2001. On that basis, the domestic auto fleet average would have to rise from 27.4 mpg in 1988 to 32.9 mpg by 1996 and 38.4 mpg in 2001. Similarly, the import auto fleet average—starting from a higher base—would have to rise from 31.5 mpg in 1988 to 37.8 mpg by 1996 and 44.1 mpg in 2001.

But the actual CAFE requirement for each manufacturer would vary according to its 1988 average. Ford, which attained a CAFE rating of 26.6 mpg on its domestic fleet in 1988, would have to raise its rating to only 31.9 mpg by 1996 and to 37.4 mpg in 2001. Toyota, on the other hand, which had a 33.0 mpg rating on its imported fleet in 1988, would have to achieve a 39.6 mpg rating by 1996 and a 46.2 mpg rating in 2001.

Neither domestic nor foreign car manufacturers have shown any enthusiasm for the Bryan bill. The Big Three contend that the bill's targets are unattainable and far more expensive than alternative means of reducing greenhouse gas emissions.[93] The Japanese car makers' complaint is that the percentage-gain requirement (as opposed to a uniform mileage standard) penalizes them for their past fuel economy gains. Importers essentially would have to widen their 4-mpg lead over domestic manufacturers (as of 1988) to 5.7 mpg by 2001. With a continued expectation of low oil prices and high demand for large cars with powerful engines, importers would find it especially difficult to meet the countervailing mandates of CAFE and the marketplace. Putting the onus on importers could work to their advantage, however, if demand for fuel-thrifty autos is rekindled by world events—as it was in the 1970s.

Bryan tried initially to attach his CAFE bill to the Clean Air Act amendments in the spring of 1990. When that attempt failed, he submitted the bill separately for consideration by the Senate. In September 1990, the Senate failed by three votes to end a filibuster by the bill's opponents, led by Michigan Sen. Donald W. Riegle Jr. (D). By preventing the bill from coming to a vote, some observers believe, the bill's opponents killed its greatest chance of passage. Iraq had invaded Kuwait only two months before, and oil prices were still gyrating wildly on the mercantile exchange.

Seven months after the end of the Gulf War, Bryan's bill was pitted against an omnibus energy bill put forward by Sen. Bennett Johnston (D-La.). The Johnston bill was the legislative counterpart of the National Energy Strategy introduced by President Bush in February 1991. The administration's plan called for the U.S. Department of Transportation to review the CAFE law—taking into account cost, safety and environmental considerations—and then set maximum achievable standards. The Johnston bill also fell victim to a filibuster, however, because it contained a provision that would have permitted exploration and drilling in the Arctic National Wildlife Refuge in Alaska.

In order to move energy legislation forward, the political compromise on Capitol Hill is a bill that includes neither Bryan's CAFE proposal nor the president's Alaskan drilling proposal. With the trade-off, the CAFE standard remains at 27.5 mpg for now. President Bush has threatened to veto an increase in CAFE if another bill surfaces. He even hinted in early 1992 that he would consider rolling back the CAFE standard—as President Reagan had done in 1986—to give Detroit's auto industry a boost.[94]

Potential for Fuel Economy Gains

At the crux of the debate over raising fuel economy standards is what kind of gains the auto industry can achieve without making cars so unappealing or expensive that consumers would prefer to hang on to their old—and less fuel-economical—ones. A 1991 industry-funded study by SRI International estimates that car makers should be able to raise gasoline mileage 9.5 percent by 2001 without downsizing their cars, reducing engine performance or creating sticker shock. Pickup trucks and minivans could become nearly 8 percent more fuel economical by 2001 without radical design changes, the SRI study found.[95]

These estimates serve as an important benchmark because, for the first time, the Big Three (as well as Honda) shared top-secret production plans and cost estimates with outside researchers. Given this disclosure, auto makers should have no qualms about a rise in the CAFE standard to 30 mpg for domestic passenger cars by 2001, and a rise in CAFE to 22.6 mpg for two-wheel-drive light-duty trucks. Such gains would remain well below the level sought by conservation advocates and some government officials, however.

Perhaps the most detailed analyses of potential auto fuel economy gains have been performed by Energy & Environment Analysis Inc. under contract to the Oak Ridge National Laboratory and the U.S. Department of Energy. In these studies (conducted in 1990) EEA concludes that domestic auto makers should be able to achieve a 32.1 mpg CAFE rating by 2001 under their current product plan, and importers should be able to reach a 34.6 mpg standard. (EEA's analysis assumes there will be no loss in automobile performance and that

Table 1

Technologies to Improve Fuel Economy

Technology	Projected Improvement Best Case	Industry	Comments
Engine changes			
Overhead camshaft	6.0%	1.0–3.5%	Already in most imported cars.
Multivalve engines	5.0	1.0–3.5	Already in most Japanese cars.
— 4 cyl. replacing 6	8.0	(–2.0)–0.5	At constant performance, otherwise
— 6 cyl. replacing 8	8.0	1.0–3.5	there is a loss of low-end torque.
Intake-valve control	6.0	1.5–3.0	Works well with multivalve engines.
Lean-burn engines	10.0	n.e.	Featured in 1992 Honda Civic VX.
Two-stroke engines	20.0	6.0–10.0	Introduction expected in mid-1990s.
Diesel engines	25.0	10–25.0	Only Germans sell diesels in U.S.
Reduced engine friction	2.0	2.0	Low-friction pistons and rings.
Transmission changes			
5-speed auto. trans.	4.7	0.5–3.0	Relative to 3-speed auto. trans.
Torque converter lockup	3.0	1.5–2.0	Already in most auto. trans. cars.
Cont. variable trans.	3.5	1.0–2.5	Featured in Subaru Justy.
Aggressive trans. mgmt.	9.0	n.e.	Creates shift "busy-ness."
Idle-off	15.0	n.e.	Requires use of a flywheel.
Other changes			
Front-wheel drive	10.0	0.5–1.0	Widely implemented since 1970s.
Weight reduction	6.6	n.e.	Per 10% reduction in weight.
Drag reduction	3.0	2.0	Per 10% cut in drag coefficient.
Advanced tires	1.0	1.0	Reduced rolling resistance.

n.e. - no estimate
Note: Percentage improvements in fuel economy are not cumulative.

SOURCE: Investor Responsibility Research Center, Office of Technology Assessment, American Council for an Energy-Efficient Economy, Ford and General Motors.

existing market trends will continue. The analysis also projects a modest rise in gasoline prices to $1.50 a gallon in 1989 dollars, and it factors in a 10 percent discount rate.)[96]

When a "maximum technology" approach is taken, EEA finds that domestic auto makers could achieve a 36.0 mpg rating by 2001, although it cautions that such an effort would entail "a heavy burden of retooling for the industry and would require unprecedented and risky changes to every product sold."[97] While the fuel-saving technologies considered under the maximum technology plan are available now, some are quite expensive and would be expected to pay for themselves only if gasoline prices rose to $1.55 or even $1.90 a gallon over the 10-year average life of a car. EEA also found that changing the fleets' performance characteristics to match those of the 1987 model year instead of the 1990 model year could lead to an additional 1.3 mpg gain. Then the domestic fleet average would be 37.3 mpg—only 1 mpg below the CAFE level sought under the original Bryan bill.

Two other analyses have been conducted on auto fuel economy that rely heavily on EEA's base assumptions yet reach somewhat different conclusions. One by the congressional Office of Technology Assessment agrees that a combined CAFE rating of 33 mpg is achievable by 2001 for imported and domestic passenger cars under the current product plan.[98] As for the maximum technology scenario, OTA believes that the improvements sought by 2001 in the EEA analysis would be more likely to occur by 2005. In other words, market penetration rates for technologies such as multivalve engines, intake valve control, five-speed automatic transmissions, continuously variable transmissions and aerodynamic designs would take four years longer to achieve than is envisioned in the EEA analysis.

Another independent study performed by the American Council for an Energy-Efficient Economy believes EEA has been too conservative in its assumptions rather than too liberal. ACEEE notes, for example, that EEA did not consider "lean-burn" engines and variable-valve timing technology, even though these technologies are commercially available. Moreover, ACEEE's analysis pegs gasoline prices at only $1.32 a gallon in 2000 (instead of $1.50 a gallon), although it uses a more favorable discount rate (7 percent instead of 10 percent) and assumes a 1987 level of vehicle performance (instead of a 1990 level). On this basis, the ACEEE analysis concludes that cost-effective current technology could enable the auto fleet to achieve an overall CAFE rating of 38.3 mpg by 2000, which matches the 40 percent gain sought in the Bryan CAFE bill.

If other technologies that "are largely in hand but still require some development" are factored in, CAFE could reach 41.9 mpg in 2000, according to the ACEEE study.[99] Such emerging technologies include aggressive transmission management, which uses computer sensors to shift gears in a manner that maximizes fuel economy rather than power, and "idle-off," which shuts off a car's engine when no power is required and uses a mechanical device (such as a flywheel) to get the vehicle moving before the engine is turned back on. While ACEEE concedes that such emerging technologies may meet customer resistance, aggressive transmission management and idle-off alone could improve fuel economy 14 percent and pay for themselves out of fuel savings.

Altogether, 14 of the technologies examined in the ACEEE study potentially could improve the average fuel economy of cars built in 2000 by 53 percent—without reducing their roominess or performance—relative to 1987 models. The net cost of these improvements, ACEEE says, would be 53 cents for each gallon of gasoline saved. If such fuel savings were realized, the nation would curb its demand for petroleum by more than 2 million barrels a day (compared with an unchanged level of fuel economy), reduce its trade deficit by $10 billion and achieve a 10 percent or greater reduction in carbon dioxide emissions by 2005, ACEEE concludes.[100]

Obviously, the auto industry and conservation advocates have a wide difference of opinion on the potential for fuel economy gains. The industry insists that a 9 percent improvement is all that is feasible by the turn of the century, while the conservationists claim that a 50 percent gain is technically and economically possible. The Office of Technology Assessment believes both analyses contain flaws. In a recent report to Congress, OTA says, "the analyses presented by several conservation groups lack an appropriate analytical foundation and, for the 1996 to 2002 timeframe, rely too heavily on unproven technologies; and those of the auto makers are skewed toward low fuel economy values by the imposition of assumptions not compatible with a strong regulatory push to higher fuel economy."[101]

Ultimately, the commercialization of fuel-saving technologies will not be dictated by theoretical studies but rather by their rate of introduction into the marketplace and the degree to which drivers embrace them. Deborah Bleviss, who conducted a comprehensive study of the fuel economy programs of the world's leading auto makers in 1988, found her research "demonstrated two things. Nobody is out of the block, ready to introduce new fuel-efficient cars. But there is a lot of research going on, and it is quite varied across regions and across companies. Generally speaking, the U.S. lags behind both its Western European and, particularly, its Japanese competitors."[102]

The remainder of this section highlights important progress being made in the quest for auto fuel efficiency improvements. Fuel-saving technologies include multivalve engines, lean-burn engines, two-stroke engines, stratified-charge diesel engines and innovative transmission systems. Modifications being made outside the powertrain include the expanded use of strong yet lightweight materials and bold, aerodynamic styling. This section concludes with a look at some prototype vehicles that combine many of these innovative features and that may be the shape of things to come.

Overhead Cam and Multivalve Engines

Traditional Otto engines are equipped with two valves per cylinder—one for intake and one for exhaust. In addition, the camshafts, which regulate the opening and closing of the valves, are placed underneath the engine cylinders rather than overhead. Two important refinements have been made to these engines over the years to make them operate more efficiently. In "overhead cam" engines, the camshafts are placed above the engine cylinders in closer proximity to the valves, creating a valve train with fewer parts and less mass than their "pushrod" counterparts. With less energy required to open and close the valves, overhead cam engines "breathe" more efficiently, raising fuel economy 3 to 6 percent.[103] Today, overhead cam engines are standard equipment in virtually all cars imported into America. Only the Big Three still sell a large number of cars outfitted with conventional overhead valve engines.[104]

The other major design change has been to outfit overhead cam engines with two extra valves per cylinder, so that each engine cylinder has two intake valves and two exhaust valves. The extra valves reduce friction between air and fuel as they pass through the engine cylinder, and they permit exhaust to exit faster. The result is a high-compression engine capable of producing 25 to 35 percent more horsepower than a two-valve engine of similar displacement. The high performance offered by these multivalve engines enables smaller four-valve-per-cylinder engines to provide the same performance as larger two-valve-per-cylinder engines (albeit at much higher engine speeds and with a loss of torque, or pulling power, at low engine speeds).

When a multivalve engine replaces a two-valve engine with an equal number of cylinders, the resulting fuel economy gain can be as much as 5 percent. The potential fuel economy gain rises to 8 or 10 percent when the multivalve engine is downsized (i.e., a 16-valve, four-cylinder engine takes the place of a 12-valve, six-cylinder engine, or a 24-valve, six-cylinder engine replaces a 16-valve, eight-cylinder engine).[105] In actual production, however, auto makers say the fuel economy gain of a multivalve engine often is reduced to 3 percent, since it is engineered to maintain the "feel" of a high-torque, "low revving" engine that can accelerate quickly.[106]

Multivalve engine technology was invented in the United States more than 80 years ago. Virtually every racing car built since 1912, in fact, has featured a double overhead cam, multivalve engine.[107] Yet multivalve engines were not introduced in passenger cars until the early 1980s, and it was Japanese auto makers—not Detroit's Big Three—who made the move. Stiff taxes levied on large engines in Japan after the oil price shock of 1979 encouraged auto makers there to find ways to give small engines higher performance. The Japanese turned to multivalve engines in search of such power—and realized a fuel economy gain in the process.[108]

That product innovation helped transform the way many people think about smaller cars. Until recently, consumers regarded subcompacts "as old answers to the last fuel crisis," explains auto industry analyst Christopher Cedergren. "The whole subcompact market has been squeezed because people want sportier cars that still get good gas mileage."[109] Multivalve engines have fulfilled this dual requirement by offering high performance *and* fuel economy—and helped Japanese auto makers take market share away from their American counterparts.

Thus, in the spring of 1989, while Ford was idling its assembly plant for the fuel-efficient but eight-valve-only Escort, the 16-valve Honda Accord was on its way to surpassing the Escort as the number one selling car in America. Meanwhile, in the 1990 model year, sales of the Toyota Camry practically caught up with Ford's best-selling mid-size car, the Taurus. One of the reasons: The 1990 Camry's 16-valve base engine delivers 115 horsepower and gets 27.7 mpg. The base engine for the 1990 Taurus, however, has a conventional four-cylinder motor with a mere 90 horsepower that gets only 22.7 mpg.[110]

As of the 1990 model year, 100 percent of the Hondas and 70 percent of the Toyotas sold in America featured multivalve engines. Similarly, 54 percent of the Nissans, 37 percent of the Mazdas and 24 percent of the Mitsubishis had multivalve engines in 1990.[111] (Now they are standard in virtually all Japanese cars.) Meanwhile, the only American-made multivalve engine in 1990 was the 16-valve "Quad Four" built by General Motors. It was sold as a $660 option under the hood of selected Pontiac and Oldsmobile models. Ford and Chrysler had no multivalve engines of their own, although they both imported multivalve engines from Japan to place under the hood of some of their sportier models, such as the Taurus SHO and the Dodge Stealth R/T.[112]

Now the Big Three are in a position of having to play catch-up on this important fuel-saving technology. Only 0.2 percent of the 9 million engines that GM built in the 1989 model year were Quad Fours. General Motors Chairman Stempel originally resisted the move to multivalve engines but now intends to aggressively expand production of them.[113] For the 1992 model year, the Pontiac Grand Am and the Buick Regal have the Quad Four as an option, and Olds Cutlass Supreme features the world's largest 24-valve, V-6 engine. GM's new Saturn Division also has a sporty two-door coupe with a 16-valve engine, but Saturn's four-door sedan still comes equipped with a conventional four-cylinder engine. For the 1993 model year, GM's Cadillac division will introduce a 32-valve, V-8 engine. But it is also devising an improved version of its conventional 3.8-liter pushrod V-8 engine, with the hope that it will deliver almost as much power (200 horsepower) as a multivalve engine of comparable size, at significantly reduced cost.[114]

Chrysler's Dodge division introduced the Stealth R/T with a 24-valve, twin-turbocharged engine for the 1991 model-year. The engine was designed in America, but it is engineered and built by Mitsubishi of Japan. Chrysler began building its own four-cylinder, 16-valve engine in late 1990 for selected 1992-model cars, such as the Eagle Talon. It has also committed $148 million to build a new 3.5-liter, 24-valve engine with 210 horsepower as an option for its new LH family of mid-size cars, which will be introduced in the fall of 1992. Chrysler expects that 80 percent of its cars will have multivalve engines by the mid-1990s.[115]

Ford plans to boost its capital spending 30 percent through 1995 to improve the design of its entire engine line-up. Much of its emphasis is on modular engines for full- and mid-size cars, which can be produced in different sizes with different valve configurations. Ford will offer a 4.6-liter, V-8 multivalve engine with 280 horsepower in the new Lincoln Mark VIII, starting in late 1992. Ford also is investing $1.2 billion to develop and build a new family of V-6 engines equipped with four valves per cylinder. Its Cleveland, Ohio, engine plant will have the capacity to produce 400,000 such engines a year by the mid-1990s.[116] In January 1991, Ford embarked on a $700 million program to expand an engine plant in Chihuahua, Mexico, to make multivalve engines for some of its four-cylinder models. Full-scale engine production at the Chihuahua plant is not expected until mid-1993, however. Once available, Ford will use the 16-valve motors to power most of the cars it sells in Europe (mainly Escorts and Sierras) for the remainder of the 1990s.[117]

The lack of a multivalve engine is one of the reasons critics offer for the comparatively low sales volume of the re-introduced Escort model, which was completely revamped in 1990. Ford did introduce an Escort LX sedan in the fall of 1991 featuring a 16-valve engine as standard equipment. "We started late," one Ford executive remarked in 1989 in reference to multivalve technology, "but we hope we can move quickly."[118]

Variable-Valve, Lean-Burn Engines

Meanwhile, some Japanese auto makers are taking multivalve technology the next step. In the fall of 1991, Honda introduced a 16-valve engine with variable-valve timing, which overcomes multivalve engines' lack of power at low speeds. By the mid-1990s, several Japanese auto makers may offer variable-valve engines routinely.

A variable-valve engine "breathes" much the way a person does, regulating its intake according to whether it is working hard or just cruising along. When the car is accelerating fast, both intake valves open wide to increase the flow of air and fuel into the engine cylinders. But under "part load" conditions—normal driving at legal speeds—one intake valve is closed to conserve fuel. (Alternatively, both intake valves can be closed early in the timing cycle to avoid pumping air across the throttle under part-load conditions.) A variable-valve lift and timing system can provide up to a 6 percent gain in fuel economy (but only if the engine is downsized to provide equal low-speed performance.)[119]

Besides having variable timing, these engines can be designed to operate with a high ratio of air to fuel. Standard engines use a ratio of about 15 parts air to one part fuel. This is just enough oxygen to burn all of the fuel—without fouling the catalytic controls for nitrogen oxides in the exhaust system. A "lean-burn" engine increases the ratio of air to fuel to 20:1—or even 25:1—which increases thermodynamic efficiency and reduces pumping losses. The resulting fuel economy gain is 10 to 12 percent over conventional two-valve engines.[120] One remaining obstacle, however, is to find a catalyst to control nitrogen oxides emissions under lean-burn conditions.

Toyota has been selling a lean-burn Carina in Europe since late 1987. But because these cars sell at a premium, they make up only one-tenth of all Carinas marketed there. Mitsubishi introduced a lean-burn engine in some subcompact models it sells in Japan in the fall of 1991. These models are priced about 5 percent higher than its regular subcompacts, yet they offer a 10 percent improvement in fuel economy at highway speeds, and a 20 percent improvement at idle speeds, according to press accounts.[121]

The lean-burn engine commanding the most attention in the United States is Honda's "VTEC-E" (which stands for Variable-valve Timing and lift Electronic Control system-Economy). Introduced for the 1992 model year, the Honda Civic hatchback VX combines variable-valve and lean-burn technology (with an air-fuel ratio of 22:1), and many other fuel saving features. The result is a car rated at 55 mpg on the highway and 48 mpg in the city—a remarkable 55 percent improvement over the EPA mileage rating for the 1991 Civic. In fact, the Civic VX's mileage rating is nearly as high as that of the Geo Metro—the most fuel-economical car sold in America. Whereas the Metro delivers only 61 horsepower, however, the Civic VX musters 92. And it goes from 0 to 60 mph in about 10 seconds, which is unprecedented in cars with super-efficient engines.[122]

Given these features, Honda's sales goals for the Civic VX in the United States are relatively modest—about 8,000 units a year, selling for about $10,000 a copy. One of the factors is that the lean-burn Civic VX fails to meet California's tough NO_x pollution standards, which will be applied nationally in 1994. Honda is working on a new catalytic converter that it hopes will enable the lean-burn VX to meet California's standard. In the meantime, Honda is selling its Civic VX in California with an exhaust gas recirculation system that maintains an ordinary mixture of 15 parts air to one part fuel in the engine's cylinders. While the loss of lean-burn fuel injection reduces fuel economy, other features—including the multivalve engine with variable-valve timing, a streamlined exterior, reduced rolling resistance (through use of special tires), low-friction piston rings and more—allow the California version of the VX to still get an impressive 51 mpg in highway driving.[123]

American auto makers have long been intrigued by lean-burn technology but felt it posed too many problems to pursue it aggressively. Ford's British subsidiary, for one, has invested heavily in research and development of lean-burn engines but has yet to design one that can meet U.S. emissions standards.[124] More fundamentally, Detroit has questioned whether there would be much demand for such a fuel-efficient engine. Even Honda almost scuttled its lean-burn program in 1989 because of cheap oil prices and falling demand for small, high-mpg cars. But after heated internal debate, the Japanese auto maker decided to move forward. "The direction [toward more fuel-economical cars] is inevitable," concluded Honda President Nobuhiko Kawamoto. "We'd rather take a small step first, instead of waiting for regulations."[125]

Two-Stroke Engines

Detroit's auto makers have been more aggressive in their pursuit of lean-burn, "two-stroke" engines, which may burst onto the American scene before the year 2000. Lacking a valve train, two-stroke engines are light and simple to operate, with 200 fewer parts than conventional four-cycle Otto engines. In fact, the two-stroke engine pre-dates the Otto engine by eight years; Etienne Lenoir patented it in 1859.[126]

Any internal combustion engine requires four steps to deliver power: intake of fuel and air, compression of the mixture, combustion (the power stroke) and, finally, exhaust. A two-stroke engine combines these four steps into two cycles. Combustion occurs as the piston moves up the cylinder, compressing the fuel-air mixture. Then the power of combustion drives the piston back down the cylinder, sucking a fresh fuel-air mixture into the chamber as the exhaust gases are flushed out.

Two-stroke engines offer considerably more horsepower for their size and weight relative to Otto engines. That is why they are commonly found in outboard motors, chainsaws and lawnmowers. As for automobiles, Saab of Sweden sold a small, two-stroke car in the United States until 1968, but it discontinued sales after the first auto emissions standards went into effect.[127]

Traditionally, two-stroke car engines mixed air and oil in the carburetor before they entered the combustion chamber. The unburned portion of the mixture spewed out of the tailpipe as a noxious blue smoke—a conspicuous symbol not only of the old, little Saabs but also of Trabants and Wartburgs manufactured in East Germany until recently. In most new two-stroke engine designs, electronically controlled, direct-injection fuel-injectors replace carburetors, making fuel combustion much more complete. The result is a relatively clean-burning engine, although the lean fuel mixture is not compatible with conventional catalysts, making it difficult to control emissions of nitrogen oxides.

While no two-stroke car engines are on the American market today, prospects for the technology have brightened in part because of optimistic claims made by an Australian inventor and real estate mogul, Ralph T. Sarich, who owns Orbital Engine Co. of Perth, Australia. In 1991, Orbital opened an engine factory in Tecumseh, Mich., with plans for manufacturing up to 250,000 engines by 1993. The two-stroke, three-cylinder Orbital engine has only half the size and weight of a conventional four-stroke engine. Lacking a valve train, the Orbital engine also sits seven inches lower under the hood, making it an ideal candidate for use in aerodynamic, low-profile, high-efficiency cars.

Sarich estimates that the Orbital engine offers a 25 percent gain in fuel economy over conventional four-stroke engines and that full-scale production could reduce engine costs by 25 percent, since it has fewer moving parts. Moreover, Sarich claims that the Orbital engine will be able to meet U.S. auto emissions standards, at least in small cars weighing 2,500 pounds and under—the main market for two-stroke engines. In early 1992, Orbital reported "promising" results from emissions tests conducted on its engine by Ford and EPA. Over 2,000 miles of travel, the engine "achieved emissions below the levels contained in the stringent California standards," the company said.[128]

Many automotive engineers are wary of Orbital's claims, however. General Motors and Ford have been experimenting with two-stroke engines for years and had failed to match the results boasted by Sarich, until the recent test. GM, for one, believes any fuel economy gain to be realized in a two-stroke engine would be in the range of only 6 to 10 percent.[129] Even so, GM and Ford have licenses to Orbital's engine technology, and GM has pledged to market two-stroke engines as early as 1995.[130] Both GM and Ford insist, however, that the engines they ultimately bring to market will be based on their own development efforts, not Orbital's. GM has designed a three-cylinder, two-stroke engine that delivers 110 horsepower yet weighs just 165 pounds and is miserly with gas.[131]

Chrysler, meanwhile, reports that it is "well into the development of a new two-stroke engine. Prototypes half the volume of a four-stroke engine have achieved up to 40 percent more horsepower."[132] One such engine prototype has found its way into a Chrysler concept car, the Neon. Like GM and Ford, however, Chrysler is concerned about the two-stroke's long-term durability and its ability to meet new emissions control standards—especially for the 10-year, 100,000 mile provision for pollution control equipment that goes into effect in 1994.[133]

Japanese auto makers also are far along in the development of two-stroke engine technology. At the 1990 Tokyo Motor Show, the world's largest auto exhibition, Toyota and Fuji Heavy Industries (maker of Subaru cars) turned the heads of American auto executives with displays of two-stroke engine prototypes.[134] Toyota has created a 24-valve, six-cylinder, two-stroke engine targeted at the luxury car market. In Japan, Toyota is testing a large two-door coupe, the Soarer, which has impressed American journalists with its high torque at low engine speeds and smooth running at cruising speeds.[135]

Finally, in Europe, Peugeot has designed its own two-stroke engine in a joint venture with the French Petroleum Institute.[136] And in a move that is sure to improve European air quality, Volkswagen has proposed a joint venture with VEB IFA-Kombinat—the maker of the lowly Trabant—to design its successor and market it throughout Europe.[137]

Diesel Engines

The diesel engine—another century-old technology—is due for another look in the 1990s because of its inherent fuel economy advantages. In heavy-duty engines and some light-duty vehicles, diesel motors are the most efficient power plants in use today. Yet in the United States diesel passenger cars have been ostracized for years. Since 1988, in fact, no domestic manufacturer has even offered a diesel car in its product line. Diesels still account for one-sixth of European new-car sales, however, and they continue to dominate the American heavy-truck market (those vehicles with a gross weight of 14,000 pounds or more.)

A diesel engine functions much like an Otto engine and can be operated in either a two- or four-stroke mode. What distinguishes a diesel engine is its high compression ratio and lean air-fuel mixture in the engine cylinder. With a compression ratio of 20:1—versus 9:1 for a standard gasoline engine—diesel fuel gets so hot during compression that it ignites spontaneously, eliminating the need for a sparkplug. At the same time, the lean fuel mixture promotes fuel savings. According to European manufacturers, diesel engines are 15 to 18 percent more efficient on average than four-valve gasoline engines. When high pressure exhaust gases are channeled through a turbine connected to the crankshaft—a turbocharger—the diesel's efficiency advantage rises to 24 or even 28 percent, while adding power and curbing diesel exhaust fumes.[138]

Like other lean-burn engines, today's diesels stratify the air-fuel mixture into distinct layers within the engine cylinder. Direct fuel injection diesels, now in production in Europe, create a rich fuel mixture at the top of the cylinder and a comparatively lean mixture elsewhere, allowing relatively complete fuel combustion. The efficiency gain offered in such direct injection, turbocharged diesels is 35 to 40 percent over conventional gasoline engines.[139]

Part of diesels' considerable fuel economy advantage over gasoline engines results from their ability to operate efficiently under part load conditions, such as in city traffic. Whereas the air-fuel mixture of conventional gasoline engines tends to be too rich for the amount of power required under such conditions, stratified-charge engines—including diesels—burn fuel efficiently when maximum power is not required. Two other important advantages of diesels are that they tend to last longer than gasoline engines and are easier to maintain. Given these advantages, auto makers could scale up production of diesel-powered cars as a means of promoting fuel economy and reducing vehicular carbon dioxide emissions.

Yet the American public has turned decidedly against diesels for a number of reasons. The engines themselves generally cost more and are noisier and less powerful than comparable gasoline engines. Moreover, since they lack sparkplugs, diesel engines can be hard to start in cold weather. Some drivers have expressed concern that fewer service stations offer diesel fuel for sale.

The soot emitted from diesels' tailpipes is another drawback of these engines. The soot (actually partially burned fuel consisting of fine carbon particles) contains polycyclic aromatic hydrocarbons, a toxin that can lodge deep in the lungs and is a suspected human carcinogen. EPA has proposed regulations to reduce such particulate emissions from urban diesel buses by

80 percent by 1994—a move that may cost diesel engine makers $70 million.[140] But while soot poses a visible and tangible threat to human health, diesel engines are cleaner than gasoline engines with respect to other emissions that humans cannot see. Tailpipe emissions of carbon monoxide and hydrocarbons, for example, are only 10 percent as much as in conventional gasoline engines.[141]

The primary emissions problem that diesel engines face is the same one that plagues all lean-burn engines—emissions of nitrogen oxides. The 1990 Clean Air Act amendments grant diesel engines a waiver from the 0.4 grams-per-mile NO_x limit set on gasoline-powered engines that goes into effect in 1994. For diesel-powered vehicles weighing less than 3,750 pounds, the NO_x limit will remain at 1.0 grams per mile. If diesels stage a comeback in the auto marketplace, however, pressures could mount to rescind the NO_x waiver. In any case, the Clean Air Act stipulates that diesels must meet the 1994 NO_x standard in effect for gasoline-powered cars by 2004.

Even though the American driving public has largely abandoned diesels in recent years—and control of NO_x emissions poses a major engineering challenge—two German manufacturers of diesel cars have not given up on the American market. Starting with the 1990 model-year, Volkswagen reintroduced a four-door, diesel Jetta sedan, and Mercedes-Benz introduced a mid-size diesel, the 250D turbo. The 1992 diesel-powered Jetta gets 40 mpg on the highway, compared with 32 mpg for the comparable gasoline-powered Jetta, yet its sticker price is only a few hundred dollars more. As recently as 1988, Volkswagen had suspended U.S. sales of its diesel-powered Golf and Jetta models, leaving Mercedes-Benz for a time as the only marketer of a diesel car in America (with its compact 190D).[142]

Mercedes has been steadfast in marketing diesel cars in the U.S. market, partly because they get such good mileage relative to the rest of its American fleet. With a CAFE fleet rating of only 21.2 mpg in 1990, Mercedes was assessed an estimated $18.6 million penalty that year for failing to meet the 27.5-mpg CAFE standard. Moreover, nearly half of the German auto maker's cars are subject to a federal "gas-guzzler" tax because they fail to achieve a minimum fuel economy standard of 22.5 mpg. (This tax is assessed on a sliding scale, ranging from $1,000 for cars getting less than 22.5 mpg to $7,700 for cars getting less than 12.5 mpg.) Accordingly, Mercedes added three diesels to its American showrooms in the 1991 model-year: the mid-size 300D and the larger 350SD and 350SDL turbos.[143] The 1992 300D has an EPA mileage rating of 25.1 mpg in city/highway driving, although it is not available in California because it fails to meet that state's strict NO_x emissions standards.

Adiabatic diesel engines: Over the short term, introduction of turbo-charged, direct-injection diesel-powered cars appears to be the best hope for reviving sales of diesels in the American market (not to mention boosting fuel economy ratings tremendously). Over the long term, creation of adiabatic diesel engines may hold even greater promise. Adiabatic engines minimize heat loss and promote still cleaner burning of diesel fuel. Engine cooling requirements normally create a cold spot, or quench zone, in the engine cylinder that prevents complete combustion. An adiabatic diesel engine eliminates this cold spot. Operating without a cooling system, the adiabatic diesel engine weighs less and has fewer parasitic power losses than comparable engines. A Ford Tempo equipped with a turbocharged adiabatic diesel engine theoretically could achieve a fuel economy rating of 80 mpg. Moreover, emissions of hydrocarbons, carbon monoxide, oxides of nitrogen and particulates could be reduced as much as 60 to 80 percent over current diesel engine technology.[144]

The main hurdle to overcome with adiabatic engines is finding a material to withstand the intense heat they generate. Ceramic materials would aid in this cause (not to mention the development of 21st-century gas turbine engines) because they are tough and can withstand temperatures above 2,500 degrees F. Neither metal nor polymer, ceramics are a unique family of solid substances that maintain strong atomic bonds even when subjected to chemical attack or extreme heat. Ceramics are very difficult to mold into the shape of engine parts, however. Even a tiny structural flaw causes these brittle substances to crack. American and Japanese scientists have identified composite ceramic materials that are moldable at temperatures of nearly 2,500 degrees F, but so far they stretch in only one direction. To be shaped into useful engine parts, ceramic materials must stretch in two directions uniformly.[145]

Cummins Engine and Adiabatics Inc. in the United States, as well as Japanese auto makers, are leading the way on adiabatic engine research. Toyota and Nissan have set up joint ventures with ceramic manufacturers.[146] Isuzu has its own ceramics research laboratory and is said to be developing an adiabatic diesel engine that works in conjunction with a system to recover energy from exhaust heat. Such an engine could provide a 30 percent improvement in fuel efficiency over today's diesel engines. More advanced ceramic engines someday may provide a 60 percent increase over today's gasoline engines, although this development is not expected for decades.[147]

Improved Transmissions

Automotive fuel economy gains can be achieved in parts of the powertrain other than the engine itself. One such area is the vehicle's transmission system, which links the horsepower of the engine to the drivetrain. Fuel economy is optimized when the transmission system combines the highest possible torque with the lowest practical engine speed. Adding extra gears enables the drivetrain to utilize more of the engine's power under varying load conditions. On that basis, four- and five-speed manual and automatic transmissions do a more efficient job of gear-shifting than three-speed automatics, usually yielding fuel savings of a few miles per gallon.[148]

Virtually all Japanese cars sold in America—including best sellers such as Toyota's Camry, Honda's Accord and Nissan's Stanza—now come equipped with four-speed automatics and five-speed manual transmissions. While the Big Three are switching from three-speed to four-speed automatics as well, many popular American cars—such as the Chevrolet Corsica, the Ford Tempo and the Dodge Shadow—still have only three-speed automatics (and five-speed manual transmissions) available. Not only does the three-gear configuration put a drag on fuel economy, it tends to make the American cars accelerate more slowly.[149]

Most cars would benefit from the addition of a fifth (and even a sixth) gear, with potential fuel economy gains of up to 2.5 percent per gear. Only manual transmission cars have five gears available at present, however. Partly for this reason, manual transmission cars traditionally have achieved a fuel economy advantage of up to 7 mpg in highway driving, and up to 2 mpg in city driving, over comparable cars with automatic transmissions.[150] Slippage and friction losses resulting from the movement of transmission fluid during gear changes also contributes to a loss of efficiency in automatic transmissions. Cars with manual transmissions minimize such losses through use of a clutch.

In recent years, American and Japanese car companies have addressed the slippage problem in automatic transmissions by adding lockup torque converters—mechanical clutches that lock in place in a given gear. (Most European cars still lack this feature, however.) The addition of lockup torque converters boosts fuel economy an average of 3 percent, but even these cars do not achieve the mileage ratings of comparable cars equipped with manual transmissions.[151]

One relatively new option for small cars is a continuously variable transmission, which uses belts and pulleys to keep the engine running at optimal efficiency regardless of a car's speed. The Subaru Justy, a subcompact car sold in America by Fuji Heavy Industries, features such a continuously variable transmission. Ford also sells a car in Europe with a continuously variable transmission.[152]

Another concept being studied by the American Council for an Energy-Efficient Economy is "aggressive transmission management." The idea here is to outfit a vehicle's automatic transmission with more gears and lower gear ratios and to use electronic controls to shift gears as soon as possible. The result is a reduction in overall engine speeds that saves fuel. In one theoretical example, ACEEE replaced a five-speed manual transmission with a six-speed automatic transmission that always selects the most energy-efficient gear. The reduction in average engine speed improves fuel economy 9 percent in city driving and 15 percent on the highway.[153] But the switch to aggressive transmission management would not be without a loss of amenity. In city driving, there could be twice as many gear changes per minute—an average of six instead of three. Moreover, when the accelerator pedal is floored, the car would be more likely to downshift—a process of declutching, engine speed-up and reclutching that takes a half-second if done well. Accordingly, there would be "shift busy-ness" that would change the feel of driving.

One final fuel-saving idea is to attach a flywheel to the transmission system to capture some of the energy normally lost when the engine is braking or idling. After the flywheel has received a sufficient charge—spinning at, say, 2,500 revolutions a minute—the engine could be turned off while the flywheel continues to run the car's accessories. When the car resumes forward motion, the flywheel could help it to accelerate and then restart the engine as more power is needed (or as the flywheel slows to below 2,500 revolutions per minute). ACEEE estimates fuel savings from such "idle-off" technology could amount to 15 percent in city driving and 2 percent on the highway.[154] But once again, there would be a change in the feel of driving. Drivers might be especially reluctant to turn against traffic with their engines turned off, even though the flywheel could perform the task safely.

Volkswagen has developed a prototype flywheel called "Glider Automatic." In addition to accelerating a car from a dead stop, the flywheel's rotor could be used in place of a conventional starter and alternator, resulting in a net reduction in vehicle weight and possibly cost. Volkswagen considered introducing an early version of the Glider Automatic in the early 1980s, but it postponed its plans to develop a second-generation technology. Nissan also is reported to be developing a flywheel device.[155]

Lightweight and Aerodynamic Designs

To raise a car's fuel economy, auto makers also have the option of using materials and designs that make cars lighter and less resistant to aerodynamic drag. Since the 1970s, cars' average weight has declined 20 percent, from 4,000 pounds to 3,200 pounds, while the coefficient of drag (Cd) has fallen 32 percent, from 0.50 to 0.36.[156] Each 10 percent reduction in a vehicle's weight yields an average fuel economy gain of 6.6 percent, assuming the engine is downsized by a similar percentage. Each 10 percent reduction in Cd improves fuel economy 2.3 percent, if an adjustment is made for the reduced power requirement.[157]

Within the engine block, lightweight aluminum, titanium and plastic polymers are being fashioned into a variety of engine components as a replacement for steel. General Motors' Cadillac Division, for one, has introduced an aluminum engine that is 100 pounds lighter than its cast-iron version.[158] GM's Saturn also features a lightweight aluminum engine. The Geo Metro has an aluminum engine that weighs a mere 157 pounds.

Use of aluminum also is starting to spread to cars' internal frames. Honda's limited edition Acura NSX, for example, is a lightweight, aluminum-intensive car; it gets 13 percent better gas mileage than if it were made from steel.[159] The price tag for the NSX is $60,000, however. Volkswagen's Audi unit may be the first to offer a truly mass-produced aluminum car—containing as much as 800 pounds of aluminum. Volkswagen signed an agreement with Alcoa in the fall of 1991 to build a car-frame plant in Soest, Germany.[160] One Japanese aluminum manufacturer predicts that aluminum's share of total vehicle weight could rise from 5 percent to 35 percent by 2000—with steel's share cut in half, from 60 to 30 percent. The net effect would be to shed 900 pounds of overall vehicle weight in Japanese cars.

From a greenhouse standpoint, the switch from steel to aluminum offers a mixed blessing. While use of aluminum would save on gasoline, refining virgin aluminum requires 10 times more electricity than refining steel. Fortunately, in the United States, aluminum manufacturers get most of their power from (noncarbon) hydroelectricity. But in the Far East, many countries import aluminum from Australia, where coal is the dominant electricity source. Hence, the net CO_2 emissions benefit there would be marginal in switching to lightweight aluminum.

Aluminum is not the only substitute for steel, however. Ford is testing a 2.3-liter, four-cylinder engine made from moldable fiber-reinforced phenolic composites. This "plastic" engine weighs only 175 pounds, compared with 300 pounds for a conventional metal engine. Besides saving weight, phenolic

> **Box 4-C**
>
> ### Cars and Auto Safety
>
> The film clip practically speaks for itself. A 4,000 pound Ford Crown Victoria smashes head-on into an 1,800 pound Subaru Justy in a government crash test. The engine compartment of the Justy compresses like an accordion, forcing its steering wheel into the chest of a dummy sitting in the driver's seat. Then a voice intones, "While smaller cars can save you gas, they could cost you something far more precious."
>
> This television advertisement was broadcast nationally during the summer of 1991 as part of a campaign by the Big Three auto makers to oppose increases in the federal law that sets corporate average fuel economy (CAFE) requirements for light-duty vehicles. The message is clear: If Congress wants to raise CAFE from the present standard of 27.5 miles per gallon, it may put thousands of lives at stake.
>
> To meet a substantially increased fuel-economy requirement, car makers insist they would have to downsize their fleets substantially. A possible 850-pound average weight reduction would lead to an additional 600 to 1,140 deaths and 2,900 to 5,300 vehicular injuries a year, according to one analysis by the government's National Highway Traffic Safety Administration (NHTSA). That makes the proposed CAFE legislation nothing more than a "highway fatality bill," remarks former Transportation Secretary (and now White House chief of staff) Samuel Skinner.[1c]
>
> Downsizing a vehicle reduces its safety for several reasons. First, smaller cars have a greater tendency to roll over as their track width—the distance between the left and right wheels—narrows. About 10,000 people are killed every year in the United States when the vehicles in which they are riding roll over. Only head-on collisions and accidents involving drunk drivers rival rollovers as the leading causes of highway deaths.[2c]
>
> Downsizing also leaves less car-body available to absorb the impact of a head-on collision. This increases the chances of occupants coming in contact with the dashboard and windshield. Also, when the front-end "crumple zone" is shortened, the car's occupants experience greater motion deceleration, putting more "G-forces" on their bodies. Moreover, cars with less mass stop more abruptly in head-on collisions because the laws of physics dictate that the deceleration of a striking car increases as its weight decreases.
>
> Despite the safety issues raised by the proposed increase in CAFE standards, there may be less to these safety arguments than meets the eye. For starters,

composites damp engine noise and vibration and resist corrosion. In full production, such engines could save 25 percent in manufacturing costs, since several parts can be combined into one piece, requiring less machining than metal. Ford and other auto manufacturers remain concerned about the durability of such engines to withstand the heat and stresses produced by internal combustion, however. Therefore, commercially available phenolic engines are still probably a decade or more away.[161] In addition, energy requirements to manufacture phenolic materials are quite high, once again raising questions about net CO_2 emissions benefits.

Other changes in car design are coming on the outside—and are clear for all to see. Bubble shapes and rounded edges have replaced boxy designs in an downsizing cars is not the only way to raise their fuel economy. Reduced engine displacement, greater use of overhead cam and multivalve engines, four- and five-gear transmissions and more aerodynamic styling can greatly improve vehicles' fuel economy without necessarily making them smaller or lighter. Moreover, in single-car accidents, the highest fatality rates occur in the middle of the weight spectrum, not in the very lightest vehicles, according to NHTSA data.[3c] Finally, any increased propensity of lighter, shorter vehicles to roll over can be offset by increasing their track width. One extra inch of track width compensates for 450 pounds of weight reduction, extrapolating from the NHTSA data.[4c]

In two-car crashes, occupants of a heavier car are much more likely to survive an accident that involves contact with a lighter car. A 3,500-pound car going just 35 miles per hour, for example, applies the same force at impact as a 2,000-pound car traveling 46 mph. This data can be interpreted two ways. A heavier car is safer for those who happen to be riding in one; for others, it is the more dangerous car on the road. As Eleanor Chelimsky, assistant comptroller general of the General Accounting Office, told Congress in 1991, "Our analysis suggests that if all cars become lighter, the increased vulnerability of lighter cars in two-car crashes would be more than offset—all things being equal—by the reduced threat to them by heavier cars."[5c]

Auto manufacturers are right to point out that some weight differences always will remain between different kinds of cars and, of course, between cars and trucks. Therefore, individual car buyers will continue to have a safety incentive to buy larger, heavier cars. Despite this, highway travelers may take comfort in the fact that fatality rates per mile of highway travel have dropped one-third since 1975 even though today's cars weigh about one-quarter less.[6c]

Car makers can do—and are doing—many things to make vehicles safer besides making them big. These include the installation of air bags, "passive" safety restraints, anti-lock brakes, energy-absorbing chassis and body parts, improved side-impact protection and roof crush protection. Such safety measures—when combined with reduced speed limits, mandatory seat-belt laws and crackdowns on drunk driving—have reduced the fatality rate 33 percent per 100 million vehicle miles traveled since 1975. In 1991, highway fatalities in the United States dropped 7 percent to 41,000, the lowest level since 1962.[7c] Fewer than half as many vehicles were on American roads 30 years ago, however, and the number of vehicle miles traveled was nearly two-thirds less. Mile for mile, car travel in the United States has never been safer than it is today.[8c]

effort to reduce aerodynamic drag. The lowest coefficient of drag on a current production model is claimed by GM's European Opel Calibra, with a Cd of 0.26.[162] In the United States, the Cd range is from 0.29 to 0.50, with an average Cd of 0.36 for new passenger cars. The congressional Office of Technology Assessment estimates that the coefficient of drag could fall to 0.24 or even 0.20 for cars manufactured in 2010.[163] GM's Impact concept car already sports a Cd of only 0.19. And one of Ford's experimental vehicles, the Probe V, has a 0.137 Cd—better than that of an F-15 jet fighter.[164]

It is not just the look of cars' exteriors that is changing; it is what their skins are made of. Glass-reinforced fiber composite materials are replacing steel on fenders, bumpers and tailgates. The vertical side-panels of GM's Saturn are made of plastic, for instance, and the plastic molds can be switched quickly, allowing future styling changes to occur with relative ease. Other vehicles, such as GM's APV minivan and selected Renault and Honda models, also have body panels made almost completely out of plastic. For the industry as a whole, plastic materials have been increasing 3.5 percent a year since 1986 as a percentage of total car weight, enabling vehicles to shed scores of pounds of weight.[165]

This weight-reducing trend seems likely to continue. Ford, General Motors and Chrysler are participating in a consortium to develop front-end chassis made from polymer-based composites. A car in which all structural body components are plastic could be 30 percent lighter than current production models.[166] In addition to boosting fuel economy, composites speed assembly because parts that ordinarily need to be welded or bolted together can be formed out of the same mold. Ford, for example, has developed one concept car in which the 400 steel parts that make up the body structure of its Taurus sedan were integrated into just five composite sections.[167] Two strikes against polymer-based composites, however, are that they cost much more to manufacture than steel and they are harder to recycle. As car designers place increasing emphasis on making car parts recyclable, questions are being raised about whether secondary markets for plastic composites will develop as they have for scrap metal. On the other hand, since many of these composite materials do not dent, scratch or rust, cars built from these materials are likely to last much longer than cars with metal frames.

The Office of Technology Assessment estimates that materials substitution could reduce the average weight of new passenger cars to 2,800 pounds by 2001—a 12 percent reduction over current models. By 2010, the total reduction could range from 21 to 33 percent, including a 3 to 8 percent savings from improved packaging of vehicles' interiors.[168] Volume-saving measures (the most important being a switch to front-wheel drive) already have achieved a 14 percent reduction in vehicular weight since the late 1970s without shrinking cars' average interior volume. Chrysler will unveil another important design innovation in the fall of 1992, when its new LH model cars are introduced. The "cab forward" design of these cars features a steeply raked windshield and rear wheels that are moved back toward the trunk. The result is a sporty-looking, aerodynamic exterior and a roomy interior that may become standard in many models by the mid-1990s.

Fuel-Efficient Prototype Cars

When engine modifications and design innovations like those described above are put together in an effort to maximize fuel economy, it is possible to make a car that gets more than 100 mpg in highway driving. At the 1991 Tokyo Motor Show (whose theme, incidentally, was "Man, Car and the Earth as One") the most talked-about vehicle was Honda's EP-X, which stands for Efficient Personal Experimental and is Honda's first-ever concept car. The EP-X is a two-seater that puts the passenger behind the driver—much like the configuration of a fighter cockpit. Its slender 1,360-pound body cuts wind resistance and is made of lightweight aluminum. Under the hood, there is an ultra-efficient three-cylinder engine. Honda did not disclose the estimated mileage rating of the EP-X; however, word spread at the auto show that it would be able to go 100 miles on a gallon of gas.[169]

A few months later at the North American International Auto Show in Detroit, General Motors executives were more outspoken about the Ultralite, a new GM concept car which they *did* say could get 100 mpg. (The four-passenger version is rated at 45 mpg in the city and 81 mpg on the highway, however.) The Ultralite has a ground-hugging, aerodynamic frame, and its 1,400-pound weight is achieved through use of lightweight carbon fibers. Such fibers cost $40 a pound, compared with 35 cents a pound for steel. Accordingly, materials for the Ultralite's body alone would cost $13,000.[170] "The Ultralite is not a production vehicle," GM President Lloyd Reuss stressed to reporters at the Detroit auto show. "It's more like a rolling laboratory....Yes, it's possible to build a car that gets 100 miles to a gallon, but the real challenge is to deliver those results economically, at a price that customers can afford to pay."[171]

The fact is, most concept cars like the EP-X and the Ultralite will never be built. But at the 1991 Tokyo Motor Show, Toyota showed off the AVX-IV, a diesel-powered prototype car that Toyota executives say could enter commercial production by the end of the decade. With an all-aluminum body, the AVX-IV is a two-seater that weighs less than 1,000 pounds and is rated at 70 to 75 mpg. In addition, its body parts have been designed for easy recyling.[172]

Previous research by Deborah Bleviss of the International Institute for Energy Conservation finds that seven car companies had built 10 prototype vehicles as of 1988 that got at least 55 mpg in city driving and 71 mpg on the highway.[173] Renault's gasoline-powered Vesta 2 actually achieved 107 mpg in highway driving tests, although the French car company has stopped testing the vehicle. Skeptics point out that seven of these concept cars are equipped with diesel engines that are currently out of favor with American drivers. In addition, some of the prototypes would not pass U.S. crash tests or emissions tests. Perhaps most important, none is in production.

Bleviss reports, however, that at least two of the vehicles—Volvo's LCP 2000 and Volkswagen's Eco Polo—were designed with production in mind. While Volvo disputes the production claim about the LCP 2000, it did design the vehicle to exceed U.S. crashworthiness tests and to approach (but not surpass) American emissions standards. The other vehicle, Volkswagen's Eco Polo, is a small, two-cylinder car with an advanced fuel-injected diesel engine that gets 62 mpg in combined city/highway driving. The German auto maker reportedly was almost ready to bring a version of this highly efficient car into production after the second oil price shock of 1979. But it shelved the plan after oil prices fell back again in the mid-1980s.[174]

Auto industry executives make the point that ultra-fuel-economical vehicles are not likely to fare well in the marketplace as long as fuel prices remain cheap. They note that the Geo Metro and the Honda Civic hatchback VX—the most fuel-economical cars now on the road—represent less than 1 percent of General Motors' and Honda's sales, respectively. Most car buyers apparently do not think that the gain in fuel savings would compensate them adequately for other sacrifices they would have to make in driving these vehicles. Nevertheless, prototype vehicles like the AVX-IV and the Ultralite demonstrate that tremendous fuel economy gains are technologically feasible, if not commercially viable, at present. Moreover, new engine technologies, drivetrain refinements and body design innovations will yield modest fuel economy gains in the years ahead that most drivers will not even notice—except, perhaps, at the gas pump.

Finally, important gains in fleet economy—such as the 40 percent CAFE rise sought by Sen. Richard Bryan—are achievable even with *no* further technological advancements. According to the Office of Technology Assessment, a 45 percent gain in fuel economy could be achieved simply by limiting new-car purchases to the dozen highest-mpg cars now sold in each weight class, and by bringing about a 12 percent reduction in the weight of the new-car fleet overall.[175] Under these circumstances, large cars weighing 4,000 pounds and over would comprise only 3 percent of the new-car fleet, instead of 12 percent today. Small cars weighing under 2,500 pounds would make up 33 percent of the new-car fleet, compared with 15 percent today. But the vast majority of new-car buyers—some 64 percent—would still choose among models weighing between 2,500 and 3,999 pounds, as 73 percent do today.

The end result in this analysis is that the "average" car would be a little more cramped (93 cubic feet of interior space versus 107 cubic feet in 1990), accelerate a little more slowly (0 to 60 mph in 15.6 seconds instead of 12.3 seconds) and be more likely to have a manual transmission than an automatic transmission. Yet in return for this loss of amenities, important gains would be made in ameliorating global warming potential and in reducing the nation's dependence on foreign oil (albeit at the risk of importing more cars from Japan). Such trade-offs are one way that American consumers might analyze the greenhouse gambit if they were to apply it to their own car-buying decisions.

Alternative Transportation Fuels

Promoting fuel-economical vehicles in an era of cheap oil is a tough sell. Today's drivers generally pay more for car insurance or engine maintenance than they spend on gasoline.[176] Those who choose a new car getting 35 miles per gallon over one getting 30 mpg save a mere $50 at the pump each year. Factoring in the higher average sticker price for the more fuel-economical vehicle (all other things being equal), the net annual savings is only $25 a year—hardly a financial incentive to conserve.[177]

While the pocketbook savings are relatively trivial, the cumulative effect of fuel conservation on the nation's energy bill is not. The United States consumes an estimated 4 million fewer barrels of oil a day than it would have without the fuel economy gains built into the light-duty fleet since 1973. That works out to more oil savings than the amount supplied daily from the Persian Gulf or Alaska's North Slope—a savings of at least $30 billion a year.[178] Even so, fuel conservation remains its own worst enemy: The greater the realized savings, the less need there is to conserve.

Given this inverse relationship, fuel economy has become less appealing to car buyers than other features of vehicles, such as performance, styling, utility, safety, reliability and comfort. Detroit is betting that consumer preferences in the 1990s will remain oriented away from fuel economy as it moves forward with production of relatively large cars and light-duty trucks with comparatively big, powerful engines. Japanese auto makers are betting much the same way, although they have a base of smaller cars from which to expand their fleet, and they have proven themselves to be market innovators in fuel-saving technologies such as multivalve and lean-burn engines.

In sum, the specter of global warming is not causing auto makers to hedge their bets very much—although they say they are not the ones who are oblivious to the threat. "We are doing what our investors expect us to do, satisfy the customer," reasons Thomas Gage, Chrysler's manager of energy and alternative fuels planning. "It is our customers, responding to the lowest fuel costs in history, who are insensitive to the global warming issue and are uninterested in additional fuel economy improvements."[179]

Gearing Up for Another Energy Crisis?

The decisive American victory in the 1991 war with Iraq has reinforced a sense of complacency regarding fuel consumption and the nation's reliance on imported oil. In reality, the United States is more dependent on foreign oil than ever before. Domestic oil production has been falling steadily in the lower 48 states since 1970, and the flow of Alaskan oil peaked in 1988. With U.S. production now on the wane, imported oil is projected to supply more than half of the nation's petroleum demand by 1995, two-thirds of the demand by 2010, and four-fifths by 2030.[180]

Meanwhile, the OPEC cartel—with nearly seven-eighths of the world's proven oil reserves—learned a valuable lesson during the 1980s: The best way to keep the driving public hooked on gasoline and diesel fuel is by making oil affordable and assuring a steady supply. "There is a tacit consensus, a sort of deal, whereby OPEC countries agree to guarantee a price for oil that does not put at risk the economic growth of major consumers," observes Pierre Terzian, editor of the Paris-based journal, *Petrostrategies*. "In return, consuming countries no longer continue their massive efforts to find energy substitutes for oil."[181]

So far, the Faustian bargain has paid off for all those with a stake in cheap oil. Nearly two decades after the first oil price shock of 1973-74, the nation's transportation system remains as dependent as ever on petroleum as a motor fuel—97 percent dependent. The linkage between oil imports and motor vehicle consumption is also clear. Nearly two-thirds of the 17 million barrels of oil used daily in America is for transportation purposes; the 7 million barrels consumed in motor vehicles roughly equals the amount imported into the United States. Other sectors of the U.S. economy have managed to decrease their use of oil 29 percent since the Arab embargo of 1973, but the transportation sector consumes 21 percent more oil now than it did back then.[182] Thus, if future tension

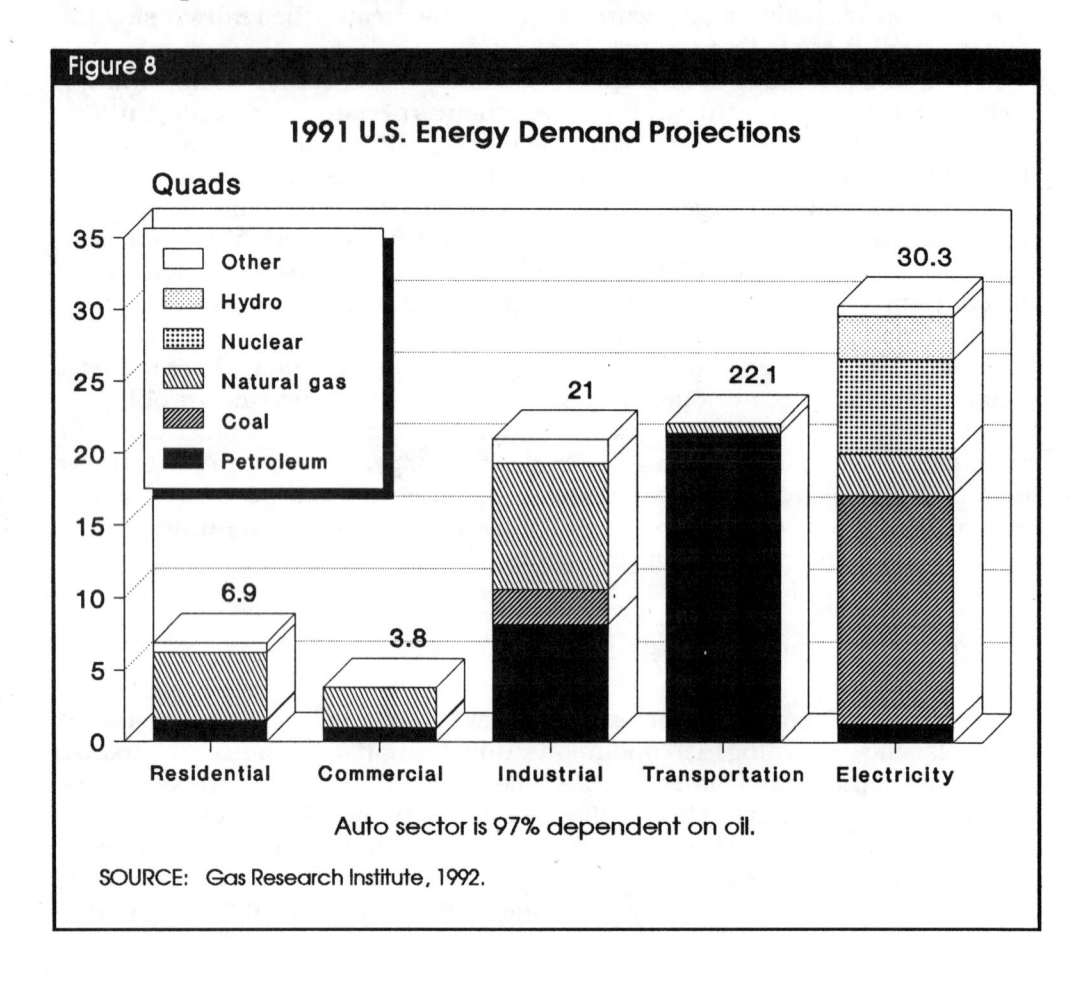

Figure 8

1991 U.S. Energy Demand Projections

Quads

- Residential: 6.9
- Commercial: 3.8
- Industrial: 21
- Transportation: 22.1
- Electricity: 30.3

Legend: Other, Hydro, Nuclear, Natural gas, Coal, Petroleum

Auto sector is 97% dependent on oil.

SOURCE: Gas Research Institute, 1992.

The Automobile Industry

in the Middle East precipitates another energy crisis, this time it would be mainly a transportation-related phenomenon.

The light-duty fleet's almost exclusive reliance on petroleum—and the growing share of oil imported from abroad—offers a strategic reason for the nation to wean itself of such dependence. The fact that motor vehicles account for roughly half of the unhealthful air hanging over American cities offers another compelling reason to weigh alternative policies. And motor vehicles' 22 percent contribution to the nation's CO_2 emissions is, of course, a third reason to reduce petroleum consumption.

But in the final analysis, boosting the fuel economy of the light-duty fleet will not solve these problems alone. It may not even make much of a dent in these problems if the number of vehicles—and the miles they are driven—continues to grow. Despite a further reduction in fuel consumption *per mile* of travel, the expected fuel savings could be wiped out by an increase in *total* vehicle miles traveled.

Since 1970, the number of vehicle miles traveled in the United States has doubled to 2.1 trillion miles, equal to an annual average increase of 3.3 percent. While changes in demographic and commuting trends are expected to halve this

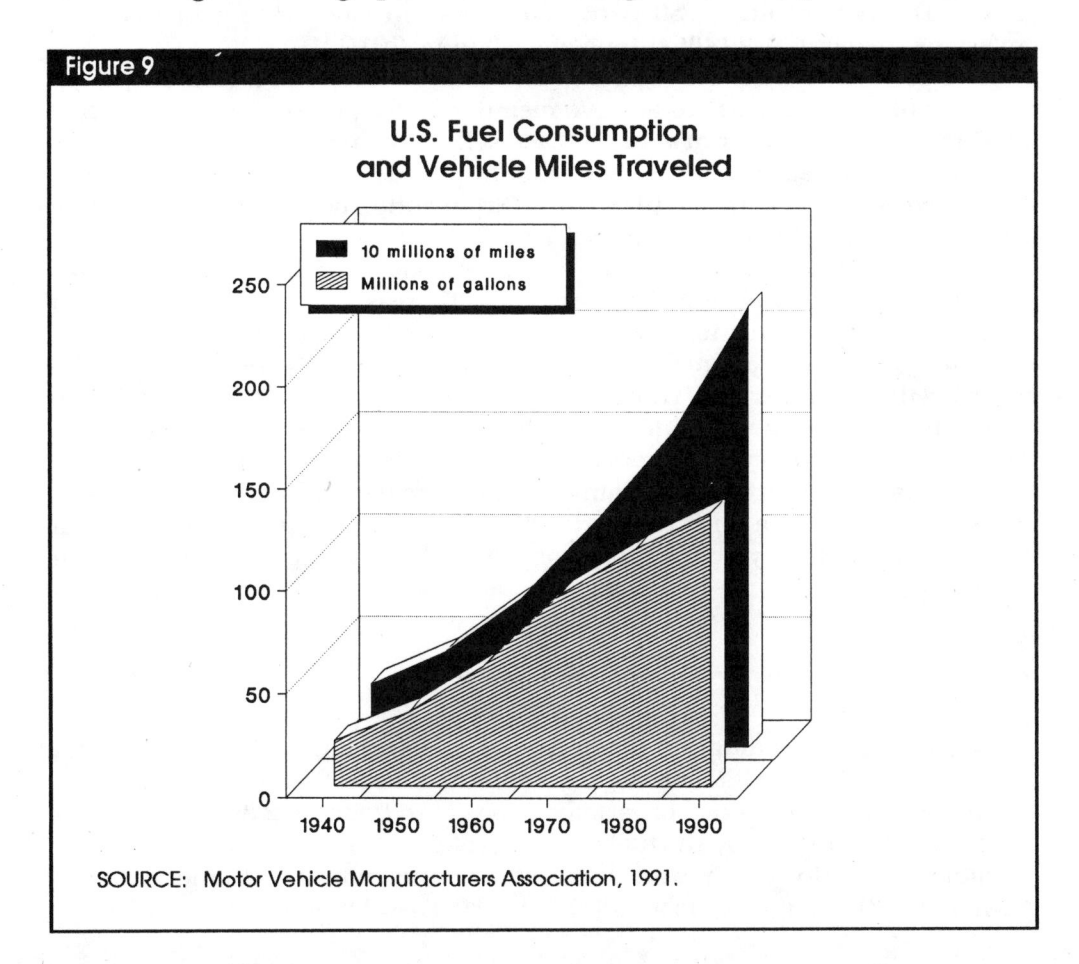

Figure 9

U.S. Fuel Consumption and Vehicle Miles Traveled

- 10 millions of miles
- Millions of gallons

SOURCE: Motor Vehicle Manufacturers Association, 1991.

rate of increase in the future, the number of vehicle miles traveled is likely to grow another 60 percent over the next quarter-century, reaching 3.4 trillion miles by 2015.[183] If the fuel economy of new American cars improves by 10 mpg over the next two decades, to 37 mpg, vehicular CO_2 emissions would still be one-third greater in the year 2015 than now, because of the increase in vehicle miles traveled, according to one analysis by the congressional Office of Technology Assessment.[184] To achieve a 10 percent reduction in such emissions over the next 25 years, the fleet average would in fact have to rise to 55 mpg by 2010—*and* vehicle miles traveled in urban areas (where more than 60 percent of driving occurs) would have to decrease by 5 percent, OTA's analysis concluded.

A report by the World Resources Institute has come to equally somber conclusions about the prospects for curbing vehicular CO_2 emissions worldwide. Since 1970, motor vehicle registrations have increased at an average rate of 16 million a year, equal to a 3 percent annual increase. Also since 1970, average CO_2 emissions per vehicle have fallen nearly 1.5 percent annually because of fuel economy gains. If these countervailing trends continue over the next 20 years, the rise in motor vehicle use would outpace the fuel economy gains. The net result: a 50 percent increase in worldwide vehicular CO_2 emissions, reaching 1.2 billion tons of carbon by 2010.[185]

Trends such as these leave automobile manufacturers in a quandary as to their greenhouse gambit. To achieve meaningful CO_2 reductions, either they must reduce the number of vehicles they sell (and encourage people to drive them fewer miles each year) or they must attain fuel economy improvements that far exceed those of the last 15 years. Either option poses serious economic consequences for the auto industry.

Yet framing the question this way assumes that virtually all motor vehicles will continue to rely on gasoline or diesel fuel. There is a third alternative: Auto manufacturers can turn to fuels that are not fossil-based. While this option raises other serious economic concerns, at least the financial burden would be shared with producers of transportation fuels. And from a greenhouse perspective, auto makers would be able to continue to sell as many vehicles as the market demands—and owners could continue to drive them as many miles as they wish—within the constraints of an increasingly congested roadway network. Therefore, a shift to alternative transportation fuels, derived from nonfossil sources, may be the most appealing option for the auto industry in the long run.

In Search of the Appropriate Fuel

Any thoughtful evaluation of alternative transportation fuels must begin with consideration of what has made gasoline and diesel fuels so popular in the first place. The practical reason is something called their "specific energy"; that is, the energy delivered by the fuel as a ratio of the fuel's weight and that of its container. An Otto engine can extract enough energy from gasoline to lift the gasoline's initial weight almost 1,000 miles.[186] Diesel fuel travels even farther

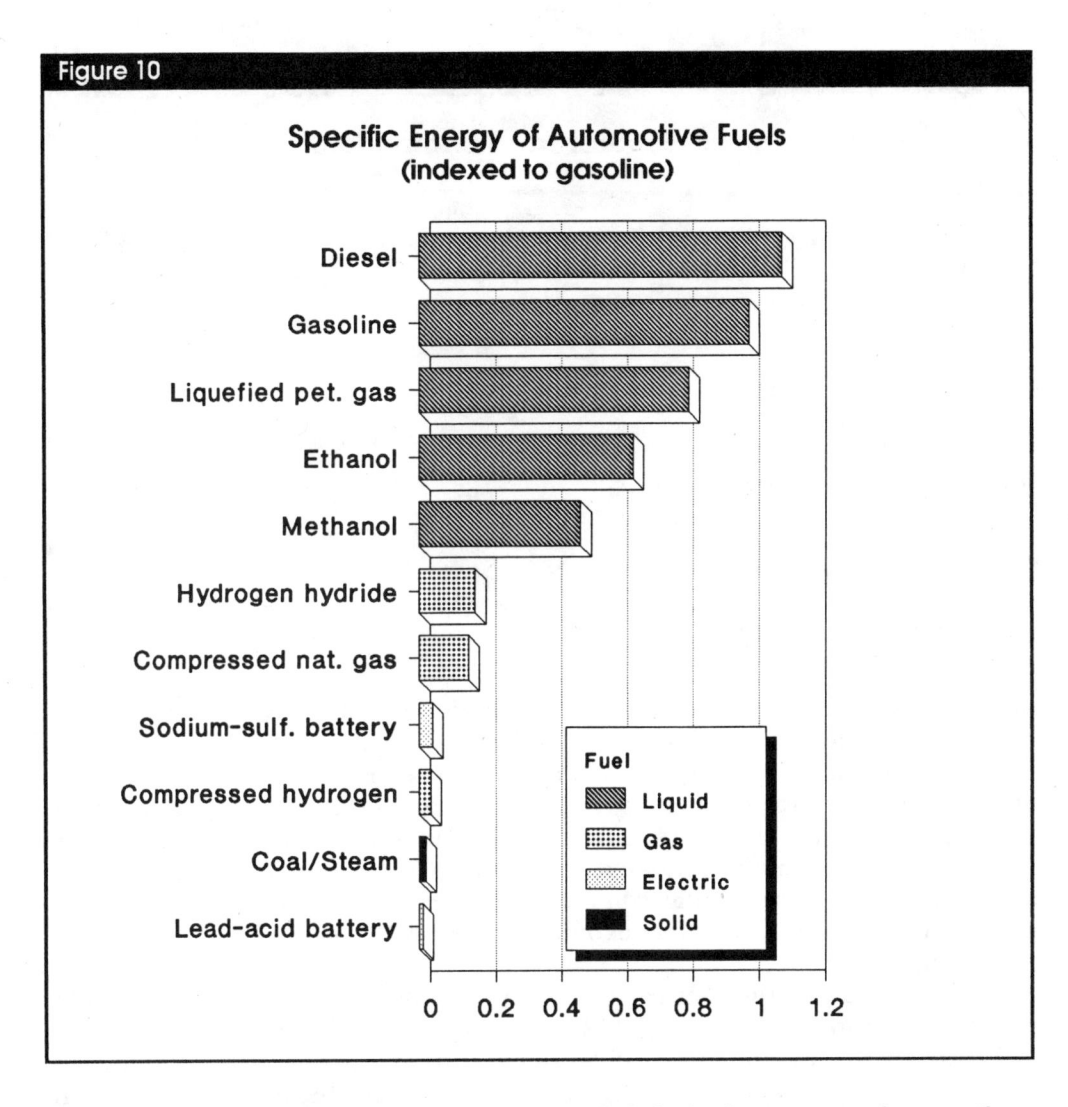

> **Figure 10**
>
> **Specific Energy of Automotive Fuels**
> **(indexed to gasoline)**
>
> | Fuel | |
> |---|---|
> | Diesel | ~1.05 |
> | Gasoline | 1.00 |
> | Liquefied pet. gas | ~0.90 |
> | Ethanol | ~0.65 |
> | Methanol | ~0.50 |
> | Hydrogen hydride | ~0.10 |
> | Compressed nat. gas | ~0.10 |
> | Sodium-sulf. battery | ~0.03 |
> | Compressed hydrogen | ~0.03 |
> | Coal/Steam | ~0.02 |
> | Lead-acid battery | ~0.01 |
>
> Fuel types: Liquid, Gas, Electric, Solid

in relation to its weight because it has a slightly higher energy density than gasoline. By comparison, ethanol fuel has only two-thirds the specific energy of gasoline, and methanol has only one-half as much. At the low end of the scale, a lead acid battery has only 1 percent of gasoline's specific energy. In other words, it delivers enough energy to lift its own weight only 10 miles.

"There are good reasons why the auto industry a long, long time ago went to gasoline and diesel," explains Michael Schwarz, an emissions planning engineer for Ford. "They are good fuels in terms of delivering movement of the vehicle in relation to the weight of that vehicle, so they provide good utility of the vehicle."[187]

Gasoline and diesel fuels have their limitations, however. In today's average car, only one-seventh of the energy in gasoline actually propels it forward. The rest escapes through the tailpipe and radiator, mostly as waste heat. In addition, the chemical composition of gasoline and diesel fuels involves large

Table 2

Vehicle Exhaust Emission Standards

Regulation		Nonmethane organic compounds	Carbon monoxide	Nitrogen oxides
		(grams per mile at 50,000 miles)		
Federal	Current	0.41	3.4	1.0
Clean Air Act	Tier I	0.25	3.4	0.4
	Tier II	0.125	1.7	0.2
California	Conventional	0.25	3.4	0.4
vehicles	Transitional	0.125	3.4	0.4
	Low emission	0.075	3.4	0.2
	Ultra-low emission	0.040	1.7	0.2
	Zero emission	—	—	—

molecules with multiple carbon-carbon bonds. Combustion of these larger molecules requires a complex series of reactions, increasing the probability of incomplete combustion. The carbon soot often seen coming out the tailpipes of diesel cars and trucks offers a vivid illustration of this incomplete combustion. In gasoline-powered cars, unburned (and invisible) hydrocarbons are the main culprit. Most modern exhaust systems capture these molecules, however, and return them to the engine cylinder for more complete combustion—or they convert them to more benign exhaust fumes with the help of a catalytic converter.

To reduce these automotive pollution problems, America's Big Three auto makers announced in June 1992 that they will work jointly to develop technologies to meet tough new clean-air standards.[188] General Motors, Ford and Chrysler will spend more than $1 billion over the next few years on pollution-control innovations, including redesigned engines and catalytic converters. Ford and General Motors already have announced that they will manufacture select 1993 models that emit only half as many non-methane organic gases as the new Clean Air Act allows in 1994. As such, these models will qualify as "transitional low emission vehicles" under California's more stringent regulations.[189]

But a ratcheting of the clean-air standards in California and nearly a dozen other states during the 1990s—and possibly nationwide after the turn of the century—is prompting researchers to look beyond the design of car engines and emission-control systems at the fuels they use. Fuels that burn cleaner than gasoline would have few or no carbon-carbon bonds—and hydrocarbons that are not highly reactive with other smog-forming compounds. Methanol, ethanol and compressed natural gas are so structured. The least-polluting combustible fuel of all would consist of simple molecules and contain no carbon whatsoever.

Hydrogen is such a fuel—the simplest and most common element of the universe. The ultimate fuel source might not even rely on combustion. Electricity stored in batteries or produced in reaction with sunlight offers such a prospect.

A final consideration with the development of alternative transportation fuels is economics. Hydrogen gas and photovoltaic electricity cost much more to manufacture than the other alternatives. Barring a technological breakthrough of some sort, hydrogen's introduction into the auto marketplace is not expected for at least a quarter-century. Electricity's introduction may come sooner, however, because of regulations in California, New York and Massachusetts which will require that a small percentage of the new-car fleet in those states be "zero-polluting" by the end of the century.

Until then, auto companies and oil producers will spend billions of dollars on the development of high-specific-energy fuels that require comparatively modest changes to Otto and diesel engines but that are intended to bring about significant improvements in air quality. These fuels include reformulated gasoline, liquid methanol and ethanol, and compressed natural gas.

Reformulated Gasoline

Petroleum refiners have tinkered with their formulas for making gasoline about as long as car companies have been tinkering with the designs of their engines. The introduction of faster and more powerful cars over the years has required gasoline to withstand higher compression in the engine cylinder before igniting. Premature ignition of the fuel mixture is known as "knocking," which robs the engine of some of its power. To keep engines with ever-higher compression ratios running smoothly, oil companies used to add more tetraethyl lead to gasoline. Octane ratings, which measure a fuel's ability to prevent knocking, rose from the low 60s in the 1920s to the low 90s by the late 1960s because of the increased lead injection.

Then in 1970 the Clean Air Act mandated the removal of lead from gasoline for health-related and environmental reasons. To maintain high octane ratings without using lead, oil companies turned increasingly to butane and aromatic hydrocarbons—primarily benzene, toluene and xylene. Today, aromatics comprise one-third (by volume) of all gasoline sold in the United States, up from 22 percent in 1971. Many premium grades contain as much as 45 percent aromatics.[190] The curious result of substituting butane and aromatics for lead in gasoline was to trade an airborne lead problem for exacerbated problems with urban smog and airborne toxic carcinogens. One important reason for amending the Clean Air Act in 1990, in fact, was to limit the aromatic content of gasoline (and thereby reduce its volatility) and curb toxic air pollution associated with these fuel additives.

The 1990 amendments did not redress another unintended side effect of the Clean Air Act, however; namely a rising level of carbon dioxide from refineries that produce high-octane and reformulated fuels. More raw fuel has been required to process petroleum and natural gas into desired end products as

(continued on p. 290)

> **Box 4-D**
>
> ### Cars, Fuels and the Clean Air Act
>
> Amendments to the Clean Air Act signed into law by President Bush in the fall of 1990 are bringing about important changes in automotive pollution control equipment and the formulation of gasoline. While these measures are sure to improve air quality in the United States, it remains to be seen whether the multibillion-dollar program will have much ameliorative effect on global warming.
>
> The original Clean Air Act of 1970 achieved tremendous reductions in automotive emissions of conventional air pollutants. Installation of catalytic converters has curbed tailpipe emissions of reactive hydrocarbons and carbon monoxide in new vehicles by 96 percent, while per-vehicle emissions of oxides of nitrogen have fallen 78 percent. Airborne levels of lead, meanwhile, have plummeted 90 percent because of the introduction of unleaded gasoline.[1d]
>
> Ozone-related smog, however, has remained a persistent air pollution problem that now ranks among the nation's most serious public health threats. Ninety-eight metropolitan areas—home to 67 million people—exceeded the federal ambient ozone standard at least three times during a three-year period ending in 1990. Ozone pollution is linked to chronic bronchitis, emphysema and other breathing problems. Several health and environmental organizations have petitioned the U.S. Environmental Protection Agency to reduce the federal ozone standard of 120 parts per billion; EPA will decide whether to revise the standard by March 1993.[2d]
>
> Ozone is a greenhouse gas. Under hot and hazy conditions—not unlike what is envisioned for a greenhouse world—tropospheric ozone concentrations can soar to alarmingly high levels. During the steamy summer of 1988, for example, urban smog levels rose 8 percent relative to the previous summer. New York City exceeded the federal ozone standard on 42 occasions—about once every other day. That same summer, Washington, D.C., violated the standard an average of once every three days. And in Los Angeles, where the smog problem is most notorious, ozone concentrations rose to unhealthful levels about once every two days—for the entire year.[3d]
>
> Ozone levels have since improved nationwide, partly because of more favorable weather conditions, but also because of reductions in the volatility of gasoline and turnover of the auto fleet. Since old cars not equipped with catalytic converters emit about 20 times more pollutants than new cars, President Bush unveiled a plan in March 1992 in which consumers living in smog-plagued cities can be paid up to $1,000 to scrap their pre-1980 vehicles. Companies doling out the cash would earn pollution credits.[4d] At the heart of the 1990 Clean Air Act, however, is a requirement that all cars sold in 1994 and later emit about 30 percent fewer reactive hydrocarbons and 60 percent fewer oxides of nitrogen—which combine to form smog—than cars sold today. The act also directs EPA to study the feasibility of further reductions in emissions beginning with the 2003 model year.
>
> The reduction of nitrogen oxides emissions is especially important given recent scientific evidence that suggests the most effective way to control smog in cities with disproportionately high ambient hydrocarbon levels is to lower the concentration of less abundant NO_x (instead of limiting the hydrocarbons directly).[5d] The Clean Air Act amendments also require that vehicles' new emission control equipment last for 100,000 miles, instead of 50,000 miles as is now the case. These measures are expected to raise the cost of most cars $200 to $600 each.[6d]
>
> The amendments also require changes in the formulation of gasoline. In 39 cities where carbon monoxide levels exceed ambient standards, gasoline for sale during

winter months must have an oxygen content of at least 2.7 percent, beginning in November 1992. In January 1995, the nine cities with the worst ozone pollution must begin to sell reformulated gasoline containing at least 2 percent oxygen content and no more than 1 percent benzene year round. Also in 1995, the overall toxic and volatile organic compound content of gasoline must be reduced 15 percent, and then be lowered 25 percent by 2000, using 1990 gasoline as a baseline. Other states may adopt these clean fuel requirements if they choose. The Clean Air Act also requires owners of vehicle fleets (amounting to 10 or more vehicles) to introduce "low emission vehicles" beginning in 1998 if they operate in one of 20 metropolitan areas that are not in compliance with the federal ozone standard. This provision is expected to boost the market for even cleaner-burning reformulated gasolines, methanol, ethanol and compressed natural gas.

Meanwhile, California—the state that pioneered auto emission controls in 1966—is staying one step ahead of federal regulations. By 2003, at least 75 percent of new vehicles sold in that state must be "low emission vehicles" that emit only one-quarter as much hydrocarbons and only one-half as much nitrogen oxides and carbon monoxide as the federal act allows. Another 15 percent of the cars sold in the state must be "ultra-low" emission vehicles, and the remaining 10 percent must be "zero emission vehicles," which effectively will mandate the use of electric motors. With annual sales averaging 1.7 million cars and trucks, California's vehicle market is larger than in all but seven nations. The District of Columbia and 11 states from Maine to Virginia also have adopted elements of California's clean air plan in place of less stringent provisions of the Clean Air Act. As a result, at least one-third of the U.S. auto market is expected to be covered by California's tough regulations.[7d]

Taken together, these clean-air requirements will foster significant capital investments by the auto and oil industries. General Motors reports that it plans to spend $1.3 billion to meet the new Clean Air Act standards as well as the first phase of the new California air quality plan. Bavarian Motor Works says it will allocate 30 percent of its research and development budget in the next four years just to meet the California standards. Japanese auto companies also are investing heavily in new emissions control technology but have not disclosed specific dollar amounts.[8d]

As for the oil industry, the price tag to retrofit refineries and storage systems for reformulated gasoline is expected to reach $20 billion to $30 billion over the decade. Atlantic Richfield reports it is investing $2 billion in its new line of reformulated "emissions control" gasolines.[9d] Amoco also says it is spending $2 billion to produce such compliance-grade gasolines.[10d] The largest share of the industry's investment is in boosting production capacity for methyl tertiary butyl ether (MTBE), an oxygenated fuel additive with an octane rating of 110. (Ethyl tertiary butyl ether, or ETBE, also can serve as an oxygenated fuel additive. It is derived from corn but costs more to produce than MTBE.) As of mid-1991, North American production of MTBE was 5.6 million tons a year. Production is expected to rise to 14 million tons a year by 1995 and possibly reach 22 million tons by the end of the decade.[11d]

The investment in auto pollution control equipment and reformulated gasolines should help clear the air of many chronic pollutants. Ambient levels of carbon monoxide and ozone, as well as emissions of air-toxic chemicals such as benzene, are expected to fall 10 to 25 percent.[12d] But no substantive greenhouse benefit may result from these changes, since any reduction in ozone-forming gases could be offset by increased CO_2 emissions resulting from the processing of reformulated gasoline, which has higher energy requirements than the refining process for ordinary gasoline.[13d]

formulas for making gasoline become increasingly complex. The refining process for high-octane unleaded gasoline consumes up to twice as many barrels of crude oil per production run as regular unleaded gasoline, for example. But carbon dioxide emissions need not increase, in a direct sense, as long as markets exist for all of the petroleum products derived from each barrel of crude.[191] Highly specialized formulas for reformulated gasoline, however, can create production imbalances and process inefficiencies that may result in less than the whole barrel being effectively utilized. Adding to the CO_2 emissions problem, substantial amounts of process energy are necessary in the distillation and/or catalytic reformation of unrefined fuel into environmentally friendly products. Therefore, refineries for "cleaner burning" gasoline emit more CO_2 into the atmosphere than refineries for ordinary, "dirty" gasoline.[192]

At least nine major oil companies now offer reformulated gasolines, usually in premium grades only. Soon the Clean Air Act will mandate the use of reformulated gasoline in all cars in selected cities. While there is no pat recipe for making reformulated gasoline, most use methyl tertiary butyl ether (MTBE) in place of some of the aromatics to maintain a high octane rating while reducing the fuel's volatility. Beyond raising process energy requirements at refineries, however, blending MTBE or other oxygenated additives (such as ethyl tertiary butyl ether, ETBE, or tertiary butyl alcohols, TBA) creates a slight fuel economy penalty in cars.[193] Fortunately, the hydrogen-carbon ratio also tends to be higher in reformulated gasolines relative to ordinary gasolines, so carbon dioxide emissions out of the tailpipe are about the same for each tank of fuel consumed.[194]

An important strategic question posed by reformulated gasolines (not to mention the low-polluting vehicles introduced by Ford and General Motors) is the extent to which their introduction will stave off the development of alternative transportation fuels. Atlantic Richfield—the first to offer an "emissions-control" gasoline in 1989—now is working on a new formula, EC-X, that could reduce the smog-forming "reactivity" of automobile exhaust by one-third.[195] If the preliminary test results hold up, EC-X could burn about as cleanly as a methanol blend known as M85. And since the manufacture of EC-X involves no proprietary technology, all gasoline marketers could sell their own version of it eventually. A $20 billion investment may be required over the next decade to retool refineries to make reformulated gasolines. Gasoline prices may rise at least 15 cents a gallon, and perhaps much more, to cover the added production costs.[196]

The long-term benefits of such a commitment to reformulated gasolines are decidedly mixed, however. While air quality should improve in cities where reformulated gasolines are introduced, the improvement may not be sufficient in some urban areas that face especially chronic air-pollution problems. Moreover, successful introduction of reformulated gasolines would extend the nation's dependence on oil; this is presumably the reason why so many oil companies are anxious to develop them in the first place.

If other fuels are even more expensive to develop—and lower the utility of motor vehicles while raising their operating costs—reformulated gasoline may be embraced as an economical and environmentally sound fuel for the future.

Figure 11

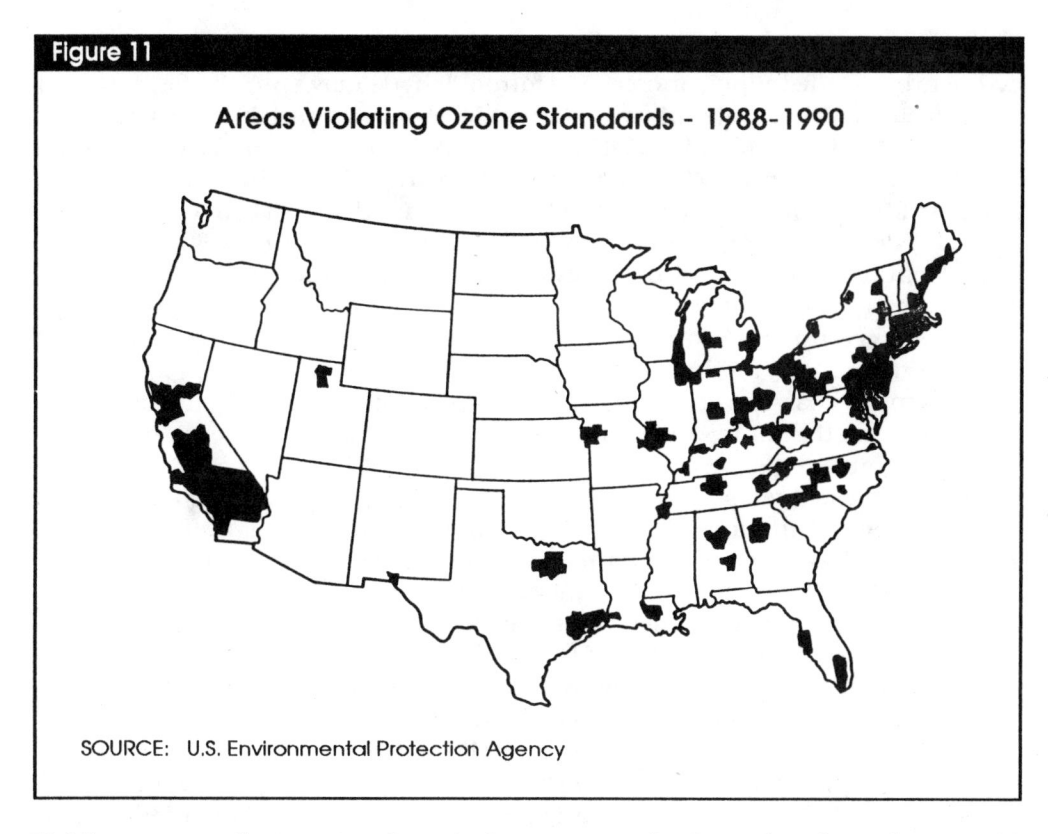

Areas Violating Ozone Standards - 1988-1990

SOURCE: U.S. Environmental Protection Agency

Yet from a greenhouse standpoint, the success of reformulated gasoline would be counted as one step forward and two steps backward. Although conventional air pollution problems would be addressed—perhaps more cost-effectively than using alternative means—carbon dioxide emissions would increase at the nation's refineries *and* the day of reckoning for a switch to nonfossil fuels would be postponed. Despite many virtues, reformulated gasoline has relatively little to offer as a greenhouse gambit.

Methanol

Before the introduction of reformulated gasolines, methanol was regarded as the leading candidate to clean up the air in smog-ridden cities. Methanol is an alcohol fuel that evaporates more slowly and burns more completely than conventional gasoline, so it emits fewer volatile hydrocarbons. Methanol also burns at a lower temperature than gasoline, thereby reducing emissions of nitrogen oxides. As an added benefit, methanol greatly reduces toxic air emissions of benzene and other suspected carcinogens, principally 1,3-butadiene and polycyclic organic compounds, when used in place of gasoline.

Despite these environmentally friendly characteristics, doubts have persisted about methanol's ultimate smog-reducing potential. The principal limitation of the fuel is that it increases tailpipe emissions of formaldehyde, a toxic and volatile gas that reacts with other hydrocarbons in the atmosphere to form ozone. While burning methanol itself reduces emissions of reactive hydrocarbons, the tailpipe increase in formaldehyde may spur the formation of smog in cities where ambient hydrocarbon levels already are high. As a result, the use of M85 fuel (a blend of methanol and 15 percent gasoline) may decrease ozone-forming potential 20 to 40 percent in some places, but increase it as much as 20 percent in others. Vehicles that run on pure M100 fuel would do a much better job of reducing ozone-forming potential—up to 80 percent in some locations—but would be of little or no smog-reducing benefit in areas where controlling NO_x is considered a more effective ozone-control strategy.[197]

From a greenhouse standpoint, the pivotal question is the feedstock used to make methanol. Methanol may be synthesized from natural gas, coal or biomass materials. The only manufacturing process that is cost-effective today is to derive methanol from natural gas (principally using a steam reforming method). Natural gas typically contains only about half the carbon found in gasoline or diesel fuel, so there is a CO_2 emissions benefit in using natural gas as a feedstock. But since methanol itself has only half the specific energy of gasoline, it takes about twice as much methanol as gasoline to go the same distance. The net result is that CO_2 emissions from natural gas-derived methanol, including production of the fuel, would be about the same as with conventional gasoline—or perhaps slightly less if the vehicle's engine were redesigned to take full advantage of an efficiency gain offered by use of methanol fuel.

If methanol were derived from coal, net CO_2 emissions would increase—perhaps substantially—since coal is a much more carbon-intensive feedstock than natural gas. Theoretically, methanol also could be derived from wood—and reduce net CO_2 emissions dramatically. But cost presents a formidable barrier to either of these feedstock options. According to the National Research Council, producing methanol from coal, using current technology, would not be economically feasible until oil prices reached $50 a barrel. Similarly, production of methanol from biomass would not be feasible until oil prices reached $70 a barrel.[198]

Barring a technological breakthrough in the conversion of biomass to methanol, this alcohol fuel holds relatively little promise for a greenhouse world. Even so, the market potential of natural gas-derived methanol cannot be overlooked, since its price at the pump would be on a par with premium unleaded gasoline, and it could reduce smog formation and CO_2 emissions—at least on the margin. The real greenhouse risk in committing to methanol is that producers might eventually turn to coal as a feedstock, assuming that wellhead prices for natural gas rise along with increased demand for the fuel. As a rule of thumb, each dollar increase in the wellhead price of natural gas (per million Btus) raises the production cost of methanol by 10 cents a gallon.[199]

Enough surplus natural gas is available around the world at present to make large amounts of methanol fuel for perhaps a few decades. Unfortunately, little of this surplus gas is found in the United States. In fact, 25 percent of the nation's methanol supply already is imported because domestic sources of natural gas are being put to higher-valued uses, such as the production of commercial formaldehyde and MTBE, the critical ingredient of reformulated gasoline. Even if the domestic methanol supply were devoted exclusively to motor fuels, it would substitute for barely 1 percent of existing U.S. gasoline and diesel fuel demand. Consequently, substantial new production capacity is essential for methanol to make any real contribution to the transportation sector.

Since natural gas supplies in the lower 48 states are not considered an economically viable feedstock for large increments of new methanol supply, production capacity would have to be built elsewhere. The North Slope of Alaska, Siberia and the Middle East are considered possible locations for large-scale methanol production, since these regions have very large reserves of natural gas yet no pipelines connected to ready markets. By some estimates, OPEC nations and former Eastern Bloc nations would be likely to provide 75 percent of an expanded methanol fuel supply market.[200]

If the United States sought to develop its own indigenous methanol fuel supply, it could rely on the nation's abundant coal resources. But in addition to the high economic costs associated with making methanol from coal, the conversion process would require two gallons of water for each gallon of finished liquid fuel. As such, large-scale production would put tremendous pressure on water resources, especially in the arid West. Moreover, in the event of global warming, it would compound farmers' difficulties in securing affordable irrigation water. And worst of all from a global warming standpoint, it would mean that the world's largest consumer of gasoline would be switching over to a motor fuel that emits even more CO_2 than gasoline.

Setting aside global warming concerns, there are other reasons to proceed cautiously with methanol fuel development. Methanol is a colorless, odorless, tasteless and water-soluble liquid. In the event of leaks from underground storage tanks, methanol would be likely to contaminate the groundwater, yet detection would be extremely difficult. When ingested or absorbed through the skin, methanol can cause blindness or other serious injury, since it has twice the toxicity of gasoline. In addition, pure methanol burns with an invisible flame, which is one reason it is usually blended with 15 percent gasoline. (Another reason is that methanol evaporates only at temperatures above 52 degrees F, which makes cold-weather starting difficult; liquid methanol will not ignite.) A bad-tasting additive could be put in methanol fuel to reduce the potential for accidental ingestion. Chemists also are looking for an appropriate trace additive to make its flame visible.[201]

Considering that CO_2 emissions would hardly be reduced if natural gas-derived methanol were substituted for gasoline—and that coal-derived methanol would increase CO_2 emissions substantially—the long-term value of the methanol option is called into question. Investors may be especially wary since expanding methanol production is a high-cost proposition. A large-scale methanol plant with 1.2 billion gallons of yearly production capacity costs more than $1 billion to build.[202] Since most new production plants are likely to be built overseas, however, at least these investment costs would not be borne domestically.

Other costs would hit closer to home. Because methanol is corrosive to some of the materials normally used to store and distribute gasoline, changes would have to be made to gas stations, pipelines, tanker trucks, ocean-going tankers and the like. The Petroleum Manufacturers Association of America estimates, for example, that new stainless steel underground storage tanks would have to be installed at a cost of $30,000 to $45,000 at each service station wanting to sell methanol fuel.[203] Fewer than 100 service stations have M85 fuel for sale at present, and three-quarters of these are in California.[204] To displace 1 million barrels a day of U.S. gasoline demand with M85, $4.1 billion would have to be spent by 91,000 service stations across the country, with another $700 million required for modifications at other fuel storage, shipping and handling facilities, according to the U.S. Department of Energy.[205]

Methanol-Fueled Vehicles

Auto makers also would face added costs when converting vehicles to run on methanol fuel. Methanol's corrosiveness may require the use of a stainless steel gas tank as well as teflon hoses and other resistant materials that come in contact with the fuel. The fuel tank also has to be extra large, since methanol contains only half the specific energy of gasoline. And since methanol (unlike gasoline) conducts electricity, fuel injectors and other engine components that carry voltage must be insulated. In mass production, the Department of Energy estimates that flexible-fueled vehicles would cost an extra $275 to build. The added cost at present is around $1,000.[206]

In March 1992, General Motors became the first auto company to offer flexible-fuel methanol vehicles for sale to the public.[207] Its Chevrolet division is building between 2,000 and 4,000 Luminas in the current model year, mostly for sale in California. A microprocessor built into the cars will detect the concentration of gasoline and methanol being supplied to the cars' fuel injectors and then make the necessary adjustments to maintain engine performance. The cars will be priced $2,000 higher than the regular Lumina, but the State of California is offering credits of up to $2,000 per car for fleet buyers. Most of the flexible-fueled Luminas are likely to be sold to corporations for use in fleets that refuel at central garages. Previously, General Motors provided 2,220 Chevrolet Corsicas and Luminas to the State of California in 1989 and 1990 that are capable of running on methanol, ethanol and unleaded gasoline.[208]

Chrysler began delivery of 2,100 Dodge Spirit and Plymouth Acclaim flexible-fuel vehicles in July 1992.[209] Unlike General Motors, Chrysler is offering these 1993 model-year cars to fleet buyers and retail customers at no extra charge. (Fleet buyers in California will still be eligible for the $2,000 rebate offered by the state.) In 1990, Chrysler became the first auto maker to announce consideration of large-scale production plans for methanol-fueled vehicles. It said it might build up to 100,000 flexible-fuel cars for the 1993 model year, although it has since pushed back that target date to 1994 and 1995 model-year cars.[210]

Ford also is in the race to introduce flexible-fuel vehicles. It plans to manufacture 2,500 flexible-fuel Tauruses after October 1992, following production of 250 in the 1991 model year and 200 flexible-fuel Econoline vans in the 1992 model year.[211] Ford has been testing methanol-fueled vehicles for more than a decade. Its test fleet includes 630 methanol-dedicated vehicles that have been in operation in the United States and Canada dating back to 1981.

But according to Roberta Nichols, who until recently headed up Ford's flexible-fuel vehicle development team, "It's safe to say from this point forward that all of the methanol vehicles that we build for demonstration fleets will be flexible-fuel vehicles. We think it's the only technology that makes sense for getting from the old fuel to the new fuel."[212]

Congress also has offered an inducement for the auto makers to introduce flexible-fuel vehicles. In rating the fuel economy of the auto fleets, flexible-fuel vehicles are granted bonus mileage credits (even though their actual mileage falls when they burn methanol or ethanol). A flexible-fuel car rated at 25 miles per gallon of gasoline, for example, is counted as a 40-mpg vehicle under the CAFE law. If an auto maker puts a sufficient quantity of flexible-fuel vehicles on the road, it may boost its CAFE rating by up to 1.2 miles per gallon. Accordingly, a car company whose fleet average is 26.3 mpg could remain in technical compliance with CAFE, even though the standard is 27.5 mpg—*and even though drivers may run their flexible-fuel vehicles solely on gasoline*. Such a provision benefits the auto makers, yet it confounds efforts to limit gasoline consumption and, hence, CO_2 emissions into the atmosphere.

In the heavy engine market, Detroit Diesel has built the first methanol-fueled engine ever certified to meet both federal clean air standards and California's more stringent regulations.[213] The modified Series 92 engine, used mainly in transit buses, emits just one-third of the hydrocarbons and carbon monoxide allowed under current federal law, and it eliminates almost all of the sooty particulates associated with older diesel buses. Development of the 277-horsepower, methanol-fueled engine is considered an important step toward the 1994 federal standard for diesel engines, which requires an 80 percent reduction in hydrocarbon emissions. As of August 1991, Detroit Diesel had received orders for about 250 of the modified Series 92 engines.[214] (Detroit Diesel is 80 percent owned by entrepreneur Roger Penske's Penske Corp. and 20 percent owned by General Motors.)

Prospects for the heavy engine market notwithstanding, methanol fuel faces an uncertain future in the auto industry. Recent developmental success with reformulated gasolines is causing oil companies, car makers and even government regulators to reconsider the value of committing to the methanol option. One telling sign was a decision in April 1992 by Chevron—one of the largest gasoline marketers in California—to suspend its programs to install methanol pumps at selected service stations in the Golden State. In announcing its decision, Chevron cited uncertainties about methanol's impacts on air quality, competition from reformulated gasolines and extremely low demand for methanol fuel. Sales have averaged only 18 gallons a day at each of 10 Chevron stations offering methanol fuel in California.[215]

From a greenhouse standpoint, the demise of methanol may be just as well. While auto makers emphasize that methanol from biomass could serve as an environmentally desirable fuel option in the long run, many environmental advocates believe that research efforts would be better directed elsewhere. James MacKenzie, a senior associate with the World Resources Institute, commented at a congressional hearing on alternative fuels in 1989 that switching to fossil-derived methanol "would have few if any positive impacts on urban smog levels, carbon dioxide emissions, national security or our balance of payments problems. Moreover, [it] would divert precious time and resources into paths that would ultimately prove to be dead ends because of global warming."[216]

The remaining alternative fuels to be discussed in this section offer greater potential benefits for a greenhouse world.

Ethanol

Biomass-derived ethanol (ethyl alcohol) eludes some of the criticisms directed at its methanol counterpart, largely because it is derived from a renewable resource. In a greenhouse world, ethanol as a replacement for gasoline could reduce CO_2 emissions dramatically—provided that the ethanol feedstock were grown, harvested and distilled with a minimum of fossil energy inputs. That is not how the ethanol production process works today, however, so ethanol's greenhouse benefits are currently negligible.

About 95 percent of American ethanol is derived from corn in an extremely energy-intensive process. The process begins with farmers who use heavy farm machinery to till the soil, plant the corn and harvest the crop, and who spray their fields with energy-intensive nitrogen fertilizer, pesticides and irrigation water. Then diesel trucks and locomotives transport the grain to processing centers, where the corn is converted to a sugar solution. Finally, ethanol is separated from water in numerous evaporation-condensation cycles—with coal usually serving as the distillation fuel. By the time ethanol is ready to enter a car's gas tank, it has created as much CO_2 (and possibly more) as occurs during the refining and burning of conventional gasoline.[217]

"Gasohol" fuel, a blend of 10 percent ethanol and 90 percent gasoline, has created a niche market for ethanol producers. In the United States, about 350 million bushels of corn—representing about 4 percent of the domestic crop—is converted into 100 percent alcohol for blending into gasoline each year. Use of gasohol in cities such as Denver has been especially helpful in curbing carbon monoxide pollution. Altogether, gasohol accounts for about 8 percent of the U.S. fuel supply—including some 920 million gallons of ethanol consumption a year.[218]

Despite the clamor for cleaner-burning transportation fuels, a surplus of domestic ethanol production capacity exists. As of 1988, 113 distillation plants had a combined capacity to process 1.2 billion gallons of ethanol a year. Seventy percent of this capacity was in Iowa and Illinois, and 70 plants (representing 15 percent of the production capacity) were idle.[219] The capital cost of building a new ethanol plant has been estimated at $143 million for 52 million gallons of annual production capacity.[220] The full cost of production from a new ethanol plant ranges from $0.85 to $1.50 a gallon, compared with wholesale gasoline prices of about $0.55 a gallon.[221] In addition, the specific energy of ethanol is about one-third less than that of gasoline, so the costs are even higher on a gasoline-equivalent basis.

Clearly, today's ethanol industry could not survive without massive government subsidies. The federal government and about one-third of the states partially exempt gasohol from gasoline taxes. The exemption from the federal tax yields a subsidy of 60 cents per gallon of ethanol (or 6 cents per gallon at the pump, since gasohol contains 10 percent ethanol). This exemption deprives the Highway Trust Fund, the main repository for gasoline taxes, of up to $1 billion a year of revenue. Archer-Daniels-Midland of Decatur, Ill., is the prime corporate beneficiary of the ethanol subsidy, since it controls about two-thirds of the U.S. ethanol market.

Despite the economic barriers confronting ethanol, its prospects have brightened with passage of the 1990 amendments of the Clean Air Act. By some accounts, mandatory use of oxygenated fuels to combat carbon monoxide problems in 39 American cities could increase demand for ethanol-derived ethyl tertiary butyl ether (ETBE) to more than 3 billion gallons of ethanol a year by 2000—or nearly three times current ethanol production capabilities. While

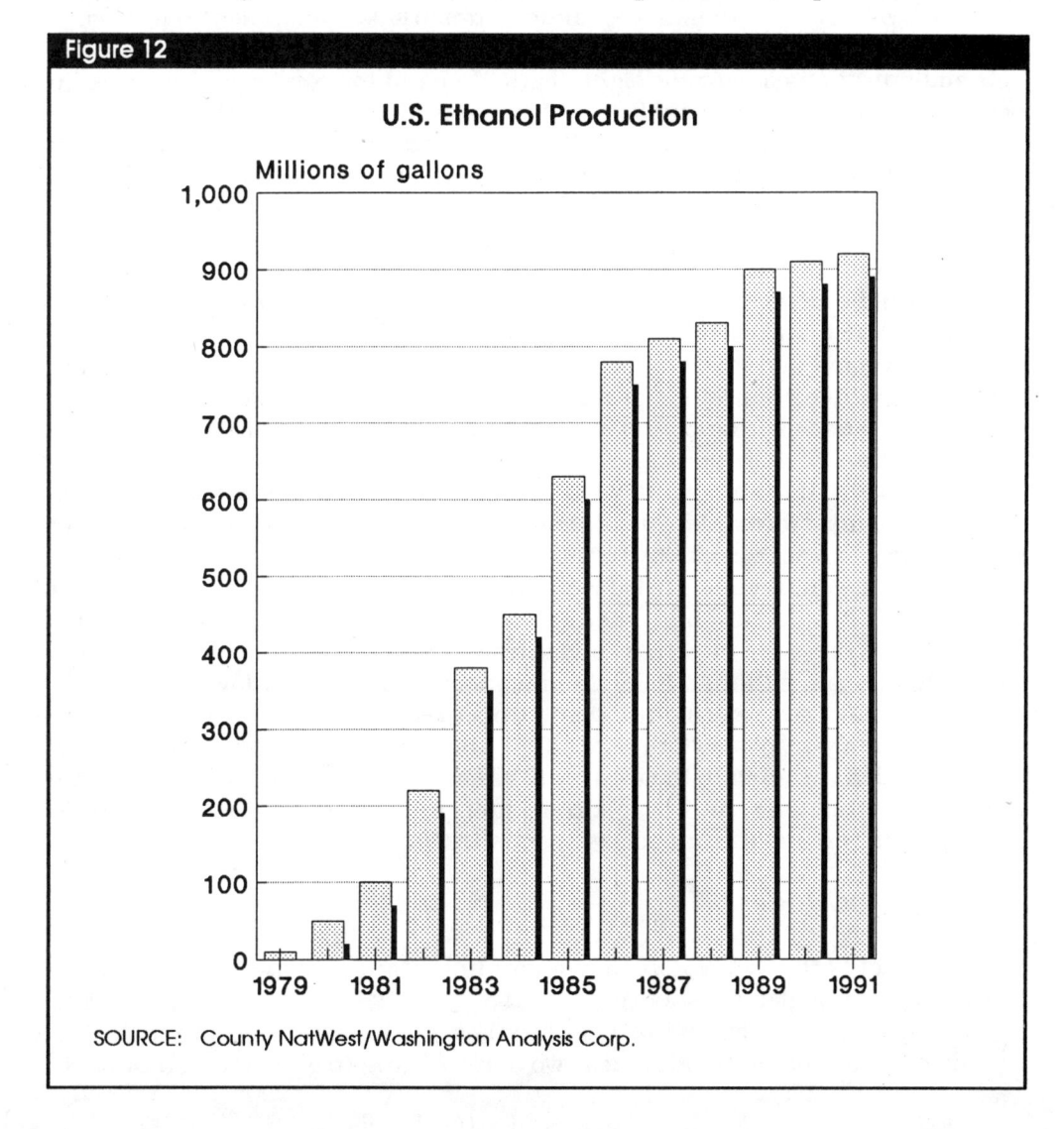

Figure 12

U.S. Ethanol Production

Millions of gallons

SOURCE: County NatWest/Washington Analysis Corp.

ethanol is effective in reducing wintertime levels of carbon monoxide, however, it also has the potential to increase summertime levels of ozone—and herein lies a problem. Ethanol increases the rate at which gasoline evaporates in gasohol mixtures. These fumes in turn release volatile organic compounds to the atmosphere, which react in the presence of sunlight during hot-weather months to form ozone, the principal ingredient of smog. Consequently, EPA is having second thoughts about the eligibility of ethanol for year-round use in smog-plagued cities, putting many ethanol expansion plans on hold.[222]

Even if ethanol supporters are successful in obtaining a waiver from EPA under the Clean Air Act, $2.3 billion in capital outlays would be required to raise ethanol production capacity to 3 billion gallons a year, according to the National Corn Growers Association.[223] A similar analysis by the U.S. General Accounting Office found that tripling corn-derived ethanol production would cost anywhere from $1.2 billion to $7 billion, depending on market development patterns.[224] Under a high-growth scenario, in which production capacity expands by nearly 300 million gallons a year for eight years, demand for corn would increase 6 percent and corn prices would rise 15 percent, GAO calculated. This would result in still higher ethanol production costs and lead to a further increase in ethanol subsidies and possibly higher costs for consumers. U.S. government farm support payments would be substantially reduced, however, since corn farmers would have an important new commercial market for their crop.

If the number for flexible-fuel vehicles also grows in the years ahead, a market could emerge to burn ethanol fuel directly (instead of turning it into gasohol or ETBE). Pure ethanol has less than one-quarter of the evaporative pressure of conventional gasoline (and about half that of methanol), resulting in sharply reduced emissions of reactive vapors. Emissions analysts at Ford estimate that use of ethanol fuel in place of gasoline could reduce transportation-related ozone pollution up to 20 percent.[225]

Ethanol also has an advantage over methanol in that it creates no tailpipe emissions of toxic formaldehyde. Incomplete combustion of ethanol does lead to formation of another aldehyde, however, known as acetaldehyde. A suspected carcinogen, acetaldehyde fosters the formation of ozone and could offset much of the smog-reducing potential of ethanol itself. Acetaldehyde also increases the presence of peroxycetyl nitrate (PAN) in the atmosphere. PAN is a greenhouse gas, although its ambient concentration is very low; it is toxic to plants and may cause eye irritation and upper respiratory problems. Improvements in catalytic converters may limit the amount of acetaldehyde that ethanol-fueled cars emit in the future. Ethanol also has cold-starting problems to overcome, however, and it absorbs water and dislodges silt, which could contaminate the fuel if transported in existing pipelines.[226]

Ethanol Feedstocks

If ethanol is to emerge someday as an important transportation fuel, the thorny policy question must be addressed of growing crops as fuel for cars, not just as food for livestock and people. It takes one bushel of corn to make 2.5 gallons of ethanol fuel, and it takes two acres of land to grow enough bushels of corn to run an ethanol-fueled automobile for a year. With proper government incentives, up to 5 billion bushels of corn could be grown for ethanol distillation purposes in the United States each year—on top of the 7.5 billion bushels of corn already grown as grain. But even such a huge increase in corn acreage would enable ethanol to replace only 13 percent of the nation's gasoline supply.[227] For that matter, the entire world corn crop, if made available for ethanol fuel conversion, would offset only one-sixth of global gasoline demand—and these calculations do not account for the tremendous amount of energy that would be required to distill the fuel.[228]

The bottom line is that corn-derived ethanol has limited potential to displace gasoline, especially in a world whose population is expected to double by the middle of the next century. Moreover, expansion of the corn crop probably would occur on lands already considered marginal for agriculture and increase the chance of crop failure in the event of adverse weather conditions. In a greenhouse world, more frequent droughts, heat waves and pest attacks could undermine the very effort to supplant gasoline with corn-derived ethanol, and such an undertaking might prove futile in a policy sense anyway, considering the tremendous CO_2 emissions associated with the production process.

Corn is not the only feedstock available to make ethanol, however. In Brazil—the world's leading producer and consumer of ethanol—sugar cane is used as a feedstock. Brazil's ethanol program began in 1975 with help from a $300 million World Bank loan; the goal was to supplant gasoline imports with a home-grown fuel source. Today, Brazil boasts 4.5 million ethanol-fueled cars and another 9 million vehicles that run on a mixture of 22 percent ethanol and 78 percent gasoline. General Motors, Ford, Fiat, Saab, Volvo, Mercedes-Benz and Volkswagen all have considerable experience building ethanol-powered vehicles for sale in Brazil.[229]

Changes to these ethanol-fueled vehicles are relatively minor. Their fuel lines and ignition systems have been modified, because ethanol fuel is somewhat corrosive, and their gas tanks have been enlarged because a gallon of ethanol fuel takes a car only two-thirds as far as a gallon of gasoline. In addition, the air/fuel ratio in the ethanol fuel injection system has been reduced to 9:1, compared with 14.5:1 in gasoline-powered cars.

Despite the proven performance of ethanol-fueled cars, however, Brazil's sugar-cane-to-ethanol program has been widely criticized by economists and environmentalists alike. Brazil's goal of improving its trade balance notwithstanding, the World Bank estimated in 1989 that government subsidies to the ethanol program generated a $2.7 billion annual deficit because of inefficient production methods and lower-than-expected oil prices.[230] Another ironic twist is that sugar cane fields are set ablaze before harvest to prevent workers from cutting themselves on the cane's razor-sharp leaves. The scorched area—extending over 10 percent of Brazil's agricultural lands—produces thousands upon thousands of tons of carbon dioxide that contribute to the global buildup of greenhouse gases.[231]

A more environmentally friendly way to produce ethanol fuel may be on the horizon. Cellulose from short-rotation herbaceous or woody crops (such as poplar, maple, cottonwood and other hardwood trees) could serve as an ethanol feedstock rather than conventional foodstuffs. The National Renewable Energy Laboratory in Golden, Colo., estimates that biomass production from wood and agricultural wastes—along with increased ethanol conversion efficiencies—already has reduced the estimated market price for wood ethanol from $3.60 a gallon in 1979 to $1.35 a gallon as of 1990. The government's goal is to decrease the subsidy-free cost of ethanol from cellulosic feedstock to 60 cents a gallon (90 cents a gallon of gasoline-equivalent) by the end of the decade.[232]

Ethanol advocate Harry P. Gregor, professor emeritus of chemistry at Columbia University, is even more optimistic about the prospects for wood ethanol. With the cost of cellulose a mere 2 cents a pound—and 14 pounds of feedstock required for each gallon of ethanol—the raw materials cost is only 28 cents a gallon, by Gregor's calculations. If permeable selective membranes made from synthetic plastics are employed, the cost of converting the cellulosic raw material to ethanol could be as low as 15 cents a gallon, Gregor estimates. That would make the total manufacturing cost for wood ethanol only 43 cents a gallon (or 56 cents a gallon of gasoline-equivalent).[233]

Best of all from a greenhouse standpoint, the energy input required for ethanol production from wood is, at most, one-third the amount required for the production of corn. If nonfossil energy sources were used to grow the wood and to convert the cellulose to ethanol, the ultimate CO_2 reduction would be 100 percent relative to burning gasoline. Even if some fossil energy inputs were included, wood ethanol would contribute only one-fifth as much CO_2 to the atmosphere as gasoline, according to biomass fuel proponents.[234] With the government already encouraging tree-planting for soil conservation and carbon sequestration purposes, the prospect of raising fast-growing trees on marginal cropland further enhances the potential for such cost-effective, environmentally sound ethanol fuel production. Even so, it may be a decade or more before such alternative means of producing ethanol are demonstrated on a commercial scale, and some remain skeptical that the optimistic cost projections for wood ethanol will be realized.[235]

Compressed Natural Gas

One of the reasons that methanol and ethanol are considered leading alternatives to gasoline is that they are liquid fuels. Compressed natural gas (CNG) is a more unconventional fuel by comparison. It requires post-factory vehicle modifications of $1,500 to $2,500 to pay for a gas-air mixer, pressure regulators, extra fuel tanks and other components.[236] But CNG fuel itself is cheaper to burn than gasoline—selling for 30 to 40 cents less than a gallon of gasoline-equivalent—and such a price advantage has spurred the development of CNG vehicles in niche markets in recent years.[237]

Italy has 300,000 CNG vehicles in operation; Russia and other former Soviet republics have 200,000; New Zealand, 130,000; Australia, 100,000; and Canada, 15,000.[238] In the United States, approximately 30,000 vehicles are equipped to run on compressed natural gas; most are owned by natural gas utilities. Typically, CNG vehicles are bi-fueled sedans, vans, school buses and light trucks with bolt-on systems supplied by after-market retrofitters. If the vehicles were designed at the factory to run exclusively on CNG, added costs per vehicle could drop to only $700 or $800, an up-front expense that could be recovered through fuel-cost savings over the life of the vehicle.[239]

Although compressed natural gas is a fossil fuel, it consists mainly of methane gas, which has one-third less carbon by weight than oil. Combustion of CNG therefore results in fewer CO_2 emissions than combustion of gasoline. Mile for mile, most CNG vehicles emit 0 to 15 percent less CO_2 than gasoline-powered vehicles.[240] If dedicated vehicles were built to optimize the use of CNG, greater CO_2 reductions might be possible.[241] It is essential, however, that few, if any, leaks occur during the transport of natural gas and the refueling of CNG vehicles—and that exhaust emissions of methane are held to a minimum—since methane has more than 20 times the global warming potential of CO_2. Even a small amount of leakage could wipe out the greenhouse benefits of using CNG as a transportation fuel.

As for conventional pollutants, CNG is considered a low emissions fuel because methane contains virtually no nitrogen or sulfur, and CNG's exhaust hydrocarbon emissions are nonreactive and do not contribute to ozone formation. Carbon monoxide emissions are reduced as well, since CNG vehicles start more easily when the engine is cold. (Gasoline-powered cars must enrich the fuel before the engine warms up, a process that increases carbon monoxide emissions.) CNG fuel does produce more "engine-out" emissions of nitrogen oxides than gasoline; however, a catalytic converter is able to clean most of this exhaust—provided that the engine is not run in a lean-burn mode. There is an incentive to "run lean" with CNG because the engine's efficiency is optimized and emissions of carbon monoxide and nonmethane hydrocarbons are reduced. But power is reduced as well (in addition to the problem of NO_x control), so lean-burn CNG vehicles may have difficulty gaining market and regulatory acceptance.[242]

Two other obstacles for CNG vehicles are the need for extra on-board storage tanks and a lack of CNG refueling stations. A CNG vehicle requires four times the fuel storage space of an ordinary car, and even then it usually travels only half as far between refuelings. Increasing the cruising range of a CNG vehicle adds weight and cost, while reducing space and performance. The heavy, pressurized storage containers also put a drag on fuel economy, which increases CO_2 emissions in turn. In the future, use of fiberglass-wrapped aluminum storage tanks could reduce the added weight of a CNG vehicle with a 300 mile cruising range to only 150 pounds over a conventional gasoline system.[243] Alternatively, liquefied natural gas (LNG) could be used to cut the space requirement in half. But liquefying natural gas increases production and delivery costs dramatically, and new LNG terminals would have to be built for bulk imports of the fuel.[244]

A broader strategic question is the availability of natural gas for use as a transportation fuel. As a rule of thumb, a CNG vehicle consumes about as much fuel in a year as a home heated with natural gas. Since compressed natural gas, unlike methanol, does not require expensive processing to be turned into a vehicle fuel, domestic gas sources could be used to augment the supply. While estimates vary, the current domestic surplus may exceed 1 trillion cubic feet per year—enough to power 25 million CNG automobiles.[245] Most U.S. supply forecasts project rising imports of natural gas during the 1990s and beyond, however, with *no* increase in natural gas usage by the domestic transportation sector. Accordingly, the amount of natural gas required to power a large U.S. fleet of CNG vehicles is almost certain to lead to even greater imports of the fuel.

CNG Fleet Vehicles

In the near term, compressed natural gas is best suited for use in fleet vehicles, especially buses, trucks and vans with access to central refueling stations. The Clean Air Act boosted the prospects for CNG in light-duty fleet vehicles (weighing 6,000 pounds or more) by requiring owners of 10 or more such vehicles in 20 polluted cities to begin switching over to alternative fuels in 1998. Thirty percent of the fleet vehicles purchased in 1998 must run on a "clean fuel"; the percentage rises to 50 percent in 1999 and 70 percent in 2000. Several states have instituted similar clean-fuel requirements. The State of Texas, for one, has passed a law requiring many state agencies, school districts, public transit authorities and even private companies to buy only alternative-fueled vehicles for their fleets after Sept. 1, 1991, and to convert at least half of their existing fleets by 1996.[246]

Some fleet-vehicle owners already have extensive experience with compressed natural gas. Federal Express has converted 70 percent of its Canadian fleet, some 550 vehicles, to run on CNG since 1982, as a result of a government program that offers tax incentives to switch away from gasoline and diesel fuel. So far, the company says, maintenance and reliability of the vehicles has been comparable to that of its gasoline-powered vehicles.[247] In the Los Angeles area, United Parcel Service has converted 20 of its delivery trucks to run on compressed natural gas, and in Oklahoma (which has also passed legislation to encourage the use of clean fuels), UPS is building two refueling stations in Oklahoma City to fuel as many as 140 of its delivery vehicles with CNG.[248] UPS's expectation is that the CNG delivery trucks will reduce emissions of ozone-forming pollutants by two-thirds compared with UPS's gasoline-powered trucks. If its test program in Los Angeles goes smoothly, UPS may convert the rest of its Los Angeles fleet of more than 2,000 trucks to run on compressed natural gas.

Compressed natural gas also appears to be well-suited for the heavy engine market. Cummins Engine has converted 110 of its L-10 diesel bus engines to run on CNG. The fuel-injected, spark-ignited engines burn so cleanly that they do not require a three-way catalytic converter to control emissions.[249] In addition, Tecogen Inc. of Waltham, Mass., has installed redesigned V-8 engines from General Motors in 10 California school buses. By the end of 1992, Tecogen estimates, demand for the reconfigured engines could reach several thousand nationwide—including 1,000 for California buses alone.[250] Ford has also pledged to begin producing trucks in 1992 that run on compressed natural gas.

The fastest growing market for CNG vehicles appears to be in vans and pickup trucks. In 1992, General Motors is building a fleet of 2,000 CNG pickup trucks for sale in selected GMC Truck Division dealerships in California, Texas, Colorado and elsewhere, with particular emphasis on fleet buyers. These pickups have three slender fiberglass-wrapped cylinders, each five feet long, tucked along their frame, giving the trucks a driving range of 150 to 200 miles. A consortium of 10 natural gas utilities provided co-funding for nearly $1 million of design and production costs associated with the vehicles. Later versions of GM's redesigned pickup trucks may have dedicated CNG engines with increased compression ratios to restore lost power, and altered combustion chambers to reduce wear on valves and pistons.[251]

In 1991, T. Boone Pickens announced that Mesa Limited Partnership, a Texas oil and gas company that Pickens controls, would invest $2 million to acquire a 70 percent stake in Clean Fuels Co. of Martinsburg, W.Va., which designs and develops car engines that burn natural gas.[252] In addition, Chrysler and the Gas Research Institute entered into an $850,000 joint project in May 1991 to determine the feasibility of introducing a CNG-powered Chrysler van for the 1994 model year.[253]

As for petroleum companies, Amoco became the first major gasoline retailer to offer compressed natural gas at four of its service stations in the Denver area in 1990. To get the gas to these stations, Amoco simply tapped into the local natural gas company's existing distribution system. Public Service Co. of Colorado—which has converted 230 of its own utility vehicles to run on CNG—spent up to $250,000 to retrofit each of the Amoco stations, with compressor equipment accounting for most of the expense.[254] (A high-speed pump cuts average refueling time of a CNG vehicle to two to five minutes.) Amoco has since installed a CNG tank and compressor at a service station on Capitol Hill in Washington, D.C., in a joint research project with Washington Gas Light Co.[255]

One of the reasons Amoco is promoting CNG is that it has more natural gas reserves than any other American company. But it is not the only gasoline retailer to outfit some of its stations with CNG pumps. Chevron and Shell Oil are installing natural gas pumps at a limited number of service stations in northern and central California. Pacific Gas and Electric, the giant California utility, is providing natural gas from its transmission lines, sharing in the costs of installing the pumps and sharing in any profits that are generated. The utility eventually expects $15 million to $20 million in annual profit from its agreement with Shell alone.[256]

The U.S. Department of Energy projects that a retail service station with four CNG pumps—each capable of filling a vehicle's CNG fuel tank in eight minutes and handling 75 vehicles per day—would cost $320,000 to build, including land acquisition costs. To displace a million barrels a day of gasoline demand would require a total capital investment of $7.6 billion at service stations, according to DOE's cost projections, plus another $1 billion to $2 billion to improve local gas distribution systems.[257] The resulting 11 percent displacement of U.S. gasoline demand by compressed natural gas should be within the bounds of what North American gas producers are able to supply.

Hydrogen

Expanded use of compressed natural gas could pave the way eventually for another gas that has been touted as the environmental fuel of choice for the 21st century: hydrogen. Like compressed natural gas, hydrogen as a transportation fuel would entail the use of large-volume storage tanks and compressors, only slightly modified gasoline and diesel engines, and gas pipeline distribution sysyems.[258] Hydrogen is unique, however, because of its simple chemistry.

Hydrogen is the most basic and abundant element in the universe. When burned, hydrogen poses no problems with carbon dioxide, carbon monoxide or reactive hydrocarbons, since there is no carbon in the fuel to begin with. In fact, hydrogen's only significant combustion by-product is water vapor that forms as it fuses with oxygen. Some oxides of nitrogen also result from the combination of atmospheric nitrogen and oxygen at high engine temperatures. But with exhaust gas recirculation, the quantities of NO_x produced by hydrogen vehicles should be far below those emitted by gasoline vehicles equipped with three-way catalytic converters.[259]

At present, hydrogen is manufactured mainly for industrial purposes, such as petroleum refining, manufacturing ammonia, methanol and fertilizer, and growing silicon crystals. The United States accounts for about half of global hydrogen consumption, and only 4 percent is used directly as fuel.[260] The engines on America's space shuttles are powered by liquid hydrogen fuel. Derived from natural gas—the cheapest feedstock—this hydrogen still costs 4.4 times more than gasoline.[261] And in terms of greenhouse gas emissions, hydrogen production from natural gas yields nearly 25 percent more CO_2 emissions than direct combustion of gasoline.[262] (Coal-derived hydrogen would produce even greater CO_2 emissions.) Accordingly, the challenge with hydrogen fuel is finding a means of reducing it to elemental form without producing copious greenhouse emissions in the process.

The solution appears to be splitting hydrogen apart from oxygen in water, in a process known as "electrolysis." At room temperature, a voltage greater than 1.23 volts (the electrical equivalent of the energy that binds hydrogen and oxygen together) will separate water into its constituent parts, with bubbles of pure hydrogen and oxygen roiling to the surface. Since collecting the pure hydrogen is easy, the hard part is finding a non-carbon energy source to provide the electric current for electrolysis to take place.

The European Community is exploring the possibility of using surplus hydroelectric power in Quebec to make the hydrogen and then transporting the fuel across the Atlantic in tankers.[263] A more practical long-term solution may be to use photovoltaics that convert sunlight to electricity. A 15 percent efficient amorphous silicon cell, built at 40 cents a peak watt, would be able to manufacture hydrogen fuel for about $2 a gallon-equivalent, according to a study by the World Resources Institute. A solar collector field 240 miles in diameter (covering 7 percent of U.S. desert land) could provide enough hydrogen fuel to equal the entire oil production of the United States, the institute found.[264]

While this analysis assumes further technological breakthroughs in photovoltaic technology and may underestimate the land requirements, the possibility remains that a land area smaller than the size of Arizona could produce enough hydrogen fuel to supply all of U.S. transportation requirements during the 21st century.[265] (Of course, collector fields could be highly dispersed rather than all in one place.) The Saudi Arabian government, in fact, has teamed up with German engineers to erect an experimental solar-hydrogen facility near Riyadh. The "Hysolar" project consists of a 350-kilowatt photovoltaic array coupled to an electrolysis plant that produces hydrogen. A similar plant—with 500 kilowatts of photovoltaic capacity—is operating in southern Germany. (BMW, the German auto maker, is a partner in the German solar-hydrogen plant.)[266]

Storing Hydrogen Fuel

Even if a practical means of making hydrogen is found, other obstacles remain to turning it into an automotive fuel. In a gaseous state, pure hydrogen takes up nearly 13 times as much room as gasoline per unit of energy and nearly four times as much room as compressed natural gas. To store the energy equivalent of 20 gallons of gasoline as hydrogen gas, storage tanks weighing 3,000 pounds would be required—taking up nearly three times the cargo space normally found in a car's trunk. Hydrogen fuel can be stored more compactly in liquid form, but it still takes four gallons of liquid hydrogen to go as far as on one gallon of gasoline. In addition, liquid hydrogen must be maintained at a temperature of -423 degrees F or it will boil away. The bulky refrigeration unit on such a liquid hydrogen vehicle would take up additional space and draw nearly as much electrical power as the energy required to operate an ordinary car.[267]

Another option is to store the hydrogen energy chemically in metal "hydride" tanks. These tanks absorb pressurized hydrogen like a sponge and release it when heated. Such hydride systems, consisting of powdered metal alloys such as titanium-iron and magnesium-nickel, weigh about 800 pounds and take up only one-third as much room as hydrogen gas cylinders (for the same amount of energy storage). In existing vehicle prototypes, hydride tanks hold about 3.5 percent hydrogen by weight—the equivalent of four gallons of gasoline—giving the vehicles a driving range of about 75 miles. The goal of ongoing research is to at least double this storage ratio to 7 percent.[268]

Daimler-Benz is among several companies pursuing metal hydride research. This German auto giant has dabbled in hydrogen fuel research since 1973, having road-tested about 20 converted station wagons and vans.[269] The station wagons are dual-fuel vehicles that run on gasoline at cruising speeds and on hydrogen when idling. The vans run on hydrogen only. In the fall of 1991, Daimler-Benz ran a full-page advertisement in several major American newspapers espousing its commitment to hydrogen fuel research. Tucked within the photograph of a cresting ocean wave, the question is posed, "Hydrogen gained from water through electrolysis: The energy carrier of the future?"

Daimler-Benz is not alone in attempting to answer this question. BMW has built several cars fueled by liquid hydrogen and plans to test a fleet of about 100 hydrogen-powered vehicles in the next couple of years.[270] Peugeot and Renault of France also are developing a hydrogen-powered car prototype as part of a government-supported "clean car" program.[271] Mazda unveiled the latest entry in the hydrogen field at the 1991 Tokyo Auto Show. Its prototype HR-X (Hydrogen Rotary Experimental) coupe runs on hydrogen gas that is stored in a rectangular fuel tank and fed into a rotary engine. Mazda billed it as the "ultimate engine of the 21st century."[272]

American auto makers have not built any hydrogen-fueled prototypes. "Hydrogen will just not be available at cost and volume that is appropriate for motor transportation for 20 or 30 years," asserts Nicholas Gallopoulas, who heads the engine research department at General Motors. Similarly, Donald Buist, Ford's director of automotive emissions and fuel economy office, estimates that gasoline prices will need to reach $4 a gallon before hydrogen becomes cost-competitive. "[We're] looking at it, sure, but it is not on the front burner," he says.[273]

Fuel cells: Ultimately, the best application of hydrogen fuel may be in fuel cells that generate an electric current chemically rather than mechanically. Fuel cells run best and most economically on pure hydrogen. (Methanol and natural gas are other options.) The process essentially is electrolysis working in reverse. As hydrogen and oxygen atoms bond together in the presence of an electrolyte, they give off an electric charge that can be transmitted and collected on a series of metal plates. The power derived from such a fuel cell could be used to spin an electric motor in a car.

The biggest obstacle with fuel cells is their high cost. America's manned space program has depended on expensive fuel cell technology since the time of the lunar missions, but more affordable commercial fuel cell prototypes for stationary electric-generating applications have appeared only recently. The installed generating cost for fuel cells has fallen to $2,000 to $2,500 per kilowatt, which is about 25 percent higher than the installed cost of a coal-fired plant outfitted with the latest pollution control equipment. Leading fuel cell developers include United Technologies, Westinghouse, and Energy Research Corp. in the United States; Mitsubishi Heavy Industries and Toshiba in Japan; and Siemens and ABB Group in Europe.[274]

Within the automotive sector, smaller companies are taking the lead. Energy Partners of West Palm Beach, Fla., is building a "proof of concept car" that runs on two hydrogen fuel cells made out of stack plates of platinum-coated Teflon. The platinum serves as a catalyst that splits the hydrogen fuel and causes electrons to flow. Energy Partners hopes to demonstrate the feasibility of a small, lightweight fuel cell for motor vehicles by the end of the decade.[275] Meanwhile, at the nonprofit American Academy of Science in Independence, Mo., a similar kind of fuel cell is undergoing testing that someday could be used in automobiles. In mass production, the $3,500 cell would be about the size of a television set and give an electric car a 300 mile range. It could also be plugged into an electric outlet and run in reverse to make hydrogen fuel, to be stored in hydrides.[276] If further testing is successful, these fuel cells could be in mass production for use in electric vehicles sometime after the turn of the century.

Electricity

While hydrogen fuel cells constitute the ultimate clean-burning, internal power plant for motor vehicles of the 21st century, other kinds of electric-powered automobiles will be on the road before this decade is out. California's new clean air standards, now being emulated in New York and Massachusetts, require the introduction of "zero emission vehicles" (ZEVs) by 1998. (Similar regulations have been proposed in New Jersey.)[277] Since electric-powered vehicles have no need for a tailpipe, they are the only ones now able to meet a zero emissions standard.

In 1998, states adopting California's clean air standards will require that 2 percent of the vehicles sold be ZEVs. This amounts to at least 25,000 vehicles, since the states already committed to the plan command about 13 percent of the U.S. auto market. By 2003, the proportion of ZEVs to total vehicle sales is to rise to 10 percent, or approximately 130,000 vehicles, in these states.

The emerging market for electric cars could bring automotive technology back to where it started. At the turn of the 20th century, battery-powered electric cars exceeded gasoline-powered cars in number by nearly a two-to-one margin, and an electric car held the land speed record.[278] In 1904, another electric car—the Baker Torpedo—became the first automobile to go faster than 100 miles an hour. Ironically, battery technology also led to the demise of the electric car, when automotive engineers figured out how to wire a lead-acid battery to a gasoline engine, allowing drivers to start their cars with a key instead of a crank. Electric cars disappeared from the landscape just about as quickly as gasoline service stations appeared.[279]

Today, electric vehicles may be on the verge of a renaissance. In 1990, for the first time in more than half a century, a major car company offered an electric car for sale. Fiat of Italy introduced the Panda Elettra, a two-passenger car that travels at speeds up to 65 mph and has a cruising range of 45 to 65 miles. With a sticker price of $20,000, however, Fiat expects to sell only 300 to 500 Elettras a year.[280] But the European market potentially is much larger. "We aim to introduce several million electric driven cars by the end of this century," European Community Transport Commissioner Karel van Mierthe announced at a 1989 gathering of electric transport experts in Bruges, Belgium.[281]

Several European auto companies are scrambling to get electric cars on the road. Peugeot is lending prototype electric versions of its 205 model on long-term trials to French companies. Bavarian Motor Works has introduced the E1, which it hails as the first "pure blooded" BMW electric car, and the E2, a larger sedan that reaches speeds of 75 miles an hour and has a 267-mile maximum cruising range.[282] Volkswagen is working on a hybrid car, the Chico, which has a small electric motor and a two-cylinder gasoline engine. The batteries give the Chico a cruising range of 12 to 13 miles of nonpolluting city driving, while the gasoline engine extends the range to about 300 miles and increases the car's top speed to 81 mph. In mass production, the Chico could sell for only $7,000, but it is not expected to be on the market until the mid-1990s.[283] In the meantime, Volkswagen's Audi unit is building a limited

Table 3

CO_2 Emissions from Alternatively Fueled Vehicles

Fuel and feedstock	Percent change from gasoline
Methanol:	
M100, improved vehicle and best coal conversion technology	+25 to +30
M85, current natural gas conversion technology	-2
M100, improved vehicle and gas conversion technology	-17
M100, derived from woody biomass	-70
Ethanol:	
Ethanol from corn using coal for process heat	-10 to +30
Synthetic natural gas from woody biomass conversion technology	-80 to -60
Natural gas:	
Compressed natural gas from domestic sources	-15 to 0
Compressed natural gas with optimized vehicle	-30 to -15
Hydrogen:	
Hydrogen derived from natural gas using steam-reforming method	+25
Hydrogen derived from solar electrolysis	-100 to -85
Electricity: (battery charging)	
Coal-fired plant	+5
Year 2000 power mix	-19
Current power mix	-23
Conventional gas turbine	-30
Best gas turbine	-45
Nuclear plant (total fuel cycle)	-91
Hydro or solar power	-100

SOURCE: M. DeLuchi and D. Sperling, University of California at Davis, 1990 estimates; and Investor Responsibility Research Center.

number of its Quattro cars with a $16,000 hybrid power system; it has an internal combustion engine for front-wheel drive and an electric battery for rear-wheel drive.[284]

Across the Pacific, Nissan is showing off its prototype FEV, or Future Electric Vehicle, which accelerates as fast as comparable gasoline-powered cars and travels up to 100 miles before recharging. What is especially remarkable about this car is that its batteries are rechargeable in just 15 minutes, compared with a minimum of three hours for most other batteries. But the quick recharge is possible only at 440 volts, which is twice the voltage capacity available in most American homes.[285] Also in Japan, Tokyo Power and Electric, the largest privately owned utility in the world, has designed the Iza, a four-seat passenger car with a top speed of 109 mph and a record-breaking cruising range of 340 miles. Because of its high cost, however, the Iza is not scheduled for production.[286]

The most talked-about electric car of all has been built by America's very own General Motors. The Impact prototype electric car is a tear-dropped two-seater that GM says could enter mass production within the next several years. While the company has not announced a production schedule for the Impact, it has selected an assembly plant in Lansing, Mich., that is capable of producing 20,000 cars a year.[287] "Our goal is to be the first automobile company since the early days of the auto industry to mass-produce an electric car," GM Chairman Roger Smith pledged in an address before the National Press Club in April 1990. "We are taking a major step toward helping our country meet its transportation needs and environmental goals."[288]

CO_2 and other emissions: That electric cars could do much to improve air quality is without question. One study conducted by the California Air Resources Board concluded that electric vehicles driven in the Los Angeles area would emit 99 percent less carbon monoxide and 98 percent fewer hydrocarbons than the gasoline-powered vehicles they replace.[289] An important consideration, however, is what power source is used to charge the batteries of electric vehicles. Presumably, emissions from these stationary generating sources would increase as emissions from mobile sources go down.

Even evaluated on this basis, the preliminary results are encouraging. Pollution emitted from the tailpipes of cars with internal combustion engines is far greater per unit of output than from the smokestacks of central-station generating plants. One analysis by the Electric Power Research Institute concluded that carbon monoxide and volatile organic compound emissions would be reduced 99 percent, and emissions of oxides of nitrogen would be cut 33 percent, if the current mix of power plants in the United States were used to recharge electric batteries.[290] Since more than half of the nation's electricity generation comes from coal-fired power plants, however, sulfur dioxide emissions would be likely to increase substantially until old plants operating without scrubbers are retrofitted or retired.

The biggest question from a global warming standpoint, of course, is what effect switching from gasoline-powered to electric-powered vehicles would have on carbon dioxide emissions. Here again, the results depend on the mix of generating sources used to recharge the batteries. Since nuclear and hydro-

Box 4-E

Assessing the 'Impact'

Electric car enthusiasts are keeping a close eye on General Motors' Impact, a battery-powered car that could enter production as early as 1994. The Impact offers the kind of performance that most electric cars have lacked. Like a sports car, it can accelerate from 0 to 60 miles an hour in eight seconds—faster than the Mazda Miata or the Nissan 300ZX. The Impact has a sleek, aerodynamic exterior, a smooth undercarriage, super-inflated Goodyear tires (with half the rolling resistance of ordinary tires) and lightweight aluminum sheels made by Alcoa. These and other design features have reduced aerodynamic drag on the Impact to half the level of most cars on the road today. As such, the Impact also is a very energy-efficient car, consuming only one-third as much energy (in the form of electricity) as a conventional gasoline-powered car.[1e]

With an 870-pound lead-acid battery pack (developed by GM's Delco Remy unit), the Impact feeds power into two induction-type motors attached at the front wheels. The motors provide the Impact with an impressive top speed of 100 mph and a cruising range of 124 miles at 55 mph. Despite these sporty features, General Motors is touting the Impact as an "ideal second car for doing errands close to home."[2e]

The big question is when the Impact will hit dealer showrooms. In the late 1970s, General Motors officials vowed to make electric cars available by the mid-1980s, by which time it was thought that gasoline prices would be topping $2 a gallon. Ten percent of GM's total car fleet was to have been converted to electric power by 1990. But when the price of gasoline failed to rise in line with projections, GM put its plans for electric vehicles in neutral. Now the company pledges to bring the Impact into production when it is convinced that consumer demand exists for 100,000 vehicles or so.[3e]

Much work must be done before the Impact can be brought to market. While the Impact has been designed to meet government safety standards, it has yet to be crash-tested—as only a few prototypes exist. It is expected that side-panel protection will have to be added to production models, increasing the car's weight and rolling resistance. Moreover, the Impact has been driven mainly under test track conditions. Its super-inflated tires—pressurized to 65 pounds per square inch—would provide a rough ride on ordinary roads and not corner well in turns. Its enclosed belly pan may also make it difficult to service the undercarriage.[4e]

The Impact's other roadblock is cost, although the economic hurdle is not as high as one might suspect. The expected capital outlay for an Impact is around $20,000—similar to other cars in its class. The monthly electric bill would be only $5 to $12 for an Impact that is driven 10,000 miles a year. (Recharging can be accomplished in two hours using an ordinary wall socket in a home's garage.) By comparison, fuel and engine servicing averages more than $50 a month for a gasoline-powered car driven 10,000 miles.

The economic downside of the Impact is that its battery pack would need to be replaced at a cost of $1,500 every 20,000 miles. (Some battery experts are skeptical the batteries would last even this long.) This need for battery replenishment raises the calculation of the Impact's operating expense to $800 or more a year, compared with less than $700 a year for a similar gasoline-powered car. General Motors says that if it can extend the batteries' life to 50,000 miles, there would be no economic penalty associated with owning an Impact.[5e]

electric plants (which do not produce carbon dioxide) already are committed to baseload power operation, coal plants generally would be next in line to provide for night-time recharging of electric vehicles. The CO_2 consequences of this are not as severe as one might think, however. According to University of California researcher Mark DeLuchi, net CO_2 would increase only 5 percent if gasoline-powered cars ran instead on batteries charged by coal-fired power plants.[291]

While this is a relatively even trade-off in terms of CO_2 emissions, it would bolster the load factor of many electric utilities, since some of their baseload plants are idle at night. One analysis of the Southern California Edison service area calculated that 1 million electric vehicles, each traveling 15,000 miles a year, would raise the utility's load factor by 4 percent—without leading to any additional need for new power plants. Nationwide, the Electric Power Research Institute estimates that enough idle baseload capacity exists to charge 50 million electric vehicles nightly without adding to existing baseload construction plans.[292]

Obviously, the more power electric vehicles derive from non-coal sources, the greater the reduction in carbon dioxide emissions relative to burning gasoline. If the current mix of U.S. power production were used (70 percent of which is generated by coal, oil and natural gas), net CO_2 emissions would drop 23 percent, according to the study by University of California researchers. If only the most efficient gas turbines were used, the CO_2 decline would be nearly 50 percent. And if nuclear or solar sources were used exclusively, CO_2 emissions from the electric motor vehicles and their power sources would be virtually eliminated.[293]

Prospects for Electric Vehicles

Someday, electric vehicles may not have to rely on external sources at all to recharge their batteries. Photovoltaic panels stitched into the cars' exteriors could do the job directly, although many consider this possibility remote. Specially designed solar vehicles have traveled across the Australian and North American continents, traveling 2,000 miles on the energy-equivalent of five gallons of gasoline. General Motors' Sunraycer, for instance, moved across the Australian outback in 1987 at an average speed of 41.6 mph—powered only by 9,500 solar cells.[294] Several aerodynamic design innovations of this low-slung experimental machine have since found their way into GM's Impact prototype electric car. As for solar cells, a few luxury cars now sport roof-mounted photovoltaic panels that operate fans to keep their interiors cool when parked in the sun.

However they get their charge, battery-powered vehicles are well-suited for a world that is increasingly electrified and with road conditions more prone to stop-and-go traffic conditions. In a battery-powered car, electricity is consumed only when the vehicle moves forward; no fuel is wasted idling in traffic or when stopped at a light. Power is even returned to the battery when the brakes are applied, as the car's momentum is used to spin the generator in reverse.

Most of today's electric vehicles are not cars, however, but battery-powered vans owned by electric utilities and other commercial users. Each of the Big

Three auto makers have electric vans on the road. While the van market overall is comparatively small, production of the Big Three's best-selling vans—the GM Vandura, Ford Econoline and Chrysler Ram models—does total about 450,000 vehicles a year. Vehma International, a developer of a General Motors-based electric van, estimates that since half of these vans are used for commercial purposes, and half of those could substitute electric battery packs for gasoline engines, there is a market for 110,000 electric vans a year.[295]

GM electric van: General Motors' "G-Van" is an electrified version of its full-size Vandura model designed by the Electric Power Research Institute. The G-Van is equipped with 36 lead-acid batteries (manufactured by Chloride EV Systems of Great Britain) and weighs 8,600 pounds—not including its 1,550 pound payload. It has a 60 mile cruising range and a top speed of 52 mph. Market analysis suggests that its payload and driving range would allow the electric G-Van to perform the tasks assigned to almost 40 percent of all conventional fleet vans.[296] But the G-Van sells for $32,000 (including $7,000 to $8,000 for lead-acid battery packs that must be replaced every 30,000 miles), which is double the price of a gasoline-powered Vandura. In 1991, Vehma International developed 100 electric-powered G-Vans in conjunction with the Electric Power Research Institute for use by electric utilities in more than two dozen cities.[297]

Ford electric vans: Ford is testing two kinds of electric vans. The ETX is based on Ford's Aerostar model. It features a sodium-sulfur battery pack (designed by ABB Group) that costs between $5,000 and $10,000 and that should last for the life of the vehicle. The ETX prototype has a top speed of 65 mph, a cruising range of 100 miles and a half-ton payload.[298] Ford's other electric van, based on the Escort minivan, has a top speed of 70 mph, a cruising range of 100 miles and a payload of 750 to 1,000 pounds (only half the payload of the gasoline-powered model). The sodium-sulfur battery pack itself weighs 1,200 pounds in the ETX and Escort. To offset some of this weight, the Escort van has an alternating-current drivetrain that weighs only half as much as a conventional direct-current drivetrain. In 1991, Ford announced plans to build up to 100 electric-powered Escort vans by 1993 for testing in the United States and Europe. Mass production could begin following completion of this demonstration program.[299]

Chrysler electric van: Chrysler's entry in the electric van market is the TEVan, based on its Dodge Caravan and Plymouth Voyager minivans. Although the TEVan is still a prototype, Chrysler has signed an agreement with the Electric Power Research Institute that may lead to commercial production of the vehicle.[300] Chrysler will build a limited number of the TEVans beginning in September 1992. With a half-ton payload, the TEVan has a top speed of 65 mph and a cruising range of 120 miles. It is outfitted with a nickel-iron battery pack manufactured by Eagle-Picher Industries. The batteries are lighter than standard lead-acid batteries yet store about the same amount of energy. They are capable of 1,200 or more deep discharge cycles up to 100,000 miles. As with the Ford and General Motors electric vans, the TEVan is charged on a 220-volt circuit—the kind used in households to run electric ovens and central air conditioning units. Drawing 40 amperes of power over eight hours for a full recharge, the TEVan consumes about the same amount of electricity as would running every electrical device in a typical house for eight hours.

Electric passenger cars: For those interested in an electric-powered passenger car, options in the United States are few and far between. One small California company called Solar Electric Engineering has been taking gasoline engines out of new cars and replacing them with electric motors and lead-acid battery packs for more than a decade. Sticker prices for these cars range up to $25,000, but the cars' range is only 60 miles between charges. Solar Electric Engineering had just 23 orders during the 1980s. Orders picked up to 28 cars in 1990 alone, however, perhaps indicative of renewed interest in electric vehicles.[301]

The LA 301: To spur the development of dedicated electric passenger cars, the City of Los Angeles (in conjunction with Southern California Edison and the Los Angeles Department of Water and Power) sponsored a competition in 1989 in which the city pledged $7 million toward supporting the development of a new electric car. No bids were received from Detroit's Big Three or Japanese auto makers. Of 18 proposals that were submitted, Los Angeles accepted one from Clean Air Transport Svenska AB, a new Swedish firm whose management team has helped develop electric prototypes in Europe. Its winning design—the LA301—is a four-passenger, three-door hatchback model equipped with 16 lead-acid batteries that can take the car to speeds of 70 mph. Like the cars built by Solar Electric Engineering, the LA301 will sell for about $25,000 and be capable of traveling only 60 miles before recharging. (A full recharge will take eight hours.)

With a small, auxiliary gasoline-powered engine, however, the LA301's cruising range can be extended to 145 miles. But exhaust from that engine will disqualify the hybrid LA301 as the kind of zero emissions vehicle that California wants introduced by 1998. Clean Air Transport hopes to raise $30 million from American investors to supplement the $7 million committed by the City of Los Angeles and $4 million in equity from Swedish and British investors. Despite a shoestring budget, production of the LA301 is expected to begin in late 1992, with sales in Los Angeles commencing in early 1993.[302] A production run of 1,000 vehicles is expected in 1993, increasing to 5,000 vehicles in 1994 and 1995.[303]

Batteries

The bane of all electric vehicles is their batteries. Batteries lose their charges far more quickly than ordinary cars run out of gas, and they also have a problem supplying power to things outside the drivetrain. A car's heater and air conditioner, for example, normally run off the engine. If batteries run these devices instead, they will lose their charge much more rapidly.[304] Most types of batteries also need periodic replacement, whereas conventional motors do not. Therefore, without further breakthroughs in the cost and performance of batteries, electric vehicles will never achieve their full potential.[305]

In January 1991, the Big Three auto makers and the U.S. Department of Energy formed the United States Advanced Battery Consortium. The collabo-

> ## Box 4-F
>
> ### Battery Research
>
> Battery researchers are focusing on various combinations of minerals in an effort to make batteries that perform better in vehicles than the ordinary lead-acid battery. Well-researched combinations include nickel-iron, nickel-cadmium and sodium-sulfur batteries. Other battery designs include nickel metal hydride and several combinations involving lithium. Because none of these batteries is in commercial production, cost and performance estimates are speculative.
>
> ***Nickel-iron:*** Nickel-iron batteries (such as those used in Chrysler's TEVan) are light and durable, and they stand up to the stress of deep discharging and recharging much better than lead-acid batteries. They also last three to four times as long as lead-acid batteries, but they cost four or five times more than lead-acid batteries per unit of output. Consequently, nickel-iron batteries are thought to be too expensive for widespread use in passenger cars. In 1990, Chrysler teamed up with Eagle-Picher Industries of Cincinnati, Ohio, to build nickel-iron battery packs for TEVans to be used as service vehicles for Southern California Edison.[1f] The project was later scrubbed when the U.S. Advanced Battery Consortium put its development effort behind other battery designs. Several European battery manufacturers also have abandoned development of nickel-iron batteries.[2f]
>
> ***Nickel-cadmium:*** Nickel-cadmium batteries are the most common rechargeable battery now on the market for low-power uses. In the summer of 1991, engineers from Japan's Nissan announced a breakthrough for vehicular applications. Its nickel-cadmium battery can be recharged to 60 percent of full capacity in only 15 minutes, instead of the typical two to eight hours of recharging time.[3f] Most batteries tend to overheat if recharged too quickly. Nissan overcame this problem by designing the battery with very low internal resistance and by making it very thin, so heat dissipates quickly. When this specially designed nickel-cadmium battery pack is placed in Nissan's FEV electric car, it can travel 100 miles at 45 mph before requiring another 15 minutes worth of recharging. But there is a catch. The fast-recharging time requires higher voltage and amperage than is available at most residences. Nevertheless, if the battery lives up to its billing, it could turn the proposition of recharging a battery overnight into more of a pit stop at an electric filling station. Proper disposal of these batteries is essential, however, since cadmium is highly toxic.
>
> ***Nickel metal hydride:*** The nickel metal hydride battery under development by Ovonic Battery Co. has a positive electrode made of nickel and a negative electrode that consists of many metals—vanadium, titanium, zirconium, nickel, chromium and several others—that soak up large amounts of hydrogen. Unlike conventional batteries, which store energy only on the surface of electrodes, the nickel metal hydride battery is able to store negatively charged hydrogen ions deep inside its

ration is significant because it represents the largest research project on which the auto makers have agreed to work jointly with the government. Over the next 12 years, the consortium is expected to spend more than $1 billion in an effort to demonstrate the feasibility of advanced battery technologies for electric vehicles.[306] In October 1991, the Department of Energy announced that it would commit $260 million in research funding over the next four metal alloy hydride. Ovonic belives that refinements in its battery might enable it to propel a compact car 300 miles on a single charge, accelerate it from 0 to 60 mph in only eight seconds, recharge in only 15 minutes and last 100,000 miles before it wears out.[4f]

Sodium-sulfur batteries: Another battery technology receiving a great deal of attention combines sodium and sulfur. Ford pioneered research into this kind of battery 20 years ago. The leading research on sodium-sulfur batteries today is being conducted by the U.S. Advanced Battery Consortium, Chloride Silent Power of England and ABB Group (the European conglomerate). ABB's battery has four times the energy density of a conventional lead-acid battery and provides two to three times the cruising range. Ford's ETX electric van operates on such a 1,200-pound sodium-sulfur battery pack.

These battery packs are unconventional for two reasons. First, they comprise some 3,000 individual cells that appear similar to flashlight batteries. Each steel-encased cell contains liquid sodium and liquid sulfur, separated by a ceramic electrolytic sleeve. Second, these batteries must be kept at a temperature of 660 degrees F in order to function, so the cells are packed into an insulated steel case. While sodium-sulfur battery packs are much more expensive than nickel-iron and nickel-cadmium batteries, they should not need replacement over the life of an electric vehicle.[5f] In addition, they require no water, produce no hydrogen gas during recharging and have very high charging efficiencies. If corrosion problems can be overcome with the sulfur electrode's casing and the cells' seals, the range of electric vehicles such as the ETX could increase to 200 miles between charges, and the battery pack itself could last for 120,000 miles or more.[6f]

Other batteries: Other advanced battery types include zinc-bromine, lithium-iron sulfide, lithium-polymer and metal-air. Lithium is a promising material because it can be sliced into paper thin layers for the negative electrode. A high-energy density material such as vanadium oxide can be used for the positive electrode, with a polymer electrolyte sandwiched in between. Such lithium batteries may achieve an energy capacity five times that of lead-acid batteries. Moreover, they can be rolled up and molded into virtually any shape that an auto designer wishes.[7f] Lithium batteries are at least five years behind the development of sodium-sulfur batteries, however.

In the long term—beyond the year 2000—metal-air batteries are promising because they combine high energy density with a mechanical (as opposed to electrical) means of recharging. The batteries' full power is restored when metal anodes are replaced, water is added and byproducts are removed. Questions remain concerning their cost, durability and complexity, however. A practical means also would have to be developed to scrub CO_2 emissions associated with metal-air batteries.[8f]

years.[307] (By comparison, the government spends only about $5 million a year on hydrogen research.)

The car makers and government agree that their collaborative effort will have to develop something better than the lead-acid battery, a technology that has been honed since the 1860s. In terms of specific energy, the best lead-acid battery stores only 60 watt-hours of energy per pound, whereas a pound of gasoline contains 6,000 watt-hours of energy. Hence, an 870-pound battery pack (like the one in the Impact) stores the equivalent of about nine pounds of gasoline, or roughly 1-1/2 gallons of fuel. That amount of power is not likely to take a car very far, no matter how efficient the rest of the vehicle is.

In May 1992, the United States Advanced Battery Consortium awarded its first major contract to Ovonic Battery Co., a subsidiary of Energy Conversion Devices of Troy, Mich.[308] Under terms of the $18.5 million contract, Ovonic will develop a nickel metal hydride battery and demonstrate its technical feasibility. Ovonic has produced the unconventional battery since 1987 for use in small electronic devices such as laptop computers and cellular phones. In vehicular testing, the battery provides a high amount of peak power, which is good for acceleration, but it also tends to wear out after only 350 to 400 discharge cycles. (A conventional lead-acid battery, by comparison, has a lifetime of 750 cycles or more.) One of the goals of the USABC testing program is to create a battery with a lifetime of 1,000 cycles, or about 60,000 miles. Stanford Ovshinsky, the founder and chief executive officer of Energy Conversion Devices, claims that the nickel metal hydride battery should be able to reach the 1,000 cycle goal, while providing a range of at least 130 miles per charge, with a recharge time of one hour or less.[309]

Researchers with the United States Advanced Battery Consortium also plan to support the development of a pilot scale plant to demonstrate the capability of sodium-sulfur batteries by 1994. In addition, they hope to develop a full-size experimental battery by 1994 using lithium-metal disulfide and lithium-polymer materials. While these batteries are more than a decade away from commercial development, they could extend an electric vehicle's cruising range to 300 miles.[310]

Someday, advances in photovoltaic cells may combine with advances in battery technology to allow electric vehicles to become stand-alone propulsion systems. Gas tanks, electric power cords—and attendant concerns over greenhouse gas emissions—could be banished to the scrap heap. While this prospect remains a gleam in a solar engineer's eye for now, one study by the Jet Propulsion Laboratory in Pasadena, Calif., did find compelling logic in the marriage of these two technologies. Photovoltaic cells mounted on the tops of electric vehicles could extend their cruising range before the vehicles have to plug into the nearest available outlet. And if the batteries are replenished with power during daylight hours, they would not go as far into their discharge cycles—possibly quadrupling the batteries' life.[311] Thus, while Sunraycer and other odd-looking, solar-powered contraptions seem like unlikely candidates to carry on the wanderlust of the automobile, they may yet emerge as the truest expression of freedom-behind-the-wheel.

Conclusions

Automotive designers and engineers already are hard at work drawing up plans for cars to be introduced after the year 2000. Whatever appearance they take on, and whatever fuel sources they use, these are sure to be the "smartest" cars ever built. On-board computers will control anti-lock brakes, active suspension systems, diagnostic dashboard displays and interior climate conditioning. Under the hood, computers will oversee most aspects of the engine's operation—the air-to-fuel ratio, the timing of combustion and the treatment of exhaust. With 20 megabytes of stored information, model-year 2000 cars will have the equivalent of a powerful desktop computer built right into them.[312]

Driving into the 21st Century

With so many advances expected by the turn of the century, one can imagine what cars might be like in 2010. On-board computers could orient drivers who are lost, suggest ways around traffic jams—or perhaps even take people where they want to go. As a first step in this direction, Toyota now offers a $2,300 option on its Crown and Soarer models in Japan that includes national road maps, highway guides and dealer information, stored on compact disks and displayed on dashboard computer consoles.[313] It may seem out of Buck Rogers, but "It is only a short leap to on-board front and rear radar used to locate surrounding traffic or on-coming curves," predicts Rich Taylor, an auto industry publicist. "Couple this with [global positioning] satellite navigation, drive-by-wire steering, brakes and throttle [control] and you have the basis for a car that drives itself. That dream could be a reality by 2010...[T]he technology exists now, it's just a matter of putting it all together."[314]

This automotive dream will come at a high price, however. The expense for all the electronic gadgetry is expected to rise from 6 percent of the value of new cars today to as much as 20 percent by the turn of the century.[315] The dream also conflicts with the vision of conservation proponents, who imagine a time when people get out of their cars and walk more, bicycle more, use more public transit and take more fast trains to get from one place to another. The human infatuation with the automobile must end, they say, in order to break the gridlock on the highways and make steady progress toward cleaner air.

The reality remains, however, that Americans rack up 50 billion more miles of motor vehicle travel each year than they did the year before. "We have built a society that is dependent on personal mobility and we are envied all over the world for that," boasts Michael Deland, chairman of the President's Council on Environmental Quality. That mobility "is consistent with our most cherished national values of independence," Deland adds. "I don't think we're going to change that in the near future."[316]

Figure 13

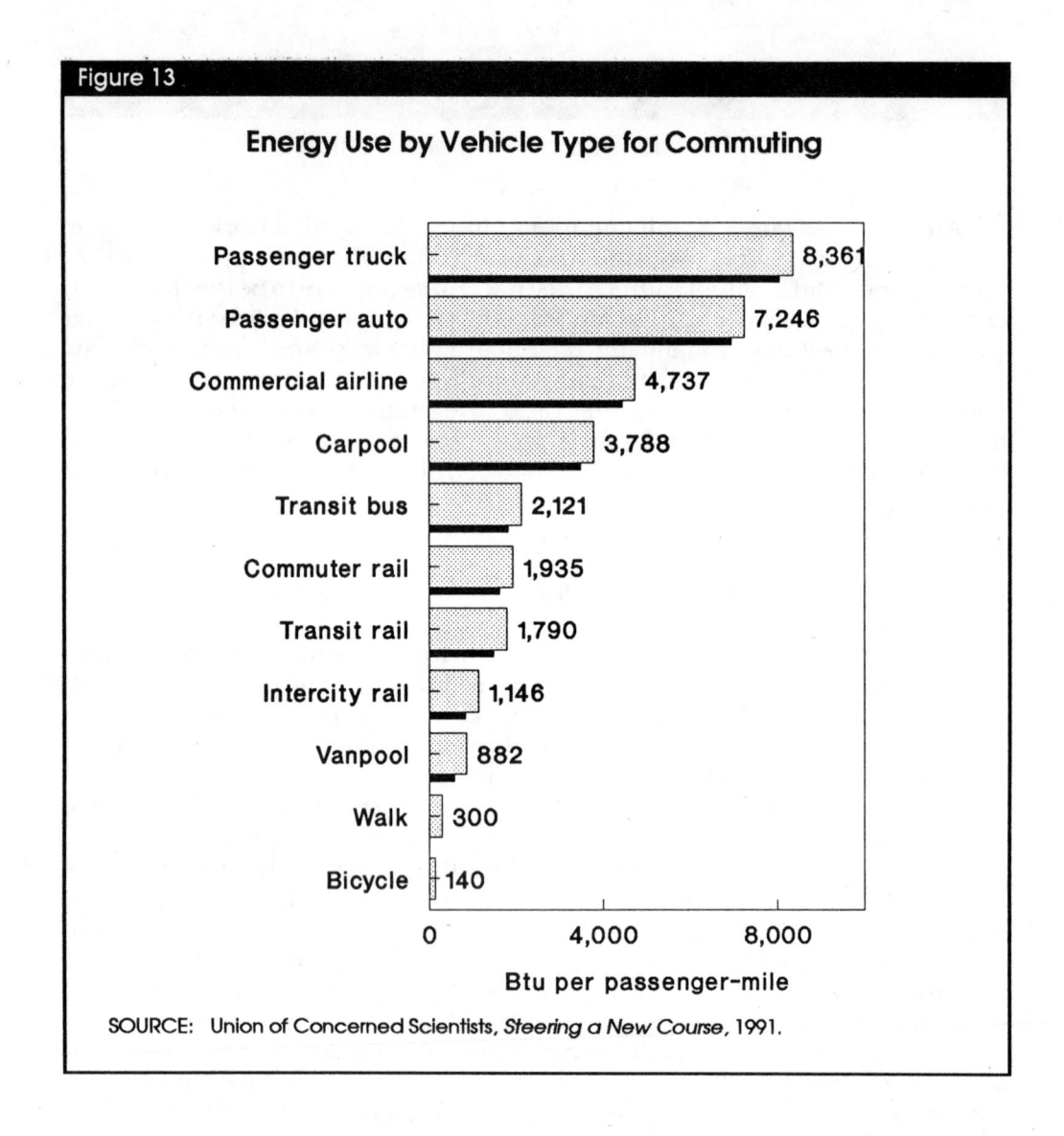

Energy Use by Vehicle Type for Commuting

Vehicle Type	Btu per passenger-mile
Passenger truck	8,361
Passenger auto	7,246
Commercial airline	4,737
Carpool	3,788
Transit bus	2,121
Commuter rail	1,935
Transit rail	1,790
Intercity rail	1,146
Vanpool	882
Walk	300
Bicycle	140

SOURCE: Union of Concerned Scientists, *Steering a New Course*, 1991.

Indeed, few things stand in the way of the private automobile and its pathway into the 21st century (except, perhaps, other automobiles). The stringent provisions of the 1990 Clean Air Act will pose a serious challenge for auto makers and gasoline producers over the next decade. But as the vehicle fleet turns over—and older, high-polluting vehicles are replaced by new vehicles with longer-lasting pollution-control devices—air quality is bound to improve. Moreover, introduction of low-emission, reformulated gasolines will help clear the air—and perhaps obviate the need for more radical fuel alternatives.

Two other obstacles loom on the horizon, however: a diminishing supply of non-OPEC oil and the prospect of global warming. For the time being, most

The Automobile Industry

Figure 14

CO$_2$ Emissions from U.S. Motor Vehicles

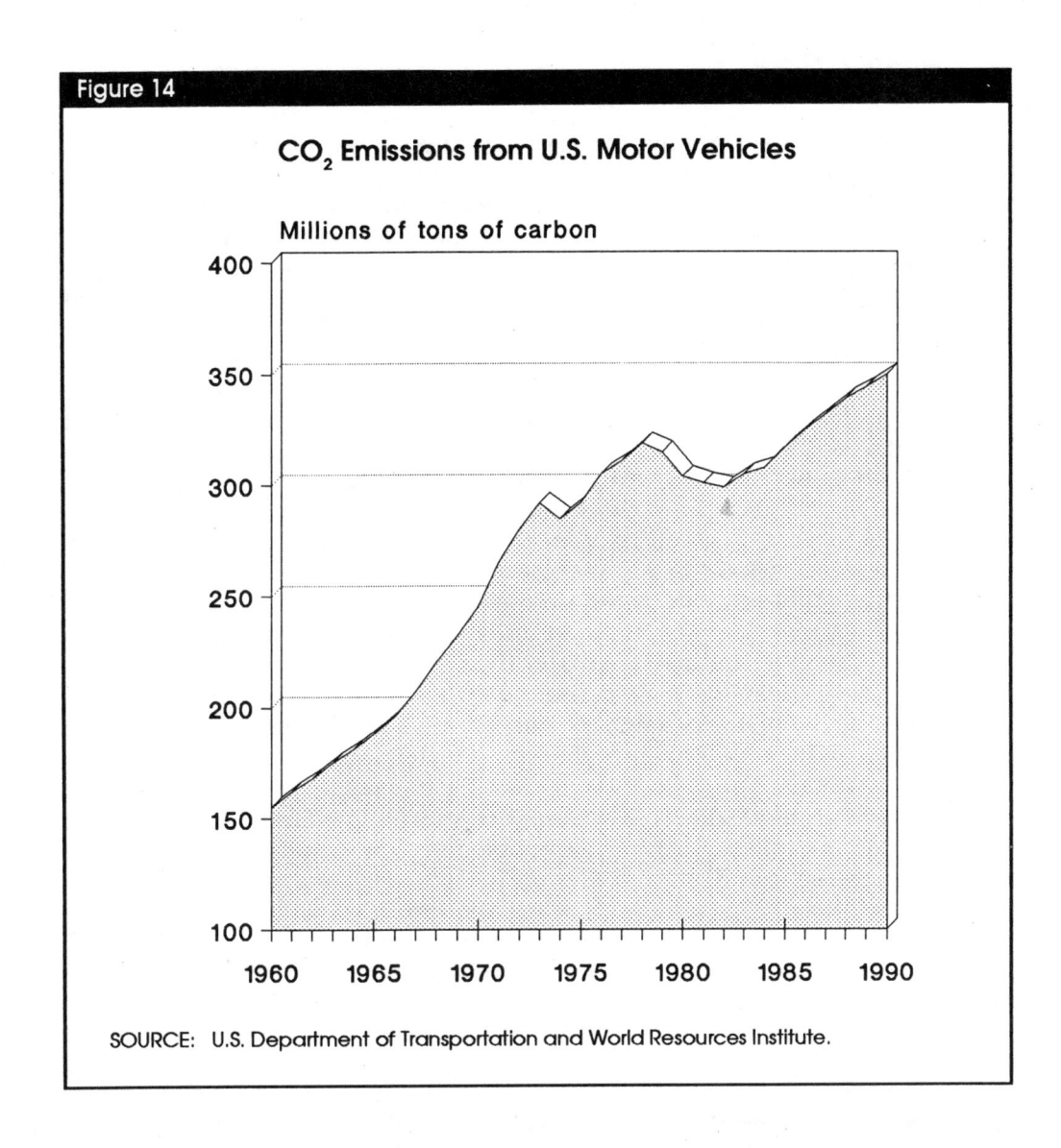

Millions of tons of carbon

SOURCE: U.S. Department of Transportation and World Resources Institute.

auto companies seem to think that these roadblocks are far enough away that there is little need to ease off the accelerator now. Signs indicate that cars built five years from now will be every bit as big, powerful and fuel-consuming as those now on the showroom floor. But the longer the auto industry careers down this path, the sooner it may collide with these barriers head-on.

Auto makers insist that consumers are the ones who have failed to look far enough ahead. To stay in business, they say, they must sell the kinds of vehicles that the driving public wants to buy. And at the moment, it is power and performance that car buyers desire, not fuel economy. As one General Motors executive explains, "We could build all Geo Metros (the most fuel-economical car sold in America), but who would buy them?"[317]

The First Move

Fortunately, there are ways to make cars and trucks more fuel-economical without sacrificing performance or radically downsizing the fleet. Although such innovations would raise average sticker prices, realized fuel savings should pay for the added costs. Therefore, the first move in the automotive greenhouse gambit can be taken now: building cars and trucks with more fuel-saving technologies and innovative design features to curb CO_2 emissions (and dependence on foreign oil).

The underlying question is, what amount of fuel-economy improvements are achievable relative to the current fleet average of 27.5 miles per gallon? The highly respected National Research Council issued a report in April 1992 that concluded a 31-33 mpg average is technically feasible for new passenger cars built in the 2001 model year—with the level rising to 37 mpg five years after that. The council figured that the savings could be achieved with an average vehicle weight reduction of 10 percent or less, although it would increase the average sticker price of cars $500 to $2,500 by the 2006 model year.[318]

While these fuel-economy gains are not as large as the 40 percent improvement sought by Sen. Richard Bryan in CAFE legislation proposed for the 2001 model year, the projected rise is greater than the level that the auto industry itself says is practicable. Some conservation advocates believe, on the other hand, that the council was much too conservative in its conclusions because it overlooked or dismissed many of the emerging technologies discussed in this chapter. They insist that Detroit is capable of achieving far greater fuel savings over a decade's time. Deborah Bleviss of the International Institute for Energy Conservation, for one, says it is "reasonable to propose standards during the next 10 years that would improve average fuel economy at a rate similar to that achieved between 1975 and 1985, when the first [CAFE] standards were in effect. This strategy would yield new automobile fleets that average 45 mpg and new light-truck fleets that average 35 mpg" by shortly after the turn of the century.[319]

The auto makers respond that future gains will not come as easily as those in the past. "We've picked all the low-hanging fruit," confides Chrysler executive Albert Slechter.[320] "Even if customers would accept the kind of 45-mile-per-gallon cars that are feasible," adds David E. Cole, director of the Office for the Study of Automotive Transportation at the University of Michigan, "the industry would face an unprecedented level of investment to effect the magnitude of change" that efficiency proponents seek by the turn of the century. "The human resource requirements would be equally daunting," Cole says. "In fact, the automotive industry realistically cannot achieve this magnitude of product change in one decade, at least not without serious risk of business failure and customer revolt."[321]

This may be so. But if an oil supply disruption or irrefutable signs of global warming appear during the 1990s, the auto industry may have little choice other than to reach for other energy-saving options still to be plucked. Such a prospect is particularly bittersweet, now that the industry has embarked on a strategy to maintain powerful engines and roomy car interiors while meeting

Figure 15

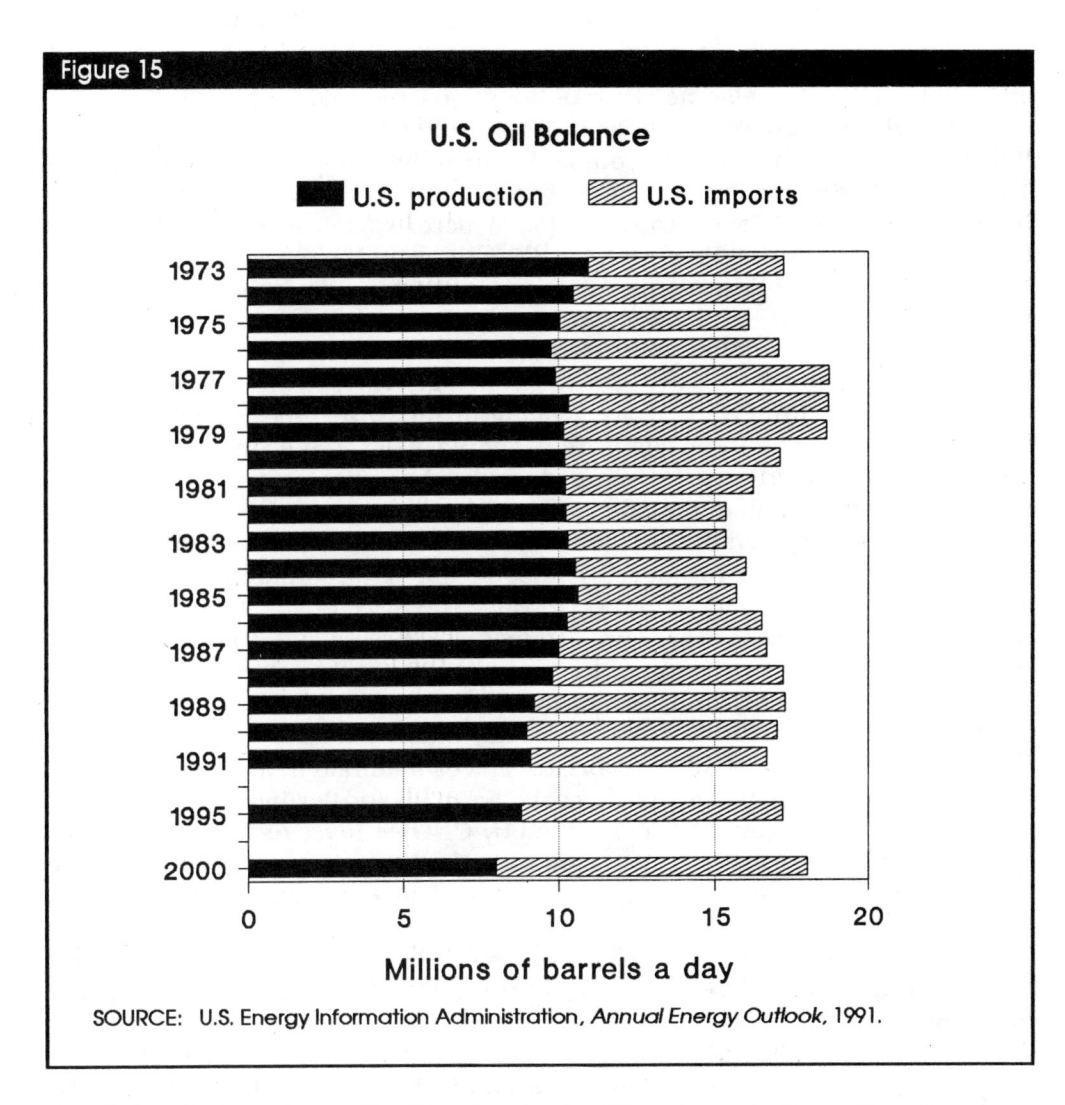

U.S. Oil Balance

■ U.S. production　▨ U.S. imports

(Millions of barrels a day, by year: 1973, 1975, 1977, 1979, 1981, 1983, 1985, 1987, 1989, 1991, 1995, 2000)

SOURCE: U.S. Energy Information Administration, *Annual Energy Outlook*, 1991.

new federal emissions and safety standards. Moreover, at a time when Japanese auto makers are slashing into Detroit's market share—and the Big Three's coffers are spilling red ink—American auto makers can ill-afford the imposition of new regulations that work more to the benefit of their overseas competitors.

The Bryan bill's call for a uniform percentage increase in each auto maker's CAFE rating (as opposed to an across-the-board *mileage* increase for all auto makers) is one way for the Big Three to gain a small advantage against the Japanese—albeit at the cost of institutionalizing the Japanese lead in fuel economy. Even so, a 40 percent improvement in average fuel economy might still exact too high a price for Detroit, since American auto makers typically require up to twice as much lead time as their Japanese competitors to revamp their production lines.[322] A 2001 target date means, in effect, that the Big Three would have only a few years in which to institute design changes that would raise their fleet averages from 27.5 mpg to 38.5 mpg. But if the target were achieved, it would take the nation a long way toward achieving a cap on its CO_2 emissions at 1990 levels—as other nations have pledged to do.

Political and economic realities being what they are, a compromise is likely that will enable domestic auto companies to achieve fuel economy gains that are more in line with their own projections. Any additional fuel savings could result in a fleet that is "some combination of more expensive, less powerful, less luxurious, and less stylish vehicles than it would have been otherwise," surmises Steven Plotkin of the congressional Office of Technology Assessment.[323] Yet the successful innovations of the automobile industry over the last 20 years—and others still to come—suggest that such sacrifices need not be as Draconian as one might fear. On the contrary, various car manufacturers (especially those from Japan) are moving ahead with production of peppy, lightweight, aerodynamic designs that should keep fuel-economical cars in demand—albeit mainly in the compact and subcompact categories in which consumer demand has diminished in recent years.

To the extent that car buyers ignore the most fuel-economical models on the market, other political measures could be taken to command consumers' attention. One possibility would be to raise the price of gasoline through an added sales tax or some other levy. To many observers, a gasoline tax makes perfect sense. "This tax is a user fee just like a turnpike toll or an admission ticket," reasons Roger Brinner, executive director of research for DRI/McGraw Hill. "The user knows the cost in advance and makes efficient economic choices; thus the driver will select an appropriate vehicle, maintain it, and choose routes and times-of-day to minimize gas-tax costs. Fairness and flexibility are built in."[324]

A CO_2-emissions analysis performed by Charles River Associates bolsters the case for such a tax.[325] The study, commissioned by the Motor Vehicle Manufacturers Association, found that a 27.5-cent-a-gallon gasoline tax (added in 5.5-cent increments over five years) would remove CO_2 from the atmosphere at a cost of only $23 a ton. By comparison, imposing a 40-mpg CAFE standard would remove CO_2 at an estimated cost of $104 per ton (although some contend this latter estimate is much too high.) The Charles River study also found that a carbon tax of only $2 a ton would achieve the same CO_2 reductions as a 40-mpg CAFE standard. The carbon tax is by far the cheapest option, the study concluded, because it compels drivers to think about fuel economy at all times—not just when they purchase a car—and because it spreads the burden of the tax among all carbon-intensive industries—not just transportation-related ones.

"If some fearless politician with enough charisma to carry it off could introduce a slowly rising tax on gasoline to, say, $4 a gallon by the end of the decade," reasons Paul MacCready, a developer of GM's Impact electric car, "the national debt would be pretty much taken care of, the economy would be revitalized, everyone's health would be better, and we'd be free of dependence on foreign oil. But," he concedes, "it's not going to happen."[326]

Elected officials are in no mood for passing higher taxes, least of all a gasoline tax. Congress rejected a proposal for a 5-cent increase in the federal gasoline tax in the fall of 1991, even though it was billed as "5 cents for America" to help repair ailing roads and bridges. In the spring of 1992, the only

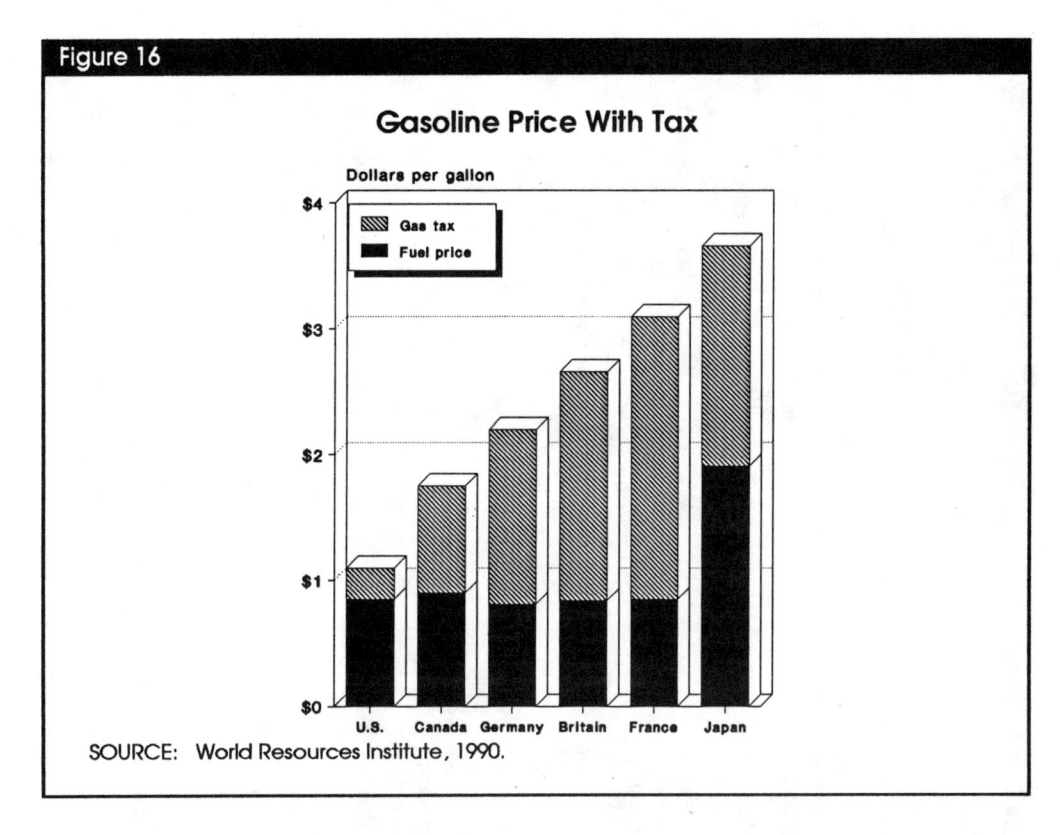

> **Figure 16**
>
> **Gasoline Price With Tax**
>
> Dollars per gallon
>
> (Gas tax / Fuel price)
>
> U.S., Canada, Germany, Britain, France, Japan
>
> SOURCE: World Resources Institute, 1990.

presidential candidate to propose such a tax, former Massachusetts Sen. Paul Tsongas (D), was roundly criticized by his opponents for doing so.

One alternative approach would be to create a revenue-neutral "feebate" system in which purchasers of "gas-guzzling" cars would pay a fee to subsidize purchasers of "gas-sipping" cars. A sliding scale could be used so that cars with the worst mileage ratings would pay the highest fees, while cars with the best mileage ratings would receive the largest rebates. That way, more customers would be attracted to fuel-economical cars, even if gas prices stayed low. People with limited incomes also might be inclined to trade in their old clunkers for new, thrifty models. Unless domestic auto makers have gas-sipping models to offer, however, they would not stand to benefit much from this arrangement.

A more radical idea is to have consumers pay for a portion of their auto insurance at the gas pump, rather than as a fixed annual premium. With such a pay-as-you-drive system, the price of gasoline would go up, say, $0.50 to $1.00 a gallon, reflecting the charge for basic automobile insurance. (The insurance would be auctioned off in blocks to private companies, with supplemental insurance available through independent carriers at the driver's option.) The rise in gasoline prices would encourage fuel conservation and tie the cost of insurance much more closely to the amount one drives. At the same time it would bring uninsured motorists into the system. For now, however, feebates and pay-as-you-drive remain theoretical concepts, no closer to implementation than a rise in the federal gasoline tax.

Table 4
Alternative Transportation Fuels

Fuel	Advantages	Disadvantages
Reformulated gasoline	Relatively low cost. Reduces ozone and toxic emissions. Compatible with existing vehicles.	Requires new refinery equipment. Pollution benefits limited/uncertain. Increases CO_2 emissions slightly.
Methanol	High octane, high engine efficiency. Reduces toxic, NO_x emissions. Possible reduction in hydrocarbons. Requires only minor changes to cars. Same CO_2 if made from natural gas.	Poison, requires special handling. Increases formaldehyde emissions. Water soluble; spills hard to detect. Likely to increase import dependence. More CO_2 if made from coal.
Ethanol	High octane, high oxygen content. Reduces carbon monoxide emissions. Domestic renewable resource. Production increases farm income. Much less CO_2 if made from wood.	High cost, requires subsidies. Gasohol raises evaporative emissions. Corrosive, hard to transport. Food/fuel competition possible. More CO_2 if made from corn.
Compressed natural gas	Low-cost, North American supply. Few reactive hydrocarbon emissions. Reduces carbon monoxide emissions. Less maintenance, longer engine life. Modest CO_2 benefit possible.	Bulky storage tanks, limited range. May increase NO_x emissions. Long refueling time, few outlets. Some loss of engine performance. Methane a potent greenhouse gas.
Hydrogen	Extremely clean burning. Fuel domestically produced. Major CO_2 benefit possible.	High fuel costs, vehicle changes. Bulky storage tanks, limited range. Current sources increase CO_2.
Electricity	No point-source pollution. Quiet operation, highly efficient. Fuel domestically produced. Major CO_2 benefit possible.	Limited range and power. Batteries take hours to recharge. Expensive battery replacement. Coal-charged battery increases CO_2.

The Right Fuels

Regardless of measures taken to improve automotive fuel economy in the near term, much of the resulting reduction in greenhouse gas emissions is likely to be wiped out in the long term by the growing number of cars and trucks on the road and the bumper-to-bumper traffic conditions they are increasingly likely to encounter. One analysis by Oak Ridge National Laboratory concludes, in fact, that even a 39-mpg standard achieved by the year 2000 would hold the nation's fuel consumption and transportation-related CO_2 emissions constant for only a decade, before they start to rise again.[327] As a result, auto makers will have to turn to nonfossil energy sources eventually to achieve any permanent reduction in the sector's CO_2 emissions.

With passage of the 1990 Clean Air Act, the first steps toward the development of alternative transportation fuels are happening now. But some of these steps may be in the wrong direction as far as the greenhouse effect is concerned. The auto industry would prefer to see the development of new fuels that require the fewest changes to conventional Otto and diesel engines (and associated fuel delivery systems). By the same token, the oil industry would prefer to manufacture petroleum-derived products that make the greatest use of existing reserves and refining capacity. On this basis, reformulated gasolines are ideal from an industry standpoint, because they should enable most vehicles already on the road to emit fewer noxious pollutants, while maintaining the preeminence of oil. But gallon for gallon, these new formulas consume more crude oil during refining and may impose a slight fuel economy penalty. As a result, "cleaner burning" gasolines will increase carbon dioxide emissions rather than decrease them.

Methanol is the auto industry's next-best option to comply with provisions of the Clean Air Act, while maintaining reliance on liquid-based fuels. But methanol derived from natural gas reduces CO_2 emissions only a little, while increasing America's dependence on foreign suppliers a lot. And methanol from coal is the worst possible choice in terms of global warming potential; it emits 30 to 100 percent more CO_2 per gallon than ordinary gasoline. Ethanol made from corn increases CO_2 emissions as well, and global climate change may diminish the capacity to grow corn at a time when nutritional needs are rising. On the other hand, methanol or ethanol made from woody biomass could serve as environmentally sound transportation fuels in a greenhouse world. But at present there is no infrastructure or cost-effective means of producing such biomass-derived motor fuels.

Among the alternative transportation fuels, compressed natural gas appears to be the most promising near-term option from a greenhouse perspective. CNG vehicles could reduce tailpipe emissions of CO_2 by perhaps 15 to 30 percent relative to gasoline-powered vehicles (and reduce conventional auto pollutants as well). For now, however, the need for bulky on-board storage containers and frequent refueling limits the role of CNG mainly to fleet vehicles. In the long run, its greatest contribution may be to pave the way for use of hydrogen gas, which emits no CO_2 and virtually no other pollutants as it burns.

Prospects for compressed natural gas and electricity as transportation fuels have been boosted by decisions of lawmakers in California and nearly a dozen other states to mandate the introduction of "low emission" vehicles. By 2003, 10 percent of the new vehicles sold in California, New York, Massachusetts and possibly New Jersey—130,000 vehicles in all—must be zero emission vehicles, which effectively mandates the introduction of electric-powered cars. The greenhouse benefits of this requirement will depend, however, on the source of electricity used to recharge the vehicles' batteries. Coal plants are generally first in line to meet such incremental demand, so there would be little or no CO_2 benefit in switching over to electric vehicles as the generating mix exists today. Over time, increased use of nuclear power and/or noncarbon renewable energy sources to recharge the batteries could reduce net CO_2 emissions substantially. Ultimately, the installation of on-board hydrogen fuel cells or solar-powered photovoltaic systems may offer the greatest potential to reduce the automotive sector's greenhouse gas emissions overall.

Since neither the government nor industry has developed a comprehensive transportation strategy to address the greenhouse effect, the likely near-term outcome is further disaggregation of the light-duty fleet. In California, for example, flexible-fueled vehicles that run on methanol, ethanol and gasoline are likely to take to the highway in increasing numbers. Elsewhere, gasohol and reformulated gasolines are sure to find their way into the fuel tanks of cars driven in smog-ridden metropolitan areas. Fleet vehicles running on compressed natural gas or electricity will find niche markets wherever refueling of these vehicles is practical, accessible and affordable.

In most places, however, the family car is still likely to run on gasoline for a long time to come; that way, it can be put to virtually any transportation-related task. In a nation of two- and three-car families, it is the *second* car in the driveway that may soon operate on a range-limited fuel—confining its use to running errands around town and commuting to work. If this is how the transition away from fossil-based transportation fuels is to begin, the rate at which it continues will depend on an ongoing assessment of the nation's air quality, the prognosis for global warming and the desire of consumers and policymakers to confront these difficult issues.

In any event, the auto industry's greenhouse gambit is not far away. The industry must decide which of the available options (including fuel economy improvements) make sense to pursue from a business standpoint—and keeping within the limits of the atmosphere to assimilate additional greenhouse gases and conventional air pollutants. While the automobile is a "hardy little beast" that is sure to survive the treacherous road ahead, the fate of particular auto companies is more in doubt. Those in the driver's seat must choose wisely among the multitude of options now before them and steer a careful, deliberate course into the 21st century.

Notes

1. Motor Vehicle Manufacturers Association of the United States Inc., *Motor Vehicle Facts and Figures '91*, Detroit, Mich., 1991.
2. *Ibid.*
3. National Research Council, Transportation Research Board, *A Look Ahead: Year 2020*, Proceedings of the Conference on Long-Range Trends for the Nation's Highway and Public Transit Systems, Special Report 220, Washington, D.C., 1988.
4. U.S. Environmental Protection Agency figures from *National Air Pollutant Emission Estimates, 1940-1989*, issued March 1991, cited in *Motor Vehicle Facts and Figures '91* (see note 1).
5. "Heat Wave Affects Production of Cars At Several Plants," *The Wall Street Journal*, Aug. 5, 1988.
6. See note 1.
7. David L. Greene, "Short Term Options for Controlling CO_2 Emissions of Light Duty Vehicles," SAE Technical Paper Series, No. 901111, Warrendale, Pa. Another 2.9 pounds of carbon dioxide emissions result from petroleum refining and distribution, so that total CO_2 emissions per gallon of gasoline produced, delivered and consumed equals 22.6 pounds. For reference, see Mark A. DeLuchi, "Emissions of Greenhouse Gases from the Use of Gasoline, Methanol, and Other Alternative Transportation Fuels," *Methanol as an Alternative Fuel Choice*, Wilfred L. Kohl, editor, The Johns Hopkins Foreign Policy Institute, Washington, D.C., 1990.
8. These figures are approximations. Fuel consumption by cars and trucks in the United States was nearly 132 billion gallons in 1989, according to the Federal Highway Administration. Assuming 20 pounds of carbon dioxide emissions for each gallon of fuel consumed (diesel fuel emits slightly more CO_2 than gasoline), cars and trucks produced 1.2 billion metric tons of CO_2 in the United States in 1989. The Energy Information Administration estimates that the United States emitted 5.4 billion tons of CO_2 altogether in 1990, based on energy consumption data.
9. Deborah Gordon, Union of Concerned Scientists, *Steering a New Course: Transportation, Energy and the Environment*, Island Press, Washington, D.C., 1991.
10. *Ibid.*
11. Intergovernmental Panel on Climate Change, *Scientific Assessment of Climate Change, Summary and Report*, World Meteorological Organization/United Nations Environment Programme, Cambridge University Press, Cambridge, Mass., 1990.
12. This percentage is based on the 100-year radiative forcing potential of each greenhouse gas. If the gases are weighted according to their average residence time in the atmosphere, carbon dioxide would account for more than 70 percent of the radiative forcing of all greenhouse gases combined. Moreover, this percentage does not consider that stratospheric cooling caused by depletion of the ozone layer may offset radiative forcing by chlorofluorocarbons. Such an adjustment would reduce radiative forcing by greenhouse gases overall and increase the relative contribution of carbon dioxide.
13. American vehicles contribute roughly half of worldwide vehicular CO_2 emissions—even though they make up only one-third of the world fleet—because they are less fuel-efficient and are driven more miles per year, on average, than vehicles in other countries.

14. Michael Renner, *Rethinking the Role of the Automobile*, Paper #84, Worldwatch Institute, Washington, D.C., June 1988. Emission figures are based on a 1986 report by the Organization for Economic Cooperation and Development, *Environmental Effects of Automotive Transport.*
15. The Intergovernmental Panel on Climate Change has not provided an estimate of global warming potential by tropospheric ozone. One report calculates that ozone accounts for 8 percent of global warming potential. See James J. MacKenzie and Michael P. Walsh, *Driving Forces: Motor Vehicle Trends and Their Implication for Global Warming, Energy Strategies and Transportation Planning*, World Resources Institute, Washington, D.C., December 1990.
16. Mark A. DeLuchi, "Transportation Fuels and the Greenhouse Effect," *Transportation Research Record 1175*, Transportation Research Board, Washington, D.C.
17. Fred Pearce, "Methane: The Hidden Greenhouse Gas," *New Scientist*, May 6, 1989.
18. U.S. Congress, Office of Technology Assessment, *Changing by Degrees: Steps to Reduce Greenhouse Gases*, OTA-O-482, U.S. Government Printing Office, Washington, D.C., February 1991.
19. Michael Schwarz, Ford Motor Co., presentation in *The Greenhouse Effect: Investment Implications and Opportunities*, proceedings of an Oct. 4, 1989, conference sponsored by the Investor Responsibility Research Center and the World Resources Institute, Douglas Cogan, editor, Investor Responsibility Research Center, Washington, D.C., 1990.
20. Calculations based on the paper by Terry G. Statt, U.S. Department of Energy, "The Use of CFCs in Refrigeration, Insulation and Mobile AC in the U.S.," paper prepared for the EPA Conference on Substitutes and Alternatives to CFCs and Halons, Washington, D.C., Jan. 13-15, 1988. The transportation industry consumed approximately 164 million pounds of CFCs in 1985, 69 percent of which was CFC-12 used in mobile air conditioning units. The total amount consumed equals approximately 485 million tons of CO_2 equivalent.
21. Bob Davis and Barbara Rosewicz, "Panel Sees Ozone Thinning, Intensifying Political Heat," *The Wall Street Journal*, Oct. 23, 1991.
22. Mark Crawford, "Environmentalists Hammer CFC Substitutes, Cite Greenhouse Effect," *Environment Week*, Jan. 31, 1991; and Douglas Cogan, *Stones in a Glass House: CFCs and Ozone Depletion*, Investor Responsibility Research Center, Washington, D.C., 1988.
23. Intergovernmental Panel on Climate Change, *1992 IPCC Supplement*, World Meteorological Organization and United Nations Environment Programme, Geneva, Switzerland, February 1992.
24. It should be noted that some partially halogenated substitutes for CFCs cause damage to the ozone layer and also are candidates for an expedited phase-out.
25. Marc Ross, Marc Ledbetter and Feng An, *Options for Reducing Oil Use by Light Vehicles: An Analysis of Technologies and Policy*, American Council for an Energy-Efficient Economy, Washington, D.C., December 1991.
26. *Ibid.* See the source cited in note 34 for a more conservative estimate of the projected savings.
27. See note 1.
28. *Ibid.*, citing data from the U.S. Department of Commerce.
29. Neal Templin, "Japan Cars Score Sales Coup in the U.S.," *The Wall Street Journal*, July 6, 1990; and Robert G. Liberatore, "An Industry View: Market Incentives," *Forum for Applied Research and Public Policy*, Spring 1990.
30. See note 1.
31. Energy Information Administration, *Monthly Energy Review*, U.S. Department of Energy, Washington, D.C., November 1991.

32. Quote appears in the article by Chris J. Calwell, "Links Between Emissions, Fuel Economy," *Forum for Applied Research and Public Policy*, Spring 1990.
33. Albert Slechter, Chrysler Corp., personal communication, Dec. 20, 1991.
34. U.S. Congress, Office of Technology Assessment, *Improving Autombile Fuel Economy: New Standards, New Approaches*, OTA-E-504, U.S. Government Printing Office, October 1991.
35. *Ibid*. The fuel savings projected are slightly less than that projected by the American Council for an Energy Efficient Economy because ACEEE freezes the fleet fuel economy standard at 27.5 mpg whereas OTA assumes an improvement to 32.9 mpg in 2001. Other critical assumptions in the OTA analysis include a 1 percent increase in miles driven for each 10 percent decrease in fuel consumed per mile—the "rebound" effect—and a 2 percent annual increase in vehicle miles traveled.
36. See note 1.
37. U.S. General Accounting Office, *Traffic Congestion: Trends, Measures and Effects*, GAO/PMED-90-1, November 1989. This estimate assumes that energy use for highway passenger transport increases at a rate of 1 percent a year, to 6.5 million barrels a day in 2005.
38. See note 9.
39. *Ibid*.
40. Bradley A. Stertz and Terence Roth, "To Western Industry, East Bloc Auto Market Is Losing Some Luster," *The Wall Street Journal*, Nov. 14, 1990.
41. See the citation in note 15.
42. Carbon dioxide emissions from the transportation sector overall are roughly equal to those of the electricity sector on a global basis. If motor vehicle transport grows faster than electricity supply, and carbon-based fuel sources remain dominant, motor vehicles may surpass electricity generation as the largest source of man-made CO_2 emissions in the early 21st century.
43. See the citation in note 15.
44. Lawrence Ingrassia and Joseph B. White, "GM Plans to Close 21 More Factories, Cut 74,000 Jobs, Slash Capital Spending," *The Wall Street Journal*, Dec. 19, 1991.
45. See note 1.
46. Bradley A. Stertz and Jacqueline Mitchell, "Auto Makers Hobble Into the New Year With Little Hope for a Robust Recovery," *The Wall Street Journal*, Jan. 7, 1992.
47. Bradley A. Stertz, "Big Three Boost Car Quality but Still Lag," *The Wall Street Journal*, May 4, 1990.
48. Joseph B. White, "Car Makers Gear Up to Turn Good Marks in Quality Poll to Marketing Advantage," *The Wall Street Journal*, July 3, 1990.
49. See note 46, which includes year-end 1991 sales figures. The percentage is even higher when one considers cars manufactured by Japanese companies and sold under American nameplates.
50. *Ibid*.
51. See note 1.
52. Paul Ingrassia, "Auto Industry in U.S. Is Sliding Relentlessly into Japanese Hands," *The Wall Street Journal*, Feb. 16, 1990.
53. Steven E. Plotkin, "The Road to Fuel Efficiency in the Passenger Vehicle Fleet," *Environment*, July/August 1989.
54. Several auto makers have argued that rising fuel prices alone prompted them to offer more fuel-economical models beginning in the late 1970s. However, one government statistical analysis concludes that fuel economy standards were at least twice as important as changes in oil prices as a "driver" of fuel economy. See David L. Greene, Oak Ridge National Laboratory, "CAFE or Price? An analysis of

the Effects of Federal Fuel Economy Regulations and Gasoline Price on New Car MPG, 1978-89," contract paper for the U.S. Department of Energy, revised Nov. 30, 1989.
55. Energy Conservation Coalition, "The Auto Industry on Fuel Efficiency: Yesterday and Today," fact sheet, Washington, D.C., 1989.
56. According to statistics provided by General Motors, the average interior volume of a 1990 new car is 5 percent less than that of a 1974 Pinto wagon, and 2 percent less than that of a 1974 Chevrolet Nova, suggesting that interior "roominess" has diminished somewhat since the early 1970s. But the congressional Office of Technology Assessment points out that average interior volume has remained virtually constant for 13 years—109 cubic feet in 1978 and 107 cubic feet in 1990.
57. See note 1.
58. *Ibid.*
59. See note 34, and David Woodruff, "Detroit's Big Worry for the 1990s: The Greenhouse Effect," *Business Week*, Sept. 4, 1989.
60. See note 34, and Matthew L. Wald, "In Cars, Muscle vs. Mileage," *The New York Times*, Aug. 16, 1990.
61. See note 3, David L. Greene, Daniel Sperling and Barry McNutt, "Transportation Energy to the Year 2020." This fuel savings estimate assumes a 5 percent discount rate. Without it, fuel savings were $200 billion in 1987 dollars.
62. See note 31, and Daniel Yergin, Cambridge Energy Research Associates, author of *The Prize: The Epic Quest for Oil, Money and Power*.
63. Lee Shipper et al., "Linking Lifestyles and Energy Use," presentation at the Workshop on Energy Policies to Address Global Climate Change, University of California, Davis, Calif., Sept. 8, 1989.
64. Matthew L. Wald, "America Is Still Demanding a Full Tank," *The New York Times*, Aug. 12, 1990. Statistics cited are from the Energy Information Administration and Cambridge Energy Research Associates.
65. See note 1.
66. See, for example, Thomas Gale Moore of the Hoover Institute in the op-ed piece, "A Hidden Culprit in Auto Imports," *The Wall Street Journal*, Jan. 14, 1992.
67. Gregory A. Patterson, "Foreign or Domestic? Car Firms Play Games With the Categories," *The Wall Street Journal*, Nov. 11, 1991.
68. "Reforming the Auto," *Environmental Action Re:Sources*, July/August 1989.
69. See note 1. Honda, for example, achieved a CAFE rating of 29 mpg in 1978, when the fleet requirement was only 18 mpg.
70. "Mazda Sedan Will Be First Japanese Vehicle Classified as U.S. Car," *The Wall Street Journal*, Feb. 7, 1992; and Warren Brown, "The Great CAFE Caper," *The Washington Post*, June 9, 1991.
71. Robert G. Liberatore, "An Industry View: Market Incentives," *Forum for Applied Research and Public Policy*, Spring 1990.
72. Melinda Grenier Guiles, "Ford's 1991 Escort Symbolizes Future; Car is Engineered by Japan's Mazda," *The Wall Street Journal*, Feb. 27, 1990.
73. Joseph B. White, "New Models, Ads Rev Up GM's Geo Sales," *The Wall Street Journal*, March 2, 1990.
74. John Flesher, "EPA: Gas Economy Is Down," Associated Press wire story, Sept. 30, 1991.
75. Deborah Lynn Bleviss, "Improving Vehicle Fuel Economy: A Critical Need," *Forum for Applied Research and Public Policy*, Spring 1990.
76. See note 25.
77. *Ibid.*
78. See note 34.

79. *Ibid.*
80. Bureau of the Census, "Truck Inventory and Use Survey," 1987 Census of Transportation, Washington, D.C., 1990.
81. See note 34.
82. See note 1.
83. *Ibid.*
84. See note 18.
85. See second citation of note 60.
86. Joseph B. White and Neal Templin, "Rise in Gas Price Finds Car Makers Backsliding on Fuel Efficiency," *The Wall Street Journal*, Aug. 16, 1990.
87. *Ibid.*
88. David E. Cole, "Fuel Economy: There Are Limits," *Forum for Applied Research and Public Policy*, Winter 1990. (Cole is director of the Office for the Study of Automotive Transportation at the University of Michigan.)
89. See note 86.
90. Joseph B. White, "GM's Cadillac Shifts to Reverse, Plan Bigger Cars for '89," *The Wall Street Journal*, Sept. 20, 1988.
91. See note 34.
92. Joseph B. White, "GM Has $1.06 Billion Loss; Ford Also Posts a Deficit," *The Wall Street Journal*, Oct. 23, 1991.
93. See, for example, "Policy Alternatives for Reducing Petroleum Use and Greenhouse Gas Emissions, Charles River Associates, Cambridge, Mass., September 1991. The study was sponsored by the Motor Vehicle Manufacturers Association of the United States.
94. See note 92, and Bob Davis and John Harwood, "Bush Plans to Offer Pollution Credits to Firms That Buy, Then Junk Old Cars," *The Wall Street Journal*, March 9, 1992.
95. SRI International, *Potential for Improved Fuel Economy in Passenger Cars and Light Trucks*, Menlo Park, Calif., July 1991.
96. C. Difiglio, K.G. Duleep and D.L. Greene, "Cost Effectiveness of Future Fuel Economy Improvements, *The Energy Journal*, January 1990.
97. Energy & Environmental Analysis Inc., *Analysis of the Fuel Economy Boundary for 2010 and Comparison to Prototypes*, draft final report prepared for Martin Marietta Systems, November 1990.
98. See note 34. This conclusion assumes there will be relatively stable gasoline prices and a general continuation of recent trends in vehicle size and power, although technologies introduced after 1995 would be engineered to maximize fuel economy over high performance.
99. See note 25.
100. *Ibid.* These figures are extrapolated from ACEEE's estimates of savings through 2005 as a result of a 40 percent increase in CAFE by 2001.
101. See note 34.
102. Deborah Bleviss, International Institute for Energy Conservation, presentation in *The Greenhouse Effect: Investment Implications and Opportunities*, proceedings of an Oct. 4, 1989, conference sponsored by the Investor Responsibility Research Center and the World Resources Institute, Douglas Cogan, editor, Investor Responsibility Research Center, Washington, D.C., 1990.
103. See notes 25 and 34.
104. See note 34.
105. See notes 25 and 34.
106. See note 34, and Samuel Leonard, General Motors Corp., personal communication, Jan. 3, 1992.

107. Rich Taylor, "Auto Tech '90," advertising supplement to *The New York Times Magazine*, May 13, 1990.
108. Dorin P. Levin, "New Japan Car Weapon: A 'Little Engine That Could,'" *The New York Times*, Nov. 26, 1989; and second citation of note 59.
109. Bradley A. Stertz, "Ford Idles Plant that Assembles Escort Models," *The Wall Street Journal*, May, 26, 1989.
110. Ford has revamped the Taurus since 1990, although the new Taurus does not have a multivalve engine either, except in the sporty SHO version. The base engine in the 1992 Taurus is a 3.0 liter, six-cylinder model rated at 24.2 mpg. Mileage ratings are from the "Gas Mileage Guides" issued each October by the U.S. Environmental Protection Agency.
111. See note 108, citing market research by TRW Inc., a maker of engine valves.
112. *Ibid.*
113. Paul Ingrassia and Joseph B. White, "GM's New Chairman: Still an Innovator?" *The Wall Street Journal*, April 4, 1990.
114. "'Multivalve' Engines Make Inroads in Detroit," *The Wall Street Journal*, May 21, 1992.
115. *Ibid.*, and Gordon Allardyce, Chrysler Corp., personal communication, Jan. 8, 1992; and "Chrysler Considers Spending $148 Million to Build New Engine," *The Wall Street Journal*, April 23, 1990.
116. See note 114, and "Ford Is Ready to Invest $1.2 Billion To Develop a New Group of Engines," *The Wall Street Journal*, Feb. 24, 1992.
117. Neal Templin and Matt Moffett, "Ford to Invest $700 Million In Mexico Plant," *The Wall Street Journal*, Jan. 15, 1991; and Robert L. Simison, "Registration Of New Cars Fall in Europe," *The Wall Street Journal*, Oct. 21, 1991.
118. Joseph B. White, "After a Brief Pause, Japanese Auto Makers Gain on Detroit Again," *The Wall Street Journal*, May 23, 1989.
119. See note 34.
120. *Ibid.*
121. Jacob M. Schlesinger, "Japan Car Firms Unveil Engines Lifting Mileage," *The Wall Street Journal*, Aug. 30, 1991.
122. Leslie Helm and Donald Woutat, "New Engine Is a Power, Fuel Economy Marvel," *The Los Angeles Times*, Aug. 9, 1991.
123. Karen Lowery Miller and Larry Armstrong, "55 Miles Per Gallon: How Honda Did It," *Business Week*, Sept. 23, 1991.
124. Richard Gould, "The Exhausting Options of Modern Vehicles," *New Scientist*, May 13, 1989.
125. See note 123.
126. See note 107.
127. See note 9.
128. "Orbital Says Engine Shows Promise in Test by Ford and the EPA," *The Wall Street Journal*, Feb. 10, 1992. For background, see Joseph B. White, "GM Gets License from Orbital to Make Fuel-Efficient, Two Stroke Car Engines," *The Wall Street Journal*, June 20, 1989; and Karen Wright, "The Shape of Things to Go," *Scientific American*, May 1990.
129. See second citation of note 106.
130. See second citation of note 128.
131. Lawrence M. Fisher, "Rehabilitating the Image of the Two-Stroke Engine," *The New York Times*, July 8, 1990.
132. Chrysler Corp. advertisement, *The Wall Street Journal*, Nov. 11, 1991.
133. Chrysler Corp. position paper, "Two Stroke Engines," undated manuscript.
134. Paul Ingrassia and Jacob M. Schlesinger, "Power, Whimsy at Tokyo Auto Show," *The Wall Street Journal*, Oct. 26, 1989.

135. See note 131.
136. See third citation of note 128.
137. See full citation in note 59.
138. See note 34, citing work performed by Energy & Environment Analysis Inc.
139. *Ibid.*
140. Rose Gutfeld, "EPA Proposes Tougher Rules For Bus Exhaust," *The Wall Street Journal*, Sept. 11, 1991.
141. See note 124.
142. "VW Once Again to Market Diesel-Engined Car in the U.S.," *The Wall Street Journal*, May 16, 1989.
143. Krystal Miller, "Mercedes to Introduce 3 Diesel Models To Help It Meet U.S. Fuel Standards," *The Wall Street Journal*, Sept. 11, 1990.
144. U.S. Environmental Protection Agency, *Policy Options for Stabilizing Global Climate: Volume II* (draft), Washington, D.C., February 1989. This report includes a discussion of the adiabatic diesel engine.
145. "Cooking Up Ceramics for Auto Engines," *The Wall Street Journal*, April 17, 1990.
146. See third citation of note 128.
147. Deborah L. Bleviss, *The New Oil Crisis and Fuel Economy Technologies: Preparing the Light Transportation Industry for the 1990s*, Quorum Books, Westport, Conn. 1988.
148. See notes 34 and 42.
149. U.S. Environmental Protection Agency, "1992 Gas Mileage Guide," Washington, D.C., October 1991; and Paul Ingrassia and Joseph B. White, "With Its Market Share Sliding, GM Scrambles to Avoid a Calamity," *The Wall Street Journal*, Dec. 14, 1989.
150. See first citation of note 149. The 1992 Lincoln Topaz, for example, is rated at 21 mpg in city driving and 26 mpg in highway driving, when equipped with a 2.3 liter engine and an automatic transmission. The five-speed manual version is rated at 23 mpg in the city and 33 mpg on the highway. The Topaz lacks a torque converter lockup device.
151. See note 34.
152. See second citation of note 59.
153. See note 25.
154. *Ibid.*
155. Deborah L. Bleviss and Peter Walzer, "Energy for Motor Vehicles," *Scientific American*, September 1990.
156. See note 25.
157. See note 34.
158. See second citation of note 59.
159. Neil Gross, "What Gets 100 MPG and Seats 6? Nothing—Yet," *Business Week*, Sept. 16, 1991.
160. Dana Milbank, "Alcoa Will Build Car-Frame Plant In Soest, Germany," *The Wall Street Journal*, Oct. 24, 1991.
161. "Plastic Engines May Aid Detroit in Weight Battle, *The Wall Street Journal*, Sept. 29, 1989.
162. See note 25.
163. See note 34.
164. See note 147.
165. Lisa Tantillo, "Auto Plastics: Into High Gear," *Chemical Week*, March 15, 1989.
166. See note 147.
167. See third citation of note 128.
168. See note 34.

169. Paul Ingrassia, "Tokyo Car Show Takes Itself Seriously," *The Wall Street Journal*, Oct. 24, 1991.
170. Neal Templin, "'Green Cars' Are Still Far in the Future," *The Wall Street Journal*, Jan. 13, 1992.
171. *Ibid.*
172. See notes 169 and 170.
173. See note 147.
174. *Ibid.*
175. See note 34.
176. See note 63.
177. See note 25. This calculation assumes that gasoline prices are $1 per gallon.
178. *Ibid.* The fuel savings estimate is based on low oil prices at $20 a barrel. The savings would be greater if oil prices were higher.
179. Thomas B. Gage, Chrysler Corp., personal communication, Jan. 7, 1992.
180. These estimates are base-case projections in the National Energy Strategy, issued by the U.S. government in February 1991. With implementation of the National Energy Strategy, which favors drilling in environmentally sensitive areas, the United States would import an estimated 45 percent of total oil demand in 2000, falling to 40 percent in 2010, before rising to 60 percent by 2030, according to the document's authors. See U.S. Department of Energy, *National Energy Strategy: Powerful Ideas for America*, first edition 1991/1992, Washington, D.C., February 1991.
181. Youssef M. Ibrahim, "OPEC Is Back and Feeling Flush," *The New York Times*, Sept. 24, 1989.
182. U.S. Department of Energy, Energy Information Administration, *Monthly Energy Review*, Washington, D.C., November 1991.
183. See note 18.
184. *Ibid.*
185. See note 15.
186. Howard G. Wilson, Paul B. MacCready and Chester R. Kyle, "Lessons of Sunraycer," *Scientific American*, March 1989.
187. See note 19.
188. Krystal Miller, "Big 3 to Cooperate On Efforts to Meet Clean-Air Rules," *The Wall Street Journal*, June 10, 1992.
189. Michael Schwarz, Ford Motor Co., personal communication, July 2, 1992.
190. It should be noted that higher octane gas generally means higher profits for oil refiners. Most premium unleaded fuels cost only 5 cents more a gallon to make but sell at the pump for 15 to 20 cents extra. The driving public's demand for high-octane gasoline is one important reason why oil companies have been able to maintain profit levels despite falling crude oil prices. Ironically, most drivers would be better off if they stuck with regular unleaded gasoline. The heavy molecular structure of aromatics makes them prone to clog fuel injectors in the 90 percent of cars that are not designed for high performance. For reference, see Edwin S. Rothschild, "The Knock on High-Octane Gasoline," *The Washington Post*, Feb. 18, 1990.
191. According to the *Oil & Gas Journal*, "Making a higher octane product can entail a crude run twice that for a suboctane product." (Quote appears in the Rothschild article cited in note 190.) University of California researcher Mark DeLuchi estimates that refining one gallon of ordinary gasoline yields about 1.5 pounds of carbon dioxide emissions. Accordingly, the refining process for extra-high-octane gasoline produces well over two pounds of CO_2 emissions per gallon. For reference, see Mark A. DeLuchi, "Emissions of Greenhouse Gas from the Use of

Gasoline, Methanol, and Other Alternative Transportation Fuels," *Methanol as an Alternative Fuel Choice*, Wilfred L. Kohl, editor, The Johns Hopkins Foreign Policy Institute, Washington, D.C., 1990.

192. This is especially true when pre-existing equipment for conventional gasoline is used and when fewer of the compounds derived from crude oil are refined into ingredients of reformulated gasoline or other petroleum products. For reference, see Matthew L. Wald, "That 'Cleaner Fuel' May Be Gasoline," *The New York Times*, Aug. 23, 1989; and U.S. General Accounting Office, "Gasoline Marketing: Uncertainties Surround Reformulated Gasoline as a Motor Fuel," GAO/RCED-90-153, Washington, D.C., June 1990.

193. Ford estimates the fuel economy penalty to be 0 to 2 percent for reformulated fuels using MTBE, ETBE or TBA. An analysis by Arco Chemical finds a 1-mpg drop in mileage ratings. Michael Schwarz, Ford Motor Co., personal communication, Jan. 2, 1992; and "Refiners Should Look to Mixing Oxygenates in Gasoline Formulas," *International Solar Energy Intelligence Report*, Oct. 19, 1990.

194. See full citation in note 106.

195. Matthew L. Wald, "Gasoline as Clean as Methanol Is Developed to Cut Pollution," *The New York Times*, July 11, 1991.

196. Matthew L. Wald, "When the E.P.A. Isn't Mean Enough About Clean Air," *The New York Times*, July 21, 1991.

197. U.S. Congress, Office of Technology Assessment, *Replacing Gasoline: Alternative Fuels for Light-Duty Vehicles*, OTA-E-364, U.S. Government Printing Office, Washington, D.C., September 1990.

198. Mark DeLuchi, Robert Johnston and Daniel Sperling, "Transportation Fuels and the Greenhouse Effect," University of California at Davis, Report UER-180, University-wide Research Group, Davis, Calif., 1987. In this study, researchers at the University of California examined CO_2 emissions from the conversion of surplus natural gas to methanol and coal to methanol. The study took into account CO_2 emissions that result from long-distance transport of methanol, methane leakage rates from pipelines and sources of electricity for methanol conversion plants—80 percent of which are coal-fired. The researchers found that net greenhouse gas emissions from the manufacture and transport of gas-derived methanol are 3 percent less than from an energy-equivalent amount of gasoline. Changing the assumptions in the model produced a range of estimates from a 20 percent reduction in net CO_2 emissions to a 30 percent increase in emissions. A similar analysis of coal-derived methanol produced a range of estimates from a 35- to 160-percent increase in net greenhouse gas emissions, with a mean increase of 98 percent. DeLuchi updated these figures for the Office of Technology Assessment in December 1990. (See note 18.) Use of an improved methanol-fueled vehicle could result in 17 percent fewer CO_2 emissions, relative to a gasoline-powered vehicle, if methanol is derived from natural gas; and 25 to 30 percent greater emissions, if the methanol is derived using the best coal conversion technology.

199. National Research Council, Committee on Production Technologies for Liquid Transportation Fuels, *Fuels to Drive Our Future*, National Academy Press, Washington, D.C., 1990.

200. Chem Systems Inc., "A Briefing Paper on Methanol Supply/Demand for the United States and the Impact of the Use of Methanol as a Transportation Fuel," prepared for the American Gas Association, September 1987.

201. For more details on the development of methanol vehicles, see Charles L. Gray Jr. and Jeffrey A. Alson, "The Case for Methanol," *Scientific American*, November 1989.

202. See report cited in note 93 for estimates of methanol capital expenditures.

203. U.S. General Accounting Office, "Air Quality Implications of Alternative Fuels," GAO/RCED-90-143, Washington, D.C., July 1990.
204. Alan L. Adler, "New Rules Could Raise Driving Expenses," Associated Press wire story, Oct. 31, 1991.
205. U.S. Department of Energy, *Assessment of Costs and Benefits of Flexible and Alternative Fuel Use in the U.S. Transportation Sector: Technical Report Five, Vehicle and Fuel Distribution Requirements* (draft), Office of Policy, Planning and Analysis, Washington, D.C., January 1990.
206. *Ibid.*
207. Joseph B. White, "GM Plans Launch Of 'Variable-Fuel' Cars in March '92," *The Wall Street Journal,* Nov. 6, 1991.
208. Charles Amann, General Motors Corp., "Highway Vehicles and Global Warming," paper presented at New Technologies, Business Opportunities and Strategies for Reducing U.S. Greenhouse Gas Emissions, sponsored by the U.S. Environmental Protection Agency, Tysons Corner, Va., June 20, 1990.
209. Matthew L. Wald, "Chrysler to Sell Alcohol-Fueled Cars Soon," *The New York Times,* April 16, 1992.
210. "Chrysler Announces True Flexible Fuel Capability at No Cost to Customers," Chrysler Corp. press release, Highland Park, Mich., Jan. 2, 1992; and Neal Templin, "Auto Makers Strive to Get Up to Speed On Clean Cars for the California Market," *The Wall Street Journal,* March 26, 1991.
211. Michael Schwarz, Ford Motor Co., personal communication, Jan. 2, 1992; also see note 207.
212. Kimberly Dozier, "Ford Leans to Flexible Fuel Vehicle As Environmental Alternative," *Environment Week,* Nov. 16, 1989.
213. Joseph B. White, "Heavy Engine Using Methanol Wins Approval," *The Wall Street Journal,* Aug. 7, 1991.
214. *Ibid.*
215. News brief, "Chevron Corp.," *The Wall Street Journal,* April 2, 1992.
216. Quote appears in the article by Dennis Wamsted, "EPA's Sunny Outlook for Methanol Clouded by OTA Analysts," *Environment Week,* Oct. 19, 1989.
217. See note 18. Estimates of CO_2 emissions by corn-derived ethanol relative to gasoline range from a 10 percent decrease to a 30 percent increase. Estimates are based on amendments by Mark DeLuchi to his earlier draft report, *State of the Art Assessment of Greenhouse Gases from the Use of Fossil and Nonfossil Fuels, With Emphasis on Alternative Transportation Fuels,* University of California, Davis, Calif., June 3, 1990. Figures recalculated on Dec. 11, 1990.
218. U.S. General Accounting Office, "Alcohol Fuels: Impacts from Increased Use of Ethanol Blended Fuels," GAO-RCED-90-195, Washington, D.C., July 1990; and Cynthia May, "GAO Says Expanded Ethanol Production Dependent On Continued Subsidies," *The Energy Report,* July 30, 1990.
219. *Ibid.*
220. See note 93.
221. S.M. Kane and J.M. Reilly, *Economics of Ethanol Production in the United States,* Agricultural Economic Report No. 607, U.S. Department of Agriculture, March 1989.
222. Scott Kilman, "Ethanol Producers Battle EPA Proposal on Clean Air," The Wall Street Journal, May 1, 1992.
223. Priscilla Smith, "Corn Growers Tout Ethanol Option," *Environment Week,* Sept. 13, 1990.
224. See note 218.
225. Tai Y. Change, Robert H. Hammerle, Steven M. Japar and Irving T. Salmeen, "Alternative Transportation Fuels and Air Quality," *Environmental Science & Technology,* July 1991. The authors are employed by Ford Motor Co.

226. Frank Pitman and Alice Agoos, "Ethanol Takes a Hit," *Chemical Week*, Oct. 5, 1989.
227. Harry P. Gregor, "Alcohol Fuel: The One for the Road," *The Washington Post*, July 9, 1989.
228. Nicholas Lenssen and John E. Young, "Filling Up in the Future," *World Watch*, May/June 1990.
229. See third citation of note 128.
230. Jonathan Kandell, "Brazil's Costly Mix: Autos and Alcohol," *The Wall Street Journal*, Sept. 28, 1989.
231. *Ibid.*
232. See note 228, which refers to the federal government's goals.
233. See note 227.
234. Lee R. Lynd, Janet H. Cushman, Roberta J. Nichols and Charles E. Wyman, "Fuel Ethanol from Cellulosic Biomass," *Science*, March 15, 1991.
235. For other expressions of doubt on ethanol, see notes 195 and 197; and Thomas W. Lippman, "Ordinary Gasoline on Way Out, but Successor Is Far From Certain," *The Washington Post*, May 8, 1990.
236. Jeff Seisler, Natural Gas Vehicle Coalition, presentation at New Technologies, Business Opportunities and Strategies for Reducing U.S. Greenhouse Gas Emissions, sponsored by the U.S. Environmental Protection Agency, Tysons Corner, Va., June 20, 1990.
237. *Ibid.*, and see full citation in note 235.
238. See note 197.
239. *Ibid.*
240. See notes 18 and 217.
241. See note 197.
242. *Ibid.*
243. M.A. DeLuchi, R.A. Johnston and D. Sperling, "Methanol vs. Natural Gas Vehicles, A Comparison of Resource Supply, Performance, Emissions, Fuel Storage, Safety, Costs and Transitions," Society of Automotive Engineers Technical Paper 881656, 1988.
244. See note 197.
245. *Ibid.* The assumptions are that the average CNG vehicle is driven 10,000 miles a year and gets the equivalent of 35 miles per gallon of gasoline.
246. See second citation of note 210.
247. Rose Gutfeld, "Steering Toward Alternative Fleets," *The Wall Street Journal*, April 12, 1991.
248. "UPS Plans Natural-Gas Stations," *The Wall Street Journal*, Jan. 31, 1992; and "UPS Trucks Will Use Gas In Los Angeles Experiment," *The Wall Street Journal*, July 12, 1990.
249. See note 236.
250. Jim Carlton, "California Clears Engine Powered by Natural Gas," *The Wall Street Journal*, Jan. 10, 1991.
251. "GM, Consortium to Build 1,000 Natural Gas Trucks," *The Wall Street Journal*, July 26, 1990. Also see note 183.
252. "Roundup," *The Washington Post*, Sept. 17, 1991.
253. "Chrysler Looks Into Gas Power," *The Wall Street Journal*, May 1, 1991.
254. Caleb Solomon, "Amoco to Pump Compressed Gas in Marketing Test," *The Wall Street Journal*, Dec. 21, 1989.
255. Cynthia May, "Amoco Introduces Reformulated Fuels, Pump Nozzles, Oil Recycling, CNG Pump," *The Energy Report*, Nov. 5, 1990.
256. Charles McCoy, "PG&E, Chevron Plan to Install Natural Gas Pumps," *The Wall Street Journal*, Oct. 10, 1991.

257. See note 205.
258. Pure hydrogen will damage certain steels, however, so inhibiting agents must be found to add to hydrogen transported through existing pipeline infrastructure.
259. See note 197.
260. Joan M. Ogden and Robert H. Williams, *Solar Hydrogen: Moving Beyond Fossil Fuels*, World Resources Insititute, Washington, D.C., October 1989.
261. Steve Nadis, "Hydrogen Dreams," *Technology Review*, August/September 1990.
262. See note 225.
263. See note 199.
264. See note 260.
265. The solar-hydrogen path faces a number of obstacles. First, the price of gasoline would have to nearly double in price before such solar-generated hydrogen fuel would be competitive in the marketplace. Second, amorphous silicon cells would have to raise their current best efficiency rating from 11 percent to 15 percent and overcome problems with degradation of the cells' efficiency in the field. Third, the cost of manufacturing those cells would have to fall seven-fold or more to reach 40 cents a peak watt. Finally, if allowances are made for access roads, cabling stations, hydrogen storage tanks and the like, the land area required for solar-derived hydrogen might be far greater than 240 miles around.
266. Israel Dostrovsky, "Chemical Fuels from the Sun," *Scientific American*, December 1991; and Peter Hoffmann, "The Hydrogen Letter," presentation at New Technologies, Business Opportunities and Strategies for Reducing U.S. Greenhouse Gas Emissions, sponsored by the U.S. Environmental Protection Agency, Tysons Corner, Va., June 20, 1990.
267. See notes 197 and 260.
268. *Ibid.*
269. Claudia H. Deutsch, "As Oil Prices Rises, the Hydrogen Car Is Looking Better," *The New York Times*, Aug. 26, 1990.
270. See note 261.
271. See second citation of note 266.
272. See note 169.
273. See note 269 for quotes from General Motors and Ford executives.
274. Amal Kumar Naj, "Clean Fuel Cells Sparking New Interest," *The Wall Street Journal*, March 19, 1992.
275. Lesley Hazelton, "Really Cool Cars," *The New York Times Sunday Magazine*, March 29, 1992.
276. "Simplifying Fuel Cells For Use in Autos," *The Wall Street Journal*, Dec. 4, 1990.
277. Beth McConnell, "Electric Car Boosters Try to Tackle Chicken-and-Egg Problem in States," *The Energy Report*, May 4, 1992.
278. See third citation of note 128. Around 1900, 38 percent of American cars ran on electricity, 22 percent used gasoline, and 40 percent were steam-powered.
279. For historical reference, see the letter to the editor by Ross F. Firestone in *The Wall Street Journal*, Feb. 8, 1990.
280. "Fiat's Electric Car, *The Wall Street Journal*, Feb. 8, 1990.
281. "Europe's Electric Cars," *The Wall Street Journal*, Sept. 25, 1989.
282. See note 275, and Ferdinand Pratzman, "The Greening of the Auto Makers," *The New York Times*, Sept. 16, 1991.
283. See note 275.
284. "How Clean Is the Plug-in Car?" *The Economist*, Oct. 13, 1990.
285. Neal Templin, "Nissan Says Its Electric Car Reduces Battery-Recharge Time to 15 Minutes," *The Wall Street Journal*, Aug. 28, 1991.
286. See note 275.

287. Jacqueline Mitchell, "GM to Discontinue the Buick Reatta, Citing Slow Sales," *The Wall Street Journal*, March 5, 1991.
288. Joseph B. White, "GM Says It Plans an Electric Car, but Details Are Spotty," *The Wall Street Journal*, April 19, 1990.
289. Lawrence G. O'Connell, "The Decade for Electric Vehicles," *Public Power*, September/October 1989. (The author is the manager of the transportation program at the Electric Power Research Institute.)
290. Taylor Moore, "They're New! They're Clean! They're Electric!" *EPRI Journal*, April/May 1991.
291. See note 18. Estimates of CO_2 emissions by coal plants range from 3 percent increase to a 10 percent increase. Estimates are based on amendments by Mark DeLuchi to his earlier draft report, *State of the Art Assessment of Greenhouse Gases from the Use of Fossil and Nonfossil Fuels, With Emphasis on Alternative Transportation Fuels*, University of California, Davis, Calif., June 3, 1990. Figures recalculated on Dec. 11, 1990.
292. See note 290.
293. See the draft report cited in note 291.
294. See note 186.
295. Matthew L. Wald, "At Last, a Practical Electric Vehicle?" *The New York Times*, Nov. 15, 1989.
296. See note 289. Estimate is based on a 1987 nationwide survey of fleet managers by Electric Vehicle Development Corp.
297. See note 290.
298. Paul J. Brown, Electrical Vehicle Association of the Americas, presentation at New Technologies, Business Opportunities and Strategies for Reducing U.S. Greenhouse Gas Emissions, sponsored by the U.S. Environmental Protection Agency, Tysons Corner, Va., June 20, 1990.
299. "Ford Getting Ready to Make Electric Van," Associated Press wire story, April 11, 1991.
300. "Chrysler to Develop Electric-Powered Van With Research Group," *The Wall Street Journal*, Jan. 7, 1991.
301. Dick Russell, "L.A.'s Positive Charge," *The Amicus Journal*, Spring 1991.
302. Robert L. Simison, "European Electric-Car Firm, Supported By Los Angeles, Plans Late '92 Output," *The Wall Street Journal*, Sept. 10, 1991.
303. See note 275.
304. Ford's electric Escort minivan requires eight kilowatts of electricity in typical city driving, for example. Using the heater adds another five kilowatts of demand on the batteries, while using the air conditioner would add six kilowatts. See, for reference, Stephen Stinson, "Electric Cars: Transportation for the Future?" *Chemical & Engineering News*, June 17, 1991.
305. A study by the Electric Power Research Institute illustrates how the cost of batteries, rather than the cost of electricity itself, is the limiting factor in the economic viability of electric vehicles. Each mile of travel in an electric vehicle consumes only 1.5 cents worth of electricity, on average, or less than half the cost of gasoline in a comparably sized vehicle. Engine maintenance adds another 1.5 cents a mile in expenses for gasoline-powered vehicles, while electric motors are essentially maintenance-free. Mile for mile, then, the electric vehicle is at least two-thirds less expensive to drive—until it comes time to replace the battery. For an electric vehicle to achieve operating expenses comparable to those of a gasoline-powered car, a $1,000 battery pack would have to last 30,000 miles, the institute's analysis concluded. This works out to the same expense as a gasoline-powered car driven 30,000 miles that gets 25 miles a gallon (at $1.25 per gallon) and pays $500 for engine maintenance. For reference, see note 289.

306. Jacqueline Mitchell, "Ford, Chrysler, GM to Cooperate On Battery Work," *The Wall Street Journal*, Feb. 1, 1991.
307. "Big Three Car Makers' Electric-Car Venture Gets Boost From U.S.," *The Wall Street Journal*, Oct. 29, 1991.
308. "Ovonic to Develop Battery For Use in Electric Vehicles," *The Wall Street Journal*, May 20, 1992.
309. Jerry E. Bishop, "New Generation of Electric-Car Batteries to Be Unveiled," *The Wall Street Journal*, May 15, 1992.
310. Taylor Moore, "The Push for Advanced Batteries," *EPRI Journal*, April/May 1991.
311. See note 301.
312. See note 107.
313. Gregory Witcher, "Smart Cars, Smart Highways," *The Wall Street Journal*, May 22, 1989.
314. See note 107.
315. *Ibid.*
316. Matthew L. Wald, "How Dreams of Clean Air Get Stuck in Traffic," *The New York Times*, March 11, 1990.
317. See second citation of note 106.
318. National Research Council, *Automotive Fuel Economy: How Far Should We Go?*, National Academy Press, Washington, D.C., April 1992. The report was commissioned by the U.S. Transportation Department's National Highway Traffic Safety Administration.
319. See note 75.
320. Albert J. Slechter, Chrysler Corp., presentation at New Technologies, Business Opportunities and Strategies for Reducing U.S. Greenhouse Gas Emissions, sponsored by the U.S. Environmental Protection Agency, Tysons Corner, Va., June 20, 1990.
321. See note 88.
322. See note 197. Japanese auto makers typically require four to five years to revamp their production lines, whereas American auto makers take up to eight years normally.
323. See note 53.
324. Roger E. Brinner, DRI/McGraw Hill, "Gasoline Tax Best Path to Reduced Emissions," *Forum for Applied Research and Public Policy*, Winter 1991.
325. See note 93.
326. For the source of the quote, see note 275.
327. See first citation of note 7.

Box 4-A

1a. Deborah Gordon, Union of Concerned Scientists, *Steering a New Course: Transportation, Energy and the Environment*, Island Press, Washington, D.C., 1991.
2a. Motor Vehicle Manufacturers Association of the United States Inc., *Motor Vehicle Facts and Figures '91*, Detroit, Mich., 1991.
3a. See note 1a and "Reforming the Auto," *Environmental Action Re:Sources*, July/August 1989.
4a. See note 2a.
5a. Scott Klinger, "Hardening of the Arteries: Study of the Infrastructure Crisis and Mass Transit," Franklin's *insight*, Franklin Research and Development Corp., Boston, Mass., March 1989.

6a. See note 2a, and Stephen C. Fehr, "More Commuters Traveling Alone," *The Washington Post*, May 29, 1992. The percentage of commuters driving alone increased from 64 percent in 1980 to 73 percent in 1990.
7a. J.A. Lindley, "Urban Freeway Congestion Problems and Solutions: An Update," *ITE Journal*, December 1989. Figures are for 1987.
8a. *Ibid.*
9a. John Yoo, "As Highways Decay, Their State Becomes a Drag on the Economy," *The Wall Street Journal*, Aug. 3, 1990.
10a. Federal Highway Administration, *The Future National Highway Program 1991 and Beyond: Urban and Suburban Highway Congestion*, working paper #10, Washington, D.C., December 1987.
11a. U.S. Department of Energy, *Long Range Energy Projections to 2010*, DOE/PE-0082, Washington, D.C., July 1988.
12a. Robert W. Stewart, "House Approves Highway Aid Bill," *The Los Angeles Times*, Oct. 24, 1991.
13a. Nancy Shute, "Driving Beyond the Limit," *The Amicus Journal*, Spring 1991.
14a. Daniel Machalaba, "States Try New Tactic to Curb Auto Traffic: Cut Highway Spending," *The Wall Street Journal*, April 8, 1992.
15a. Daniel Machalaba, "Longtime Symbols of Decay and Delay, Commuter Railroads Undergo a Revival," *The Wall Street Journal*, Oct. 1, 1991.
16a. See note 2a. The CO_2 calculation is an IRRC estimate.
17a. See note 1a.
18a. *Ibid.* For example, a car cruising at 30 miles per hour gets twice the gas mileage of a car crawling along at 10 mph, so the slower vehicle emits twice the amount of carbon dioxide over the length of the trip.

Box 4-B

1b. Fred Westbrook and Phillip Patterson, U.S. Department of Energy, *Changing Driving Patterns and Their Effect on Fuel Economy*, paper presented at the Society of Automotive Engineers Government/Industry Meeting, Washington, D.C., May 2, 1989.
2b. Unless otherwise noted, projections of future on-road fuel economy are based on the paper cited in note 1b.
3b. See note 1b.
4b. U.S. Congress, Office of Technology Assessment, *Improving Automobile Fuel Economy: New Standards, New Approaches*, OTA-E-504, U.S. Government Printing Office, October 1991.
5b. R. McGill, *Fuel Consumption and Emission Values for Traffic Models*, FHWA/RD-85/053, Oak Ridge National Laboratory, Oak Ridge, Tenn., May 1985.

Box 4-C

1c. "White House Stiffens Its Opposition On Fuel Efficiency," *The Wall Street Journal*, Aug. 17, 1990.
2c. John H. Cushman Jr., "Auto Roll-Overs Are New Target of a U.S. Push," *The New York Times*, Jan. 14, 1992.
3c. U.S. Department of Transportation, National Highway Traffic Safety Administration, "Effect of Car Size on Fatality and Injury Risk," unpublished paper widely distributed to congressional committees in the spring of 1991.

4c. Eleanor Chelimsky, "Automobile Weight and Safety," GAO/T-PEMD-91-2, U.S. General Accounting Office, Washington, D.C., April 11, 1991.
5c. *Ibid.*
6c. *Ibid.*
7c. See note 2c.
8c. Motor Vehicle Manufacturers Association of the United States Inc., *Motor Vehicle Facts and Figures '91*, Detroit, Mich., 1991. The highway traffic fatality rate in 1989 (when there were 47,000 such fatalities) was 2.2 per 100 million miles of vehicle travel, compared with a fatality rate in excess of 5 per 100 million miles of vehicle travel in 1962.

Box 4-D

1d. U.S. Environmental Protection Agency, "National Air Quality and Emissions Trends Report, 1989," EPA-450-/4-91-003, February 1991.
2d. Jonathan M. Moses, "EPA to Review National Standard For Ozone Smog," *The Wall Street Journal*, March 2, 1992.
3d. J. Raloff, "Summer Smog: Not Just an Urban Problem," *Science News*, July 8, 1989; and Hilary French, *Clearing the Air: A Global Agenda*, Worldwatch Institute, Paper #94, Washington, D.C., January 1990.
4d. Bob Davis and John Harwood, "Bush Plans to Offer Pollution Credits to Firms That Buy, Then Junk Old Cars," *The Wall Street Journal*, March 9, 1992.
5d. National Research Council, *Rethinking the Ozone Problem in Urban and Regional Air Pollution*, National Academy Press, Washington, D.C., December 1991.
6d. Richard D. Wilson, "Motor Vehicles and Fuels: The Strategy," *EPA Journal*, January/February 1991.
7d. Matthew L. Wald, "9 States in East Plan to Restrict Pollution by Cars," *The New York Times*, Oct. 30, 1991. (Two other states and the District of Columbia have since adopted California's rules.)
8d. Neal Templin, "Auto Makers Strive to Get Up to Speed On Clean Cars," *The Wall Street Journal*, March 26, 1991.
9d. "Arco Unveils Gasoline For California Market That Burns Cleaner," *The Wall Street Journal*, Sept. 7, 1990.
10d. Cynthia May, "Amoco Introduces Reformulated Fuels," *The Energy Report*, Nov. 5, 1990.
11d. Andrew Wood, "Clean Air Drives the MTBE Race," *Chemical Week*, July 31, 1991; and Thomas C. Hayes, "Exxon Plans Additive for Cleaner Fuel," *The New York Times*, Sept. 10, 1991.
12d. David Hanson, "Engine Emissions Prove Hard to Lower," *Chemical & Engineering News*, Jan. 7, 1991.
13d. Mark A. DeLuchi, "Transportation Fuels and the Greenhouse Effect," *Transportation Research Record 1175*, Transportation Research Board, Washington, D.C., 1989.

Box 4-E

1e. Richard W. Stevenson, "G.M. Displays the Impact, An Advanced Electric Car," *The New York Times*, Jan. 4, 1990.
2e. Rick Wartzman, "GM Unveils Electric Car With Lots of Zip But Also a Battery of Unsolved Problems," *The Wall Street Journal*, Jan. 4, 1990.

3e. Joseph B. White, "GM Says It Plans an Electric Car, but Details Are Spotty," *The Wall Street Journal*, April 19, 1990.
4e. Taken from Charles Amann's abstract, "Highway Vehicles and Global Warming," paper presented at New Technologies, Business Opportunities and Strategies for Reducing U.S. Greenhouse Gas Emissions, sponsored by the U.S. Environmental Protection Agency, Tysons Corner, Va., June 20, 1990.
5e. See note 2e.

Box 4-F

1f. Paul C. Judge, "Race On for an Electric-Car Battery," *The New York Times*, July 18, 1990.
2f. U.S. Congress, Office of Technology Assessment, *Replacing Gasoline: Alternative Fuels for Light-Duty Vehicles*, OTA-E-364, U.S. Government Printing Office, Washington, D.C., September 1990.
3f. Neal Templin, "Nissan Says Its Electric Car Reduces Battery-Recharge Time to 15 Minutes," *The Wall Street Journal*, Aug. 28, 1991.
4f. Boyce Rensberger, "New Battery Required For Autos of Future," *The Washington Post*, May 25, 1992.
5f. Dick Russell, "L.A.'s Positive Charge," *The Amicus Journal*, Spring 1991.
6f. Howard G. Wilson, Paul B. MacCready and Chester R. Kyle, "Lessons of Sunraycer," *Scientific American*, March 1989.
7f. Gary Stix, "Electric Car Pool," *Scientific American*, May 1992.
8f. See note 2f.

Chapter 5
The Electric Utility Industry

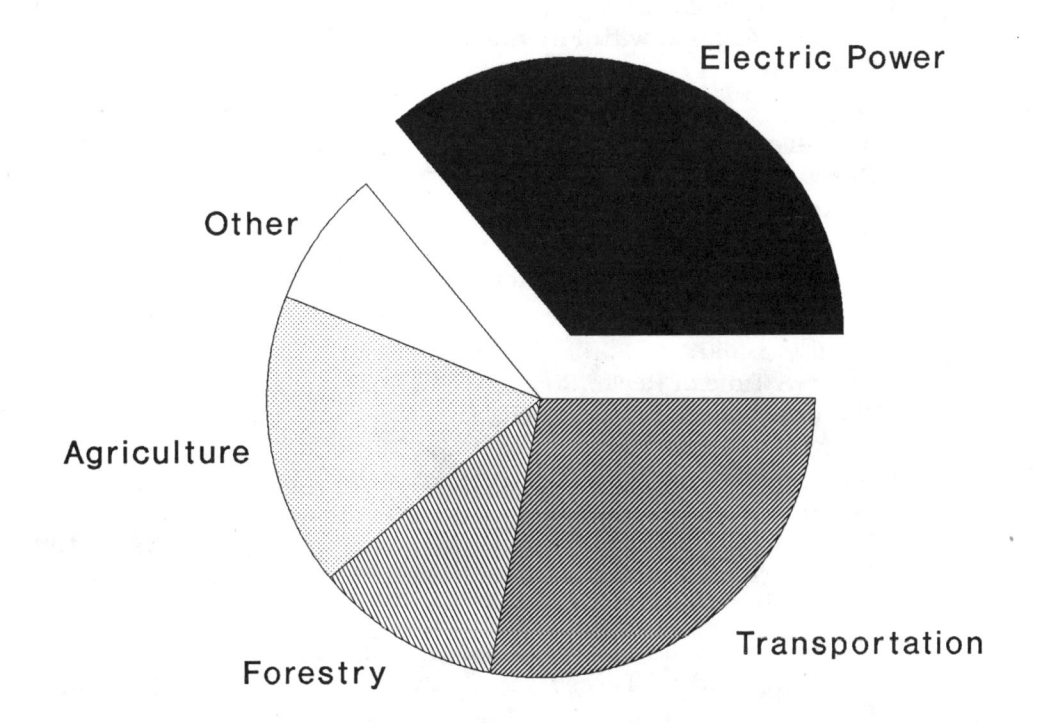

Relative contribution to global warming potential.

Contents of Chapter 5

Introduction
    Dawn of the Electronic Age 347
    Setting A Global Example 351

Effects on Power Generation
    Energy for the Electronic Age 356
    Electricity in a Greenhouse World 359
    Studies of Impacts on Utilities 363
    Sorting through the Options 366
    Utilities Respond with Fire and 'ICE' 370
    Taking Action 372

Energy Efficiency
    'Unselling' Electricity 375
    Surveying Utility Programs 380
    Making Efficiency Pay 381
    The Coming Boom in DSM 383
    Demand-side's Potential 386

Alternative Generating Sources
    What to Build? 390
    Coal: A Time of Reckoning 394
    Natural Gas: A 'Bridge' Fuel 401
    Reviving Nuclear Power 406
    A Market for New Power 408
    Progress in Renewables 422
    Envisioning a Solar Future 428
    Finding the Right Mix 430

Conclusions
    Factoring in Externalities 435
    Assessing Carbon Taxes 438
    Europe Moves Forward 441
    America Holds Back 442
    The Energy Gambit 445

Boxes
    Box 5-A: National Energy Strategy 354
    Box 5-B: Electric Utilities in the 'Greenhouse Summer' 360
    Box 5-C: Sizing Up Energy Savings 378
    Box 5-D: Creating a Market for Energy Efficiency 384
    Box 5-E: Capturing Carbon Dioxide Emissions 398
    Box 5-F: Nuclear Power in Retreat 410
    Box 5-G: Focus on Fusion 418

Notes 453

Introduction

Throughout the Industrial Age, energy policy has pursued a simple goal: to acquire fuel supplies at the lowest possible cost. Adherence to this policy has been a proven formula for economic success. Those nations able to acquire and consume the most fuel generally have had the fastest-growing economies, attained the highest standards of living and secured the brightest prospects for the future.

Resource constraints did not enter into this energy formula until recently. The earth was thought to contain such an endless bounty of fuel that human demand could not possibly overreach it. Now it is apparent that the earthly store of fossil energy that took a billion years to create may be unearthed and consumed in only a matter of centuries. Yet demand for fossil fuel continues to grow even as the supply dwindles.

Proven fossil energy reserves amount to 10 trillion barrels of oil and oil-equivalent. At present rates of consumption, the supply should last for another 170 years.[1] While soaring demand in newly industrialized countries may deplete this store more quickly, enhanced recovery methods in combination with conservation measures and efficiency gains could stretch out the supply for an even longer period. In any event, the world's fossil fuel reserves should not disappear for at least several more generations to come.

A more imminent shortage is emerging on the back end of the carbon cycle, however. The earth is running out of places to put the combustion byproducts of fossil fuels—except in the atmosphere, where carbon dioxide is reaching unprecedented levels. If the remaining store of fossil fuel is burned over the next few hundred years, the level of atmospheric CO_2 will rise by a full order of magnitude, and a severe greenhouse warming will be virtually assured.[2]

The fundamental constraint on future energy consumption, therefore, is no longer predicated on the amount of fossil fuel still underground. The limit now resides in the atmosphere, the oceans and the terrestrial biosphere, where "sinks" for carbon dioxide are rapidly filling up. The upshot for energy policy is that procuring additional fuel supplies at the lowest possible cost has become a wanting objective. Indeed, how much fossil energy remains to be extracted is largely beside the point. Long before the last pound of fuel is pulled from the ground, the rest of the planet will have run out of ways to assimilate the extra carbon dioxide and other greenhouse gases without inexorably changing the climate.

John Holdren, a professor of energy and resources at the University of California at Berkeley, has described the coming transition in energy policy this way: "The transition is from convenient but ultimately scarce energy resources to less convenient but more abundant ones, from a direct and positive connection between energy and economic well-being to a complicated and multidimensional one, and from localized pockets of pollution and hazard to impacts that are regional and even global in scope."[3]

In other words, the conventional energy formula for economic success is being turned on its head: The faster fossil fuels are burned, the more the economy is taxed, the farther the standard of living declines and the dimmer the outlook for the future becomes.

Dawn of the Electronic Age

The energy transition now underway stems from technological innovations of the last 200 years. At the beginning of the Industrial Age, wood was the primary source for heat and power. While smoky steam engines worked well for cranes and forges, most small industrial establishments still depended on the strength of human muscles to turn mechanical cranks, wheels and levers. Then in the late 19th century, Thomas Edison hit upon a revolutionary idea. "If the enormous energy latent in coal could be made to appear as electrical energy by means of a simple transforming apparatus," he said in 1887, "the mechanical methods of the entire world would be revolutionized."[4]

Edison probably had only an inkling of the enormity of his idea. His Pearl Street power station, opened in 1882, generated electricity solely for the purpose of lighting lamps in the financial district of Manhattan. Yet within a decade, most of urban America was illuminated with electric lights—and 55 of the 58 largest U.S. cities operated electric trolley car systems as well. By 1920, half the wheels of American industry turned via electric current, instead of using water wheels, steam turbines and internal combustion engines.[5]

By 1950, one-seventh of the nation's entire primary energy demand was devoted to making electricity. Electricity's share reached one-quarter of the total in the early 1970s, and it surpassed one-third of the total in the early 1980s. By the end of the 1990s, electricity should account for 40 percent of U.S. primary energy demand and, by 2025, it may represent fully one-half of the nation's energy requirements.[6] Should electric vehicles stage a comeback—gaining a large share of the auto market now dominated by petroleum fuels—electricity could account for more than two-thirds of U.S. primary energy demand eventually.

Clearly, the Industrial Age is giving way to a new Electronic Age as the 21st century approaches. Electricity has put power at people's fingertips virtually anywhere they choose to work and live. It has automated assembly lines and made mass production possible. Now it is the driving force behind such "high-tech" devices as computers and lasers. But the fuel source Edison praised more than a century ago—coal—remains the dominant source of power generation today, and herein lies a problem.

While electricity appears clean to the end user, coal as a generating source is especially dirty. Those who mine coal know that deposits usually are laden with sulfur and often infused with methane gas. The methane is released to the atmosphere as the coal is mined; the sulfur is vented later when the coal is burned. Sulfur dioxide and sulfate aerosols resulting from coal combustion contribute to acid rain and other air quality problems, unless smokestack emissions are thoroughly scrubbed. From a global warming perspective,

The Electric Utility Industry

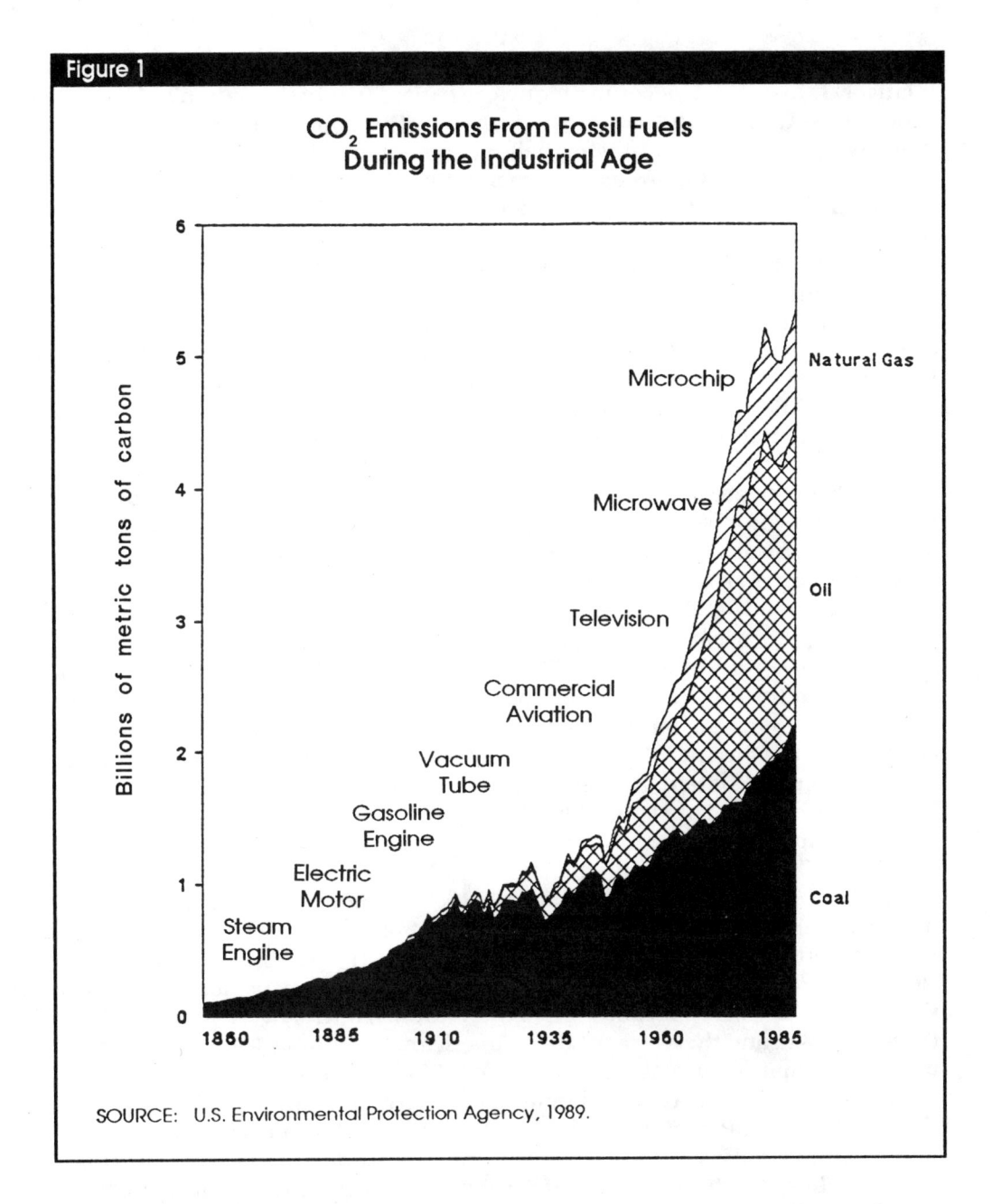

Figure 1

CO₂ Emissions From Fossil Fuels During the Industrial Age

[Chart: Billions of metric tons of carbon vs. years 1860–1985, showing stacked areas for Coal, Oil, and Natural Gas. Labels along the curve mark technological milestones: Steam Engine, Electric Motor, Gasoline Engine, Vacuum Tube, Commercial Aviation, Television, Microwave, Microchip.]

SOURCE: U.S. Environmental Protection Agency, 1989.

however, coal's greatest liability is its voluminous emissions of greenhouse gases—methane from the coal seams and carbon dioxide and nitrous oxide from combustion.

Of all the conventional fuels used to generate electricity, coal is the most rich in carbon. Each kilowatt-hour of coal-derived electricity puts more than 2 pounds of carbon dioxide into the air. By comparison, oil emits fewer than 1-3/4 pounds of CO_2 per kilowatt-hour, and natural gas (burned in a conventional gas turbine) emits about 1-1/3 pounds.[7] Yet coal, because it is plentiful and relatively inexpensive, accounts for nearly three-fifths of the nation's electricity generation—and nearly half of the world's. The combined share of oil and natural gas represents a much smaller percentage of the fuel mix. The other principal generating sources—hydroelectricity and nuclear power—emit no greenhouse gases. But they, too, account for a small share of the fuel mix relative to coal.

Accordingly, when all of the carbon and noncarbon generating sources are added together, each kilowatt-hour of electricity produced in the United States emits an average of 1.4 pounds of CO_2 into the atmosphere—reflecting the dominance of coal. (A kilowatt-hour of electricity will illuminate 10 100-watt light bulbs for an hour.) Considering that the United States produced more than 2.8 trillion kilowatt-hours of electricity in 1991, the magnitude of the CO_2 emissions from the nation's electricity sector is clear. American power plants released more than 500 million metric tons of carbon (in the form of carbon dioxide) into the atmosphere in 1991. That tonnage represents 7.5 percent of CO_2 emissions from fossil-fuel combustion worldwide. As a percentage of total man-made emissions of greenhouse gases, CO_2 production by American power plants accounts for about 4.5 percent of the world total.[8]

With demand for electricity *and* coal's share of the fuel mix expected to keep on growing, American utilities may emit 665 million metric tons of gaseous carbon by 2005 and nearly a billion metric tons by 2015.[9] Carbon dioxide emissions in the electricity sector could *double* in a quarter-century, in other words, if the current forecast holds up. A call for holding CO_2 emissions constant in the face of this forecast seems daunting—and a recommendation for a 20 percent cut in CO_2 emissions is all the more ambitious. Yet the Intergovernmental Panel on Climate Change calculates that a 60 percent reduction in CO_2 emissions would be necessary to keep the atmospheric content of CO_2 from rising further.[10]

Such huge reductions in carbon dioxide emissions would not come easily in the American utility sector, but the problem is compounded in the developing world, where electrification is just beginning in earnest. Whereas the average North American consumes roughly 11,000 kilowatt-hours of electricity a year, the average African consumes fewer than 400 kWh; the average Asian, fewer than 700 kWh; and the average South American, fewer than 1,400 kWh. China, whose per-capita electricity consumption is 375 kWh annually, plans to exploit its abundant coal reserves and double its electric generating capacity by the year 2000. Chile, Pakistan and Portugal are among dozens of other developing nations that are also counting on coal to meet expanded power requirements.[11]

Accordingly, industrialized and developing nations alike will have to change their plans for even greater use of coal if an international CO_2-reduction program is to be implemented. Should the United States alone take the hypothetical step of shutting down its coal plants—and building no new ones— the projected doubling time for worldwide CO_2 emissions would be delayed by only five years or so.[12] Since World War II, in fact, the relative contribution of the United States and other industrialized nations in creating CO_2 emissions has diminished as the economies and populations of developing nations have

Figure 2

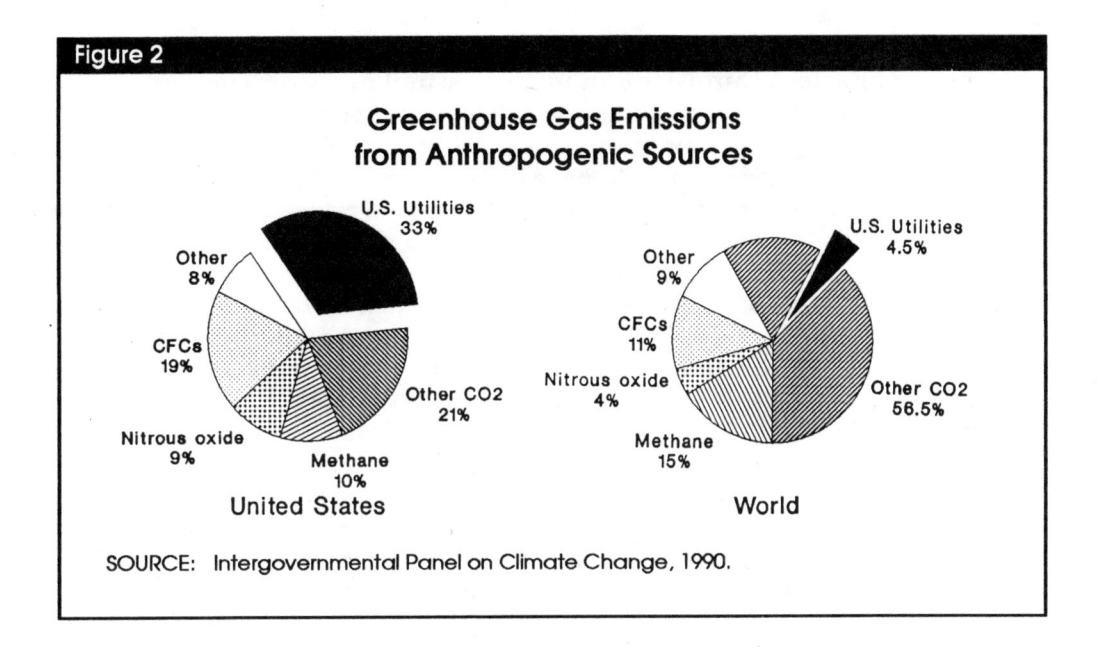

Greenhouse Gas Emissions from Anthropogenic Sources

United States:
- U.S. Utilities 33%
- Other 8%
- CFCs 19%
- Nitrous oxide 9%
- Methane 10%
- Other CO2 21%

World:
- U.S. Utilities 4.5%
- Other 9%
- CFCs 11%
- Nitrous oxide 4%
- Methane 15%
- Other CO2 56.5%

SOURCE: Intergovernmental Panel on Climate Change, 1990.

blossomed. In 1950, the United States, Canada and Western Europe accounted for 70 percent of the world's CO_2 emissions. Now their combined share is below 40 percent of the total. China's share of world CO_2 emissions, meanwhile, has risen from less than 1 percent in 1950 to more than 10 percent today. Similarly, the CO_2 share for the remainder of the developing world has increased from 6 percent to nearly 20 percent.[13] However, the United States remains the largest single emitter of greenhouse gases by a wide margin.

Setting a Global Example

Several conclusions can be drawn about the changing relative share of CO_2 emissions from nations around the world. First, it points out the futility of international actions to address global warming if developing nations fail to participate in the process. Second, it highlights the awkward position of industrialized nations, who must ask developing nations to "do as I say, not as I do" when it comes to modernizing their economies. While past development efforts involved quantum leaps in energy consumption, new energy policies must find alternative approaches. New generating technologies, in particular, will have to be deployed affordably around the world if they are to serve as practical substitutes for fossil energy sources that are still relatively cheap and abundant.

Perhaps the most important energy-related message, however, is that the future belongs to the efficient. While past energy policies put an emphasis on securing plentiful, low-cost fuel supplies, future strategies must stress the value of using those stocks wisely. Large-scale energy consumption can no longer assure a nation of its ability to compete in the global marketplace. The

United States produces four times as many goods and services as the struggling economies of Eastern Europe, for example, per unit of energy consumed. Yet it manages to produce only half as many goods and services as the powerful Japanese economy on a per-unit basis. Developing nations seeking to modernize their own economies know which energy model to follow.

While the United States lags behind Japan and Western Europe in energy efficiency, it does have the advantage of having more potential energy savings still to reap. By some estimates, aggressive conservation and efficiency measures could cut U.S. primary energy demand nearly in half.[14] Yet decoupling energy demand from economic growth complicates the task of setting new energy priorities. The difference between 1 percent growth and 2 percent growth in annual energy demand—compounded over 50 years—amounts to more energy than is presently consumed. Therefore, not only are future sources of energy now in question, so is the amount of power ultimately required.

Policies instituted to combat the greenhouse effect may clarify the situation to a considerable degree. Since the cleanest fuels are ones that are never burned, so-called "demand-side" measures are likely to take precedence in the years ahead, shaping the supply requirements of all other fuels. At the same time, developers of noncarbon energy sources will jockey to win the hearts, minds and pocketbooks of energy consumers who at present remain heavily dependent on fossil fuels.

Oil, coal and natural gas provide 38, 27 and 19 percent of the world's primary energy requirements, respectively.[15] Since coal emits more CO_2 than any other conventional fuel source per unit of energy produced, it would be dealt a heavy blow by greenhouse gas controls. Use of oil also would have to be curtailed, since oil-burning (especially in the transportation sector) accounts for more CO_2 entering the atmosphere than any other source. Natural gas, on the other hand, is one fossil energy source that may enjoy a rising market share in a greenhouse world. As a relatively clean-burning fuel—with half the carbon content of coal—it may serve as a bridge fuel until the world crosses over to nonfossil energy sources.

Concerns about the greehouse effect should not eclipse the coming Electronic Age in any case. Since generation of electricity derives simply from the flow of electrons, no greenhouse gases need be emitted. Atoms can be split or fused to generate heat and create steam for electricity, for example, or rushing water can be used in place of steam. The heat beneath the earth can be tapped to generate power, and the force of the wind can be harnessed. Even the rays of the sun can be concentrated on solar collectors to produce heat or stimulate the flow of electrons on semiconductor materials.

In short, the real "energy crisis" facing policymakers today is one of too many choices rather than too few. If only a couple of ways of making energy existed, the direction of a new policy would be plain for all to see. Instead, there are myriad supply sources—fossil, nuclear, renewables, etc.—all vying for shares of an energy market where efficiency measures compete as well. Sizing up the costs of these options will be critical to determining the direction of future energy policy: the costs of generating equipment, the costs of competing fuels and the net costs of energy savings. Yet the most important cost is also the hardest to measure: the

Figure 3

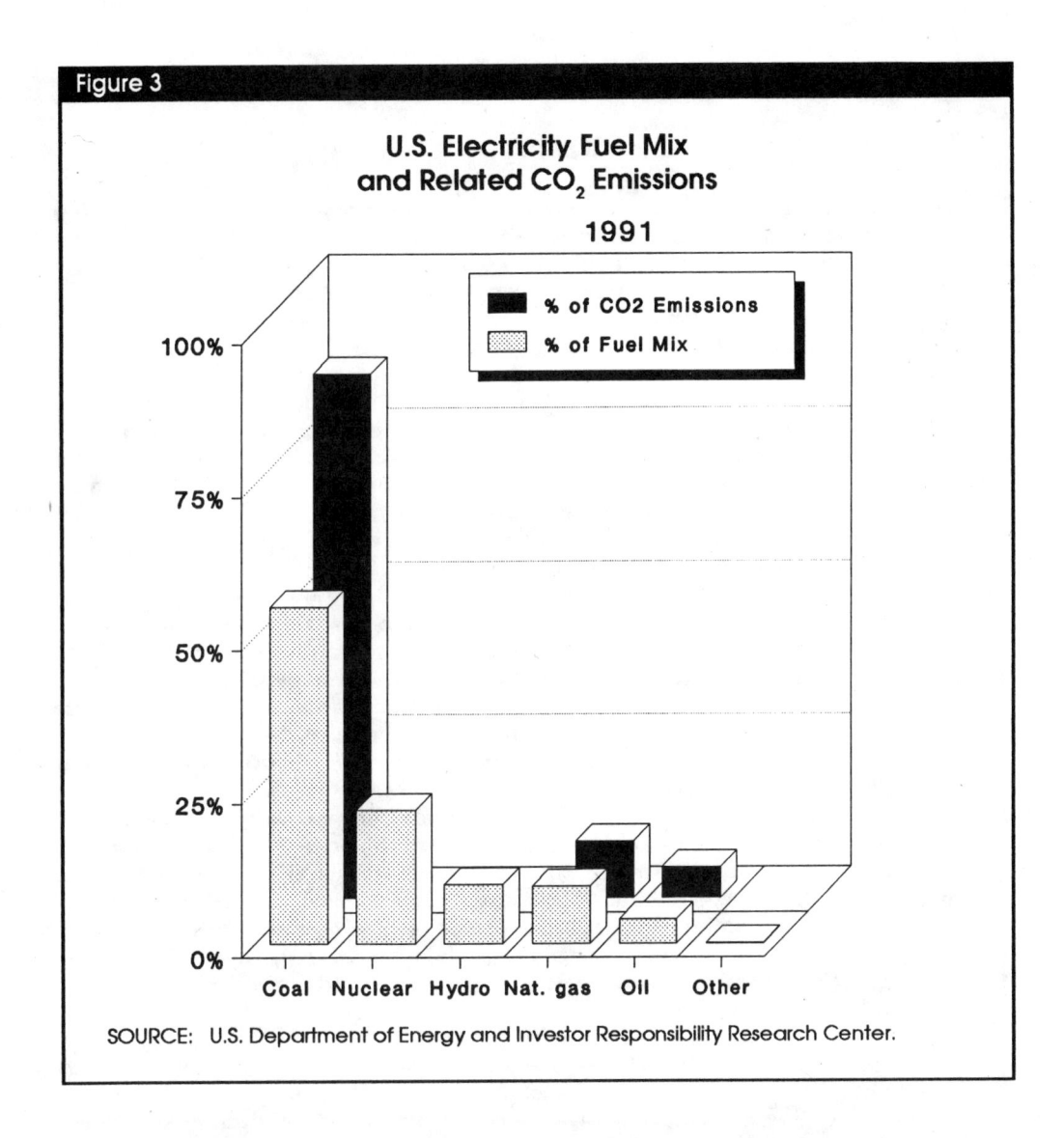

U.S. Electricity Fuel Mix and Related CO_2 Emissions

1991

[Bar chart showing % of CO_2 Emissions and % of Fuel Mix for Coal, Nuclear, Hydro, Nat. gas, Oil, and Other]

SOURCE: U.S. Department of Energy and Investor Responsibility Research Center.

costs to the environment from different kinds of energy development.

The remainder of this chapter is devoted to exploring alternative energy strategies for electric utilities, which will be at the nexus of many vital energy policy decisions for the Electronic Age. The discussion is broken into four main sections. Ways in which global warming could affect the operation of power plants and demand for electricity are considered first. Demand-side measures that promote efficient use of electricity are examined next. Then each of the future electric generating technologies is evaluated in turn. Finally, the critical policy questions are addressed.

The basic unanswered question is whether the world can afford to develop an energy strategy that holds global warming in check—or whether it can afford not to. It is at the core of the greenhouse gambit.

> **Box 5-A**
>
> ### National Energy Strategy
>
> Since the time of the Arab oil embargo in 1973, the U.S. government has enacted three national energy plans. The most recent of these plans—the National Energy Strategy—was unveiled by the Bush administration in February 1991. Each of these plans has focused on America's rising dependence on foreign oil, but they have done relatively little to solve the problem. The United States consumes a greater percentage of imported oil now, in fact, than it did in 1973.[1a]
>
> America's first strategic plan to end reliance on imported oil was President Nixon's "Project Independence," launched in 1974. Domestic oil drilling incentives were an important element of this plan, but its cornerstone was the expansion of nuclear power. By the year 2000, the nation was to derive half of its electricity from nuclear plants. Most of Nixon's energy plan never went on-line, however. The share of nuclear power in America's electricity mix today stands at only 20 percent, and it is not likely to rise any higher before the end of the century.
>
> President Carter declared "the moral equivalent of war" on energy shortages. Donning a cardigan sweater for a fireside chat to the nation, Carter conveyed an austere message that personal sacrifices would be necessary in order to win this energy war. The strategic thrust of his "National Energy Plan" was to develop a range of alternative energy sources, especially synthetic fuels made from tar sands, oil shales and other carbon-intensive sources that America has in abundance. His Synthetic Fuels Corp. never made a dent in the nation's dependence on imported oil, however, because it failed to bring the price of synthetic fuels down to a level that would allow them to compete with conventional energy sources.
>
> President Reagan concluded the best national energy strategy was not to have one. Placing his faith in market forces, he watched the price of oil plummet from a high of nearly $50 a barrel in 1981 (expressed in 1990 dollars) to a low of little more than $10 during his second term. The historic decline demonstrated just how well the market can respond to price signals. As the price of oil rose during the 1970s, utilities switched from oil to cheaper coal, while car buyers traded in their gas guzzlers for thriftier models, and home owners weatherized around their doors and windows to save on their energy bills.
>
> These and other conservation measures enabled the nation's primary energy consumption to remain flat between 1973 and 1985, even while the economy grew 34 percent. While no president—least of all President Reagan—set out explicitly to accomplish this efficiency goal, it was perhaps the most important energy policy achievement of the post-embargo era. The longstanding, lock-step relationship between energy consumption and economic growth was severed—at least temporarily. (During Reagan's second term, primary energy demand increased once again at the same rate as overall growth in the economy.)
>
> President Bush's recent National Energy Strategy reflects elements of all of the energy policy initiatives that have come before it.[2a] It embodies President Reagan's free-market philosophy, making market forces "a keystone of the strategy." Yet it also includes "a national commitment to greater efficiency in every element of energy production and use"—something that was lacking during the Reagan years. The National Energy Strategy also sees a major role for coal in America's energy future, as did President Carter's plan, with billions of dollars earmarked for demonstration of "clean-coal" technologies. And like President Nixon's plan, the National Energy

Strategy seeks to reinvigorate the nuclear option, aiming to nearly double the nation's nuclear generating capacity over the next 30 years.

Bush's program even envisions an expanded role for renewable energy, proposing to have these sources provide fully one-sixth of the nation's electricity requirements by the year 2020. But like the plans preceding it, the National Energy Strategy mainly has a goal of reducing America's oil imports, especially from the Persian Gulf. Since few domestic reserves remain untapped, it seeks permission for oil companies to enter Alaska's Arctic National Wildlife Refuge and portions of the Outer Continental Shelf that have been off-limits to drilling.

Taken together, the National Energy Strategy's 100 proposals offer something for nearly everyone—except those who believe that combatting global warming should be the central aim of the plan. Given its concern for "achieving greater energy security" while relying on free enterprise, the National Energy Strategy opposes taxes on fossil fuels (yet offers further tax breaks for domestic oil and natural gas producers), opposes higher miles-per-gallon standards on automobiles, and fails to set any national goals for efficiency improvements, despite the plan's stated "commitment" to energy efficiency.

As for government research and development funding, the strategy commits more than a billion dollars annually toward the development of new clean-coal technologies and other fossil fuels, yet comparatively little for renewable energy and conservation. For fiscal year 1992, for example, the Department of Energy requested only $164 million for renewable energy programs. Altogether, about 5 percent of the federal government's energy research and development dollars are earmarked for renewables and conservation and, in real terms, support levels have dropped 80 percent since 1980.[3a]

At its core, the new National Energy Strategy keeps the faith in market forces to make the right energy decisions. It assumes that the market already has exploited any cost-effective conservation measures and that further efficiency gains must await higher energy prices. While acknowledging that environmental "externalities" could be reflected in energy prices to influence market preferences, it purposely "rejects influencing the markets with tax policies or aggressive product mandates that narrow or direct consumer choice. The administration believes that forcing radical departures in fuel use and mix upon the economy would have crippling and unacceptable consequences," the National Energy Strategy says.

The Bush administration dismisses the possibility, however, that continued heavy reliance on fossil fuels could lead to irrevocable changes in climate that have "crippling and unacceptable" consequences of their own. In that respect, the National Energy Strategy is no different from the other presidential initiatives that came before it. They all have regarded depletion of fossil energy reserves—rather than the ongoing, related buildup of greenhouse gases—as the overriding, long-term threat to the nation's security.

In Congress, efforts to enact the National Energy Strategy's proposals into law have stalled, largely because of opposition to drilling in the Arctic National Wildlife Refuge. A new bill, stripped of the more controversial provisions, is working its way through Congress in 1992. As for the nation's dependence on imported oil, it is expected to continue rising—passing the 50-percent mark by 1995—if the National Energy Strategy is not adopted in its entirety.[4a]

Effects on Power Generation

By definition, electricity is the fuel of choice for the Electronic Age—and the Electronic Age has arrived. Today, each major customer class (except transportation) consumes less oil, coal and natural gas than it did back in 1973, yet their collective demand for electricity has risen by 50 percent.[16] The switch has raised electricity's share of U.S. primary energy demand by 10 percentage points; now it accounts for 36 percent of the total. Meanwhile, the nation's CO_2 emissions have risen only 7 percent since 1973, despite a 50 percent expansion of the gross national product.

Electrification of the U.S. economy has been an important factor in keeping a lid on the nation's CO_2 emissions while fostering economic growth.[17] Many end-uses of electricity are highly efficient, enabling energy demand to fall as electricity displaces the use of other fuels. Even a coal-burning power plant—with its attendant generation and transmission losses—can bring about a net reduction in CO_2 emissions when its power is substituted for direct combustion of fossil fuels in such activities as making steel and glass. These qualities make electricity the power for the Electronic Age.

Power for the Electronic Age

Each customer class has had its own reasons for turning to electricity in recent years, and some are climate-related. In the residential sector, the advent of affordable air conditioning has played a significant role in the nation's move to the South and West. In the fastest growing regions of the country—California, Texas, Florida and the desert Southwest—air conditioning now is regarded more as a consumer necessity than a modern convenience. If global warming proceeds as forecast, people throughout the nation may come to value air conditioning in much the same way. Already, seven out of 10 new homes feature central air conditioning.

As electricity demand in the residential sector has increased more than 60 percent since 1973, demand in the commercial sector has actually doubled. The introduction of computers and advanced electronic equipment has raised electricity requirements in the commercial sector in two significant ways. First, such equipment draws on electricity as a power source. Second, the equipment produces waste heat, so cooling requirements have risen as well. By the year 2000, computers may draw as much power as lighting fixtures in the commercial sector.[18]

Other factors have been at work in the industrial sector. Use of electricity has grown at a more moderate pace of 37 percent since 1973, reflecting a transition away from heavy industry, less need for space conditioning and greater reliance on electric motors, which have become more efficient over the years. Electricity also has had a deserved reputation among industry as an

Figure 4

Consumer Demand for Electricity

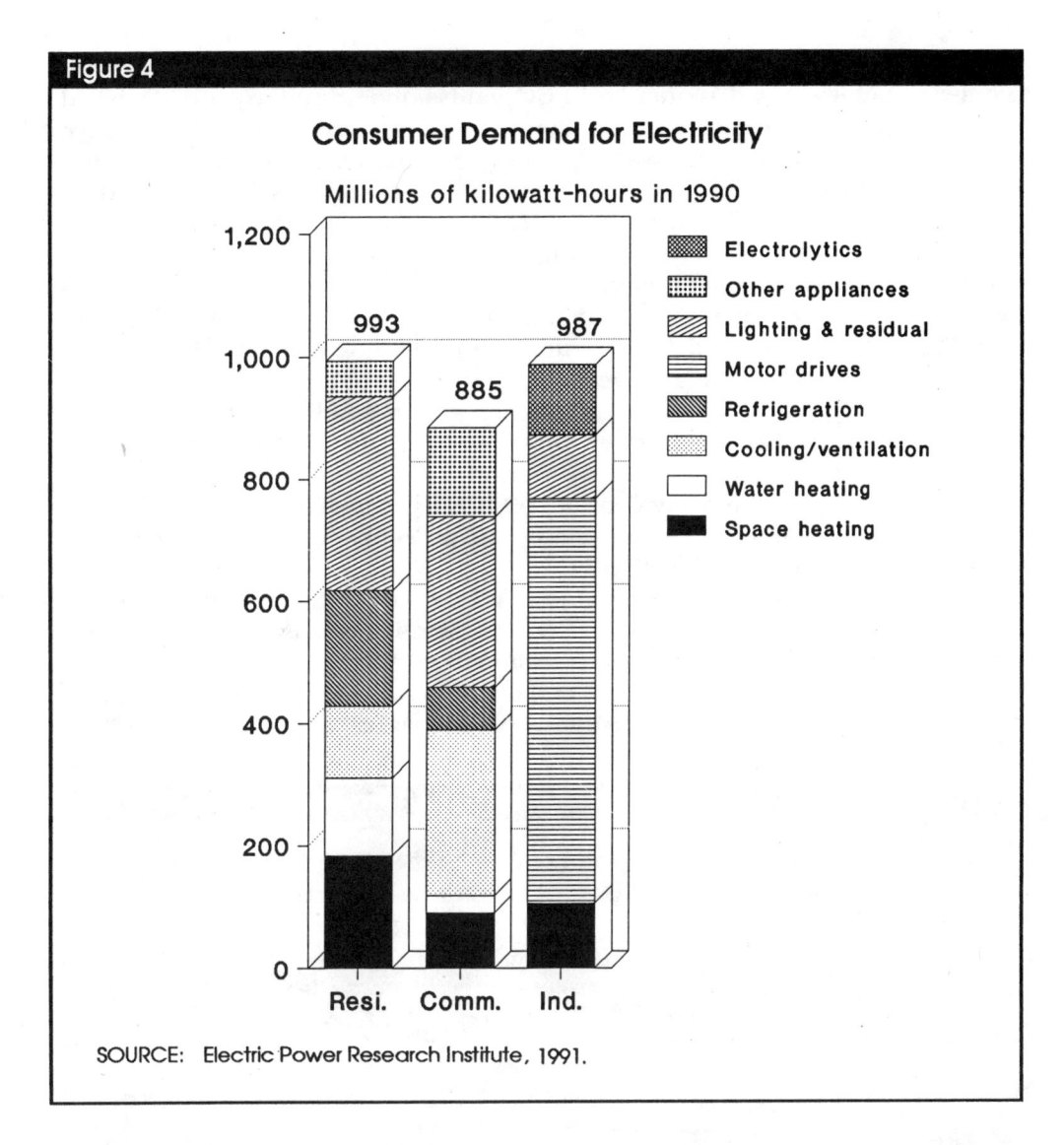

Millions of kilowatt-hours in 1990

- Electrolytics
- Other appliances
- Lighting & residual
- Motor drives
- Refrigeration
- Cooling/ventilation
- Water heating
- Space heating

Resi.: 993
Comm.: 885
Ind.: 987

SOURCE: Electric Power Research Institute, 1991.

expensive, prodigal fuel. On average, two-thirds of the primary energy used to make electricity is rejected as waste heat right at the generating plant. Tacking on transmission losses, only 30 percent or less of the primary energy normally gets beyond the factory gate.[19]

Even so, industry's electricity demand continues to grow as combustion-driven manufacturing processes are converted to more efficient electricity-based ones. This is an important and largely positive development from a greenhouse standpoint, provided that certain efficiency and economic criteria are met. The key determinant in the switchover is whether an electricity-driven process offers sufficient gains in end-use efficiency and energy savings that the inherent inefficiencies of power generation are overcome. For example, an electricity-driven manufacturing process must be at least 2.1 times more energy-efficient than an alternative industrial process that burns *coal* directly in order to achieve a net reduction in CO_2 emissions. Similarly, an electrical process must be at least 2.4 times more energy-efficient than an alternative process that burns *oil* directly—and 3.4 times more efficient than one using *natural gas*—to reduce CO_2 emissions overall.[20] (This analysis assumes that an average kilowatt-hour of electricity generation emits 1.35 pounds of CO_2.)

Five major electrical manufacturing processes pass such a CO_2-efficiency test: electric arc furnaces for making steel; electromagnetic induction heating and conditioning of various other metals; electric melting, conditioning and annealing of glass; infrared curing and drying of automotive paints, textiles, paper, plastics and metals; and freeze concentration of liquid waste materials (as an alternative to heat-producing evaporation and distillation processes).[21] The industry-funded Electric Power Research Institute figures that electricity used to displace fossil fuels in these and other selected manufacturing processes could save the equivalent of 250,000 barrels of oil a day *and* reduce the industrial sector's CO_2 emissions by 17 percent. Yet overall electricity demand would increase only slightly.[22]

Electricity also has the potential to make inroads into the transportation sector and achieve a net reduction in CO_2 emissions. Electric trains rival diesel trains as the most energy-efficient way to ship freight over land. Electric trolleys hold a wide advantage over gasoline- and diesel-powered buses in terms of CO_2 emissions. Even electric vehicles are considered a viable alternative to gasoline-

Table 1

Effects on Electricity Demand of Demand-Side Management (with Incentives) and Beneficial Electrification

	Electricity Consumption (billion kilowatt-hours)	Primary Energy Use (quads)	CO_2 Emissions (billions of tons)
1990			
Baseline	2,865	30	2,100
2000			
Baseline	3,485	37	2,548
Efficiency Improvements	-210	-2.1	-153
Beneficial Electrification	+386	-1.4 to -2.7	-71 to -175
Net Effect (percent change)	+4%	-9 to -13%	-9 to -13%
2010			
Baseline	4,044	42	2,957
Efficiency Improvements	-462	-4.6	-335
Beneficial Electrification	+663	-4.8 to -7.0	-264 to -438
Net Effect (percent change)	+5%	-22 to -28%	-20 to -26%

SOURCE: Electric Power Research Institute, 1990.

powered cars, not so much because of an inherent CO_2 advantage, but because they reduce other smog-producing emissions associated with internal combustion engines. Aggressive use of electric cars, trolleys and trains could reduce CO_2 emissions in the transportation sector by 8 percent, one energy consulting firm has determined, if the power were derived from the existing mix of generating plants.[23]

Thus, while the Electronic Age assures a greater role for electricity in virtually all energy-consuming sectors, policies to ameliorate the greenhouse effect could boost demand for electric power even higher. The Electric Power Research Institute concludes that use of beneficial electric technologies could raise the overall demand for electricity by 663 billion kilowatt-hours in 2010—16 percent above its base-case forecast—while reducing net carbon emissions by 120 million metric tons relative to its base case. Although this analysis makes optimistic assumptions about the market penetration rates of various electric technologies, it suggests that greater electrification of the U.S. economy could reduce primary energy demand by up to 7 quads relative to the base case for 2010.[24] (Greater incentives for demand-side management programs could reduce primary energy demand by another 4.6 quads and hold the overall increase in electricity demand to 5 percent above the baseline forecast.)

Electricity in a Greenhouse World

Of course, global warming itself could affect the demand for power and the reliablility of electric service in the years ahead. While regional forecasts of climate change remain highly circumspect, it is reasonable to assume that rising temperatures would drive air-conditioning loads up in the summer and heating loads down in the winter. Yet there is a lively debate as to whether the greatest temperature changes would occur during the summer or winter months. A similar debate rages about climate change's likely effects on day- and nighttime temperatures.

These are important considerations for electric utilities. They already encounter major fluctuations in the demand for electricity—on both a daily and a seasonal basis—and they dispatch their power plants to accommodate these natural ebbs and flows. Climate change would make it harder for utilities to rely on historical meteorological data when making such planning decisions. Global warming could increase the utilization, or "capacity factor," of power plants if electricity demand increases during periods of traditionally slack demand, or it could spawn shortages of power if demand increases when generating systems already are running near peak capacity.

Demand for electricity usually is lowest at night, when people are asleep, and greatest during the day, when people are busy at work and at school. On that basis, a rise in nighttime temperatures would be good news for most utilities, since it would increase air-conditioning demand at a time when many power plants lie idle. For some utilities, however, a rise in nighttime temperatures also might decrease demand for electric heating during cold-weather months; their plants' capacity factors actually might go down.

> **Box 5-B**
>
> ## Electric Utilities in the 'Greenhouse Summer'
>
> The "greenhouse summer" of 1988 provided a taste of what may lie ahead for America's electric utility industry. At least 56 utility systems across the country experienced record demand for electricity during that summer's extended heat wave. In August alone, consumers' electricity bills rose an average of 15 percent—an extra $449 million—mainly because of higher demand for air conditioning.[1b] Some utilities saw old peak demand records shattered by margins of up to 10 percent. In the Midwest, American Electric Power hit a peak not forecasted to occur until 1996. Detroit Edison activated remote control devices to shut off 55,000 residential air conditioners and 100,000 electric water heaters for brief intervals in order to hold down the system's peak demand. Across the nation, industrial customers on interruptible contracts saw their power supplies cut for hours at a time as utilities struggled to keep up with surging demand.[2b]
>
> In some respects, these customers were the lucky ones. Others had their electrical service disrupted without warning (and without the benefit of discounted, interruptible rates). Power outages and voltage reductions plagued electric utility systems from coast to coast. In one torrid day in Los Angeles, more than 400 electrical transformers blew up, overburdened by record demand for air conditioning. A downtown area of Seattle endured a three-day blackout.[3b] Utility workers at Consumers Power in Michigan had to hose down substations to keep them from overheating. And in portions of New England and the mid-Atlantic states, utilities instituted 5 percent system-wide voltage reductions—"brownouts"—in order to keep parts of their systems from "blacking out" altogether.
>
> To be sure, heat was a problem. But a shortage of water posed an even greater challenge. One coal plant in Arkansas had to shut down for lack of cooling water. Many other coal-fired plants experienced supply disruptions because water levels dropped too low for coal-laden barges to ply the nation's interstate waterways. Nuclear plants also had their share of water-related problems. Northern States Power had to reduce output from its Monticello nuclear plant by 25 percent because its water discharges into the Mississippi River threatened to raise the water temperature above permissible regulatory limits. At the Tennessee Valley Authority's Sequoyah nuclear plant in Tennessee, the temperature of cooling water at the intake pipe reached 83 degrees F, exceeding the design basis for that plant's emergency cooling system.[4b]
>
> Hydropower production was affected most severely. At the start of the summer of 1988, the Tennessee Valley Authority already had lost $110 million in hydroelectric output because of drought persisting since the beginning of the year. By mid-July, TVA's hydroelectric generation—as well as that of the Southeast Power Authority and The Southern Co.—was running 40 to 50 percent below normal. New England Power Co. also reported a 50 percent loss in generation from its hydro plants on the Connecticut River.[5b] Meanwhile, Manitoba Hydro, which normally exports power to the United States, actually had to import electricity from the Dakotas, Wisconsin and Minnesota because of low water levels at its dams. And in the Pacific Northwest, the Bonneville Power Administration cut off power exports to California for the first time in 15 years because of low stream flows.

Daytime temperature increases would present a bigger problem for most utilities. Added load requirements at times of peak demand could stretch generating systems to the limit. Since the "greenhouse summer" of 1988, in fact, summertime heat waves already have resulted in several "brownouts" and "blackouts" in portions of New England, the mid-Atlantic states, metropolitan Chicago and Seattle.[25]

The potential seasonal effects of climate change are harder to generalize. Utilities operating in regions susceptible to hot, humid weather may experience their greatest increases in demand during extended summertime heat waves. If that happens to be the season when electricity demand already is at its peak, reliability of service could suffer. So-called "winter peaking" utilities, on the other hand, may benefit from increased air-conditioning demand, since summer is a time when their generating systems are relatively underutilized. Milder winters may reduce heating demand for such winter-peaking utilities, however, offsetting some or all of the summertime sales gains.

There is also a possibility, of course, that climate change may simply make the weather more unpredictable. If recent years are any indication, summertime heat waves may become more intense *and* wintertime cold spells less frequent yet more severe. Hence, it is the uncertainty associated with climate change—as much as the expected magnitude of the change—that gives utility planners pause. A rising frequency of heat waves, cold outbreaks and severe storms could severely tax the reliability of utility generation, transmission and distribution systems. An increase in the full range of weather-related events—hurricanes, tornadoes, wind storms and ice storms—could wreak havoc on utilities' power poles and transmission lines, in particular, which would increase service and maintenance costs in turn.

Power plants would not be immune to the effects of climate change, either. Higher air and water temperatures can reduce the efficiencies at which power plants operate. A rise in sea level, moreover, would create flooding problems at central-station power plants sited along the nation's coastlines. In most locations, however, the likely problem would be access to too little water rather than too much.

Cooling water is critical for the operation of thermoelectric power plants. Electricity generation ranks second behind agriculture, in fact, as the largest user of water in the United States. While irrigation accounts for virtually all of the water consumed in the country (i.e., not returned to its source after use), thermoelectric power generation has been the fastest-growing consumer of water in recent decades. Between 1960 and 1980, the amount of water consumed at U.S. thermoelectric plants increased sixteenfold, even though power generation from such facilities only slightly more than doubled. The primary reason for this dramatic increase has been greater reliance on evaporative cooling towers in place of once-through cooling systems.[26] Now power plants require more than 130 billion gallons a day for cooling purposes, representing two-fifths of all water withdrawn daily from rivers, lakes and streams around the country.[27]

Some general circulation models indicate that mid-continental, mid-latitude regions may become hotter and drier as a result of climate change, causing a reduction in stream flows. A 3.5 degree F rise in temperature combined with a 10 percent drop in precipitation could reduce stream runoff 24 to 40 percent, depending on the size and characteristics of the watershed. Even with a 10 percent *increase* in precipitation, stream runoff could fall by 18 percent as a result of the increased heat and evaporation.[28]

Higher water temperatures associated with lower stream flows might diminish water quality and make it harder for utilities to comply with water pollution discharge regulations. The temperature of cooling water at the intake pipe could pose a problem as well. During the drought summer of 1988, for example, Westinghouse researchers discovered that water temperatures on some rivers rose to levels that exceeded the design basis for its nuclear plants. Concern that higher inlet water temperatures might foil the reactors' cooling systems prompted Westinghouse to change the design parameters of new reactors and put operating restrictions on some of its existing plants.[29]

Water also is critical to hydroelectric power generation, of course. Since hydro plants—like nuclear reactors—emit no greenhouse gases, more hydroelectric generation may be sought in a greenhouse world. Yet during the hot, dry summer of 1988, production from hydro plants fell 21 percent and was winnowed to its lowest level in 15 years because of low water flows on many of the nation's rivers and streams.[30] An extended drought in California has compromised much of that state's hydro capacity as well, which in "normal" water years provides more than one-fifth of the Golden State's generating requirements.

Global warming may pose especially vexing problems for rivers where there are competing water uses. In California and the hydro-dependent Pacific Northwest, several types of salmon and other anadromous fish species already are threatened with extinction because spawning fish must scale giant fish ladders at dam sites as they swim upstream—and young fry must avoid being sucked through hydro generating turbines on their maiden voyage to the sea. Drought-induced low-water flows and reduced dissolved oxygen content would make the rivers even more perilous for the fish. If hydroelectric generation is sacrificed to bolster their chances for survival—as is now being contemplated in California and in the Pacific Northwest—more expensive generating sources would be called upon to make up the difference. Most of this replacement power is fossil-fueled.

To address emerging hydro problems in the Pacific Northwest, as well as chronic air pollution problems in southern California, the Bonneville Power Administration announced an innovative power-swapping agreement with Southern California Edison in 1991. BPA agreed to export 200 megawatts of additional power from its hydro plants on the Columbia River to SCE's service area during the summer months, so that SCE will have to start up its own fossil-fired power plants less often during that especially smog-prone season. In the fall and winter months, SCE will return the "borrowed" power to BPA by increasing generation at its own fossil-fueled plants, enabling BPA to pass more water over its dams (rather than through its turbines) as young salmon are making their way to the sea.[31]

Such an agreement illustrates the kind of environmentally driven power swap that may become more common among utilities in the event of climate change. If the greenhouse summer of 1988 is any indication, however, BPA may have difficulty holding up its end of the bargain already struck with SCE. In 1988, reduced stream flows limited BPA's summertime export capabilities, while sweltering heat in southern California created especially severe smog and extremely high demand for power. If BPA were forced once again to suspend its hydro exports because of low-water flows on the Columbia River, SCE would have to buy other power on the spot market or crank up its own oil- and gas-fired generating units, which would exacerbate local air quality problems (not to mention the global greenhouse effect). Accordingly, such innovative power swaps will work only to the extent that the weather serves as a cooperative third party in the agreement.

Finally, excessive heat and more frequent drought could affect electricity requirements by raising demand for irrigation water, which must be pumped from underground, and possibly demand for desalinated water to augment the water supply. In California, Southern California Edison and the Los Angeles Department of Water and Power already have evaluated a plan to build a giant $1.5 billion sea-water desalination plant south of Tijuana, Mexico, that would produce up to 100 million gallons of fresh water daily.[32] The desalinated water would serve members of the Los Angeles Metropolitan Water District and the San Diego County Water Authority, while an adjacent power plant would provide as much as 500 megawatts of electricity also for sale in southern California. (Waste heat from the natural gas-fired turbines would be used in the distillation process to remove the salt from sea-water.) SCE has concluded that less expensive desalination options are available and no longer plans to participate in the project.[33] If the project goes forward anyway, Bechtel would design and engineer the plant, and Coastal Corp. would supply the natural gas.

Studies of Impacts on Utilities

The long, hot summer of 1988 sparked new interest in the global warming phenomenon. A flurry of greenhouse-related bills appeared in Congress, which in turn prompted the U.S. Department of Energy, the Edison Electric Institute and the Electric Power Research Institute (EPRI), among others, to launch their own investigations of the possible costs of complying with greenhouse gas controls. Most of these policy-oriented studies have been completed only recently. By contrast, the most comprehensive studies of the possible *direct* effects of climate change on electric utilities were conducted before and during the fateful summer of 1988.

ICF, an environmental consulting firm in Fairfax, Va., began an evaluation in 1985 of global warming's possible impacts on electricity demand and the reliability of service. The U.S. Environmental Protection Agency, the Edison Electric Institute, EPRI and the New York State Energy Research and Development Authority commissioned the work, which provided case studies of the potential effects of climate change on the New York Power Pool and a hypothetical southeastern utility. ICF completed its initial work in December 1987.[34] It subsequently broadened the scope of its research for a report that EPA presented to Congress in October 1988.[35]

Using the general circulation model developed by NASA at the Goddard Institute for Space Studies, ICF calculated that electricity demand on a nationwide basis might increase 14 to 23 percent—above and beyond a base-case forecast—by the middle of the next century as a result of global warming.[36] The additional generating requirements would amount to 400,000 megawatts, much of peaking capacity to meet greater daytime air-conditioning loads. Although the forecast varies by region, each state on average would require the equivalent of 16 additional oil- or gas-peaking units (rated at 250 megawatts each) as well as eight additional coal or nuclear baseload units (rated at 500 megawatts each) to accommodate the extra demand brought about by global warming through 2055. The capital investment required to build these additional plants easily could reach $12 billion to $16 billion per state (in 1992 dollars).

Climate change would be likely to affect regional demand for electricity in different ways, however. In most of the Southwest, southern Great Plains and Southeast, the ICF study projected that the increase in generating demand would exceed the national average, since air-conditioning loads would become virtually continuous all year round. Consequently, requirements for new generating capacity in these regions would be 20 to 30 percent higher than if no warming occurred. The northeastern and northwestern corners of the country, on the other hand, might see no relative increase in electricity demand—or even a 10 percent decline—because wintertime peak heating loads would fall by a greater amount than summertime peak air-conditioning demand would rise. The states where this "wash effect" might occur are Maine, New Hampshire and Vermont in the East, and Washington, Oregon, Montana and Wyoming in the West, ICF concluded. Peak demand in the 41 other continental American states would be likely to rise in the event of climate change. Moreover, hydroelectric generation in states like New York might decline 8 to 10 percent by 2015 as a result of reduced river flows—exacerbating a possible power supply shortage.[37]

As part of its analysis, ICF also made an interesting comparison of the costs and benefits of having utilities engage in a greenhouse gambit. In its two case studies, ICF estimated how much utilities would be likely to spend on new capacity if they took early steps to prepare for a warmer climate, relative to what they would spend if they waited until after the climate had changed. A utility anticipating the change in climate would install efficient baseload generating capacity to provide for increased air-conditioning loads and reduced hydroelectric output. In the long run, it would save money by not having to build short lead-time plants that are more expensive to operate.

In one case study, ICF estimated that the cost for a New York utility would amount to $102 million (in 1985 dollars)—if it took near-term measures in anticipation of climate change—and $174 million if it waited until much of the warming had occurred (assumed in the model to be 1.5 degrees F by 2015). In the other case study, a southeastern utility with larger capacity requirements would pay $212 million for taking near-term measures, compared with $267 million if it waited until after the warming. The potential savings from acting

Figure 5

Effect of Warming on Generating Capacity

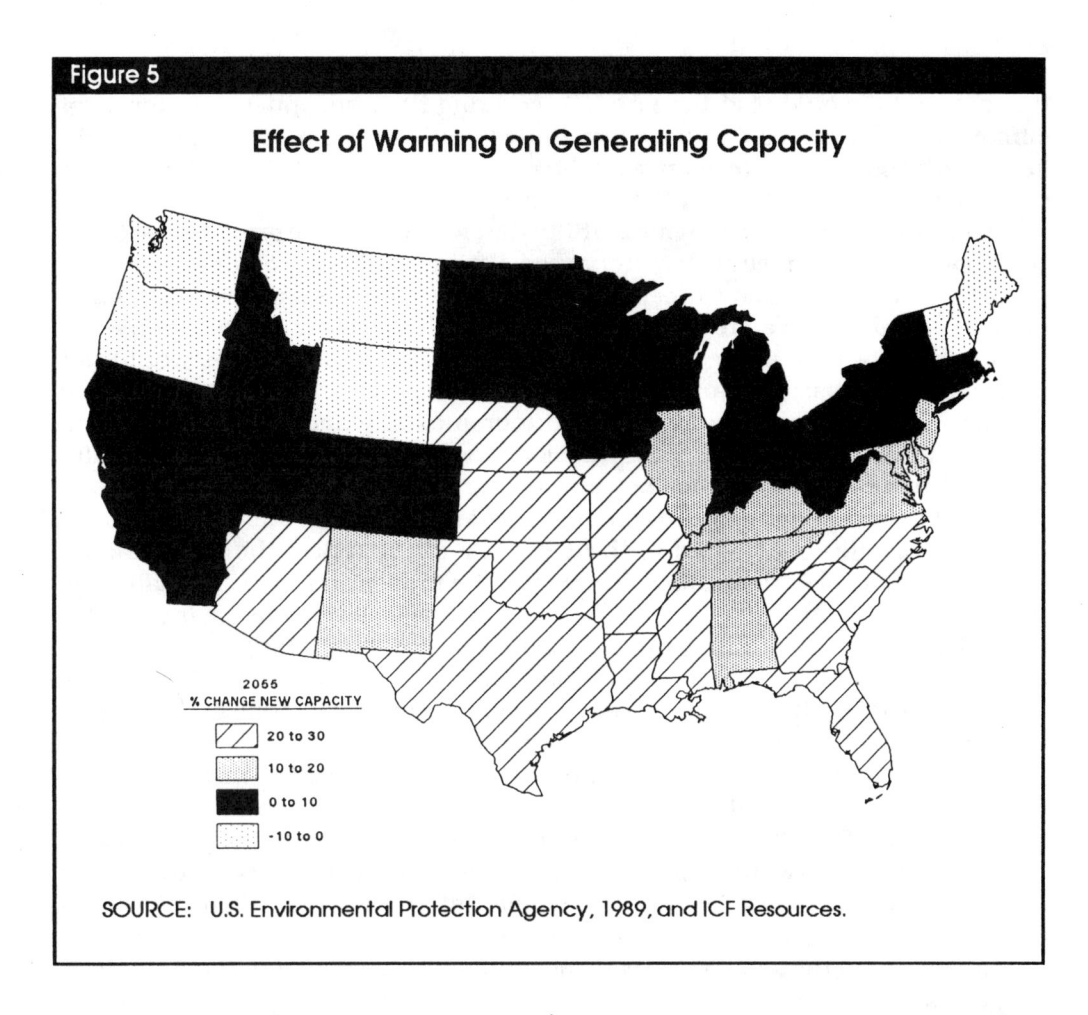

2065
% CHANGE NEW CAPACITY
- 20 to 30
- 10 to 20
- 0 to 10
- -10 to 0

SOURCE: U.S. Environmental Protection Agency, 1989, and ICF Resources.

sooner rather than later, therefore, is $72 million in the case of the New York utility and $55 million for the southeastern utility.[38]

The key to determining whether this gambit is worth taking is to calculate the penalty that would result from wasted expenditures in the event the climate did not change after all. In the case of the southeastern utility, the extra generating capacity and needless fuel switching would impose an $11 million penalty if the climate did not change, compared with $55 million in savings if it did. Thus, the risk-to-reward ratio for taking early action would be a favorable 1:5. A similar analysis for the New York utility yielded an even more favorable risk-to-reward ratio. Given the seemingly manageable cost of such insurance, ICF concluded in its study, "Utility planners should start now to consider climate change as a factor affecting their planning analysis and decisions. Large impacts are not imminent, but the importance of climate change impacts for utility planning are likely to increase over time."[39]

Ironically, the gambit considered in ICF's utility case studies involved building coal-fired, combined-cycle power plants in anticipation of climate change, rather than waiting to build oil- and gas-fired combustion units later on. Either way, fossil fuels would be utilized for the new power plants, despite the fact that it would lead to more CO_2 entering the atmosphere. In terms of climate change, such a response by utilities would signal a positive feedback: Fear of impending climate change leads them to burn more of the very fuels that contribute to the warming in the first place.

In reality, it is hard to imagine a utility that would take the greenhouse effect so seriously as to enact anticipatory measures to address potential warming problems, and then go on to deepen its commitment to the most carbon-rich of all fuels—coal. From a greenhouse perspective, it would make more sense for the utility to build a nuclear or hydro plant—since those sources would not increase CO_2 emissions—or turn to other renewable technologies that have shorter lead-times and could be added incrementally according to emerging climate change trends. The most judicious move of all might be for the utility to invest more in energy efficiency and demand-side management programs, since high-risk capacity questions might be avoided altogether.

The strategies considered in the 1987 ICF study reflect the conventional thinking among electric utilities, however. The nuclear option essentially has been counted out because no American utility has ordered a nuclear plant for more than a decade, and no utility is seriously contemplating a new reactor order in the 1990s (with the possible exception of the Tennessee Valley Authority and perhaps one or two others). Large-scale renewable energy development is not considered a viable option, either, because these technologies account for less than 1 percent of the current generating mix (excluding hydropower) and are still regarded as being too costly and too intermittent in power output. Finally, utility executives have doubted that conservation and efficiency measures would be able to reduce demand by the margins necessary to forestall construction of new power plants. By default, then, coal-fired electricity emerges as the industry's preferred base-load generating option for the future—even though coal already accounts for 85 percent of the industry's CO_2 emissions.

Sorting Through the Options

In 1992, ICF completed another study commissioned by the Edison Electric Institute that delves further into the implications of global warming for future generating supplies and demand-side management programs.[40] This time, however, ICF did not attempt to gauge the direct effects of global warming on electricity demand. Rather, it focused on the cost implications of policies to achieve a stabilization of carbon dioxide emissions at 1990 levels (or reducing those emissions by 20 percent) in comparison with two base-case scenarios. Each scenario makes different assumptions about the development potential of nuclear and renewable resources in relation to coal and natural gas, and each makes different assumptions about demand-side management potential. As a result, one scenario projects much higher costs of compliance with CO_2 control measures than the other, illustrating the importance of these built-in assumptions.

The higher-cost scenario assumes relatively little change from current policies and trends: No new nuclear plants would be built; up to 100,000 megawatts of renewable resources are assumed to be available, but at high development costs; natural gas reserves are relatively limited; and demand-side management programs have only a minor impact on electricity requirements, reducing average demand growth from 2.5 percent to 2.3 percent a year. Under these circumstances, the nation's demand for electricity would grow by more than 75 percent over the next quarter-century, and the amount of coal-fired generating capacity would increase from approximately 300,000 megawatts today to 560,000 megawatts in 2015. Generation from coal would increase from 58 percent of the fuel mix today to 65 percent of the total by 2015. As a result, carbon emissions in the utility sector would double over the next quarter century—to 1 billion metric tons annually by 2015.

The next step in the recent ICF analysis was to consider what would happen if policies were instituted to stabilize CO_2 emissions at 1990 levels or reduce emissions 20 percent from those levels. In either case, coal would take a tremendous hit. To stabilize CO_2 emissions in the utility sector by 2005, and keep them at 1990 levels thereafter, generation from coal would drop by 75 percent. If a further 20 percent reduction in CO_2 emissions were required by 2015, coal generation would drop by a total of 87 percent. Coal plants would operate at a low capacity factor of 29 percent in 2015, and they would make up only 21 percent of the fuel mix, compared with 58 percent today.

The final step in the ICF study was to estimate how much such controls would cost. In this higher-cost scenario, no new coal plants would be constructed regardless of whether the call is for CO_2 stabilization or a 20 percent CO_2 reduction. Since no new nuclear plants would be built, either, and only limited amounts of high-cost renewables would be developed, dependence on natural gas would grow immensely. Demand for natural gas as a generating source would soar by 500 percent, even under the CO_2 stabilization policy. By 2015, natural gas would provide 57 percent of the nation's generating requirements, compared with just 13 percent today. Such a surge in demand would exceed existing natural gas pipeline capacity by 30 percent, and delivery prices would more than triple to $7.50 per million Btu. Largely because of higher prices for gas-fired generation, the projected annual cost of complying with the CO_2 stabilization goal would be $24 billion in 2005 and as much as $90 billion by 2015. A 20 percent reduction in CO_2 emissions from 1990 levels would impose an even greater annual cost—$52 billion in 2005 and $118 billion in 2015.

The lower-cost scenario developed by ICF is built on more optimistic assumptions about the availability of generating sources other than coal and natural gas, and it forecasts greater penetration of demand-side management programs. The resulting efficiency gains limit growth in electricity demand to only 1.6 percent a year, instead of 2.3 percent in the higher-cost scenario, so that the total increase in electricity demand through 2015 is held below 50 percent. In addition, new nuclear plants are assumed to provide up to 20 percent of the capacity additions over the next 25 years, and up to 300,000 megawatts of renewable resources are assumed to be available at affordable development costs. Finally, domestic natural gas reserves are projected to be 15 percent larger in this lower-cost scenario.

Table 2

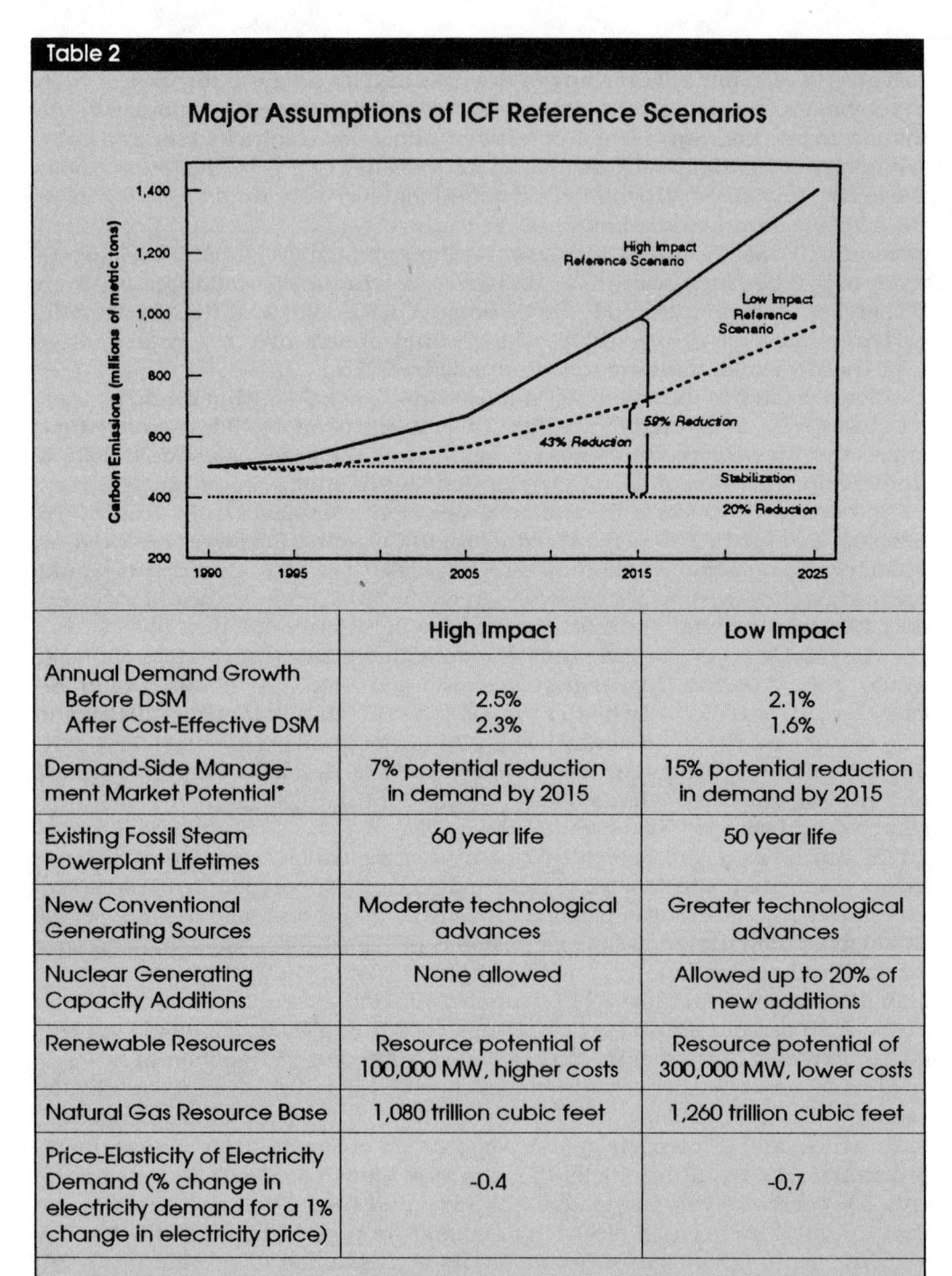

Major Assumptions of ICF Reference Scenarios

[Chart: Carbon Emissions (millions of metric tons) from 1990 to 2025, showing High Impact Reference Scenario rising to ~1,400; Low Impact Reference Scenario rising to ~900; with markers for 59% Reduction, 43% Reduction, Stabilization, and 20% Reduction]

	High Impact	Low Impact
Annual Demand Growth Before DSM After Cost-Effective DSM	2.5% 2.3%	2.1% 1.6%
Demand-Side Management Market Potential*	7% potential reduction in demand by 2015	15% potential reduction in demand by 2015
Existing Fossil Steam Powerplant Lifetimes	60 year life	50 year life
New Conventional Generating Sources	Moderate technological advances	Greater technological advances
Nuclear Generating Capacity Additions	None allowed	Allowed up to 20% of new additions
Renewable Resources	Resource potential of 100,000 MW, higher costs	Resource potential of 300,000 MW, lower costs
Natural Gas Resource Base	1,080 trillion cubic feet	1,260 trillion cubic feet
Price-Elasticity of Electricity Demand (% change in electricity demand for a 1% change in electricity price)	-0.4	-0.7

* DSM potential reductions are over and above reductions achieved by standards and naturally occurring efficiency improvements.

SOURCE: ICF Resources, *Assessment of Greenhouse Gas Emissions Policies*, 1992.

With electricity demand growing more slowly—and a wider range of fuels providing new increments of supply—the utility sector's carbon emissions do not rise nearly as much under this lower-cost scenario. Instead of doubling, carbon emissions rise only 40 percent by 2015, to 709 million metric tons. Consequently, a policy to stabilize CO_2 emissions at 1990 levels, or to reduce emissions 20 percent from those levels, becomes much easier to achieve. With reduced growth in electricity demand, there is no need for new coal plants through 2005, and only a small amount of fuel switching to natural gas is required by then to achieve CO_2 stabilization. Accordingly, the projected annual cost of achieving a stabilization target in 2005 amounts to only $5 billion, compared with $24 billion in the higher-cost scenario.

After 2005, a great deal of nuclear and renewable generating capacity is assumed to come on-line in the lower-cost scenario, regardless of whether there is a program to control CO_2 emissions. By 2015, nuclear generation is forecast to nearly double from current levels. At the same time, renewable resources would generate nearly as much electricity as nuclear power, representing an even greater percentage increase from today's levels. If a program were enacted to stabilize CO_2 emissions by 2015, reliance on nuclear and renewable generating sources would increase still further. Instead of generating 20 percent of the nation's electricity requirements in the lower-cost base-case scenario, nuclear and renewables would provide 34 percent of the generating mix to stabilize CO_2 emissions in 2015. In case of a 20 percent reduction in CO_2 emissions by 2015, nuclear and renewables would generate 39 percent of the nation's electricity and account for two-thirds of the CO_2 reductions necessary for electric utilities to meet this tougher policy goal.

The bottom line is that natural gas shoulders a much lighter generating burden in this lower-cost scenario. Even with a 20 percent cut in CO_2 emissions, natural gas's share of the fuel mix would rise to only 27 percent in 2015, instead of 57 percent in the higher-cost scenario. And once again, lesser dependence on natural gas would have a tremendous bearing on the total cost of achieving the CO_2 targets. The projected cost of reducing CO_2 emissions by 20 percent in 2015 is $27 billion annually in this lower-cost scenario, instead of $118 billion. The projected cost of stabilizing emissions in 2015 is $13 billion annually instead of $90 billion.

There is good news and bad news in these cost estimates. The good news is that steps can be taken to reduce the projected cost of CO_2 controls by 70 to 80 percent—and reduce the projected increase in electricity prices from 30 percent above the base-case forecast levels to only 10 percent above those levels. The bad news, as noted by Edison Electric Institute, which commissioned the ICF study, is that: "For the [low-cost] case to be more likely to occur than the [high-cost] case, a number of policy changes would be needed, including greater incentives to reduce the costs of demand-side management and energy efficiency measures; greater efficiency improvements in conventional power plants; the reemergence of nuclear energy; greater development of renewables; and even greater expansion of the natural gas resource base."[41]

The authors of the ICF study are upbeat in their conclusions, nevertheless: "In summary, the [low-cost] scenario provides a framework under which carbon

Figure 6

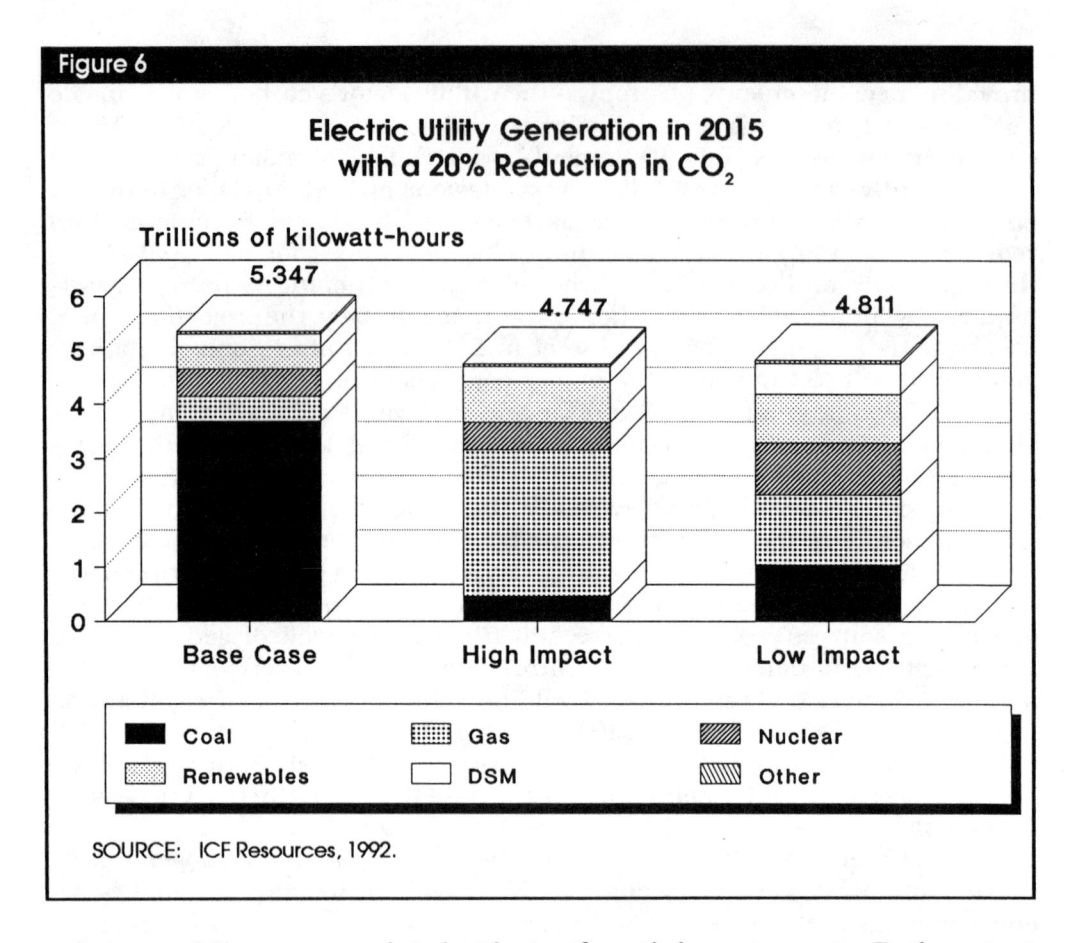

Electric Utility Generation in 2015 with a 20% Reduction in CO_2

Trillions of kilowatt-hours

- Base Case: 5.347
- High Impact: 4.747
- Low Impact: 4.811

Legend: Coal, Gas, Nuclear, Renewables, DSM, Other

SOURCE: ICF Resources, 1992.

policies could be accommodated with significantly lower impacts. To the extent that the elements of this framework do not now exist, either because of regulatory, financial or other barriers, there is a significant opportunity for policymakers to shape a future setting that has fewer carbon emissions independent of other greenhouse gas policies."[42]

Utilities Respond with Fire and 'ICE'

Electric utilities are highly ambivalent about whether to seize such an "opportunity...to shape a future setting that has fewer carbon emissions." As the nation's largest emitter of carbon dioxide, accounting for 36 percent of the U.S. total, the electric utility industry officially favors a "no regrets" policy on global warming, whereby CO_2 emissions will be reduced if such actions can be justified on grounds in addition to climate change. Yet the long-range forecast is for further electrification of the U.S. economy—with coal leading the way for new base-load capacity. Consequently, the electric utility industry's share of U.S. carbon emissions may reach 45 percent over the next quarter-century.[43]

Because of this large—and growing—contribution, proponents of action on global warming believe American utilities are a requisite choice for a carbon abatement strategy. If this industry does not lead on combatting the greenhouse effect, they argue, no other industries are likely to follow. Utility executives insist, on the other hand, that their industry must not be singled out for emission controls. Since American utilities are responsible for only 7.5 percent of the world's CO_2 emissions, and only 4.5 percent of the total greenhouse gas budget, even a wildly successful carbon abatement program—in which the domestic industry's CO_2 emissions fell by, say, 90 percent—would shrink the worldwide greenhouse gas budget by less than 4 percent. Accordingly, American utility officials believe their efforts must be part of a multilateral approach to the global warming problem, which involves many industries in many countries that emit many kinds of greenhouse gases.

To hone this message, the Edison Electric Institute has convened a Global Climate Change Issues Group among its member companies. The group points out that while electricity demand is rising, CO_2 emissions per unit of gross national product have been falling for many years. Accordingly, consumers should be encouraged to continue switching from selected combustion-driven processes to more efficient electric-driven alternatives. In addition, the group has noted the great potential for efficiency improvements within the electricity sector that should lead to further reductions in CO_2 emissions. Nevertheless, the greatest concern of utility chief executives remains the potentially high and comparatively certain costs of CO_2 abatement relative to the largely uncertain and unquantifiable costs of global warming. The greatest dissension within the industry, in fact, appears to be on this fundamental question of whether global warming is a problem in need of practical solutions.

In the spring of 1991, a group of two dozen electric utilities and coal-related interests launched a print and radio advertising campaign to cast doubt on the global warming theory. The Information Council on the Environment (or "ICE") asked in its advertisements: If the Earth is getting warmer, why are certain regions of the globe actually cooling off? The ads went on to claim that weather records for Minneapolis, Minn., Albany, N.Y., and the state of Kentucky show a decline in average temperatures in recent decades, instead of an increase. "Facts like this simply don't jibe with the theory that catastrophic global warming is taking place," the ads concluded. Members of ICE include The Southern Co., one of the largest coal-burning utilities in the nation; the Western Fuels Association, which markets coal to electric cooperatives in the West; and the National Coal Association. The Edison Electric Institute also contributed funding to support polling of consumers' reactions to the ads.[44]

The ICE campaign drew fire from many quarters and even created a rift within the utility industry. "I'm not sure it's appropriate to deal with a subject such as global warming with what I would call a slick ad campaign," commented Mark DeMichele, president of Arizona Public Service Co., in an industry trade journal. "My feeling is that there are a number of things you can do, such as increasing energy efficiency, that make sense whether the Earth is getting warmer or not," he said.[45]

Some climatologists further criticized the ICE advertisements for offering selective information that overlooked the preponderance of evidence of warming from temperature stations around the globe. The most highly regarded temperature records at the National Climatic Data Center in Asheville, N.C., do not even support the claim that temperatures in places such as Minneapolis are in fact falling.[46] Two scientists—themselves skeptics of global warming claims—later asked ICE to withdraw their names as science advisers to the group.

While the ICE campaign appears to have melted down, the effort is indicative of the desire by some utilities and fuel suppliers to sway public opinion against calls for action on the greenhouse effect. As DeMichele of Arizona Public Service has observed, "The greenhouse effect is not fully understood, and we can urge Congress to wait by claiming 'my scientific studies can beat your scientific studies.' But Congress is driven by public moods, not by in-depth studies. Scientists don't re-elect senators, voters do. And many voters—particularly baby boomers who are raising families and worrying about quality-of-life issues—don't want excuses. They want action."[47]

Taking Action

In 1991, three electric utilities did announce that they would take steps to reduce greenhouse gas emissions in an effort to appease the global warming threat. As such, they became the first (and so far the only) major American companies to make such publicly stated commitments. Two of the utilities—SCEcorp and New England Electric System—are investor-owned. The third utility, the Los Angeles Department of Water and Power, is one of the nation's largest municipally owned utilities. (A fourth utility, Pacific Gas and Electric, has pledged to meet its load growth requirements through the end of the century with a combination of renewable resources and demand-side management programs, but it has not committed to specific CO_2 reduction targets.)

Southern California Edison, a subsidiary of SCEcorp and the nation's second largest investor-owned utility, was the first to offer such a plan. In the spring of 1991, SCE pledged to work toward a target of reducing its CO_2 emissions by 20 percent—from 31.8 million tons in 1988 to 25.9 millions by 2010. SCEcorp Chairman John Bryson explained at the time of the announcement, "Taking prudent, reasonable, economical steps to reduce CO_2 emissions is warranted by current scientific understanding of the potential for global warming. This 'no regrets' approach means that, whether or not CO_2 emissions are eventually determined to cause global warming, Edison will not be sorry it took early action."[48]

Two-thirds of SCE's expected reduction will come from the institution of demand-side management programs. The rest will be the result of adding 400 megawatts of non-carbon, renewable power plants to its generating system; converting 1,500 megawatts of existing capacity to combined-cycle operations, which run more efficiently; and phasing out the use of oil as a fuel source. SCE said the savings would be achieved without significant costs to its electric customers.

Concurrent with SCE's announcement, the Los Angeles Department of Water and Power, one of the nation's largest municipally owned utilities, pledged to reduce its own CO_2 emissions from 17.7 million tons in 1989 to 14.1 million tons in 2010—also a 20 percent reduction over the period. The reductions would be achieved mainly through demand-side management programs and greater use of renewable energy sources, although it also plans to fund tree-planting efforts to sequester carbon from the atmosphere. The Los Angeles utility estimated that the cost of implementing its plan would be $12 million annually through 2000.[49]

Six months after these announcements on the West Coast, New England Electric System, based in Westborough, Mass., came out with its own plan to reduce or offset 20 percent of its CO_2 emissions—which total 15 million tons annually—by 2000. NEES will invest up to $20 million a year in renewable power plants and about $100 million a year in demand-side management programs. It also plans to increase purchases of electricity from independently owned cogeneration and hydro facilities.

NEES also will solicit proposals for more exotic offset strategies. It may offer payments of up to $2 per ton of sequestered carbon (or carbon-weighted equivalent) to those who plant trees, recover methane, recycle CFCs, use coal ash in cement, or enhance marine algal growth to offset CO_2 emissions from NEES's fossil-burning power plants. "We think we can bring about a great deal of environmental improvement at a modest price," said John W. Rowe, the utility's president and chief executive officer, as the abatement program was unveiled.[50] Announcements such as these are significant because, for the first time, corporate chief executives have acknowledged the value of engaging in a greenhouse gambit—taking early steps to counter the threat from global warming.

Yet when one considers the particular circumstances of these utilities, it is clear they had relatively little to lose in making these commitments—and a public relations bonanza to gain. Both SCE and the Los Angeles Department of Water and Power, for example, operate in the most heavily regulated air quality management region in the nation. Because of intractable smog problems in and around Los Angeles, the development of new fossil-based generating sources in that region has been virtually precluded anyway. When these companies examined their regulatory compliance strategies in light of these local regulations and the 1990 Clean Air Act, they recognized that their CO_2 emissions were going to decline in any case.

New England Electric's situation is similar. Once heavily dependent on oil as a fuel source, NEES has made a concerted effort to convert much of its existing capacity to coal, which has saved on fuel costs but raised compliance issues with the Clean Air Act. Recently, NEES decided to increase its reliance on natural gas and renewable fuels to round out its generating mix. Moreover, it has boosted its spending on conservation and efficiency programs, largely because Massachusetts regulatory authorities allow the company to earn a handsome return on its demand-side investments.

As a result, NEES and the two West Coast utilities have been able to tell the world of their lauded plans to curb CO_2 emissions, even though existing programs and strategies were mainly cast in a new light. While some may dismiss these machinations as a clever public relations ploy, the announcements achieve three important objectives nonetheless. First, they lend credibility to the global warming issue by acknowledging the potentially serious nature of the problem—and utilities' role in contributing to it. Second, the announcements put subtle pressure on other utilities (and perhaps companies outside the utility industry) to make public their own commitments to curb emissions of greenhouse gases. Third, and perhaps most important, they demonstrate that a rise in CO_2 emissions in the utility sector is not a fait accompli. On the contrary, each of the companies stressed that it expects to offset 20 percent of its CO_2 emissions without raising electricity prices appreciably or lowering the quality of service.

Whether the 1991 announcements are the start of a trend remains to be seen. The evolving state of knowledge concerning global climate change and other factors of more immediate consequence to the industry surely will be important factors. In any event, electric utilities are likely to remain on the hot seat when it comes to global warming. Fossil-fired power plants are the largest sources of carbon dioxide emissions, and their emissions are easily monitored, making them obvious targets for regulatory controls. In addition, an oversight structure is already in place to check on the industry's compliance with new emission controls—if and when they are enacted. (Significantly, the new Clean Air Act requires electric utilities to begin reporting their CO_2 emissions in 1993 as a baseline for possible future reductions.)

As representatives of a "public service" industry, electric utility officials also know they are hard pressed to oppose environmental proposals that garner wide support among the public. Consequently, elements within the industry seem likely to continue sewing doubt about the scientific evidence underlying the greenhouse debate, making it harder for the public to form a consensus on the issue. A broader spectrum of utility executives is sure to highlight the possibly adverse consequences of CO_2 abatement programs on electricity consumers and the economy in general.

If the announcements by SCE, NEES and the Los Angeles municipal utility are any indication, however, a rising number of utilities may begin to stress the value of demand-side management programs and noncarbon-generating sources in securing a brighter future for their industry and the world. The greater the role these demand and supply components play in utilities' business plans, the more the industry can say that electrification of the economy is helping to alleviate the greenhouse effect rather than exacerbating it. Such demand and supply considerations are examined in detail in the next two sections of this chapter.

Energy Efficiency Potential

The electric utility industry is in a vaunted position as the world enters the Electronic Age. Demand for its product is rising because electricity is the power of choice for computers, appliances and other eletromechanical devices that are the wave of the future. Electricity may have even more to recommend it in a greenhouse world. It could beat the heat by powering air conditioners and heat pumps, turn sea-water into desperately needed drinking and irrigation water, and provide highly efficient, noncombustive power for industry and transportation.

The dilemma appears to be in how to generate tomorrow's electricity. If coal and, to a lesser extent, natural gas and oil are the primary fuel sources, they will exacerbate the greenhouse effect by adding more carbon dioxide, methane and nitrous oxide to the atmosphere. Consequently, nonfossil options like solar and/or nuclear power seem destined to become the dominant long-term supply choices. And yet, framing the debate in this way shows only half of the picture. There is a demand-side component to consider as well. Capacity planners must determine the extent to which conservation and efficiency measures are able to serve in place of new power plants before they make their generating decisions.

'Unselling' Electricity

Historically, demand for electricity has grown by leaps and bounds. Between 1950 and 1973, the annual rate of demand growth was 8.2 percent, doubling the nation's generating requirements every nine years or so. Since 1973, however, electricity demand has risen only 2.7 percent a year on average. Reduced economic growth, structural shifts in the economy and electricity price increases have contributed to this slowdown, but more efficient use of electricity has been the greatest factor of all. Had electricity use per unit of GNP remained constant at 1973 levels, generating requirements would have exceeded 4,000 billion kilowatt-hours (kWh) in 1990; instead, they were only 2,700 billion kWh.[51]

Now U.S. electricity consumption is not expected to reach 4,000 billion kWh until at least 2010 (concerns about the greenhouse effect notwithstanding.) The National Energy Strategy put forward by the Bush administration forecasts that electricity demand will rise 2.5 percent a year through 2000, and grow at a reduced rate after that, reflecting slower expansion of the population and the economy after the turn of the century. Even so, the National Energy Strategy projects that the nation's demand for power will rise by nearly 30 percent—or roughly 750 billion kilowatt-hours—during the 1990s alone, necessitating a capital investment of $100 billion to $200 billion in new generating equipment.[52]

The Electric Power Research Institute forecasts an even slower rate of electricity growth over the period, with annual demand rising 1.8 percent through 2000, and 1.5 percent from 2000 through 2010.[53] Two vital assumptions are embedded in EPRI's forecast. First, electricity consumption in 2000 is projected to be 8.5 percent lower than what it would have been without the advent of government-mandated appliance efficiency programs, technology improvements and price-induced conservation measures. Second, demand-side management programs instituted by utilities are expected to reduce year-round electricity demand by an additional 3.0 percent in 2000 (and summer peak demand by an additional 6.7 percent) relative to the amount of demand that would have prevailed with the government programs, technology improvements and consumer conservation initiatives acting on their own.

All told, utility demand-side management (DSM) programs should forestall 100 billion kilowatt-hours of electricity demand in 2000—equivalent to the output of 450 100-megawatt combustion turbines—according to EPRI. Given

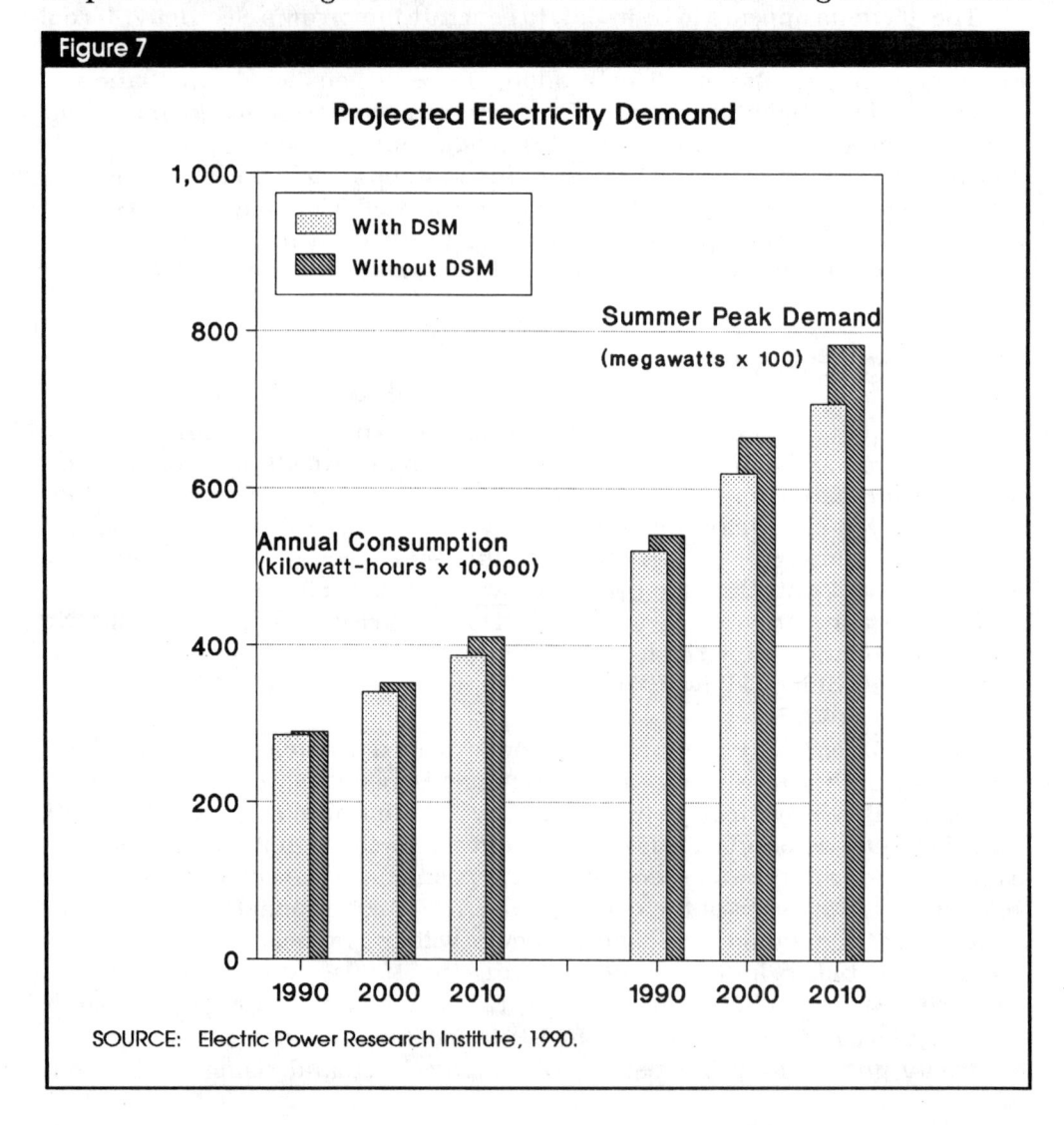

Figure 7

Projected Electricity Demand

SOURCE: Electric Power Research Institute, 1990.

favorable regulatory incentives, EPRI believes that utilities would invest even more in their DSM programs, enabling electricity savings to double to 200 billion kilowatt-hours in 2000.[54]

"The vision of an industry spending money to 'unsell' its own product is in part what makes the DSM story so captivating," extols Veronika Rabl, EPRI's DSM program manager. "Intentionally reducing electricity sales seems to contradict the very essence of U.S. business philosophy. However, this is only counter-intuitive if one thinks of utilities as being solely in the commodity business of selling kilowatt-hours," Rabl says. "If instead one remembers that customers don't buy kilowatt-hours but buy the services that electricity powers, and if one recognizes these energy services as a utility's products, then DSM makes perfect sense. From this perspective, energy efficiency programs are not a means of unselling a utility's product but an opportunity to enhance the value of electricity and to broaden the utility's business."[55]

In essence, DSM has become a way for utilities to increase their overall share of the energy market by emphasizing the efficiency advantages that electricity has to offer. While the net result is that electricity sales increase relative to the amount of generation that would have occurred without customers switching from other fuels, DSM programs at least hold the increase in generation to more manageable levels. More important from a greenhouse perspective, the combination of fuel switching and DSM programs is able to shrink primary energy demand (and, hence, carbon dioxide emissions) relative to baseline projections, making the strategy suitable as a greenhouse gambit.

A recent EPRI report on the potential for saving energy and reducing CO_2 emissions with electricity illustrates this important point.[56] In EPRI's base case, demand for electricity (and carbon emissions) increases 22 percent between 1990 and 2000, and 41 percent between 1990 and 2010. Substitution of beneficial electric technologies for combustion-driven technologies raises electricity demand by an additional 11 percent in 2000 (or by a total of 33 percent relative to 1990 levels). In conjunction with DSM, however, the net generation increase is held to 26 percent in 2000. Similarly, for 2010, beneficial electric technologies increase electricity growth 64 percent (relative to 41 percent in the base case). But in conjunction with DSM, the increase in generation is held to 47 percent in 2010, only a few percentage points above the baseline.

If today's mix of generating plants were still used in 2010, carbon emissions within the utility sector would increase by the same amount as the sales increase (i.e., 47 percent as a combination of demand growth, electrification and DSM programs). In all likelihood, utilities' greater reliance on coal would bring about an even larger increase in carbon emissions. As the utility sector's carbon emissions grow, however, *primary* energy use and *total* carbon emissions would fall relative to base-case projections—shrinking by 9 to 13 percent in 2000 and 20 to 28 percent in 2010. Expressing these carbon emissions relative to 1990 levels, substitution of beneficial electric technologies in combination with DSM would hold the total carbon increase to 6 to 11 percent in 2000 and 4 to 12 percent in 2010 (excluding changes in primary energy demand not related to

(continued on p. 380)

Box 5-C

Sizing Up Energy Savings

Energy efficiency experts believe there are myriad ways to save energy by substituting efficiency innovations for direct consumption of fuel. The Electric Power Research Institute has identified at least 50 technically feasible means of saving electricity. The Rocky Mountain Institute, a private research group, has come up with more than 1,000 specific electricity-saving measures, nearly all of which, it says, are cost-effective today. Many technological innovations have appeared only recently. Keeping up with the state-of-the-art "is important," stresses Amory Lovins, the Rocky Mountain Institute's research director, "because over the last five years the quantity of electricity you can save has roughly doubled and its real cost has fallen roughly threefold."[1c] Following are examples of energy-saving technologies now available to electricity consumers.

Lighting: Lighting is the largest source of electricity consumption in the residential and commercial sectors, accounting for nearly one-third of their total demand. Simply replacing inefficient lights with highly efficient ones could reduce the nation's CO_2 emissions by 4 percent. Compact fluorescent lamps consume only one-quarter as much electricity as standard incandescent bulbs. Similar savings are possible with a new electronic bulb that uses a radio signal to excite gases and cause a phosphor coating inside the glass to glow. These more efficient lights cost $10 to $20 but they also last up to 14 years and pay for themselves many times over through energy savings.[2c] In the commercial sector, use of such high efficiency lamps as well as high pressure sodium lamps and metal halide lamps and installation of high efficiency ballasts, occupancy sensors and timers, reflectors, current limiters and daylighting controls can reduce average lighting requirements by up to 40 percent.[3c]

Windows: Approximately 5 percent of all the energy consumed in America is used to offset heat losses and gains through windows—equivalent in energy to the flow of the Trans-Alaska Pipeline.[4c] A single pane of glass has an insulating value of R-1; double panes raise the value to R-2; and triple panes, to R-3. Treating the windows with low-emissivity coatings (which transmit visible light inside a room while letting less heat out) raise insulating values still further. "Superwindows" achieve an R-8 insulating value by taking three panes of glass with low-emissivity coatings and adding invisible argon gas (which has a lower heat conductivity than air) in the spaces between the panes.[5c] As a replacement for single-glazed windows, such superwindows save enough energy in an electrically heated home to reduce CO_2 emissions from a coal-fired power plant by nearly 10 tons a year.[6c]

Refrigeration: In non-electrically heated homes, old refrigerators generally are the largest source of electricity demand. Refrigerators built in the early 1970s consume an average of 1,725 kilowatt-hours (kWh) a year. The average for models built in 1980 is 1,280 kWh, and the average for those built in 1990 is only 920 kWh. On that basis, replacing a typical 1973 refrigerator with an efficient new model will reduce CO_2 emissions from a coal-fired power plant by more than a ton a year.[7c] Studies by the Department of Energy indicate that electricity consumption in new refrigerators could drop to 650 kWh and still be cost-effective. More stringent appliance efficiency standards scheduled to go into effect in 1993 should assure that this level of performance is achieved.[8c] Congress approved appliance efficiency legislation in 1987 that sets minimum standards for 13 categories of household appliances, including refrigerators, water heaters, air conditioners, heat pumps and fluorescent lighting. As the appliance stock turns over, these standards should save consumers an estimated $44 billion and cut peak electricity demand by 30,000 megawatts.[9c]

Heating and cooling: Thermal performance improvements—added insulation, weatherstripping, caulking around drafty windows and the like—can reduce whole-house electricity consumption by 25 percent for heating and cooling purposes.[10c] More efficient heating and cooling equipment can cut electricity consumption still further. The Electric Power Research Institute estimates that new central air conditioning units consume 25 percent less electricity on average than those now installed in most homes, and new room air conditioners consume 20 percent less electricity. Heat pump water heaters used in place of conventional electric resistance water heaters can reduce electricity consumption by 50 percent. Solar collector panels used to pre-heat water for electric resistance water heaters can reduce electricity demand by an average of 46 percent, EPRI estimates.

Electric resistance space heating is a very inefficient use of energy, however. While high-efficiency heat pumps can reduce electricity consumption by up to 50 percent relative to conventional electric resistance heaters, high-efficiency oil and gas furnaces generally are a better bet, especially in northern climates. Replacing an electric resistance space heating system with a gas furnace in a northern home (with 1,850 square feet of living space) reduces CO_2 emissions from a coal-fired power plant by 23 tons a year—making it the single greatest step one can take to reduce CO_2 emissions in the home.[11c]

Electricity requirements for water heating in the commercial sector can be cut by 50 percent, as in the residential sector, when heat pump water heaters replace conventional electric resistance water heaters. Adding a heat recovery system to the heat pump can reduce total water heating demand by up to 67 percent in commercial establishments such as supermarkets.[13c] Heating and cooling systems that run on natural gas, thermal energy storage systems, and intelligent control systems that monitor electricity pricing schedules and anticipate changes in heating and cooling loads based on outdoor temperatures are other cost-effective means of reducing electricity bills.

The biggest absolute reduction in energy consumption in the commercial sector occurs when electric motors are equipped with variable-speed drives.[12c] Such motors are suitable for equipment with changing load profiles, including air-conditioning chillers, ventilation systems and water pumps. When less than full capacity is required, the motors' speed can be adjusted electronically by changing its input voltage and frequency.

Industrial savings: Two-thirds of electricity demand in the industrial sector also involves motor drives. As in the commercial sector, use of variable-speed motors yields electricity savings of 20 to 50 percent.[14c] Because high-efficiency, variable-speed motors are more expensive to purchase, however, they account for only 5 percent of the industrial market at present. If industry replaced its existing motor stock with such high-efficiency alternatives, generating requirements would drop by 80,000 to 190,000 megawatts, the equivalent of 80 to 190 large nuclear reactors.[15c]

Tremendous efficiency gains also are made possible by a switch *to* electricity in specific industrial applications, including: resistance heating and melting of metals, ceramics, silicon and glass; radio-frequency drying and heating of textiles, foods, lumber, paper, metals and plastics; plasma refining for cutting, welding and heat treating various materials; ultraviolet curing of inks and dyes; and industrial separation processes such as membrane separation, freeze concentration, solvent extraction and critical fluid extraction.[16c]

electricity use). If utilities managed to reduce coal-fired generation as they increased electricity's share of primary energy demand, it would be possible to hold total carbon emissions at or below 1990 levels even as the overall economy continued to grow.

Surveying Utility Programs

Electric utilities face a number of formidable obstacles in their quest to promote more efficient uses of electricity. Chief among these are educating the public about the value of energy-saving technologies, bringing the technologies to market and creating a regulatory structure that rewards customers and utilities alike. Recent signs suggest that these obstacles will be overcome. But until they are, utilities' spending on new generating equipment seems likely to remain at least an order of magnitude above their spending on DSM programs.

To get a sense of utilities' early interest in DSM, the Investor Responsibility Research Center conducted an extensive survey in 1986 and 1987. Of the 123 utility respondents to the survey, 85 percent had instituted formal conservation and/or load management programs over the previous 10 years. Spending on DSM programs totaled approximately $1 billion in 1985 for the industry as a whole, and for the 1986-1995 period the surveyed utilities estimated that 20,000 megawatts of capacity additions would be avoided as a result of their energy-saving expenditures.[57]

More recent estimates by the Edison Electric Institute indicate that the savings rate may pick up somewhat during the 1990s. Thomas Kuhn, president of the institute, forecasts that 25,000 megawatts of capacity savings will be realized by the turn of the century, compared with 20,000 megawatts of actual savings during the 1980s.[58] To put these savings in perspective, the current utility generating base exceeds 730,000 megawatts, and it is expected to reach 800,000 megawatts or more by the turn of the century—with utilities spending $30 billion or more a year on new generating equipment.

Many energy analysts believe the potential exists for far greater electricity savings. Some claim it is even possible to eliminate electricity demand growth altogether with technologies that already exist. Others are not swayed by such optimistic projections, however. "It is impossible not to be skeptical," comments former New York Public Service Commission chairman Alfred Kahn. "If there are such enormous opportunities out there, why aren't consumers taking advantage of them?"[59]

Market failure and lack of public information may be partly to blame for the apparent gap between potential and realized energy savings. After all, consumers will not invest in efficiency measures if they do not know they exist or where to get hold of them. But Kahn retorts: "If the [problem] is lack of information, why aren't hordes of businessmen advertising the benefits of more efficient equipment or insulation?"

Part of the answer may be that the energy efficiency industry is still in its adolescent phase. Compact fluorescent lamps are generally available only in markets where electric utilities are actively promoting these products, for example. Limited market penetration of heat pump water heaters and ground-coupled heat pumps also illustrates the problem of nascent supply infrastructure. The marketing and distribution problem is such that the U.S. Environmental Protection Agency has instituted a "Green Lights" program to encourage major corporations to install more efficient lighting and ballast systems, since engineers at these companies apparently failed to appreciate the potential for tremendous energy savings on their own. Next, EPA plans to institute a "Golden Carrot" program, whereby refrigerator manufacturers will bid for the exclusive right to manufacture a super-efficient model that participating utilities will market in their service areas. The selected manufacturer will receive up to $30 million from the utilities in a "winner take all" prize, and customers who buy the refrigerator will receive a rebate from their utility.

Programs such as these hurdle several market barriers at once: the lack of information and market development, the high initial cost of the energy-saving equipment and the "payback gap" that exists between energy consumers and larger institutions. Market studies indicate that consumers want energy-saving measures to pay for themselves in only two or three years—equivalent to a 33 to 50 percent return on their investment—while utilities are willing to spend up to 10 years to build a power plant and then settle for a 13-percent-or-less return on their investment once the plant is in the rate base. This payback gap skews the selection process toward capacity additions.

Utilities can narrow the payback gap by financing the purchase of energy-saving devices on behalf of their customers. Not only are they able to buy the devices in bulk at wholesale prices, they can finance the purchases at their lower costs of capital, since they already are huge borrowers in financial markets. Skeptics like Alfred Kahn question the need for such an intermediary role, however. "If the asserted reason [for a lack of customer interest in energy-saving measures] is the unwillingness or inability of consumers to finance such investments, how [do you] reconcile that explanation with the more than $2.5 trillion of home mortgage debt—and installment debt of more than $700 billion?"

Split incentives between those who construct buildings and those who live in them may be a final part of the efficiency puzzle. While builders have an incentive to keep the initial price of housing and commercial space low, buyers benefit from up-front measures taken to reduce long-term energy costs. Similarly, landlords have an incentive to outfit rental housing with inexpensive and inefficent appliances, provided that their tenants are the ones paying the utility bills. Those who make the capital investments, in other words, need not be concerned with the operating costs associated with their purchasing decisions.

Making Efficiency Pay

From a utility's perspective, the biggest obstacle to the success of energy-saving programs is a rate structure that has encouraged more electricity consumption, not less. "In the current scheme of regulation," former Maine utility commissioner David Moscovitz observed in 1988, "utilities make money in only one way—selling kilowatt-hours. Utilities lose money when customers engage in conservation. They likewise lose money when they *encourage* customers to engage in conservation. This money-losing proposition is not significantly improved by any of the conservation cost recovery or incentive mechanisms now in use."[60]

Results of the IRRC demand-side management survey in 1986 and 1987 tended to support this conclusion. Two-thirds of the survey respondents reported that their primary sales objective was to *increase* electricity sales during off-peak hours or at all times, even though seven-eighths of these same utilities had formal DSM programs in place. Just seven utilities in the IRRC survey accounted for 70 percent of all of the expenditures on DSM programs, and just five utilities claimed 50 percent of the projected savings in power plant construction through 1995.[61] Most other utilities focused their efforts on performing simple energy audits for customers, recommending the use of efficient electrical appliances (such as heat pumps) in place of oil- and gas-fired alternatives, and sponsoring public service advertisements about the virtues of saving energy. As Peter Bradford, chairman of the New York Public Utilities Commission, remarked in 1988, "Most utilities...have adopted measures that convey an enthusiasm for conservation without actually cutting [their] sales."[62]

Regulatory reform is sweeping the nation, however, and the outlook for DSM programs in the 1990s appears to be brightening. Regulators are finding ways to decouple utilities' profits from their electricity sales. In many cases, they now can earn a fair return on their efficiency investments, just as they do on power plant investments. While some of these measures may spur an increase in electricity on a per-kilowatt-hour basis, ratepayers should not object as long as their own monthly electric bills go down.

The Electric Power Research Institute reports that, as of August 1991, at least 32 states allowed utilities to recover money spent on DSM programs, typically by treating the allocation as a rate-based investment rather than as an expense. At least 13 states have severed the link between kilowatt-hour sales and utilities' profits altogether. In some of these states, the ratemaking formula is independent of sales; in others, utilities are allowed to recover lost kilowatt-hour revenues after conservation and efficiency measures are in place. Finally, at least 19 states now offer pure incentives for utilities making efficiency investments. Some provide a higher rate of return on demand-side investments than on supply-side investments. Others reward utilities with bonuses for achieving certain conservation goals.[63]

In Massachusetts, for example, utilities' rate of return on DSM programs rises by one full percentage point after half of a pre-determined level of electricity savings has been achieved. Further bonuses are offered as the utility approaches its overall savings goal. In California, some utilities get to keep 15 percent of the monetary savings that result from efficiency investments made on behalf of their customers (but which customers might not have made on their own). In New York, utilities get to keep 5 to 20 percent of such "shared savings."

Niagara Mohawk Power in upstate New York has distributed a basic conservation packet—including a low-flow showerhead, a compact fluorescent light bulb and a water heater "jacket"—for customers that want to save on their electric bills. The resulting loss of electricity demand has reduced the utility's electricity sales by an estimated $72 a year, but it has also saved the utility an estimated $40 a year in avoided fuel and capacity costs. The difference—$32 a year—is made up through a moderate increase in the rates of residential customers, plus a $6 yearly charge paid by those customers who install the energy-saving devices. Niagara Mohawk gets to keep $5 of this charge as profit, so that its yearly return on investment in these devices works out to 15 percent. Participating customers do not mind the increase in rates and the $6 yearly charge, since they are saving $34 a year on their own electric bills.[64]

The Coming Boom in DSM

With arrangements such as these, utilities appear more inclined than ever to use DSM programs as a strategy for meeting future generating requirements. The American Council for an Energy-Efficient Economy says that 16 utilities have responded to recent regulatory incentives by nearly tripling their projections for electricity savings in the 1990s. Nine other utilities lacking these incentives have raised their savings projections by 70 percent, the council reports.[65] EPRI estimates that the electric utility industry as a whole spent a record $1.5 billion on DSM programs in 1991—a 50 percent increase since 1985—with some utilities devoting 3 to 5 percent of their gross annual revenues to demand-side programs.[66]

The commitment to DSM is especially strong on the nation's coasts, where utilities are counting on conservation and efficiency programs to cover a large percentage of new resource additions. In the Pacific Northwest, the Northwest Power Planning Council expects DSM programs to provide 1,500 megawatts of that region's 2,300 megawatts of incremental capacity requirements during the 1990s. The giant Bonneville Power Administration may nearly double its planned DSM spending by 2003, raising its expenditures for the period from $1.4 billion to $2.7 billion.[67]

The California Energy Commission is counting on DSM programs to an even greater extent, with three-quarters of California's electricity resource additions expected to come on the demand side rather than on the supply side during the 1990s.[68] Pacific Gas and Electric, the nation's largest investor-owned utility, will commit up to $2 billion for conservation and efficiency programs over the next 10 years—having spent $150 million on such programs in 1991. As an incentive, PG&E gets to keep 15 cents out of every dollar that its customers save by participating in the programs.[69]

The Sacramento Municipal Utility District, meanwhile, is building a 600 megawatt "conservation power plant" in the 1990s. Like the DSM programs instituted by other utilities, SMUD's conservation plant consists of the performance of energy-saving tasks: weatherizing buildings, replacing inefficient electrical appliances, installing energy-efficient lighting, even planting shady trees on sunny neighborhood streets. The energy savings come at an average price of 3 cents a kilowatt-hour—well below what customers would pay for electricity generated at a new conventional power plant.[70]

(continued on p. 386)

Box 5-D

Creating a Market for Energy Efficiency

Not long ago, energy planners assumed that demand for electricity was dictated by market forces and largely beyond their ability to control. All critical decisions came on the supply side; namely, how many power plants to build and which fuels to burn. Now energy planners realize that measures can be taken to control the demand for electricity as well as the supply. So-called "integrated resource planning" methods are sweeping the nation, whereby the costs and benefits of limiting electricity demand are evaluated in relation to investments in new supply. The Clean Air Act Amendments of 1990 require public utility commissions to adopt such integrated resource planning methods if utilities are to receive financial incentives for investments in demand-side programs (and not be penalized for any resulting loss of sales). As of August 1991, at least 31 states had adopted integrated resource planning, and another 10 were reviewing implementation options.[1d]

With energy efficiency now put (more or less) on an equal footing with power generation, the basic principle of supply and demand is coming into balance in the electricity market. A properly functioning market also requires that consumers be well-informed, however. In an effort to better inform the public of their choices, the government now requires fuel-cost labeling on new appliances and mileage stickers on new vehicles. Most electric utilities also perform comprehensive energy audits for customers who want to find ways to cut down on their fuel bills.

Other market obstacles may require more innovative solutions, however. One persistent problem involves divergent interests between builders and building dwellers. Builders seek to limit how much they spend on insulation, windows and energy-consuming appliances, while dwellers want to limit how much they spend on their monthly energy bills. One way to encourage construction of more energy-efficient buildings is to charge extra hook-up fees on less efficient buildings and offer rebates on new buildings that are more energy-efficient than average. The arrangement can be made revenue-neutral, so that fees collected on wasteful buildings are passed through as rebates for the efficient buildings. At least seven states are experimenting with such sliding-scale "feebate" systems.[2d] In effect, contractors and landlords who are unwilling to invest in conservation and efficiency measures end up subsidizing the investments of others who do.

In the future, banks could get in on the act by offering preferential terms on mortgages for energy-efficient homes. Buyers would qualify to borrow additional money (perhaps at lower interest rates) and amortize the added debt over the term of the mortgage. Homeowners' monthly energy bills would fall as a result of the realized energy savings, and the bank's risk of mortgage default also would decline. In addition, such energy-efficient homes should have greater value in the resale market, benefiting homeowners and mortgage lenders alike. A joint government task force of the U.S. Department of Energy and the Department of Housing and Urban Development is working now to establish national guidelines for mortgages on energy-efficient homes.[3d]

Eventually, the most popular way to spur greater efficiency in energy markets may be to create a trading mechanism in which rights to energy savings are bought and sold on the open market. "Increasingly..., utilities are moving to market not just [energy-saving] 'negawatts,' but to *make markets in* negawatts," says Amory Lovins of the Rocky Mountain Institute; "that is, to make saved electricity into a fungible commodity subject to competitive bidding, arbitrage, derivative instruments, secondary markets, and all of the other attributes of copper, wheat and sowbellies."[4d]

Such a trading concept will come into play in 1995 as a result of the new Clean Air Act, which allows utilities to buy and sell credits for sulfur dioxide emissions from power plants. The Chicago Board of Trade plans to list the auction price of these permitted trades beginning in 1993—creating a futures-related market.[5d] In essence, one utility will be able to buy another utility's right to pollute. By the same principle, one energy provider should be able to reap the benefit of another provider's ability to save energy.

Several contracts already have been signed in the western United States in which one utility has agreed to pay another to save electricity, with the two splitting the savings at a mutually advantageous price. Puget Sound Power & Light of Bellevue, Wash., for example, actually sells efficiency measures in utility service areas of nine other states. More than a dozen electric utilities have set up conservation subsidiaries altogether.[6d] Gas utilities have been especially pernicious bounty hunters for electricity savings. They finance customers' switch from electricity to more efficient natural gas appliances; their "finder's fee" is a share of the energy savings.

Congress is looking at the trading concept in terms of CO_2 emissions. Reps. Jim Cooper (D-Tenn.) and Mike Synar (D-Okla.) are behind a plan that would require utilities to freeze their CO_2 emissions but allow them to buy offset permits from other utilities in the event they wanted more fossil-energy capacity.[7d] A utility wanting to build a new coal plant, for instance, could buy CO_2 permits from another utility that switches fuels from coal to natural gas or renewables, or which launches efficiency programs that reduce energy demand overall.

The Environmental Defense Fund—which is credited for the emissions trading allowance featured in the Clean Air Act—thinks the trading concept eventually should be broadened to include all industries and be applied internationally. Forest interests in Brazil could be paid to plant trees to offset utilities' CO_2 emissions in America. Natural gas developers in Eastern Europe could be paid to plug methane leaks in their pipelines. Coal companies in Nova Scotia could be paid to trap coal-seam methane emissions. Even junkyard owners could be paid to capture CFCs that otherwise would escape from old refrigerators and abandoned automobiles.

Emissions trading proponents believe that applying these market principles is the best way to overcome the problems of market failure. Since the right to pollute would carry an explicit price, there would be an incentive to ferret out ways to reduce emissions at the lowest possible cost, unlike traditional command and control approaches. Auctioning the permits also would overcome a potential problem of taxing pollution: that polluters might pay the tax and go right on polluting. Setting an upper limit on emissions would ensure that the control targets are met—provided that the emissions are monitored. As long as ways are found to abate greenhouse gases at costs lower than the price of the permits, the market for emissions trading would remain alive and well.

"I submit the efficiency revolution is here," declares efficiency advocate Amory Lovins. "We are in an era of costly energy and relatively cheap efficiency. The customers will figure this out and want to buy less energy and more efficiency, and it is a good idea to sell customers what they want before someone else does. The only questions are who is going to sell the efficiency and whether the traditional energy suppliers want to choose participation in the efficiency revolution or obsolescence."[8d]

Back on the East Coast, New York utilities expect conservation and efficiency programs to provide almost half of that state's new "power" requirements through 2007. Expenditures on DSM have virtually doubled in New York in the last three years, reaching an estimated $195 million in 1992.[71] Consolidated Edison alone plans to spend $4.2 billion on DSM programs through 2008. The metropolitan New York utility has trimmed its peak demand forecast for 2008 by 23 percent, which should forestall construction of any new central-station power plants over the period.[72]

In neighboring New England, utilities have earmarked between $1 billion and $2 billion for DSM programs over the first half of the 1990s, with several thousand megawatts of capacity savings expected.[73] By 2005, efficiency programs should offset 8 percent of that region's forecasted annual electricity demand and 12 percent of its summertime peak demand.[74] New England Electric, for one, has increased its DSM budget from just $24 million in 1988 to $85 million in 1991.[75] The Massachusetts utility earned a 12 percent return on the $65 million it invested in DSM programs in 1990. Customers did not object to NEES's $7.8 million profit because their own electric bills were expected to fall by more than $150 million in the process.[76]

Savings from demand-side programs are not limited to American utilities, of course. Ontario Hydro, Canada's largest utility, will invest $5.72 billion (U.S. dollars) in DSM programs through 2014 to save an estimated 10,000 megawatts of electricity demand. The utility now expects to defer construction of any large-scale power plants until at least 2009, seven years later than it had estimated in a previous resource plan.[77]

Demand-side's Potential

If the regulatory incentives now in place in California, the Pacific Northwest and the Northeast spread to the rest of the country, EPRI believes it would be possible for American utilities to double the amount of savings to be realized from DSM programs—from 25,000 megawatts to 50,000 megawatts—during the 1990s. (These savings are in addition to 20,000 megawatts of savings realized by DSM programs in the 1980s.) The 200 million kilowatt-hours of electricity savings in 2000 would negate about 41 million tons of carbon emissions, equivalent to about 8 percent of the industry's present emissions.[78]

But the savings need not stop there. If the appropriate incentives are widely adopted, EPRI estimates that DSM programs could reduce U.S. electricity consumption 11.4 percent below its baseline forecast for 2010. Moreover, if the most energy-efficient equipment available today were to replace all inefficient equipment stock by 2000, it would be theoretically possible for electricity demand to fall below the baseline by a total of 24 to 44 percent.[79] (These savings are in addition to the 8.5 percent savings expected from government appliance efficiency standards and market-driven technology improvements.) EPRI cautions, however, that the "maximum technical potential" of DSM programs is not likely to be achieved because not all of the measures are cost-effective at today's electricity prices, and there are physical constraints that limit the

Figure 8

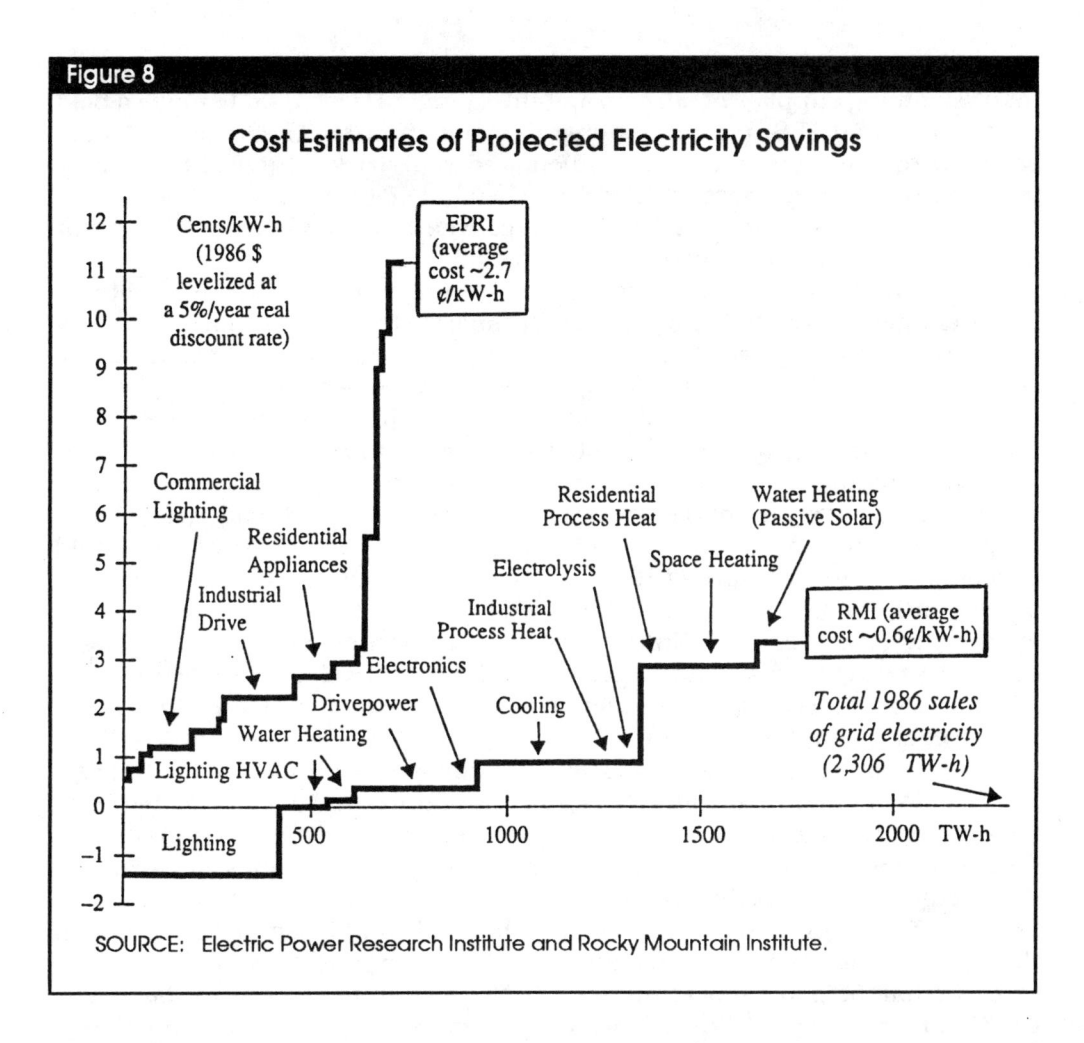

Cost Estimates of Projected Electricity Savings

Cents/kW-h (1986 $ levelized at a 5%/year real discount rate)

EPRI (average cost ~2.7 ¢/kW-h)

RMI (average cost ~0.6¢/kW-h)

Total 1986 sales of grid electricity (2,306 TW-h)

Labels on curve: Lighting; Lighting HVAC; Water Heating; Industrial Drive; Commercial Lighting; Residential Appliances; Drivepower; Electronics; Industrial Process Heat; Cooling; Electrolysis; Residential Process Heat; Space Heating; Water Heating (Passive Solar)

SOURCE: Electric Power Research Institute and Rocky Mountain Institute.

applicability of energy-efficient equipment. Nevertheless, the aggregate cost of instituting DSM to its maximum potential would be only 2.7 cents a kilowatt-hour—far below the costs of generation from new power plants.[80]

Amory Lovins of the Rocky Mountain Institute believes that the long-term potential for electricity savings is far greater than even EPRI suggests. Lovins figures that the nation could get by on 75 percent less electricity per unit of economic output.[81] Many efficiency measures would more than pay for themselves through realized energy savings. The rest of the options would impose net costs, Lovins says, but only a few would be more expensive than the cost of generation from new power plants. The average price for all of the energy savings would be less than a penny per kilowatt-hour, Lovins figures. Under these circumstances, three power plants could be retired for each new plant that is brought into service during the 21st century.

Many energy analysts regard Lovins's forecasts as heresy. But Lovins has confounded the experts before. In his 1976 article, "Energy Strategy: The Road Not Taken," in the *Foreign Affairs* journal, Lovins posited that unrealized energy savings offered a much cheaper and more accessible pool of energy "reserves" than remaining supplies of oil, coal and uranium.[82] He went on to make a bold prediction: that U.S. primary energy demand through 1990 would remain below 100 quads, contrary to government and industry forecasts of 135 quads. In actuality, primary energy demand reached only 84 quads in 1990—a scant 10 percent increase since 1976—whereas conventional wisdom suggested that demand would rise 80 percent over the period.

The resulting savings in energy-related capital and operating expenditures has amounted to $150 billion a year. But, claims Lovins, "We still have a long way to go. The energy now wasted—that is, energy that we use when we could use cheaper efficiency to do the same things as well or better—is still costing us about $300 billion a year, or slightly more than the entire [U.S] military budget...."[83]

While Lovins's projections of potential energy savings still exceed others' projections by a wide margin, a consensus is emerging that significant efficiency improvements are attainable and cost-effective. The National Academy of Sciences, for one, has identified numerous energy-saving measures that would improve the efficiency rate of the economy by 1 percent or more a year and impose little or no net cost to the economy. As part of a comprehensive abatement program, "the United States could reduce its greenhouse gas emissions by between 10 and 40 percent of the 1990 level at very low cost," the academy concluded. "Some reductions may even be at a net savings if the proper policies are implemented."[84]

Another study by researchers at the American Council for an Energy-Efficient Economy, Oak Ridge National Laboratory and Lawrence Berkeley Laboratory comes to similar conclusions. Their analysis compares the potential for energy savings relative to the amount of energy that would be consumed if the energy-intensity of the economy remained frozen at 1990 levels. Without further efficiency gains, primary energy demand would rise to 106 quads in 2000 and 131 quads in 2010, and the nation's carbon emissions would increase by more than 50 percent over the period. By comparison, a forecast issued by the U.S. Department of Energy—which incorporates current trends in energy efficiency—projects that U.S. primary energy demand will reach only 96 quads in 2000 and 107 quads in 2010, with the nation's carbon emissions rising 25 percent over the period.[85]

Under a "high efficiency" forecast developed by the laboratory researchers, U.S. primary energy demand would grow to only 86 quads in 2000 and 88 quads in 2010—with carbon emissions remaining roughly constant. The nation's fuel bill would climb from $530 billion in 1990 to $850 billion in 2010, reflecting the net costs of these energy-saving measures and rising energy prices overall. But the bill in 2010 would still be $100 billion less than the amount projected in the DOE forecast—and $300 billion less than the amount projected in the frozen energy-intensity forecast.[86] In other words, the high-efficiency energy policy that keeps the nation's carbon emissions from rising also is the least expensive of the three options considered here.

The electricity sector would share in these energy savings. Electricity consumption in 2010 would be reduced by 780 billion kilowatt-hours relative to the DOE forecast—equivalent to the annual output of 150,000 megawatts

Figure 9

EPRI Estimates of DSM Potential

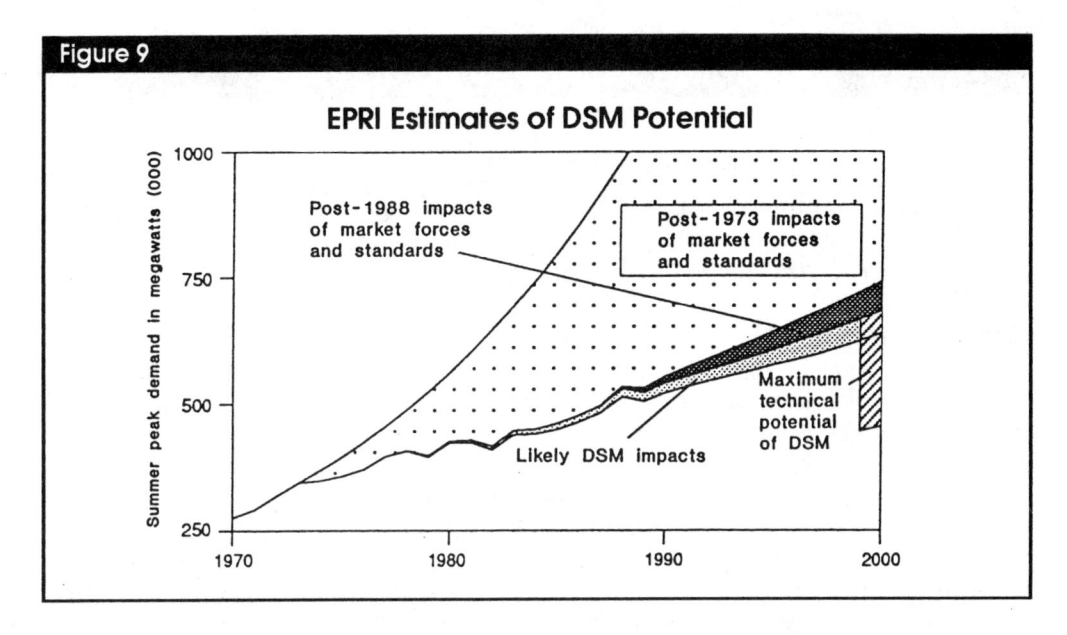

[Graph showing Summer peak demand in megawatts (000) from 1970 to 2000, with regions labeled: Post-1988 impacts of market forces and standards; Post-1973 impacts of market forces and standards; Likely DSM impacts; Maximum technical potential of DSM]

worth of power plants. Nevertheless, overall power demand would continue to grow—albeit at a manageable pace of 0.9 percent a year—because the economy would consume a rising share of energy in the form of electricity.

Thus, while continued efficiency gains are an *essential* component of a carbon abatement strategy, they probably are not *sufficient* to reduce CO_2 emissions overall. As the laboratory researchers noted in their study, "Our savings estimates indicate that the energy-efficiency initiatives can cut CO_2 emissions substantially from otherwise projected levels, but only slightly reduce the absolute level of CO_2 emissions between 1990 and 2010."[87] Should absolute reductions in CO_2 emissions become necessary, the nation would have to either commit itself to even greater energy savings than now seem practical or shift away from more carbon-intensive fuels to greater reliance on natural gas, nuclear and/or renewable energy sources.

A similar outlook prevails internationally—only the challenge is compounded by the likelihood of tremendous population growth and economic expansion in developing nations. By the middle of the 21st century, the world's population is expected to double and per capita income may nearly triple, so that the size of the world economy is almost six times larger than it is today. Therefore, even if conservation and efficiency measures cut the present rate of growth in electricity demand by one-third, and the present rate of growth in total energy consumption falls by one-half, total electricity demand and primary energy demand would still increase 4.7 times and 2.5 times, respectively, by 2060 because of the underlying economic expansion.[88] While greater reductions in energy demand growth may be possible, it does not appear possible to eliminate worldwide growth in electricity demand altogether. Accordingly, a vast array of supply-side options must be considered to meet the world's burgeoning electricity requirements without exacerbating the greenhouse effect. These generating options are considered in the next section.

Alternative Generating Sources

Utilities thinking about building new power plants are well advised to consider first the vast and largely untapped potential for saving energy. The National Energy Strategy estimates that conservation and efficiency measures have the potential to eliminate 45,000 megawatts of U.S. capacity requirements between now and 2010, with an additional 45,000 megawatts of savings possible by 2030. Such "least-cost" and "integrated resource planning" methods would trim the nation's generating requirements in 2030 by 7 percent—yet leave more than 500,000 megawatts worth of additional power plants still to build.[89]

Many energy analysts think the National Energy Strategy downplays the potential for saving energy far too much. As noted in the previous section, researchers at some of the nation's leading energy laboratories believe that suitable government polices and incentives could reduce additional generating requirements by 150,000 megawatts over the next 20 years alone. Aggressive demand-side programs implemented by utilities could supplant one-fifth to one-half of future capacity additions—perhaps more.[90]

Despite the increasing sophistication of integrated resource planning methods, forecasters still cannot determine precisely how much of the potential energy savings will be captured in the years ahead. Utility planners tend to err on the side of caution, assuming that realized energy savings will never quite catch up to their maximum potential and that electricity demand—as a proxy for economic growth—will rise incessantly. Under these circumstances, the power supply seems almost destined to grow; it is only a matter of whether it will grow by a lot or a little. Even if the forecasts are wrong, however, and demand for electricity does not grow at all, old power plants will continue to wear out and need replacement. Accordingly, the fundamental question is not whether new capacity will be needed, but what kinds of power plants to build and how many of them.

What to Build?

The Bush administration believes the time has come for the electric power industry to launch a major new round of power plant construction. Its National Energy Strategy calls for doubling the nation's nuclear and renewable generating capacity over the next 40 years and—more ominously for global warming—increasing the nation's coal-fired capacity by two-thirds. In the next decade alone, the National Energy Strategy foresees that "$100 billion to $200 billion in new capital investment will be needed...to meet the nation's growing electricity needs."[91] Through 2010, it estimates that between 190,000 megawatts and 275,000 megawatts of additional capacity will be required, 85 percent of which is "base-load" capacity. Such a forecast bodes well for the coal industry—already the dominant fuel supplier for base-load power plants.

Figure 10

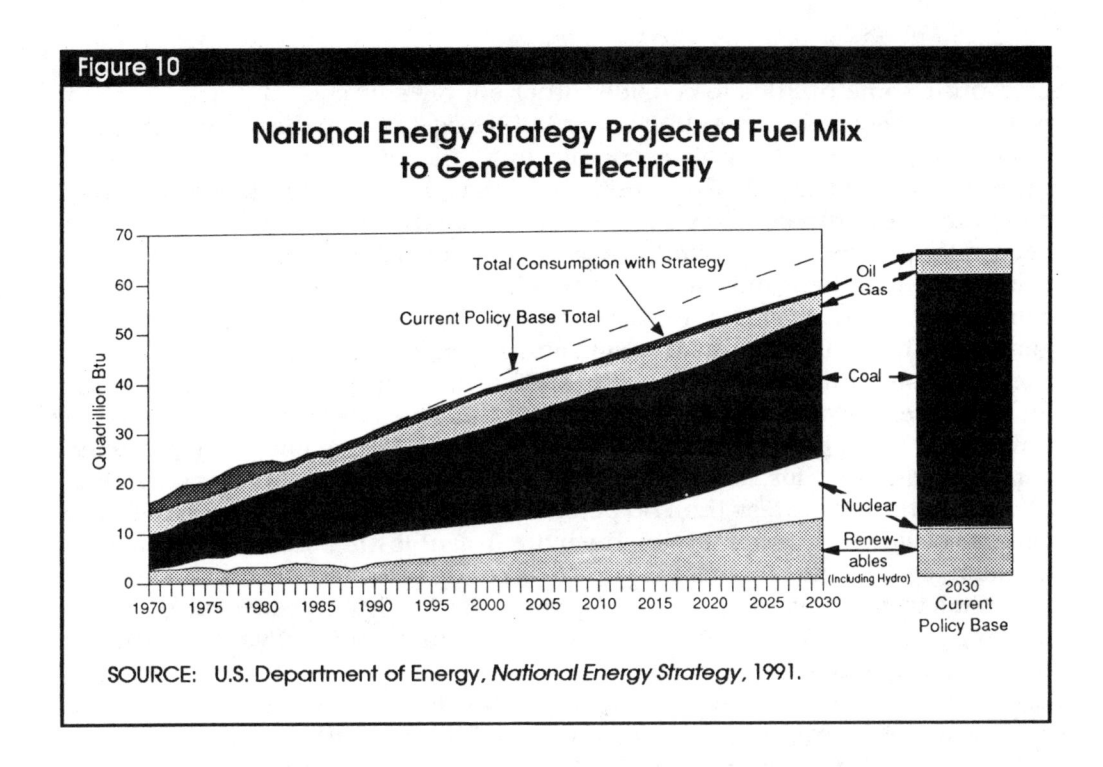

National Energy Strategy Projected Fuel Mix to Generate Electricity

[Chart showing Quadrillion Btu from 1970 to 2030, with fuel mix including Oil, Gas, Coal, Nuclear, and Renewables (Including Hydro). Shows Total Consumption with Strategy and Current Policy Base Total lines. Rightmost bar shows 2030 Current Policy Base.]

SOURCE: U.S. Department of Energy, *National Energy Strategy*, 1991.

Competing fuel suppliers are jumping on the environmental bandwagon, however, to decry excessive reliance on coal. The U.S. Committee on Energy Awareness—which spends millions of dollars a year to promote nuclear energy in print and television advertisements—now stresses that nuclear power emits "no greenouse gases" or other pollutants to the atmosphere. The solar lobby has long made similar claims about non-carbon renewable generating sources, although it has had to depend more on a sympathetic media than on paid advertising to get its message across.

The natural gas industry, meanwhile, has been taking its claims of environmental friendliness all the way to the bank. As the "prince of carbon fuels," natural gas contains virtually no sulfur or particulates. And when it is burned in plants using new turbine technology, natural gas emits 85 percent fewer nitrogen oxides and 60 less carbon dioxide than coal plants in use today. Partly for this reason, natural gas has emerged as the fuel of choice for utilities (and industry) in need of near-term generating capacity. The majority of new power plants brought on-line between 1992 and 2001, in fact, will be natural gas-fired, according to the government's Energy Information Administration.

The specter of global warming has done less to boost the prospects of natural gas than passage of the Clean Air Act amendments of 1990. The new law requires that utilities cut their smokestack emissions of sulfur dioxide and nitrogen oxides in half by the year 2000. To meet this requirement and still burn coal, utilities face the prospect of switching to lower-sulfur grades of coal, retrofitting older plants with scrubbers or introducing new "clean-coal" power plants. All of these measures would raise the cost of electricity generation, prompting some utilities to consider other supply sources. But with nuclear power and renewable generating options still largely out of the picture, natural gas has appeared as the only economically viable alternative to coal.

Such a constrained set of options clearly has left many utilities in a quandary. It normally takes them seven to 10 years to select, construct and license large-scale power plants. Yet at the start of the 1990s, utilities had committed to building only one-quarter of the 104,000 megawatts of additional generating capacity slated to come on-line by 2000. Independent power generators have stepped into this void and are expected to provide up to half of the new capacity additions during the 1990s. But a large gap remains between forecasted electricity requirements and new plants actually under construction. As of 1990, ground had been broken on only 37 percent of the projected capacity additions for the decade, and of those plants under construction, about one-third were less than 50 percent complete.[92] The longer the industry waits to build more power plants, the more urgent it will become to embark on a crash construction program—or to step up demand-side management programs to quell electricity demand growth. By 2015, 7 percent of the nation's existing generating capacity will be due for retirement, and fully 50 percent will be at least 40 years old.[93]

Putting this all into perspective, one can imagine a scenario in which replacement of worn-out power plants in combination with increased electrification of the economy and direct effects of global warming creates a market for, say, 400,000 megawatts of new generating capacity over the next quarter-century. One might imagine further that efforts to curb CO_2 emissions might lead to a moratorium on construction of new coal plants. The electric power industry could revive the nuclear option to provide the added generating requirements—building 400 large (1,000 megawatt) nuclear plants by 2015. But the rate of reactor completions would have to be rapid-fire: one reactor finished every three-and-a-quarter weeks, which is nearly twice the rate of completions during the 1970s and early 1980s when the first generation of nuclear plants came on-line. A more realistic calculation would consider, however, that new reactor prototypes generally are rated at 600 megawatts or less and are at least a decade away from demonstration. That means an even greater number of reactors would have to be completed in a shorter amount of time—almost one reactor per week between 2000 and 2015. The cost of such a Herculean construction program could easily top $1 trillion.

A scramble to build renewable energy facilities over the next 25 years presumably would be even more frenetic, given that these plants rarely exceed 100 megawatts in size. The very fact that renewable energy plants tend to be small, decentralized and intermittent in power generation is a basic reason why utilities have shied away from these resources. High capital costs associated with most renewables have been another limiting factor. Consequently, renewable energy development to date has fallen mainly in the hands of independent producers. Since the mid-1980s, the fledgling renewable energy industry has installed generating capacity at the rate of about 1,000 megawatts a year, equivalent to the size of one large nuclear plant.[94] To provide 400,000

The Electric Utility Industry

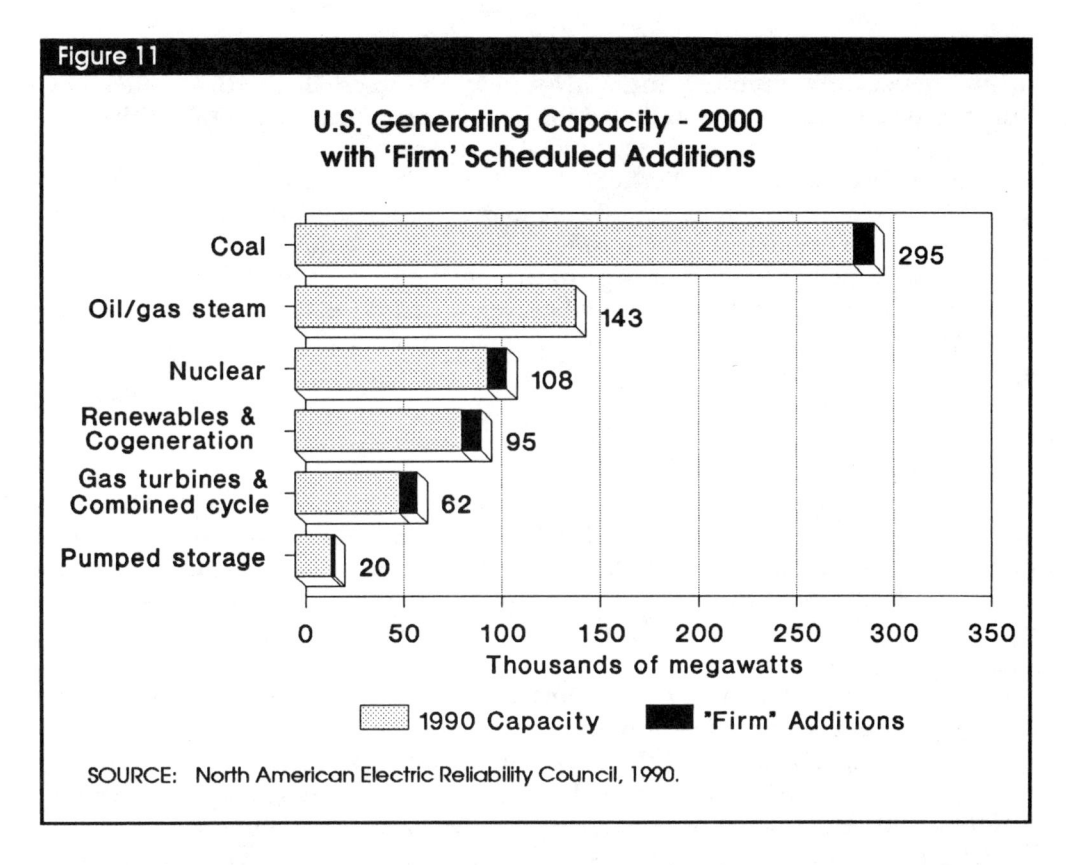

Figure 11

U.S. Generating Capacity - 2000 with 'Firm' Scheduled Additions

Source	Thousands of megawatts
Coal	295
Oil/gas steam	143
Nuclear	108
Renewables & Cogeneration	95
Gas turbines & Combined cycle	62
Pumped storage	20

Legend: 1990 Capacity / "Firm" Additions

SOURCE: North American Electric Reliability Council, 1990.

megawatts of new capacity requirements by 2015, 15 times as many installations would be required each year. Fortunately, the lead times for renewables are quite short, rarely exceeding a year or two. Even so, the independent power industry would have to mature very fast or utilities would have to take a much greater interest in renewables for these sources to meet the generating challenges of the early 21st century.

Another possibility would be for the nation to rely on natural gas for the incremental power requirements. Here, too, utilities could enlist support from independent producers, whose 23,000 megawatts of cogenerated power—installed mainly by industry over the last 20 years—represents 4 percent of the nation's generating supply. A combination of small industrial cogeneration plants (usually less than than 50 megawatts in size) and larger utility plants (up to 250 megawatts in size) could provide America's new power needs through 2015. But once again the price tag would be high. A study by the ICF consulting firm, noted earlier, found that demand for natural gas might soar 500 percent by 2015 if the policy goal were to stabilize CO_2 emissions and the only other generating alternative was limited amounts of high-cost renewables. The resulting surge in natural gas demand would overwhelm existing pipeline capacity and cause delivery prices to more than triple. Such extensive reliance on expensive natural gas would add as much as $90 billion to the nation's annual fuel bill.[95]

In the final analysis, no one energy source appears capable of supplying all future generating requirements. Nor does it seem likely that efficiency improvements will obviate the need for new power plants. As long as there is a growing economy in an Electronic Age, there will be rising demand for electricity. The question is whether future generating requirements will be close to today's level—or much larger. If conservation and efficiency measures hold future electricity demand growth to a minimum, the industry can afford to be choosy. If capacity requirements grow rapidly, however, then the industry will be inclined to tap energy sources wherever it can develop them. Hence, the National Energy Strategy's call for more nuclear power, more renewable energy and, yes, more oil, natural gas and coal.

Coal: A Time of Reckoning

There is probably no escaping the fact that coal will remain an important component of the nation's—and the world's—generating supply for some time to come. Coal accounts for 90 percent of all known conventional fossil fuel reserves worldwide. At the current rate of consumption, estimated economically recoverable reserves of coal would last 239 years—twice as long as the combined reserves of oil and gas.[96] The United States possesses more than 240 billion tons of recoverable reserves—23 percent of the world's total. On a percentage basis, the United States has a greater endowment of the world's coal than Saudi Arabia does of the world's oil. The U.S. coal supply would last another 260 years at the present rate of domestic consumption.[97]

The 1973 oil embargo prompted utilities to increase their reliance on coal and uranium as sources of power and to cut back on the use of oil and natural gas. Between 1973 and the end of 1991, American utilities reduced their consumption of oil by nearly two-thirds and their use of natural gas by nearly one-quarter. Today, these fuels account for only 4 percent and 9 percent of the nation's generating mix, respectively. To keep pace with rising demand for electricity—which has grown 50 percent since 1973—nuclear generation has surged more than 530 percent. At the same time, however, coal-fired generation has practically doubled off of a much larger existing capacity base. Even now, coal outproduces nuclear power by a margin of 2.5 to 1 in the United States.[98]

In fact, U.S. coal-burning power plants, by themselves, emitted 22 percent more CO_2 in 1991 than all U.S. fossil energy plants in 1973 combined. Coal's share of the nation's generating mix has risen from less than 44 percent in 1973 to 50 percent in 1981 and to 55 percent in 1991. By 2000, coal's share is expected to exceed 60 percent. And according to the National Energy Strategy, coal's portion of the nation's fuel mix could reach 75 percent by 2030, if the moribund nuclear power industry is not revived in the meantime. (With implementation of the National Energy Strategy, coal's share would be held to 49 percent of the generating mix in 2030.)

It is mainly because of coal, then, that American utilities' CO_2 emissions have continued to rise, despite the major decline in oil and natural gas consumption since 1973 and the meteoric rise in carbon-free nuclear power.

Looking to the future, there is a strong chance that America's dependence on coal will increase further—concerns about the greenhouse effect notwithstanding. ICF estimates that domestic coal consumption in the utility sector will rise from 690 million metric tons in 1990 to at least 1.1 billion tons—and perhaps as much as 1.55 billion tons—in 2015, depending on the rate at which demand for electricity grows and the extent to which nuclear power, renewables and natural gas are developed as alternative generating sources. Coal's share of the nation's generating mix in 2015 could be several percentage points higher or lower than its current share of 55 percent, but coal is likely to continue to produce more than half of the nation's generating requirements in any case. (The Gas Research Institute, by comparison, forecasts that coal's share of the generating mix will dip to 48 percent in 2000 and then rebound to 51 percent by 2010.)[99]

Should policies be enacted to combat the greenhouse effect, however, the outlook for coal might change entirely. ICF envisions that no new coal plants would be built under a control program to stabilize CO_2 emissions at 1990 levels. Rather than having as much as 560,000 megawatts of coal-fired generating capacity by 2015, the U.S. total would remain below 300,000 megawatts. Generation from coal plants would drop 75 percent relative to an ICF base-case forecast that assumes heavy reliance on coal. Ironically, with a CO_2 stabilization target, *more* electricity could be derived from coal if alternative generating sources and demand-side programs were developed to a greater degree than is assumed in this base-case forecast. With more noncarbon supplies available to meet reduced demand for electricity, far less switching away from coal would be necessary to meet the CO_2 stabilization target. ICF concludes in its analysis that it would be in the coal industry's own interest to support alternative energy development and energy efficiency programs should CO_2 controls be enacted.[100]

Worldwide, prospects for coal are similar. The Earth's population now derives about 28 percent of its primary energy needs from coal—and more than two-fifths of its electricity requirements.[101] The former Soviet Union has about as much coal in reserve as the United States, and China has half as much. Altogether, the world's coal reserves are expected to outlast the world's oil reserves by well over a century.[102]

A consensus forecast of world energy consumption suggests a 50 or 60 percent increase in total energy use over the next 20 years, with coal leading the way. China—already the world's largest consumer of coal—may nearly double its consumption, and by 2010 the developing world may eclipse Western Europe in terms of total energy consumption. The net effect is that CO_2 emissions from fossil fuels may increase 50 to 60 percent worldwide by 2010, with coal surpassing oil as the leading source of emissions—accounting for about two-fifths of the CO_2 emitted from all fossil fuels.[103]

In terms of greenhouse gases, however, coal's emissions are not restricted to CO_2. Coal burning also releases nitrous oxide into the atmosphere, a long-lived greenhouse gas that accounts for an estimated 4 percent of global warming potential. In addition, methane gas escapes from coal seams and surrounding rock strata when coal is mined. These releases account for 5 to 10 percent of

Table 3			
Fossil Energy Reserves			
Region	Oil (billions of barrels)	Coal (billion tons)	Natural Gas (trillion cubic feet)
Middle East	661.6 66.1% >100 years	0.2 -- --	1,319.1 30.1% >100 years
North America	41.7 4.2% 10.2 years	249.2 23.9% 260 years	266.0 6.1% 12.3 years
OECD Europe	14.5 1.5% 9.0 years	98.6 9.5% 166 years	178.6 4.1% 25.7 years
Non-OECD Europe	58.8 5.8% 15.1 years	315.4 30.4% 306 years	1,776.4 40.4% 58.9 years
Asia & Australasia	44.1 4.4% 18.3 years	303.6 29.1% 196 years	299.3 6.8% 48.2 years
Africa	60.4 6.0% 24.5 years	62.2 6.0% 336 years	310.3 7.1% >100 years
Latin America	119.8 12.0% 43.1 years	11.4 1.1% 278 years	238.4 5.4% 69.2 years
World	1,000.9 100% 43.4 years	1,040.5 100% 239 years	4,378.1 100.0% 58.7 years

Key: Line 1 - Proven reserves
Line 2 - % of world reserves
Line 3 - Life of reserves at current production rate

SOURCE: British Petroleum, *Statistical Review of World Energy*, 1992.

all estimated methane emissions attributable to human activities. Within the United States, where coal mining is especially intensive, the resulting releases may amount to as much as 10 or 20 percent of anthropogenic methane emissions.[104]

Coal mine operators traditionally have vented most of this methane to the atmosphere to prevent a dangerous, potentially explosive buildup of the gas. Some operators are beginning to capture this gas, however, and burn it as fuel. Such an undertaking is not only profitable but also preferable from a greenhouse standpoint, because methane converted to CO_2 contributes far less to global warming potential. About 18 percent of U.S. coal-seam methane emissions are captured at present.[105] Questions remain as to who owns the methane found in many coal seams, however—those who own the mineral rights or those who own the land. Until this legal matter is resolved, many coal mine operators will remain disposed to venting the gas directly to the atmosphere.

Coal plants themselves can reduce CO_2 emissions by making sure that the maximum amount of heat released during coal combustion is converted into electricity. Improved plant operation and maintenance of existing coal plants can raise their efficiency levels by up to four percentage points.[106] A modern, well-maintained coal plant outfitted with scrubbers is about 34 percent efficent. Further efficiency gains can be achieved in newly designed coal plants. Fluidized-bed combustors and integrated coal-gasification, combined cycle plants can achieve efficiency ratings of 38 percent or higher.[107]

'**Clean-coal' plants:** A fluidized-bed combustor burns pulverized coal suspended on a cushion of air to promote more complete combustion. Limestone or dolomite injected into the bed also reacts with sulfur oxides formed during combustion to yield a dry calcium sulfate—virtually eliminating sulfur emissions from the smokestack. More than 1,000 megawatts of existing coal-fired plants are being retrofitted with atmospheric fluidized bed combustors as part of the government's Clean Coal Demonstration Program. Newly built, larger-scale units may be ready for use by the mid-1990s. More efficient pressurized fluidized bed combustors have greater technical obstacles to overcome, and few plants are expected to be operating before the turn of the century.[108]

Another way to raise the efficiency of coal combustion is to gasify the coal and burn it in a combined cycle plant. In an integrated gasification, combined cycle (IGCC) plant, coal is converted into a synthetic gas, with virtually all of the sulfur in the coal removed as hydrogen sulfide gas. The rest of the gaseous mixture, consisting mainly of hydrogen and carbon monoxide and lesser quantities of methane and carbon dioxide, is burned in a combustion chamber to drive a gas turbine. Then, in the combined cycle, exhaust gases exiting the gas turbine flash water into steam to drive a steam turbine. In a more advanced concept, the steam turbine is replaced by a modified (humid air) combustion turbine, raising the operating efficiency of such a plant above 40 percent. About 4,000 megawatts of coal-burning IGCC plants are presently planned for installation in the United States, Europe and Asia.[109]

Both IGCC plants and fluidized bed combustors lack "scrubbers," which is one of the reasons why they are more efficient than most conventional coal plants. But like other coal plants, these "clean-coal" facilities still emit copious amounts of carbon dioxide. Limited by their efficiency advantages, clean-coal plants serving in place of existing coal plants would have the potential to reduce utilities' CO_2 emissions by less than 10 percent (all other things being equal).[110]

> **Box 5-E**
>
> ### Capturing Carbon Dioxide Emissions
>
> In 1990, for the first time ever, U.S. coal production topped 1 billion tons. While this was good news for the coal industry, it may turn out to be bad news for the environment. Each ton of coal emits 2.3 tons of carbon dioxide when it is burned. Once released, CO_2 persists in the atmosphere for 50 to 200 years. Thus, a lasting legacy of 1990's coal production record will be reflected in the atmosphere for decades or perhaps centuries to come.
>
> Utilities typically have dismissed any attempts to capture CO_2 from coal-burning power plants, mainly because of the sheer volume of gases involved. The one place thought large enough to accommodate "scrubbed" CO_2 emissions is the deep ocean, which already represents a huge sink for CO_2. It is not known how long CO_2 would stay put in the deep ocean, however, before it works its way back to the surface. Moreover, there is the practical matter of transporting all that CO_2 from the coal plant into the ocean—at least 100 miles offshore and 1,500 feet deep. One suggested method is to liquefy the CO_2 and pump it through pipelines.
>
> While each step of the carbon-scrubbing process appears technically feasible, the system has never been tested in its entirety. In 1984, researchers at the Brookhaven National Laboratory made a rough calculation of what it would cost to remove 90 percent of CO_2 emissions from all fossil-fueled power plants operating as of 1980. The process of removing, recovering, liquefying and disposing of the CO_2 would consume about 16 percent of total electric generating capacity in coal-burning regions, the researchers estimated, and it would raise electricity production costs by as much as 120 percent.[1c]
>
> A more recent study by Dutch researchers has identified a means of burning coal and capturing CO_2 emissions that may raise electricity costs by only 30 percent.[2c] The technology used would be an integrated gasification, combined cycle (IGCC) power plant, employing a two-step process: coal-gasification to generate heat for electricity, and physical absorption of the CO_2 to capture 88 percent of the carbon emissions. The captured CO_2 would be dried, compressed and transported through pipelines to abandoned natural gas and oil wells, where it would be injected under

In addition, coal-fired fluidized bed combustors produce substantially more nitrous oxide (N_2O) emissions than conventional coal plants, because they are designed to operate at low operating temperatures to control emissions of nitrogen oxides (NO_x)—a precursor of acid rain. The resulting increase in N_2O, a potent greenhouse gas, exceeds N_2O emissions from conventional coal plants by up to 30 times, counteracting the decrease in CO_2 emissions.[111] Moreover, calcium carbonate sorbents, commonly used in fluidized bed combustors to react with sulfur, also release CO_2, further eroding the greenhouse benefits of these plants.

Thus, while clean-coal plants would reduce emissions of sulfur and nitrogen oxides by a substantial amount, they would not reduce CO_2 emissions by very much, and N_2O emissions in fluidized bed combustors would increase. It is even possible that clean-coal plants may increase the propensity for global warming, considering recent evidence about the negative feedback effects of great pressure. Such wells in the United States are thought to be capable of storing up to 80 years of CO_2 at present rates of emissions from coal-fired power plants. The CO_2 recovery and compression process would consume about 13 percent of the IGCC plant's electricity production, raising generation costs by 25 percent. Pumping and disposal operations would tack on additional costs of 5 to 10 percent.

Robert Williams of the Center for Energy and Environmental Studies at Princeton University has taken this novel concept one step farther. Considering that the gas created in the coal-gasification process is composed mainly of hydrogen and carbon monoxide, the gas could be used to derive pure hydrogen rather than to generate electricity. The cost of such hydrogen would be only 60 to 65 cents for a gallon of gasoline-equivalent, Williams estimates. The cost of extracting and sequestering the CO_2 from this gas would add only 12 cents more a gallon (assuming that the abandoned wells are less than 60 miles away), making hydrogen fuel price-competitive with gasoline.

"Thus, ironically, coal, the dirtiest of fuels, might be used as a feedstock for initiating a shift to hydrogen, the cleanest of fuels," Williams concluded in a 1990 analysis.[3e] "Coal could probably provide the basis for a hydrogen economy for only a few decades, until the oil and gas well CO_2-sequestering capacities are exhausted," he added. "It is probably not too fanciful to suggest that if the coal industry were to become engaged this way, it might eventually lose interest in coal, as it evolves into a 'hydrogen industry,' for which there would be a growing diversity of feedstocks over time."

Researchers at the Electric Power Research Institute add a dose of reality to this high-minded idea, however. "Many billions of cubic feet of pure hydrogen are routinely produced in the world's oil refineries, at costs that are a fraction of electrolytic reduction," they point out. "Nevertheless, there are no indications that a transition from conventional end-use systems to hydrogen based systems has been of practical interest. There are no developments now visible that are likely to remove these economic and technical barriers," they add, "although the obvious merits of hydrogen production producing only water as a byproduct provides a tantalizing target."[4e]

sulfate aerosols.[112] Conventional coal plants lacking scrubbers spew tens of thousands of tons of sulfur dioxide gas into the atmosphere every day. Within hours, these gases turn into sulfate aerosols—either droplets or particles—that contribute to atmospheric haze and (possibly) the formation of clouds. The net effect is to shade the planet from incoming sunlight, keeping the surface of the Earth cooler than it would be otherwise. Clean-coal plants—by eliminating most sulfur emissions—would reduce this shading effect yet continue to release large amounts of carbon dioxide and other greenhouse gases into the atmosphere. Accordingly, the net effect of replacing old, "dirty" coal plants with "clean" new ones may be to exacerbate the greenhouse effect.

Eventually, a generating technology may be developed whereby an electrochemical process takes the place of the combustion process to convert gasified coal into electricity. A molten carbonate fuel cell, for example, could replace the combustion turbine in an IGCC plant and avoid the heat cycle limitations of

Figure 12

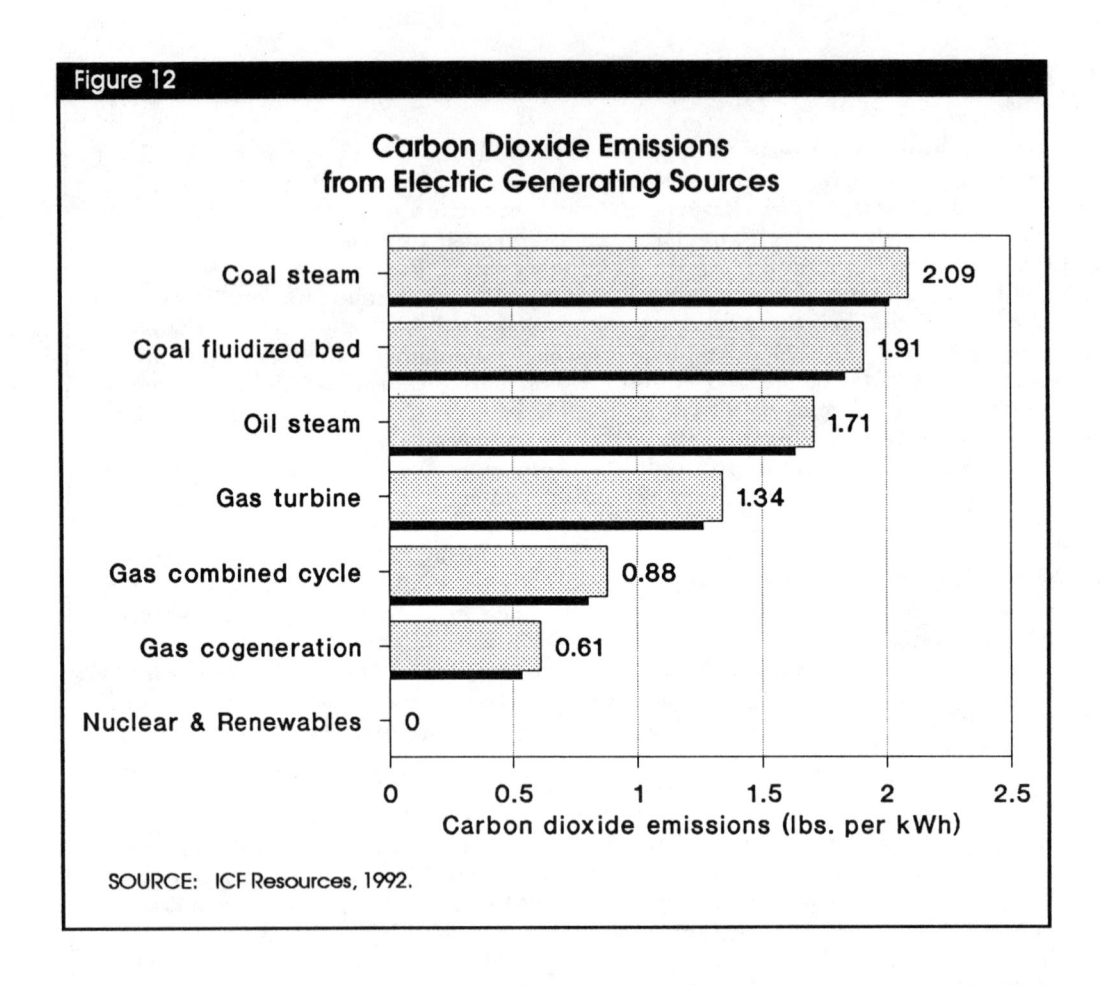

Carbon Dioxide Emissions from Electric Generating Sources

Source	Carbon dioxide emissions (lbs. per kWh)
Coal steam	2.09
Coal fluidized bed	1.91
Oil steam	1.71
Gas turbine	1.34
Gas combined cycle	0.88
Gas cogeneration	0.61
Nuclear & Renewables	0

SOURCE: ICF Resources, 1992.

burning the gasified coal. Coupled with a steam turbine, a molten carbonate fuel cell could raise the operating efficiency of an IGCC plant to 46 percent. Coupled with a humid-air combustion turbine instead, the fuel cell could raise the efficiency of the power plant to perhaps 55 percent.[113] EPRI researchers believe that "a commercially successful fuel cell could offer the most efficient, cleanest, and most versatile means of converting gasified fossil fuels to electricity, reducing carbon and pollutant emissions to about two-thirds of present practice."[114] Such a technology is still decades away from realization, however. EPRI researchers estimate that it would take 35 to 50 years to install 75,000 to 125,000 megawatts of molten carbonate fuel cell equipment and entail $80 billion to $150 billion in research, engineering and deployment costs.

The only conceivable way to curb CO_2 emissions from coal beyond the 33 percent reduction offered by the molten carbonate fuel cell would be to capture CO_2 directly as the coal is processed or burned. The flue gas in a typical power plant contains 12 to 15 percent CO_2 by volume. Chemical absorption scrubbing technologies are available to recover most of this gas, but the processes identified to date would raise the cost of electricity generation from 35 to 120 percent. Moreover, questions remain as to where to deposit the billions of tons of CO_2 after it has been captured.

While energy producers have long appreciated coal's virtues as an abundant, low-cost and widely distributed energy resource, a time of reckoning may be fast approaching for coal. It is precisely because coal is such a widely used fuel—and the prevailing trend around the world is to consume even more of it—that any serious attempt to address global warming must begin with efforts to curtail its use. Since electric utilities account for 80 percent of coal consumption in the United States, and power generators abroad account for 50 percent of coal use, they are likely to be hit first—and hardest—in the event of CO_2 emission controls.

Natural Gas: A 'Bridge' Fuel

If coal is the fossil fuel with the most to lose in case policies are enacted to ameliorate global warming, natural gas is the fossil energy source that stands the most to gain. Natural gas already is nipping at the heels of coal in terms of global energy consumption. Demand for natural gas has doubled worldwide since 1970, making it the fastest growing of the major fuel sources. Although consumption in the United States actually is below the levels of 20 years ago, natural gas demand is projected to grow more rapidly than any other fuel source through 2000, with the exception of renewable energy supplies.[115]

While natural gas is by no means pollution-free, it contains only trace amounts of sulfur and particulates, and it releases less than half as much carbon dioxide and nitrogen oxides (NO_x) as coal when burned in a conventional steam turbine. Natural gas's greater virtue may be that it is highly efficient in converting stored energy into electricity. Since the 1960s, tremendous advances have been made in the design of aeroderivative gas turbines, essentially aircraft jet engines mounted on skids. Such turbines fueled by natural gas are capable of achieving conversion efficiencies of nearly 40 percent.

Steam-injected aeroderivative gas turbines are even more efficient. The turbine's hot exhaust gases—in addition to generating electricity—flash water into steam, which is fed back into the combustion chamber to increase energy production and further reduce NO_x emissions. If the air is cooled between stages of compression, even greater efficiencies and lower NO_x emissions result. (Efficiency rises because less power is required to compress the cooler air, while NO_x emissions fall as a result of lower combustion temperatures.) An intercooled, steam-injected gas turbine can achieve a 47 percent efficiency rating or higher.[116]

When turbines fueled by natural gas are operated in a combined-cycle mode, efficiency ratings of greater than 50 percent are possible. In this case, the steam created by the gas turbine's exhaust is channeled into a connected steam turbine. The extra-high efficiency rating and the low carbon content of natural gas enables a combined-cycle plant to release just 40 percent as much CO_2 per kilowatt-hour produced as a conventional coal-fired plant. Better yet, producing power with a combined-cycle plant costs about one-quarter less than with a new coal plant.

Some engineers believe that combined-cycle plants may attain efficiency ratings of 55 percent by the mid-1990s. However, natural gas used in a cogeneration mode achieves the greatest system efficiencies overall. Here a gas turbine's waste heat is utilized in thermal applications in addition to generating electricity; efficiency ratings of 60 to 70 percent are possible. Industry has installed more than 23,000 megawatts of cogeneration capacity, mostly over the last two decades. The congressional Office of Technology Assessment estimates that it is technically and economically feasible for industry to increase cogeneration capacity by some 60,000 megawatts in the years ahead.

Even though natural gas-fired electricity offers many benefits—low capital costs, short lead times, high operating efficiencies and reduced emissions—utilities have been slower to order new gas-fired power plants than their industrial counterparts. Utilities are particularly wary of possible interruptions in the delivery of natural gas. During the record cold winter of 1977, regulation-induced shortages of natural gas caused schools and businesses to close in 22 states. Congress responded by putting restrictions on the use of natural gas in new power plants to preserve what was then perceived to be a limited supply. While most of these regulations have since been repealed, gas producers remain reluctant to enter into the kind of long-term delivery contracts with utilities that coal suppliers are eager to provide.

Ironically, the predicament facing natural gas producers today is a glut of supply—rather than a shortage—which is squeezing their profit margins. Given the nation's concern over rising oil imports, stringent new clean air laws and continued federal tax credits for certain natural gas exploration costs, it seems incongruous that the market for natural gas should be so depressed. The specter of global warming is yet another factor that should buttress the fortunes of the natural gas industry. Calculations show that replacing half of existing U.S. coal capacity with gas turbines over the next 20 years would by itself cut the nation's CO_2 emissions by 16 percent.

A spate of warm winters in recent years (perhaps related to the greenhouse effect) has contributed to the industry's woes, however. Gas producers normally make two-thirds of their annual profits in the November-through-February period, when heating demand soars and gas prices rise. But in the winter of 1991-92, warm weather and sluggish heating demand led natural gas prices on the spot market to drop to a record low of slightly more than $1 per thousand cubic feet. Since many producers consider their break-even price to be $1.50 per thousand cubic feet, they have been inclined to limit production rather than sell at a loss.[117]

Conserving the supply of natural gas may not be such a bad idea from a global warming standpoint. Natural gas is a versatile fuel that is suitable for cooking, heating, power generation, transportation and other end-uses. Many energy analysts believe it is the appropriate "bridge" fuel to spell energy consumers until noncarbon sources—including hydrogen—are more widely available. The challenge is in maintaining orderly demand for natural gas in the meantime. Should energy consumers in several sectors decide all at once to increase their dependence on natural gas, its price could surge with tremendous repercussions for the economy.

Figure 13

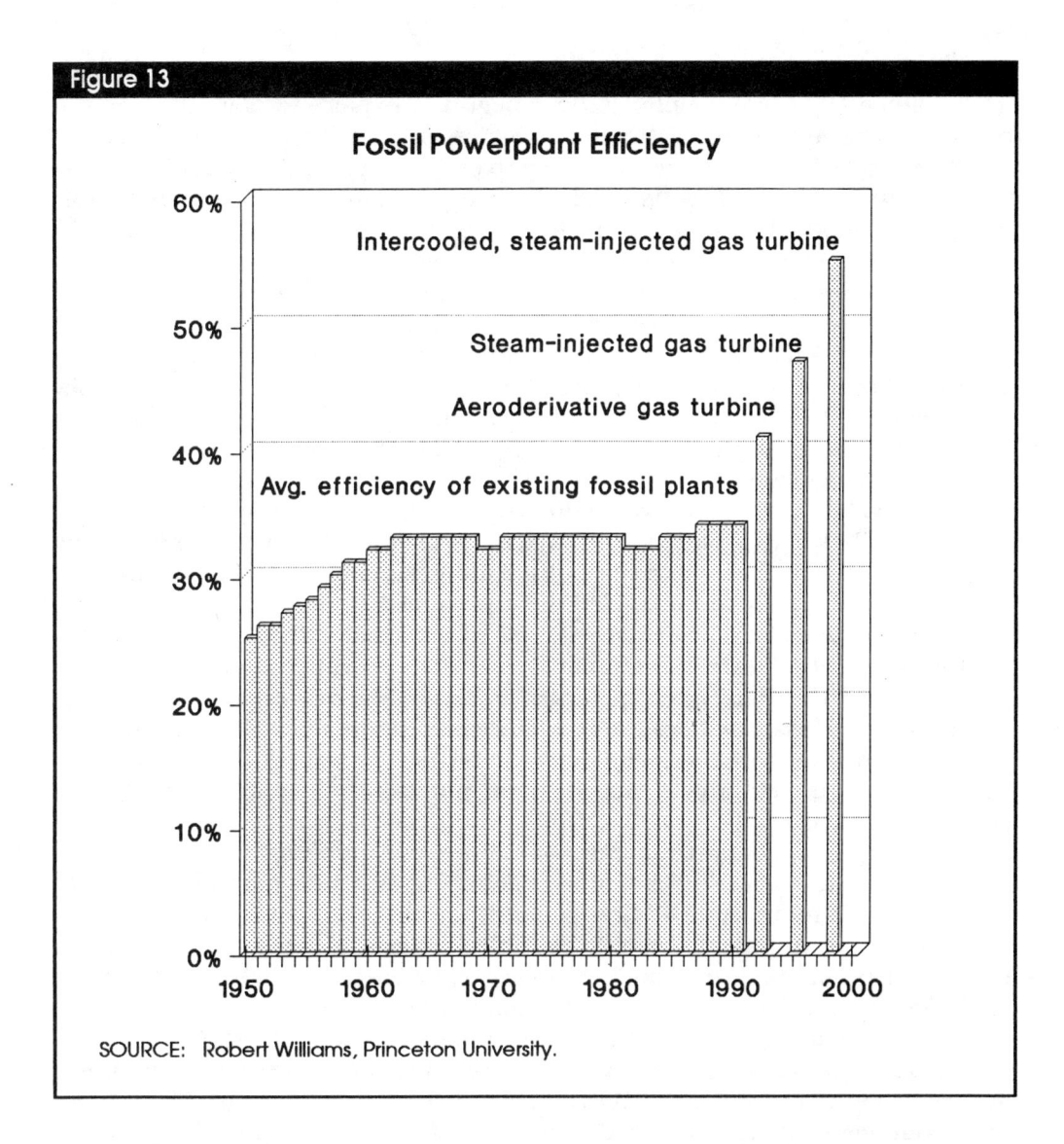

Fossil Powerplant Efficiency

- Intercooled, steam-injected gas turbine
- Steam-injected gas turbine
- Aeroderivative gas turbine
- Avg. efficiency of existing fossil plants

SOURCE: Robert Williams, Princeton University.

While delivery shortages in the 1970s created a perception that natural gas is in short supply, in reality some 4,400 trillion cubic feet of known gas reserves lie underground in the Soviet Union, the Middle East, Africa and South America, and off the shores of Norway, Indonesia, Texas and many other places. In fact, natural gas is more plentiful and widely distributed than oil, accounting for about 40 percent of the world's proven hydrocarbon reserves.[118] Roughly half of the discovered gas is in remote areas that make it uneconomic to transport at today's prices, though. Moreover, the United States has only about 4 percent of the world's proven reserves, and another 4 percent resides in neighboring Canada and Mexico. To a large extent, then, the United States will have to depend on overseas suppliers if it chooses natural gas to bridge the gap between conventional and alternative energy sources.

Proponents of natural gas point out that estimates of proved reserves have tripled since 1970, as companies have begun to explore for gas in areas not related to oil development. Estimates of recoverable reserves also are increasing as advanced technologies are developed to tap into "unconventional" reserves in low permeability formations. In the United States alone, 1,000 trillion cubic feet of natural gas is thought to reside in such formations, and another 400 trillion cubic feet is trapped in coal beds. Recent estimates suggest that 450 trillion cubic feet of this unconventional gas may be recovered, ultimately, with help from advanced technology.[119]

Fuel cells: Efficient applications of natural gas also are likely to extend the life of the resource. Fuel cells, in particular, offer an exciting prospect, since they can produce electricity far more efficiently than any available combustion technology. Fuel cells produce a low-voltage trickle of electricity as hydrogen and oxygen molecules bond to form water. A single fuel cell is capable of producing about one volt of electricity. The more fuel cells that are stacked together, the greater the electricity production. Tokyo Electric Power is testing a $30 million electrical plant that consists of 18 stacks of fuel cells, each 11 feet high and 4 feet square. The 11 megawatts of electricity produced is enough to serve 4,000 homes.[120] Residual heat from fuel cells also can be used in thermal applications.

Because fuel cells are clean, quiet (there are no moving parts), modular and non-polluting, they have great potential to serve in urban areas. Natural gas is an obvious feedstock because it is so rich in hydrogen. Coal, wood, crop residues and other biomass feedstocks can be used as well, or hydrogen can be derived through electrolysis, an electrical process that splits water into its constituent parts.

Not surprisingly, natural gas companies are pushing hardest for the commercialization of fuel cells. United Technologies, a U.S. manufacturer, has U.S. orders for more than 30 first-generation fuel cells—all from natural gas utilities. Other American developers include Westinghouse and Energy Research Corp., of Danbury, Conn. Overseas competitors include Mitsubishi Heavy Industries, Toshiba and nearly a dozen other companies in Japan as well Germany's Siemens AG and ABB Group of Zurich, Switzerland.[121]

At $2,000 to $3,000 per kilowatt of installed generating capacity, fuel cells still cost 25 to 50 percent more than new coal-fired plants and two to three times more than conventional gas-fired power plants. Fuel and maintenance costs for fuel cells range from 3 to 4 cents per kilowatt-hour.[122] As more fuel cells are manufactured, however, their installed costs are expected to drop to $1,500 per kilowatt or below. Arthur D. Little, the Cambridge, Mass., consulting firm, estimates that fuel cells "reasonably" could provide 4,000 megawatts of generating capacity annually by 2000, representing a market with sales of $6 billion a year.[123]

If early demonstrations prove successful, electric utilities could order far greater amounts of fuel cells in the early 21st century. Such a development might constitute the most appropriate use of a fossil fuel in a greenhouse world. A fuel cell deriving its hydrogen from natural gas consumes 30 to 40 percent less fuel per kilowatt-hour than natural gas burned in a conventional steam turbine. Hence, the fuel cell puts out 30 to 40 percent fewer carbon dioxide emissions

The Electric Utility Industry 405

as well. If the future belongs to the efficient, then the fuel cell holds one of the keys to 21st century power generation.

As for the natural gas industry itself, it, too, appears to be repositioning strategically for the 21st century. The natural gas industry traditionally has aligned itself with other fossil producers on energy policy matters; half of U.S. gas reserves are in fact owned by multinational oil companies. In the spring of 1992, however, the American Gas Association conducted its first joint study with the trade associations representing the energy efficiency and renewable energy industries. The purpose of the study was to formulate a "plausible, market-driven energy future for the United States."[124]

In *An Alternative Energy Future*, government subsidies and regulatory biases that favor carbon-intensive fossil fuels would be eliminated. Integrated resource planning methods would be vigorously expanded. And government research and development efforts would be reallocated away from high-carbon

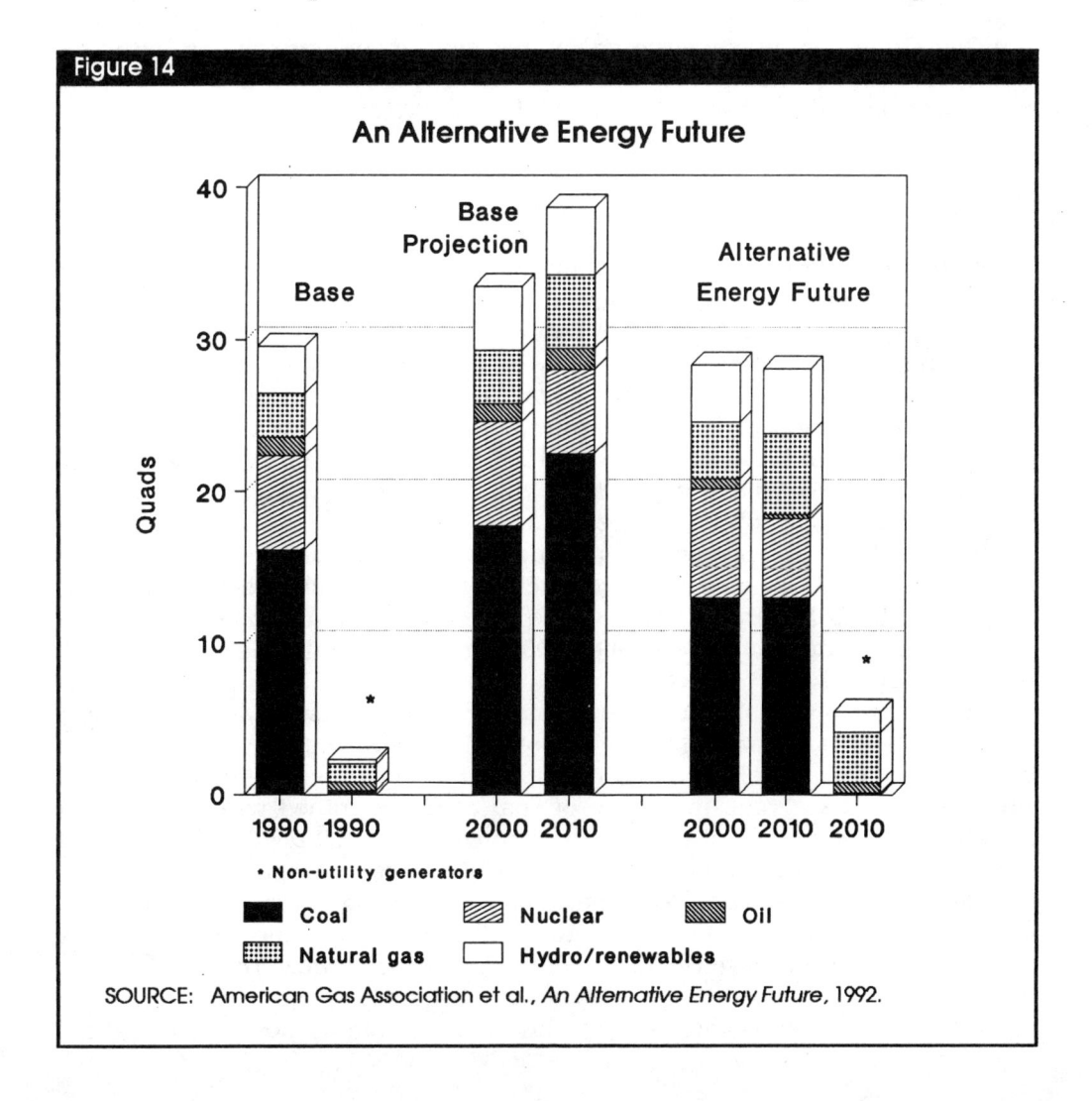

Figure 14

An Alternative Energy Future

* Non-utility generators

■ Coal ▨ Nuclear ▧ Oil
▦ Natural gas ☐ Hydro/renewables

SOURCE: American Gas Association et al., *An Alternative Energy Future*, 1992.

fuels and nuclear power to natural gas, renewables and end-use efficiency equipment. The study concluded that these changes would enable total energy consumption to remain constant at 1990 levels, while the nation's carbon dioxide emissions would fall 12 percent below 1990 levels. Natural gas's share of primary energy demand would rise from less than 24 percent in 1990 to nearly 33 percent in 2010, and renewables' share (including hydropower) would rise from less than 7 percent to nearly 13 percent.

The combined share of oil and coal, meanwhile, would fall from 63 percent in 1990 to 48 percent in 2010; the coal industry alone would lose 44,000 jobs relative to a base-case projection. Those losses would be more than offset by the creation of 215,000 additional jobs in energy conservation services, renewable energy development and natural gas production, however. Best of all from a consumer standpoint, the nation's energy bill would be only $599 billion in 2010, instead of $736 billion in the base-case forecast—a projected savings of 19 percent.

This study may presage the policy direction of the natural gas industry. While the industry has yet to call for the institution of programs to address global warming, the rationale is provided in *An Alternative Energy Future*.

Reviving Nuclear Power

While no one can be certain what energy trends will carry into the 21st century, there is almost sure to be continued reliance on coal, greater use of natural gas, further development of renewable technologies and more efficient use of energy of all kinds. Nuclear power, by contrast, is one energy source whose future is highly uncertain. Today, nuclear power provides nearly 17 percent of the world's generating requirements. But very few new reactors are being ordered, and between 1991 and 1992 total installed nuclear generating capacity declined for the first time since the industry got its start in the 1950s. In the United States, only one nuclear plant is under active construction; it was ordered in 1974. If more orders fail to materialize, nuclear power's share of generation in the United States will decline steadily, from nearly 22 percent today to perhaps 16 percent in 2000, 13 percent in 2010 and then, with a wave of plant retirements, only 1 percent by 2030 (equivalent to about six large plants' worth of generation).[125]

"The result is an unbalanced electricity generation system dominated by a single fuel"—coal, according to the National Energy Strategy. "Apart from the potential reliability problems inherent in overreliance on any single fuel, it is unlikely that these vast quantities of coal could be mined without causing substantial environmental stress," the strategy warns. Indeed, the specter of increased carbon emissions from coal is one of the most compelling reasons for reviving the nuclear option. As a noncarbon-generating source, nuclear power offsets approximately 150 million tons of U.S. carbon dioxide emissions annually that would have resulted from burning fossil fuels instead.[126]

If the number of operating nuclear plants begins to decline, however, a rise in carbon dioxide emissions seems almost inevitable as power demand in-

Figure 15

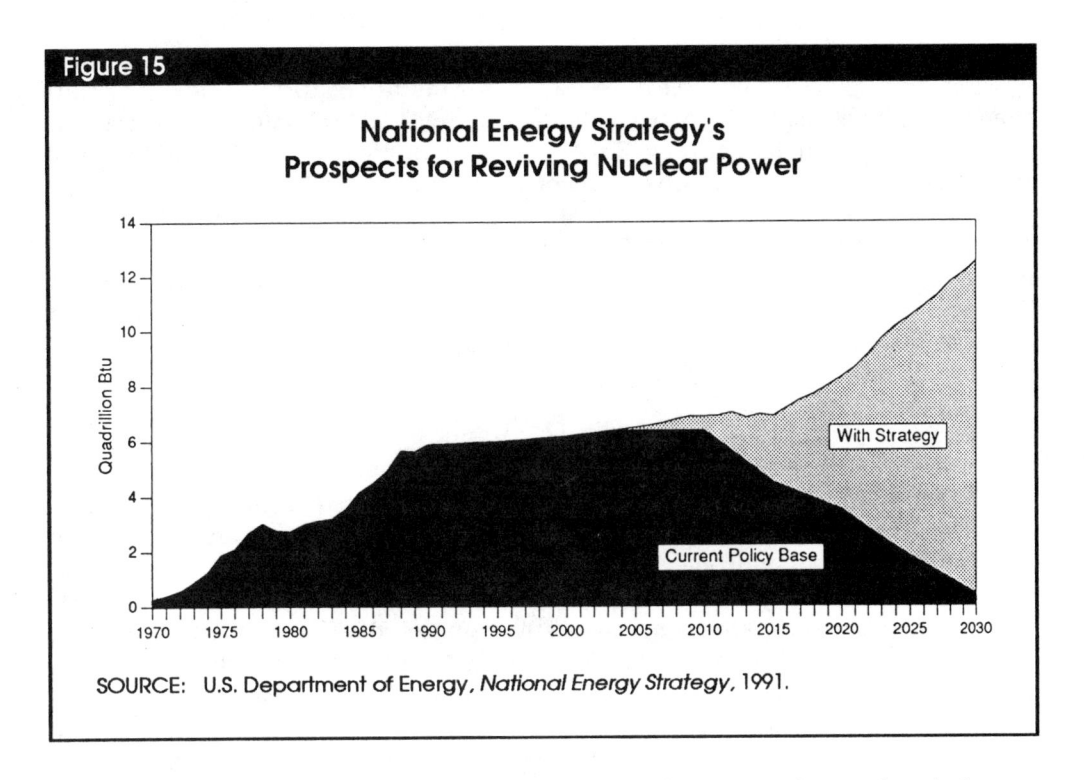

National Energy Strategy's Prospects for Reviving Nuclear Power

[Chart: Quadrillion Btu vs. Year (1970–2030), showing "Current Policy Base" and "With Strategy" scenarios]

SOURCE: U.S. Department of Energy, *National Energy Strategy*, 1991.

creases—that is, of course, unless the nation's fuel mix takes a decided turn away from coal toward natural gas and renewables, with energy efficiency keeping the overall rise in demand to a minimum. The National Energy Strategy discounts this latter possibility and recommends doubling the nation's nuclear generating base from about 100,000 megawatts today to nearly 200,000 megawatts by 2030. Instead of flickering out, then, nuclear power would retain 22 percent of a much larger generating base in 2030. At the same time, the projected costs of nuclear generation would fall from 9.9 cents per kilowatt-hour for plants brought on line since 1980 to only 6.6 cents per kWh (in 1990 dollars), the National Energy Strategy projects. That would make nuclear power competitive in price once again with electricity from other conventional supply sources.[127]

Many energy analysts agree that more nuclear power in the nation's fuel mix would make it easier and perhaps less costly to comply with future CO_2 control measures. An analysis by ICF for the Edison Electric Institute found that unconstrained development of nuclear power theoretically could cut the costs of CO_2 abatement by two-thirds—from $118 billion annually to $37 billion annually—if the program's goal was to reduce the utility sector's CO_2 emissions by 20 percent over the next quarter-century. That would mean adding 250,000 megawatts of additional nuclear capacity, however, and extending the operating lives of existing reactors. ICF concedes this is not a realistic scenario because of prevailing financial, regulatory and environmental barriers. "Nonetheless, in the face of greenhouse gas policies, relatively inexpensive nuclear power would be a valuable resource," the consulting firm emphasizes.[128]

Many obstacles stand in the way of such a nuclear renaissance. A new generation of reactors has yet to be tested, for instance, and it is not clear that they would be able to overcome the financial, political and safety concerns now associated with nuclear energy. Proliferation of nuclear technology also would heighten the potential for diversion of uranium for military purposes. The greatest obstacle to nuclear's revival remains economic in nature, however. Many opponents of nuclear power insist there are other less costly means of meeting future generating requirements. Since investment capital, ultimately, is in scarcest supply, they say it makes sense to invest in these other least-cost options first and then turn to nuclear power only as a last resort.

The Economics of Nuclear Power

The economics of nuclear power have turned increasingly sour since the mid-1960s, when the first commercial reactors went on-line. Many of the older nuclear plants rival hydroelectric power as the cheapest source of electricity in the United States. By contrast, most of the newer plants are among the most expensive generating sources. Recently completed nuclear plants cost five times more to build than the inflation-adjusted averages of reactors finished in the mid-1960s. From the time of groundbreaking to the time of plant completion, cost estimates have risen an average of 200 percent. And once in operation, most nuclear plants have performed at only 75 percent of their expected capacity factors.

Meanwhile, incurred operational and maintenance costs have risen three to four times above the level predicted as recently as the late 1970s.[129] The government's Energy Information Administration reports that operating and maintenance costs for American nuclear plants soared an average of 12 percent a year between 1974 and 1984. Reactor components have required repairs and replacement far more frequently than was expected. Moreover, the Nuclear Regulatory Commission has completely rewritten its safety standards since the Three Mile Island accident in 1979—forcing costly retrofits at both operating reactors and ones under construction. Since 1987, it has been more costly just to *operate* an average nuclear plant than to operate a coal-fired plant, even though nuclear fuel is much less expensive than coal.[130]

The fivefold rise in capital costs has posed an even bigger financial obstacle for nuclear power in the last 20 years. The newest reactors are about three times more expensive to build than comparable coal plants outfitted with scrubbers. As these pricey reactors enter the rate base, utility commissioners have been extremely harsh in their treatment of costs. According to a 1989 study by Theodore Barry & Associates, state commissioners have disallowed an average of one-seventh of the installed cost of recently completed U.S. nuclear plants. For a unit with a $3.5 billion price tag—the going rate for today's large reactors— such a penalty works out to a half-billion dollars. Few utilities can bear such a loss without requiring dividend cuts (or worse) of their shareholders.

Rising costs and attendant safety concerns also have been a blow to the nuclear industry internationally. According to the International Atomic Energy

Table 4

Nuclear Power Plants in Operation and Under Active Construction
January 1, 1992

	Operating # of units	Operating MW	Under Construction # of units	Under Construction MW
AMERICAS				
United States	110	99,523	1	1,150
Canada	20	13,993	2	1,762
Argentina	2	935	1	692
Mexico	1	654	1	654
Brazil	1	626	1	1,245
Cuba	0	0	2	816
Subtotal	**134**	**115,731**	**8**	**6,319**
WESTERN EUROPE				
France	56	56,808	6	8,460
United Kingdom	37	11,506	1	1,175
Germany	21	22,408	0	0
Sweden	12	9,817	0	0
Spain	9	7,067	0	0
Belgium	7	5,500	0	0
Switzerland	5	2,952	0	0
Finland	4	2,310	0	0
Netherlands	2	508	0	0
Subtotal	**153**	**118,876**	**7**	**9,635**
EASTERN EUROPE				
Czechoslovakia	8	3,264	6	3,336
Bulgaria	6	3,538	0	0
Hungary	4	1,645	0	0
Slovenia	1	632	0	0
Romania	0	0	5	3,125
Subtotal	**19**	**9,079**	**11**	**6,461**
COMMONWEALTH OF INDEPENDENT STATES				
Russia	28	18,893	4	2,900
Ukraine	14	12,095	0	0
Lithuania	2	2,760	0	0
Kazakhstan	1	335	0	0
Subtotal	**45**	**34083**	**4**	**2,900**
ASIA & REST OF WORLD				
Japan	42	31,994	10	9,012
South Korea	9	7,220	2	1,900
India	9	1,814	5	1,100
Taiwan	6	4,890	0	0
South Africa	2	1,842	0	0
China	1	288	2	1,860
Pakistan	1	125	0	0
Subtotal	**70**	**48,173**	**19**	**13,872**
TOTAL WORLD	**421**	**325,942**	**49**	**39,187**

SOURCE: International Atomic Energy Agency and Worldwatch Institute.

> **Box 5-F**
>
> ## Nuclear Power in Retreat
>
> Thirty years ago, nuclear power seemed destined to be the next great source of power for the world. If not "too cheap to meter," it was to be the most economical energy choice, serving as the inevitable successor to oil and coal. Today, more than 420 nuclear reactors are in operation around the world, but many of these are among the most expensive power plants ever built. Meanwhile, the battle to wrest fossil fuels from the top of the generating heap is far from over. Coal and oil still account for nearly four times as much of the world's generating capacity as nuclear power—and the momentum that nuclear power once enjoyed is gone.
>
> The United States has not placed an order for a nuclear plant since 1978, and all reactor orders since 1974 have been canceled. Similar de facto moratoriums exist in other countries with nuclear programs, including Belgium, Germany, Spain and Switzerland. In Sweden, a national referendum in June 1990 called for the shutdown of that country's 12 nuclear plants by 2010, even though these reactors make up 45 percent of the nation's generating capacity and are among the most efficiently run nuclear plants in the world. The Italian parliament also approved a measure in 1990 to dismantle that country's three once-operating nuclear units.
>
> Austria, Greece and the Philippines shut down their sole reactors before the plants ever generated electricity commercially. Denmark, Norway and The Netherlands are among other nations that have pledged never to develop nuclear power in the first place. In Latin America, the three countries with operating nuclear plants—Argentina, Brazil and Mexico—have put plans for future reactors on hold because of high construction costs and poor operating performance at four existing plants.[1f]
>
> At the moment, the former Soviet Union has the most ambitious expansion plans for nuclear power. As of year-end 1991, 45 reactors were operating and 25 more were under development, mainly in Russia and the Ukraine. But the explosion at Chernobyl in 1986—and mounting public opposition since the age of *glasnost*—is causing a major reassessment of its nuclear program. The pace of construction has slowed considerably pending safety reviews, and many projects have been suspended or canceled. Several other eastern European nations also have shut down Soviet-designed nuclear plants or canceled plants under construction, because of safety concerns following the Chernobyl accident. Fifteen Chernobyl-type reactors remain in service, however.[2f]
>
> One nation often praised for the success of its nuclear program is France. It possesses the second largest number of reactors after the United States—57 as of July 1992—and it is the world's most nuclear-dependent nation. But with three-quarters of the country's electricity now coming from the atom, France's reactor program also is winding down. Whereas a decade ago new plants were coming on-line at the rate of five a year, now France has only five reactors still under construction. The pace of reactor completions has slowed to an average of one every three or four years in France. With the domestic power market saturated, France

Agency, 421 nuclear plants were in operation in January 1992, 10 fewer than in January 1989. Another 76 plants were under construction, although work has been suspended on about one-third of these plants. In 1990, for the first time in 35 years, no new construction of nuclear plants began; only three plants began construction in 1991. Today, the backlog of reactor orders is at its lowest is selling more of its nuclear electricity at a discount to neighboring countries. In addition, its sole reactor vendor, Framatome S.A., has teamed up with Germany's Siemens AG in an effort to bolster the sagging export market for nuclear generating equipment. Their new export company, Nuclear Power International, is designing a new, medium-sized reactor for sale to developing countries—where interest in nuclear power also has waned considerably following Chernobyl.[3f]

Across the English Channel, the British House of Commons nixed a plan in 1989 to sell off the government's fleet of reactors as part of Prime Minister Margaret Thatcher's utility privatization scheme. Potential buyers of the government-owned utility system balked at the prospect of having to pay the decommissioning costs of seven aged nuclear plants, and the buyers showed little more enthusiasm for Britain's newer reactors—generally regarded as unreliable and uneconomic. In the end, "The House of Commons was not prepared to give the unprecedented guarantees sought by London's financial district to underwrite the high cost of nuclear power," *The Wall Street Journal* reported of the proposed reactor sale.[4f]

With nuclear programs in decline in other parts of the globe, Japan stands out as one nation still committed to an ambitious construction effort. The Ministry of International Trade and Industry wants to nearly double the number of operating plants in Japan to 81 by 2010—raising reliance on nuclear power from 27 percent to 43 percent of the nation's generating mix. In support of this goal, the Japanese government is allocating nearly $300 million annually toward research and development of conventional reactors (about 50 percent more than the U.S. government spends) and nearly $800 million a year for new advanced reactors (about four times what the U.S. government spends).[5f]

Even in Japan, however, opposition to nuclear power is growing. Japanese utilities suffered a spate of nuclear-related incidents in 1991, including one near Kyoto where a failed valve and a broken cooling water tube triggered a contained release of radioactive emergency cooling water. By world operating standards, this was no big nuclear event. There have been more than 400 recorded instances of water tubes breaking in reactors around the world, and thousands of instances of stuck and failed valves. (An average U.S. nuclear plant contains 40,000 such valves.) But in Japan—where the nuclear industry is heralded as being the safest in the world—it marked the first time there had been such an incident involving loss of coolant.[6f]

The recent turn of events could make it more difficult to site new reactors in Japan. While 10 reactors are under construction, no planned reactors have received local permission for plant siting since 1986.[7f] If the outlook there does not improve, industry officials concede it will be impossible to build even half of the 40 plants sought by the Japanese government over the next 20 years.[8f]

A similar predicament prevails worldwide in terms of nuclear power responding to the global warming challenge. If a new round of construction does not begin soon, it may be too late for nuclear power to play a meaningful role in efforts to curb carbon dioxide emissions over the next quarter-century.

level in more than 15 years.[131]

In order to meet future electricity demand and hold down greenhouse gas emissions, nuclear advocates believe the pace of reactor orders must pick up dramatically—and soon. Writing in a special energy edition of *Scientific American*, German physicist Wolf Hafele recommends the yearly addition of

40,000 megawatts of nuclear generating capacity over the next 40 years, raising the total number of reactors in operation to 2,500.[132] If orders continue their recent downward slide, however, the actual number of operating reactors may top out at just under 500 shortly after the turn of the century—and start to decline after that. In terms of generating capacity, the figure would rise only 10 percent above today's level—to 360,000 megawatts. This is less than one-tenth the amount of nuclear energy forecast by the International Atomic Energy Agency in 1974.

Reactor vendors have a special interest in seeing this slide reversed. A promotional brochure issued by Westinghouse offers the following rationale: "To avoid an even greater reliance on foreign oil—with its threats to our national security and our international balance of trade—or an almost exclusive dependence on coal—with its implications for acid rain and the 'greenhouse effect'—the United States needs to have nuclear power available as a realistic option."[133]

So far, this argument seems to be falling on deaf ears. The United States has 110 reactors in commercial operation—nearly twice as many as any other nation—but only three of eight planned additional units are likely to be completed.[134] By the time the last of these plants comes on-line, the first in a wave of plant retirements will be ready to begin. Between 2000 and 2015, 47 U.S. nuclear plants will reach the end of their 40-year design lives. Although licensing extensions may permit some of these plants to continue operating for longer periods, new reactor orders must materialize soon if nuclear power is to hold its own in this country.

The uranium fuels market is equally depressed. Spot uranium prices have plummeted from a record $43 a pound in 1980 to less than $9 a pound recently, trading at their lowest levels in nearly two decades. A rebound in the price of uranium is not expected for at least several more years, partly because of sales of uranium from the military stockpiles of the former Soviet Union, China and the West. At the start of the 1980s, most of the major U.S. oil companies, including Amoco, Atlantic Richfield, Chevron, Mobil and Tenneco, were involved in uranium exploration and production. Now not one of them is.[135]

Building Safer Reactors

While the deteriorating economics of nuclear power have hamstrung efforts to build new plants in the United States and around the world, the underlying concern of nuclear opponents remains the issue of plant safety. To allay these fears—and clamp down on rising reactor costs as well—the nuclear industry is looking forward to the introduction of new plants with so-called "passive" safety designs. In the event of an accident, these reactors would use non-mechanical processes such as natural convection of heat and gravity feeding of emergency coolant from overhead tanks to keep the reactor stable. Such passive safety features reduce the possibility of operator error and obviate the need for many of the "defense-in-depth" systems included in today's nuclear plants. The elimination of external, redundant safety equipment enables passively safe reactors to be designed with only half as many valves, pumps, heat exchangers and piping, and up to 80 percent fewer control cables, as in conventional nuclear plants.

Some of the plants now being studied offer radical departures from lightwater reactor designs. Helium gas or liquid sodium can be used as a coolant instead of water, for example. Such plants also boast a negative temperature coefficient, which slows the rate of fission in the reactor core when vessel temperatures rise abnormally high (as would be the case in a loss-of-coolant accident). Consequently, gas-cooled and liquid-metal-cooled reactors would not require the massive concrete containment structures shrouding most conventional nuclear plants, according to the reactors' proponents, because a core meltdown is considered virtually impossible.

The lack of redundant safety equipment and other economizing features may reduce the installed cost of new, passively safe reactors. Savings also can be achieved through use of modular, standardized components that are assembled at the factory rather than at the plant site. In the United States, preapproval of such standardized designs by the Nuclear Regulatory Commission might allow new plant start-ups within five years of a reactor order. Today, it usually takes 10 to 14 years to license and complete a nuclear plant.[136]

The advanced reactor designs that are the farthest along are based on lightwater reactor technology. Westinghouse expects to receive NRC design certification for its AP600 (advanced passive 600-megawatt) reactor by 1994. General Electric hopes to receive similar certification for its own 600-megawatt, passively safe boiling water reactor by 1997. General Electric also is looking forward to NRC certification of a new 1,350-megawatt boiling water reactor design in 1995. This large reactor incorporates some of the same passive safety features as in the 600-megawatt models. Because of its size, however, fewer parts can be factory-assembled. Moreover, its passive safety features are comparatively limited, so human intervention would be required shortly after a serious accident in order to regain control of the reactor. (The smaller reactors are designed to keep the reactor stable for at least three days before a runaway situation could start to develop.)[137]

The Japanese government approved orders for two of the 1,350-megawatt plants in 1991—marking the first time in 15 years that General Electric has received reactor orders of any kind.[138] Tokyo Electric Power hopes to bring the reactors on-line in 1996 and 1998, 140 miles northwest of Tokyo. Hitachi Ltd. and Toshiba are joint venture partners in the project.

Meanwhile, nuclear vendors in Germany and Japan as well as the United States are pursuing the commercialization of modular high-temperature gas-cooled reactors, which have been in existence since the 1950s. Innovative features of these plants include the use of helium gas as a coolant and fuel rods made from tiny thorium or uranium pellets, which in turn are encapsulated in several layers of ceramic material. The protective ceramic coating allows the fuel rods to withstand core temperatures up to 3,600 degrees F, so they would not be likely to melt down even if the helium cooling system were to fail. Additional protection is provided by graphite blocks surrounding the fuel rods in the reactor core. The blocks serve as a high capacity heat sink and as a repository for granulated boron carbide, which can be dumped into the reactor core to absorb neutrons during an emergency shutdown. Unlike the Chernobyl reactor, which also was based on a graphite-moderated reactor concept, the high-temperature gas-cooled reactor has a negative temperature coefficient. This would enable the reactor to remain stable in a loss-of-coolant accident, rather than causing a sudden spike of power that could blow the reactor apart.

In the United States, General Atomics Corp. is heading the effort to develop modular high-temperature gas-cooled reactors. It already has a contract from the U.S. Department of Energy to build a model plant in Idaho, although it will be used to produce tritium for nuclear weapons instead of generating power. General Atomics' commercial plant design consists of four reactors, each rated at 134.5 megawatts, which provide steam for two turbine generators. Consumers Power, a Michigan electric utility owned by CMS Energy Corp., has expressed interest in acquiring such a gas-cooled nuclear plant for generating purposes after the year 2000. Coastal utilities also may take an interest in gas-cooled reactors, because they also produce tremendous amounts of process heat that would be ideally suited to operate desalination plants.[139]

'Breeding' Solutions—and Problems

The advanced reactor design that is the farthest from commercialization is the liquid-metal-cooled reactor. General Electric is working on a Power Reactor Inherently Safe Module (or PRISM), for which it hopes to receive NRC certification by 2005. The PRISM reactor would link together a series of 155-megawatt reactors in "power packs" to supply increments of capacity up to 1,395 megawatts—the size of the largest conventional reactor. Each of the reactors would be buried underground and submerged in a pool of liquid sodium. Natural convection of heat away from the reactor would keep the liquid sodium at temperatures several hundred degrees below its boiling point of 1,600 degrees F. The fuel in the PRISM reactor would be a metallic alloy consisting of plutonium, uranium and zirconium. Such a configuration "breeds" nuclear fuel by converting nonfissile uranium-238 into fissile plutonium-239.

The breeding capability of a liquid-metal-cooled reactor is important because the worldwide supply of raw uranium is limited. With 360,000 megawatts of existing and planned nuclear capacity, global uranium supplies are expected to last only 100 years if there is no fuel reprocessing. (By the same token, the nuclear fuel supply would last only 20 years if the reactor base were quintupled, as some have recommended.) This potential shortage makes the use of breeder reactors a virtual necessity for the long-term expansion of nuclear power, since breeders are able to extract 100 times more energy from the same amount of uranium as conventional light-water reactors.[140]

Liquid-metal-cooled plants also are valuable from the standpoint that they can "breed" plutonium from the unused uranium found in the spent fuel of conventional light-water reactors. This recycling process would limit the waste material being sent to geologic repositories to comparatively short-lived fission waste byproducts, thereby reducing the volume of waste and the burden

The Electric Utility Industry 415

Figure 16

Advanced Liquid Metal Reactor Power Plant Design

SOURCE: General Electric Co.

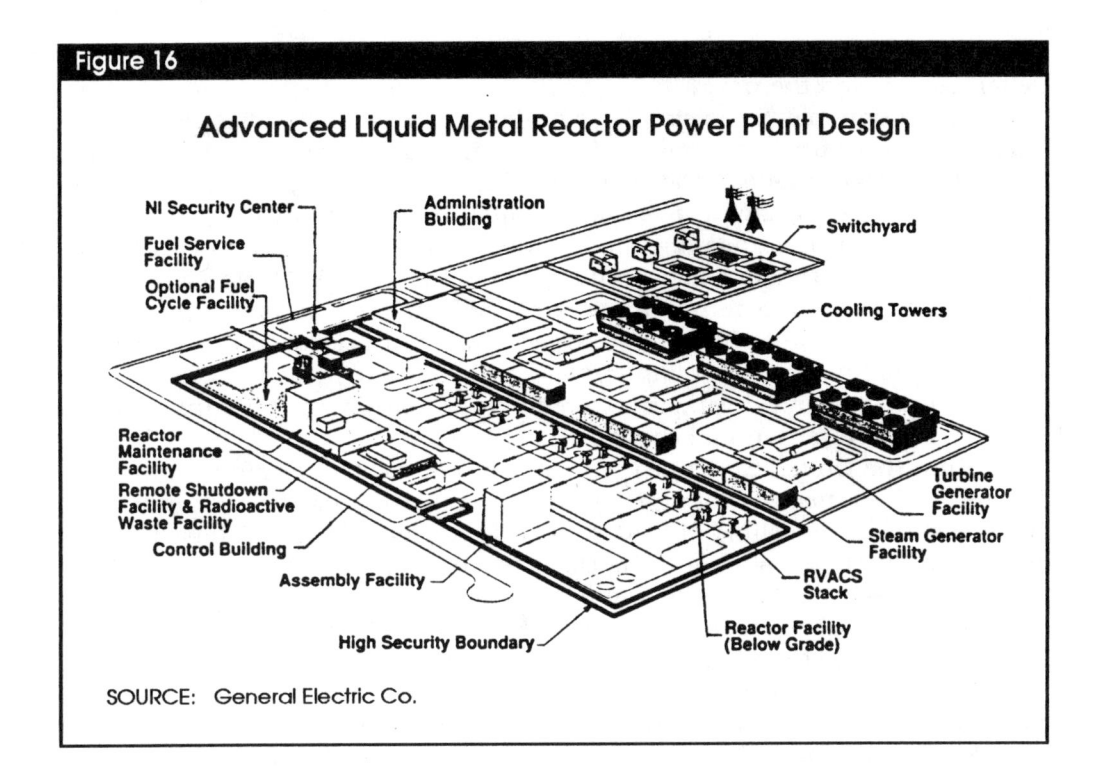

imposed on future generations. It would also lessen the prospect of having to mine and process progressively lower quality uranium ores—activities that result in CO_2 emissions.

To date, no permanent repositories for high-level waste have been sited anywhere in the world; only temporary storage sites have been selected.[141] Yet spent nuclear fuel is accumulating at an annual rate of about 20 metric tons per conventional 1,000-megawatt reactor. That means about 8,000 tons of high-level waste eventually will need to be dealt with every year, given the number of reactors in operation and on order. Since a single waste repository is not expected to hold more than 70,000 tons of waste material, a new disposal facility would have to be built somewhere in the world every nine years, assuming the spent fuel is not reused in breeder reactors. (Once again, a fivefold increase in nuclear generating capacity would require building new repositories at the much faster rate of one every two years.) Thus, development of breeder reactors would address two important problems simultaneously: the potential shortage of raw uranium and the limited storage capacity of nuclear waste repositories.

But breeders themselves require the use of fuel reprocessing plants to chemically extract the long-lived transuranium nuclides from the short-lived fission products contained in spent fuel. At present, there are only four such reprocessing plants in operation (one each in England, France, India and Japan), and five more are under construction. (Russian and Chinese reprocessing capabilities are unknown.) These existing and planned reprocessing facilities are capable of handling only half of the nuclear waste already being generated around the world.[142]

The solution may be as simple as building more reprocessing plants. But this raises the troublesome prospect of nuclear proliferation, since bomb-grade plutonium can be made at these plants. (India, for one, separated plutonium from its reprocessing facility to explode a nuclear weapon in 1974.) The Carter administration turned against the breeder/reprocessing option with its endorsement of the Nuclear Non-Proliferation Act passed by Congress in 1979. The act called for the termination of all civilian nuclear fuel reprocessing activities in the United States and urged other countries to do the same. A handful of nations are pressing forward with nuclear fuel reprocessing programs, however, and diversion of bomb-grade plutonium remains a major concern of the international community.

Within the United States, the moratorium on civilian waste reprocessing has heightened the sense of urgency surrounding the creation of a permanent high-level waste repository. By 2010, 66 U.S. nuclear plants will have reached their limits of on-site storage. The U.S. Department of Energy has picked the site for the nation's first permanent repository, Yucca Mountain, 110 miles northwest of Las Vegas, Nev. But the State of Nevada is not pleased with its selection, and the Nevada legislature has passed two resolutions outlawing a high-level nuclear waste dump within the state's boundaries. A legal battle now is being waged in Congress and the courts. In 1991, the U.S. Supreme Court upheld a decision by the Ninth Circuit Court of Appeals ordering Nevada to issue the necessary permits.[143] Meanwhile, the opening date for the nation's first high-level repository has been pushed back from 1998 to 2010, and debate continues as to whether Yucca Mountain or any other site would remain geologically stable for 10,000 years as the radioactive material decays.[144]

Disposal of spent fuel is not the only unresolved question concerning nuclear power. There is also the matter of plant decommissioning. Since no large commercial reactor has reached retirement age, there is no historical precedent on which to base the cost of plant decommissioning. Yankee Atomic, a small reactor brought on-line in 1960 at a cost of $39 million, will cost an estimated $247 million to decommission, according to its owners, a group of New England Utilities. Estimates of decommissioning costs at larger reactors range from $250 million to more than $1 billion.[145]

The Outlook for Nuclear Power

While uncertainties about permanent waste disposal and reactor decommissioning cast a long shadow over nuclear power, recent licensing reform has brightened the prospects for a restoration of nuclear plant orders. Utilities until recently needed two permits from the NRC—a construction permit and an operating permit—to generate electricity for sale from nuclear plants. This two-step licensing process gave the public the right to intervene in regulatory proceedings right up through the time of a plant's start-up. In some instances, opponents' delaying tactics raised the costs of financing as high as the actual costs of plant construction, or even higher. Some nuclear utilities have been forced to default on their loans or file for bankruptcy protection. In 1992, however, Congress approved major reforms for nuclear plant licensing, authorizing the NRC to issue combined construction and operating permits for utilities ordering new reactors. If opponents of the new plants believe that significant safety issues are raised during the plants' construction, now they must persuade the NRC to intervene in the licensing process on their behalf.[146]

The advent of standardized plant designs and one-step licensing should alleviate much of the risk associated with building nuclear plants in the United States. But these actions alone do not guarantee that the new plants will become safe or profitable investments. Only two utilities, the government-owned Tennessee Valley Authority and American Electric Power (an investor-owned utility holding company in Columbus, Ohio), publicly expressed much interest in ordering new nuclear power plants before the enactment of the recent licensing reforms.[147] Whether Wall Street now will step forward to underwrite the costs of new nuclear plant investments—having been chastened by its recent experience—remains to be seen. The Department of Energy has suggested that consortia of equipment firms, construction companies and utilities work together to build new nuclear power stations as means of spreading the financial risks. Toward this end, Congress has approved regulatory changes that would allow utility and nonutility companies to gain access to utility transmission lines, become wholesale generators and operate power plants in more than one state without being subject to the Public Utility Holding Company Act.[148]

Ultimately, the nuclear power industry must regain the public's trust if the industry is to move forward. "Rate shock" from new reactors entering the rate base, the accidents at Three Mile Island and Chernobyl, and recent government admissions about the public's exposure to radiation from nuclear weapons facilities has heightened the public's uneasiness about the atom. Accordingly, many citizens will be highly skeptical of claims that new nuclear plants can be made safer and more affordable.

Moreover, the track record for experimental new reactors is not especially encouraging. High capital costs and low reliability of gas-cooled reactors, for instance, prompted the nuclear power industries of England and France to abandon early development of this technology in the 1950s and 1960s in favor of light-water reactors. More recently, a prototype gas-cooled reactor in the United States—Fort St. Vrain, operated by Public Service Co. of Colorado—as well as two gas-cooled reactors in Germany were closed because of similar cost and reliability concerns.[149]

Even safety problems cannot be dismissed entirely with advanced reactor designs. The sodium coolant used in liquid-metal reactors, for example, is highly reactive with air and water. (Indeed, some of the nuclear industry's worst accidents have occurred at experimental facilities where liquid sodium was being tested: Fermi I in Michigan and the Experimental Breeder Reactor I in Idaho.) As a result, chemical fires and explosions cannot be ruled out at these plants.[150] Nor can one exclude the possibility that plutonium would be diverted to make a nuclear bomb—perhaps the ultimate safety risk.

> **Box 5-G**
>
> ### Focus on Fusion
>
> Even though fission reactors will be the only source of atomic-generated electricity for long into the future, researchers dream of the day when fusion energy takes nuclear power the next technological step. Fusion reactors are at least 40 or 50 years away from commercialization, yet the prospect of an inexhaustible supply of cheap energy—much like the fire burning in the Sun—is too tantalizing to ignore even now. Since the mid-1970s, the U.S. Department of Energy has devoted a large portion of its research and development budget to fusion power. The $323 million it spent on fusion research in fiscal year 1990, in fact, was more than it allocated for all fission reactor development programs that year, and it dwarfed spending on energy conservation and renewable technologies by a wide margin.[1g] Other industrialized nations are similarly enthusiastic about the prospects for fusion power. Collectively, the 21 industrial market nations belonging to the International Energy Agency spent more money on fusion research in 1989 than on energy efficiency and renewables research combined.[2g]
>
> Here is how a fusion reactor would work. Atomic hydrogen particles of deuterium and tritium would be fused together at temperatures exceeding 220 million degrees F to create helium and energetic neutrons. The neutrons would escape through a wall of the fusion chamber into a surrounding "blanket" that absorbs heat from the fusion process to create steam for electricity generation. Meanwhile, the helium nuclei would stay inside the chamber, bouncing into other deuterium and tritium atoms and triggering additional fusion reactions. If enough of these reactions take place, the fusion process would become self-sustaining; the tremendous heat generated by the fusions would eliminate the need for injecting additional energy from outside the chamber.
>
> The hope is that a fusion fire would burn for as long as new deuterium and tritium fuel is fed into the reaction chamber. There is no need to worry about eventually running out of such fuel, because deuterium is found in sea water, and tritium can be generated on site as part of the fusion process. Best of all, as far as the greenhouse effect is concerned, a mere 2.5 pounds of deuterium-tritium fuel could replace 9,000 tons of coal in generating 1,000 megawatts of power.[3g]
>
> To date, no experimental fusion reactor has produced more energy than it consumes in raising temperatures to start the atomic reaction. In the past 15 years, however, fusion power production in hollow metal doughnut-shaped devices, called tokamaks, has improved by a factor of more than a million. Two magnetic-confinement reactors—one in Princeton, N.J., and another near Oxford, U.K.—are close to reaching the energy break-even point. British scientists took a major step in November 1991 when they produced a burst of more than a million watts of power in deuterium-tritium reaction lasting about two seconds. The amount of energy released was 90 percent of what would be required to reach the break-even point.[4g]
>
> The U.S. Department of Energy would like to have an operating demonstration fusion plant in place by about 2025, but Congress is balking at the high price of

The greenhouse effect may provide the impetus to put these risks aside and press ahead with further development of nuclear power. But even if a decision were made tomorrow to replace all coal-fired plants with nuclear plants over the next 40 years—a step that would reduce worldwide greenhouse gas emissions research. In fiscal year 1991, it slashed almost $50 million from the magnetic-confinement research budget—instead of raising it by the nearly $100 million that DOE had requested. DOE subsequently canceled the $1.8 billion Burning Plasma Experiment, which was supposed to be able to heat plasma to the point where the energy produced by fusion of deuterium and tritium nuclei could sustain a nuclear reaction. Other demonstration projects are underway, however. One is the Tokamak Physics Experiment, involving a $400 million machine that should be capable of maintaining a heat plasma for at least 1,000 seconds, instead of only a second or two as is the case today. While the experiment will not achieve the densities needed for ignition, it will provide valuable data on the performance of tokamaks.[5g]

The United States also is participating in the International Thermonuclear Experimental Reactor project, the aim of which is to build a fusion reactor that generates more energy than it consumes. The joint effort involves the former Soviet Union, Japan and the European Community as well as the United States; each group is contributing about $40 million annually. While this fusion reactor will not produce electricity, researchers hope to gain a better understanding of the the physics of ignition and sustained fusion burning as well as the fundamental nature of the transport of particles and heat across lines of magnetic force (a process that reduces the temperature of the burning plasma and causes the fusion reaction to peter out). The international consortium hopes to commission the $7.5 billion tokamak device by 2005, serving as a forerunner of a prototype fusion power plant to be built by 2025.[5g]

"Cold fusion" also continues to draw its share of interest and controversy within the scientific community. Two scientists at the University of Utah, Stanley Pons and Martin Fleischmann, created a great stir in early 1989 when they claimed to have created "fusion in a bottle."[6g] In a simple table-top electrolysis experiment, the scientists inserted a palladium rod, saturated with nuclei of deuterium and encircled by a platinum wire, into a flask of "heavy" water (water containing an extra oxygen atom). When electric current was run through the platinum wire, more energy was released into the water as heat than had been injected into the wire as electricity, the scientists reported. They maintained that the added heat must have come from energy released by fusion of deuterium nuclei inside the palladium rod.

Most scientists have since dismissed these claims. The U.S. Department of Energy terminated research on cold fusion in late 1989 and the State of Utah ended its support of such research at the University of Utah in 1991, after granting $5 million previously. In 1992, however, the Electric Power Research Institute announced that it will spend $12 million over three years to support cold fusion research at SRI International in Palo Alto, Calif. SRI researchers have repeated an experiment about a dozen times in which the amount of heat produced through electrolysis of heavy water (one or two watts) represents up to 10 percent more energy than is injected into the experiment electrically. EPRI—which spends $400 million a year to fund electricity-related research and development—plans to continue funding the program unless it is determined that the source of the heat being produced is not nuclear in nature.[7g]

by 20 to 30 percent—the scale of nuclear construction would be an order of magnitude larger than the world has ever seen. According to an analysis by energy researchers Bill Keepin and Gregory Kats, a nuclear plant would have to be completed every one to three days over the next four decades, at a total price of nearly $9 trillion.[151] While this is an extreme scenario, to be sure, it points out that expansion of nuclear power is not likely to be able to stabilize or reduce CO_2 emissions all on its own. Moreover, it raises a question as to whether other generating strategies would be more practical, affordable and safe in a world of limited resources. The next section examines the renewable generating options with this question in mind.

A Market for New Power

Renewable energy sources are perhaps the closest thing to a cornucopia for the world's energy requirements. Unlike fossil energy and nuclear power, renewable fuels are, by definition, practically limitless. Moreover, most renewable sources (biomass being the major exception) do not emit much, if any, greenhouse gases. The sunlight striking the earth each day contains 15,000 times more energy than the world's present energy requirements.[152] The wind skimming above the earth's surface, the latent heat trapped underground and the waves upon the ocean possess tremendous power as well. If the means were found to harness even a tiny portion of this energy, the solution to global warming and the world's "energy crisis" would be at hand.

At present, however, renewable fuels account for less than one-fifth of the world's primary energy demand and an even smaller percentage of electricity demand.[153] Most renewable energy today is derived from damming rivers and burning plants and trees—which create perhaps as many environmental problems as they solve. With generally high capital costs and low capacity factors, renewable technologies are expected to supply no more than half of the nation's generating requirements even 40 years from now. Some suggest that the contribution will be limited to 12 percent or less if low-cost energy storage devices are not developed in the meantime.[154] A true "solar society" may be achieved eventually, but it is not likely to come in our lifetimes.

A concerted effort to combat the greenhouse effect is one development that may speed the commercialization of renewables. Proponents of solar technologies must be careful not to wish too hard for a global warming trend, however, as changes in climate could affect renewable energy production in several deleterious ways. Hydro generation and biomass production could be parched by drought, for example. An increase in cloud cover could diminish the output of solar thermal and photovoltaic devices. Reduced temperature differentials between the poles of the earth and the equator could even reduce the wind flows that determine the output of wind farms. (On the other hand, more rainfall, fewer clouds and greater wind speeds could bolster production from renewables.) In addition, renewable technologies must harness energy from the sun and wind as they are available, so it is also critical that the natural ebbs and flows of these resources remain in synch with diurnal and seasonal demand for electricity.

A more immediate concern of renewable electricity producers is demonstrating that their technologies can serve as viable alternatives to conventional generating options. Electric utilities remain the dominant buyer of power plants, and they generally regard renewable facilities as too expensive, too small-scale and too unreliable to warrant much of a share of the nation's generating mix. Hydropower supplies roughly 10 percent of U.S. electric utilities' annual generating requirements at present, mainly from large-scale dams. All of the other renewable technologies—geothermal, biomass, wind and solar—supply only about 0.4 percent of the utilities' total.[155]

Utilities have a long and well-established relationship with manufacturers of large central-station power plants. A switch to renewable electricity systems would mean working with smaller, younger companies in many instances. In addition, development of renewable technologies, some of which are amenable to on-site generation, would entail fundamental changes in the operating structure of utilities. It is even possible that widespread deployment of decentralized renewable systems would undermine the economies of scale that serve as the rationale for utilities' monopoly franchises. Nuclear power, the other principal noncarbon generating source, does not pose such a dilemma for utilities.

The hiatus in nuclear plant orders has given renewable generating technologies the upper hand in the market for new and emerging power sources, nevertheless. Congress facilitated this advance by enacting the Public Utility Regulatory Policies Act of 1978 (PURPA), a law that exempts independent developers of renewable power and cogeneration facilities from traditional utility regulations. At the same time, PURPA requires utilities to purchase electricity from these developers at "avoided cost" prices. Utilities that were contemplating investments in new coal or nuclear capacity, or that relied heavily on oil and gas to meet peaking power needs, suddenly had a new option: buying power from nonutility producers to displace their own marginal costs of production.

Since 1978, independent producers have erected approximately 8,000 megawatts of renewable energy plants and nearly three times as much cogeneration capacity (most of it natural gas-fired) that has been sold into the grid under PURPA. In states with high avoided cost prices, opening the power market to independent producers has led to dramatic changes in the composition of the fuel mix. In California, for example, utilities now obtain about 12 percent of their power from independently owned renewable energy and cogeneration facilities. Largely because of these purchases of wind, solar, hydro, biomass, geothermal and cogenerated electricity, the state's CO_2 emissions from the electric power sector fell 17 percent during the 1980s.[156]

On the East Coast, a handful of utilities in New England and the mid-Atlantic states also have contracted with independent producers to meet the bulk of their incremental capacity needs. Virginia Power, a utility subsidiary of Dominion Resources Inc., signed contracts for 2,100 megawatts of independent power in 1989. These resources will make up 20 percent of that utility's generating capacity once the new plants are built. Since the majority of contracted capacity in this case happens to be coal-fired cogeneration, however, the CO_2 benefit will be limited to the efficiency gains of burning coal in cogeneration plants rather than in conventional steam turbines.

Nationwide, about 24,000 megawatts—or 4 percent of the nation's generating capacity—was owned by independent producers as of 1988, according to the North American Electric Reliability Council. The council, which forecasts power demand in the United States and Canada, also estimated that another 28,000 megawatts of nonutility capacity was planned or under construction, mainly gas-fired cogeneration. Through 2000, the council expects that one-quarter to one-half of all capacity brought on-line in the United States will be owned by independent producers.[157] Some believe that the nonutility portion may be even higher, now that Congress is set to approve major reforms to the Public Utility Holding Company Act.

While the rising contribution from independent producers is clearly depressing the demand for new central-station power plants—and reducing the carbon-intensity of the nation's generating mix—renewable energy is by no means about to monopolize the power market overall. On the contrary, a 1989 study by the Investor Responsibility Research Center found that orders for renewable energy facilities were themselves poised for a decline after several years of rapid growth. The 108 independent power companies surveyed by IRRC planned to bring on-line a record 1,250 megawatts of renewable generating capacity in 1989. But their projections of future plant completions fell to 750 megawatts in 1991 and to only 200 megawatts by 1993.[158]

To an extent, these fledgling companies have become victims of their own success. In 27 states where auction systems have been established to make independent producers compete for the right to supply new capacity, the bids have amounted to eight times more capacity, on average, than the utilities sought.[159] The abundance of contract offers—mainly from industrial cogenerators—has eased utilities' fears of impending power shortages. At the same time, falling oil and gas prices and the success of demand-side management programs have reduced utilities' own costs of generating peak power. As a result, the avoided-cost contracts being offered today to independent producers are not nearly as lucrative as those made before 1985.

Another frustration for renewable power producers has been the on-again/off-again attitude of state and federal legislators toward renewable energy tax credits, complicating efforts to obtain project financing. (Energy legislation working its way through Congress is expected to put renewable energy tax credits back on again in 1993.) Moreover, an 80 percent cut in federal support for renewable energy research and development during the 1980s dampened high expectations for rapid technological advancement of these alternative energy sources. Consequently, production costs at many renewable facilities remain higher than utilities' own marginal costs of generation.

Steady Progress in Renewables

The early 1990s has been a time of retrenchment and consolidation for renewable power producers. The world's largest photovoltaics company, Arco Solar, was sold to a German firm, Siemens AG, in 1990. The world's largest solar thermal company, Luz International of Los Angeles, Calif., filed for bankruptcy protection in 1991. The nation's wind industry—the fastest-growing supplier of renewable power during the 1980s—planned to erect no new major windfarms in 1992. The geothermal industry, which added more megawatts of generating capacity during the 1980s than any of the other renewable industries, has brought on-line fewer than 100 megawatts since 1990.

Despite adverse political and economic trends and lack of interest among most utilities, renewable power suppliers still can point to steady advances in their technologies. If the progress continues—and concerns about global warming continue to mount—utilities may take another look at renewables and decide the time has come to diversify their coal-dominated fuel mix. Volume orders from utilities could enable manufacturers of renewable power systems to achieve better economies of scale and lower the cost of financing installations.

Photovoltaics: Photovoltaics offer a good example of the chicken-and-egg problem plaguing many renewable energy vendors. Photovoltaics produce electricity when sunlight stimulates the flow of electrons from specially treated semiconductor materials. The price of such solar power has been much too high for a large market to develop; but without a large market, manufacturing costs are not likely to drop very fast. Despite this obstacle, photovoltaic companies did achieve a 70 percent reduction in manufacturing costs during the 1980s, to about $4.50 per peak watt, or approximately 30 cents per kilowatt-hour.[160] At this rate, photovoltaics may serve as an economically viable energy source in some off-grid applications, particularly where the alternative is to run diesel-fueled generators. But further cost reductions—and more orders—will be necessary to increase the tiny market for photovoltaics, which now amounts to only 50 megawatts a year (mostly in stand-alone applications).[161]

Oil companies that once dominated research and development efforts have much less presence in the photovoltaics business today. With the sale of Arco Solar to Siemens in 1990, Amoco remains the only American oil company with solar cells on the market, sold through its Solarex subsidiary. (Mobil and British Petroleum, which purchased Standard Oil of Ohio, still have active photovoltaic research and development programs, however.) Exxon withdrew from the race to commercialize photovoltaics in 1984, even though it was the third largest manufacturer of solar cells at the time. Other large American companies that dropped out of the photovoltaics business during the 1980s include Eastman Kodak, General Electric and Westinghouse.

A few major companies remain keenly interested in the photovoltaics market, however. Martin Marietta is engaged in research of solar cells made from copper indium diselenide, for example. Dow Chemical purchased a minority interest in AstroPower Inc. of Newark, Del., in November 1991 "to assist Dow in acquiring in-depth knowledge of the photovoltaics business and exploring opportunities in photovoltaics," according to a press statement issued jointly by the two firms.[162] AstroPower, a small start-up company with $4 million in estimated annual sales, is developing polycrystalline silicon thin-film technology that has demonstrated efficiency ratings of nearly 10 percent. (Two of its arrays, rated at 20 kilowatts-peak each, have been erected in California at the Photovoltaics for Utility Scale Application, managed by Pacific Gas and Electric.)

One of the most heartening recent developments concerning photovoltaics came when Texas Instruments and Southern California Edison unveiled a module design innovation in the spring of 1991. The new solar cells—developed by Texas Instruments after six years of closely held research—measure four inches on a side and are arranged in 10-foot-square panels. Each panel consists of 1.5 million tiny spheres of silicon embedded in aluminum foil. For $2,000, one solar panel is capable of generating one-third of a typical home's peak electricity demand. Texas Instruments and SCE expect to have the panels on the market in 1993. Some industry observers believe the spheral cells, presently 10 percent efficient, could reduce the price of photovoltaic electricity to only $1 to $2 per peak watt, or 14 cents a kilowatt-hour. At that price, solar electricity would be only twice as expensive as generation from many conventional power sources.[163]

Also in 1991, DOE announced a plan that calls for the installation of about 900 megawatts of U.S.-manufactured photovoltaic systems in the United States and another 500 megawatts abroad during the 1990s.[164] If successful, this would increase the cumulatively installed photovoltaic generating base by an order of magnitude. It remains unclear how much financial support the federal government intends to lend to this ambitious development effort, however. (The Bush administration has requested $63.5 million in federal research and development funding for photovoltaics in fiscal year 1993.) Over the longer term, say, by the middle of the 21st century, photovoltaics might be able to supply tens or even hundreds of thousands of megawatts of affordable electricity around the world—but only if technological advances continue.

Wind power: Wind power is another renewable technology that has made great technological strides in recent years and has the potential to become a major source of generation in the decades to come. Unlike photovoltaics, however, wind power is competitive in price with conventional generating technologies today—at least in places where the wind blows hard. Despite a hiatus in wind farm installations in 1992, more than 15,000 wind turbines have been erected in three especially windy passes of California since 1981. Together, these turbines represent 1,500 megawatts of installed generating capacity.[165] Refinements in turbine design and better understanding of local wind regimes have pared maintenance expenses and boosted capacity factors demonstrably, driving down wind power's average generating costs from 50 cents a kilowatt-hour in 1981 to about 7.5 cents a kWh today.

Because the vast majority of installations have been in California, one might assume that other parts of the country are not well-suited for wind power development. This is not the case. About 90 percent of the wind power potential in the United States is in the Great Plains, and many states in the East have as much wind resource potential as California in terms of meeting in-state generating requirements.[166]

In what could be an important breakthrough for wind power, plans were announced in the fall of 1991 to site the first two major American wind farms outside of California. Northern States Power, a utility based in Minneapolis, Minn., will invest $8 million in a 10-megawatt windfarm along windswept Buffalo Ridge. Construction is scheduled to begin by late 1995.[167] Buffalo Ridge, extending for 100 miles in southeastern Minnesota, offers the potential for 4,500 megawatts of wind power development eventually—three times the amount of wind power now installed in California.[168]

Figure 17

U.S. Wind Resource Potential

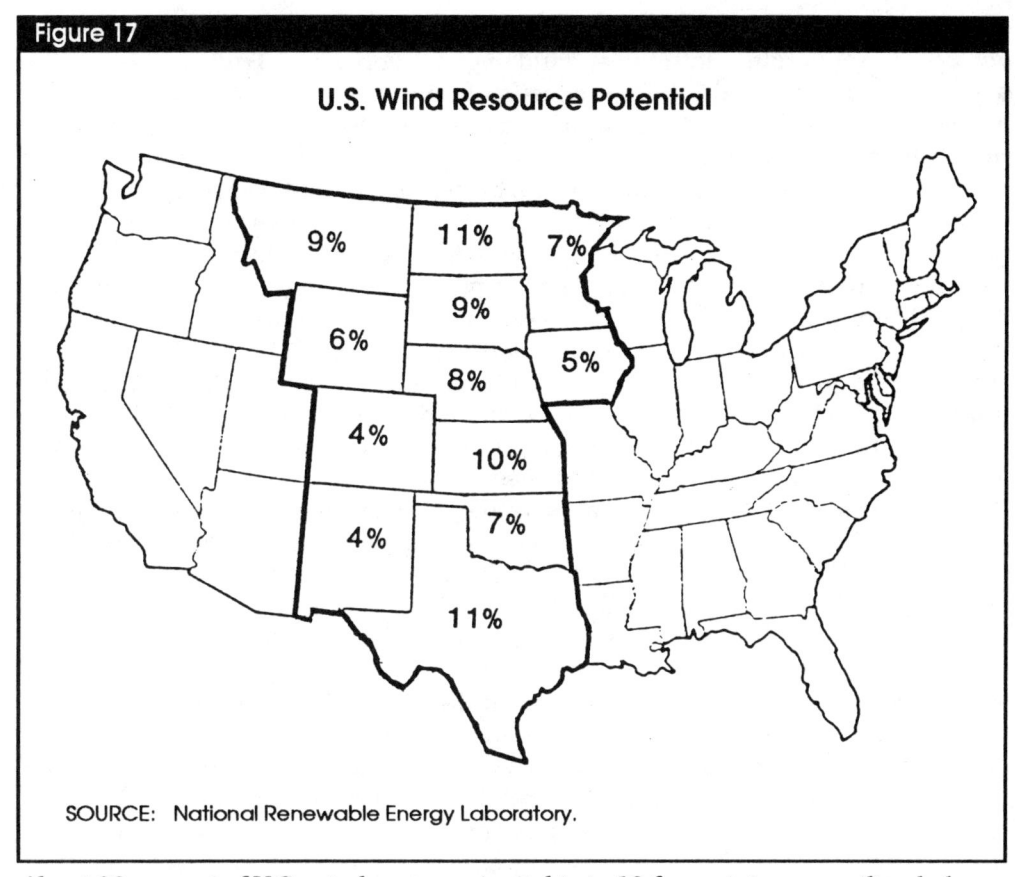

[Map of the United States showing wind resource potential percentages by state: Montana 9%, North Dakota 11%, Minnesota 7%, Wyoming 6%, South Dakota 9%, Iowa 5%, Nebraska 8%, Colorado 4%, Kansas 10%, New Mexico 4%, Oklahoma 7%, Texas 11%.]

SOURCE: National Renewable Energy Laboratory.

About 90 percent of U.S. wind power potential is in 12 farm states, as outlined above.

In neighboring Iowa, Iowa-Illinois Gas & Electric of Davenport, Iowa, has formed a joint venture with U.S. Windpower of San Francisco, Calif., the nation's largest wind turbine manufacturer. Beginning in 1994, thousands of wind turbines will be interspersed among the crops of midwestern farmers who lease wind development rights to their land. After an initial investment of $200 million to $225 million, the Iowa project may be expanded to 500 megawatts in five years, making it the size of a new coal plant or modular nuclear plant.[169] At the same time, the project will herald the introduction of a new generation of wind turbines that may produce electricity for only 5 cents a kilowatt-hour—about as cheap as any competing generating source.

U.S. Windpower's new turbines, rated at 300 kilowatts each, feature a variable-speed drivetrain that enables the blades to spin faster as wind speeds increase. Most of today's wind turbines have to brake the drivetrain in order to maintain a constant level of output transmitted into the grid. This braking leads to wear and tear on the machine and increases maintenance costs. U.S. Windpower's new design places a converter between the turbine and the utility's power line. The converter allows electrical output to increase as wind speed increases, rather than placing stressful torque on the drivetrain. The new U.S.

Windpower turbine will operate over a greater range of wind speeds as well. It will begin to generate electricity at wind speeds of only 8 or 9 miles an hour and achieve maximum production at speeds of 25 miles an hour or greater. This represents a 4-mile-an-hour improvement over its current machines and makes it practical to use in areas with only moderate wind resources.[170]

Gerald Anderson, president of U.S. Windpower's parent company, Kenetech, believes the 1990s will be a transition period in which utilities begin to take wind power seriously. After 2000, wind power could surge if utilities start ordering wind farms instead of conventional power plants. Although some might accuse him of tilting at windmills, Anderson believes that wind power could contribute up to 10 percent of the nation's electricity requirements by 2010.[171] A more conservative forecast made by researchers at the Electric Power Research Institute is that 100,000 megawatts or more of wind power capacity could be installed in the United States by the middle of the 21st century.[172] Globally, the installed capacity could be several times greater.

Solar thermal: Solar thermal technology is yet another renewable energy source that has made significant headway in recent years. Unlike photovoltaic power, solar thermal troughs, dishes and heliostats convert sunlight into heat before generating electricity. Since 1984, Luz International has sited nine power stations in California's Mojave desert employing solar parabolic trough technology. Costs of generation from these facilities have dropped from 23 cents per kilowatt-hour in 1984 to only 8 cents per kWh at Luz's most recent installations. Altogether, Luz has 354 megawatts of solar thermal capacity in operation and power contracts for another 320 megawatts.[173] Luz has fallen on hard times, however. Its decision to seek bankruptcy protection in 1991 came in the face of higher-than-expected property tax assessments and investor uncertainty over the extension of solar energy tax credits. The solar electric generating plants were exempted from the filing, though, and Luz is searching for a buyer.[174]

Meanwhile, Southern California Edison and two California municipal utilities announced plans in 1991 to move ahead with solar "power tower" technology. In this case, sun-tracking heliostats are used to concentrate heat on a central receiver, which is filled with molten salt and connected to a heat exchanger. A 10-megawatt demonstration facility will be completed in 1994 as a successor to a prototype plant that operated from 1982 to 1988. Larger commercial facilities capable of producing up to 200 megawatts of power eventually may generate electricity at a cost of 12 to 15 cents per kilowatt-hour, Southern California Edison estimates.[175]

Geothermal: Some renewable generating technologies have track records dating back well before the 1980s. Geothermal power plants, which tap heat from beneath the earth's surface, have been operating at Pacific Gas and Electric's complex at The Geysers in California since 1960. Most geothermal resource development has taken place after 1970, however. In the United States, installed geothermal capacity rose from a mere 78 megawatts in 1970 to approximately 2,700 megawatts today. More than 5,600 megawatts of geothermal power plants are in operation worldwide, in places ranging from Iceland to El Salvador, The Philippines and New Zealand.[176]

While most geothermal facilities tap into dry steam reserves such as The Geysers, many new plants use new single- or double-flash technology or binary cycle systems to exploit more abundant liquid-dominated reserves. The development of these technologies will be critical to expanding geothermal's long-term potential, since dry steam reserves are limited and not truly renewable. At The Geysers, for instance, electricity production is diminishing faster than expected from nearly 2,000 megawatts of installed generating capacity. In 1992, Pacific Gas and Electric will retire four geothermal plants that were installed at The Geysers between 1960 and 1968.[177]

Elsewhere in California, geothermal fields that have yet to be fully explored might yield 6,000 megawatts of additional power. And throughout the American West, the U.S. Geological Survey estimates that known steam and hot-water geothermal reservoirs could provide a total of 23,000 megawatts of power over the projected 30-year lifetime of the fields. Lower-temperature hydrothermal reservoirs, moreover, could provide up to 150,000 megawatts of power if continued progress is made in single- and double-flash technology and binary cycle systems.[178]

Geothermal generating capacity potentially could increase by several orders of magnitude if more economical means are found to draw heat from areas where subsurface dry rock has been heated by molten magma. According to scientists at Los Alamos National Laboratory in New Mexico, about 2 percent of the underground land mass of the United States has a temperature gradient of about 125 degrees F per mile. By pumping water deep into this crust and returning the heated water to the surface, it may be possible someday for hot dry rock technology to provide abundant baseload capacity at a cost of 9 or 10 cents a kWh.[179]

Biomass: Biomass is a renewable energy source that has been used since the dawn of human civilization. By some estimates, biomass fuels account for nearly 15 percent of world energy use, and one-third of energy use in developing countries. Of course, biomass is used mainly as a cooking and heating fuel, not as a fuel for power plants. In poor countries with limited energy resources, deforestation and/or desertification often accompanies heavy reliance on biomass fuels.

Nevertheless, prospects for generating electricity with biomass are very promising in nations with large quantities of biomass residues. The United States has more than 4,000 megawatts of wood-burning power plants in operation (although the electricity these plants produce is mainly for in-house uses at pulp and paper mills).[180] Biomass residues could be gasified and burned in aeroderivative gas turbines, like those used for natural gas and gasified coal. Because these turbines are highly efficient, relatively more energy is produced per unit of biomass fuel consumed, thereby reducing CO_2 emissions. Someday, hydrogen derived from biomass could be used in even more efficient fuel cells.

Today, sugar cane residues are readily available as a biomass fuel in 80 developing nations where sugar cane is grown. If all of these residues were gasified and burned in steam-injected gas turbines, these nations could generate half as much power as they currently derive from all generating sources, according to Princeton University's Robert Williams, an expert on biomass gasifier-gas turbine technology.[181] Moreover, the electricity generated from gasified biomass fuel would be less expensive than if diesel fuel or oil were used. The electricity could even be economically competitive with nuclear, coal and hydropower, Williams maintains, when super-high efficiency gas turbines are used. Researchers at the Electric Power Research Institute estimate that biomass might eventually supply one-quarter of the fuel for electricity by the middle of the 21st century.[182]

Hydro: Finally, there is hydropower, a renewable resource that has produced electricity since the late 19th century. As the most mature of the renewable technologies, hydropower generates 20 percent of the world's electricity, which is more than nuclear power but less than coal or oil.[183] In the United States, hydropower represents 10 percent of the utility industry's fuel mix, producing more than twice as much electricity as oil yet not quite half as much as nuclear power. Relative to the other renewables, however, hydropower produces 25 times as much of the industry's electricity.[184]

Considering how long this technology has been around, hydropower has experienced impressive growth in recent years. Hydroelectric production worldwide grew 40 percent from 1977 through 1987, while in the United States growth was more modest at 15 percent.[185] Many hydro sites remain to be exploited, especially in the developing world. Eventually, hydro generating capacity could reach four times the current level.[186] In many places, however, further hydro development would inundate large land areas, displace human and animal communities and wipe out carbon-sequestering forests. Drought-induced low-water flows also could affect hydro production adversely. For these reasons, it is not likely that all economical hydro sites will be developed in the years ahead.

Within the United States, hydro development also is likely to be constrained by environmental factors. Nevertheless, a combination of turbine upgrades and plant expansions at existing hydro sites, plus the development of small-scale hydro sites and selected larger sites could increase the nation's generating capacity from about 75,000 megawatts today to 125,000 megawatts or more by the middle of the 21st century.[187] As a result, hydropower is expected to remain the largest source of renewable electricity production for many decades to come, even if its rate of growth is slower than for the other renewable technologies.

Envisioning a Solar Future

The amount of power that could be extracted, ultimately, from renewable energy sources is enormous. Researchers at Battelle Pacific Northwest Laboratories estimate, for example, that 25 percent of current U.S. generating capacity—some 170,000 megawatts—could be provided by wind farms installed on the windiest 1.5 percent of the continental United States.[188] The vast majority of this wind-blown area is barren grazing land in 12 contiguous western states. Wind turbines mounted high above the ground would allow farming and ranching activities to continue much as before. Yet the value of the land would rise substantially as it serves in a new dual capacity. Land prices in California's Altamont Pass already have soared from $400 to $2,000 an acre because of royalties paid to land owners by wind developers.[189]

Photovoltaic and solar thermal installations on sun-baked desert land also could have a tremendous impact on generating supplies and land values. Kenneth Zweibel, a photovoltaics expert with the National Renewable Energy Laboratory in Golden, Colo., estimates that photovoltaic arrays, assembled in a square stretching 67 miles on a side (equal to less than 0.25 percent of total U.S. land area), could produce as much electricity as the entire nation now consumes.[190] Of course, there would be no need to gather all of these modules in one place. Photovoltaic arrays could be dispersed on rooftops across the country—as Texas Instruments and SCE plan to do—and no ground space would be taken up.

For the average American home, a 20-foot by 20-foot array consisting of 12 percent efficient modules would be capable of meeting all of its power requirements.[191] A rooftop array in cloudy Boston would have to be larger than one in, say, sunny Phoenix—and back-up power sources or batteries would have to be used at night. Even so, a properly configured system with photovoltaics at its base could revolutionize the way electricity is generated and transmitted—and reduce CO_2 emissions at the same time.

While virtually all renewable sources have siting limitations of one sort or another, enough different options are available around the globe to permit a dramatic curtailment in the use of carbon-based fuels. Researchers at the Worldwatch Institute, an environmental think tank in Washington, D.C., have examined ways in which different regions could exploit indigenous renewable resources as part of an international effort to combat the greenhouse effect.[192] Northern Europe could replace most carbon-based generating capacity with a mixture of wind, biomass, solar and hydropower, for example (with natural gas imported from Russia providing the rest of its power needs), while the Mediterranean, northern Africa and the Middle East could depend more on direct sunlight. Africa could derive 75 percent of its current generating requirements by burning sugar cane residues in high efficiency gas turbines, and geothermal energy along the Great Rift Valley could serve as a substantial source of new power in the African sub-continent.

The Worldwatch researchers imagine that Pacific Rim countries could rely extensively on geothermal energy as well. Virtually the entire country of Japan, for example, lies over an enormous heat source that one day could meet much of that country's generating requirements. Besides geothermal energy, southeastern Asian nations could depend on wood, agricultural wastes and solar power to meet the bulk of their power needs. (And once again, natural gas could fill the remaining void.)

Finally, the United States—a country endowed with a wide assortment of renewable resources—could derive 30 percent of its electricity requirements from sunshine, 20 percent from hydropower, 20 percent from wind power, 10 percent from biomass, 10 percent from geothermal, and 10 percent from natural-gas-fired cogeneration by 2030, according to the Worldwatch researchers. Their analysis hinges on an assumption, however, that countries taking cost-effective steps to conserve electricity would reduce overall generating requirements below today's levels. It remains to be seen whether demand-side programs will achieve this outcome, and whether such high penetration of renewables in a nation's power supply is practical and affordable.

Finding the Right Mix

While one cannot rule out the possibility that virtually all energy requirements—not just electricity—may come from renewable resources someday, few expect that vision to become a reality anytime soon. A 1990 white paper prepared by five national laboratories for the U.S. Department of Energy estimates that only 15 percent of the nation's primary energy supply is likely to be derived from renewables by 2030, assuming that the nation's energy priorities do not change. Were the Department of Energy to double its research and development budget for renewables, these environmentally preferred technologies could meet 30 percent of the nation's primary energy requirements, the interagency white paper found.[193]

The Bush administration has not been compelled to increase funding support for renewables in light of these findings. In fiscal year 1992, only 7 percent of the $3 billion that DOE requested for civilian research and development activities was appropriated for renewables, and only 8 percent of the total was allocated for energy efficiency programs. For fiscal year 1993, DOE increased its civilian R&D budget request to $3.2 billion, but renewables and energy efficiency programs were awarded the smallest budgetary gains. In round numbers, the combined funding request for renewables and energy efficiency programs in FY 1993 amounts to less than $500 million, while nearly $700 million is sought for nuclear power (including fusion research) and $1.1 billion is requested for fossil-fuels related research (including the clean-coal demonstration program). The fossil energy R&D budget request is in fact the largest such appropriation since 1982. In percentage terms, DOE is seeking a 28 percent increase in coal-related R&D spending and an 18 percent increase in nuclear R&D spending, while seeking only a 3 percent increase in renewables R&D spending.[194]

Such budget priorities are consistent with the administration's view that renewable energy development will be limited under the National Energy Strategy. According to the strategy's baseline forecast, renewables' share of the nation's primary energy supply is expected to rise from 8 percent today to 18 percent in 2030. The strategy's goal is to raise renewables' share to 21 percent instead.[195] Such a three-percentage-point increase remains well below the 15 percent gain thought possible in the 1990 government white paper (whereby renewables' share of the nation's primary energy supply would jump from 15 to 30 percent in 2030).

Within the electricity sector, the National Energy Strategy seeks to have renewables account for about one-fifth of the nation's generating supply by 2030. Hydropower would generate 8 percent; biomass's share (including municipal solid waste) would be 4 percent; geothermal, 3 percent; wind, 2 percent; and all other renewable electricity sources, including solar thermal

Figure 18

Federal Funding for Energy Research

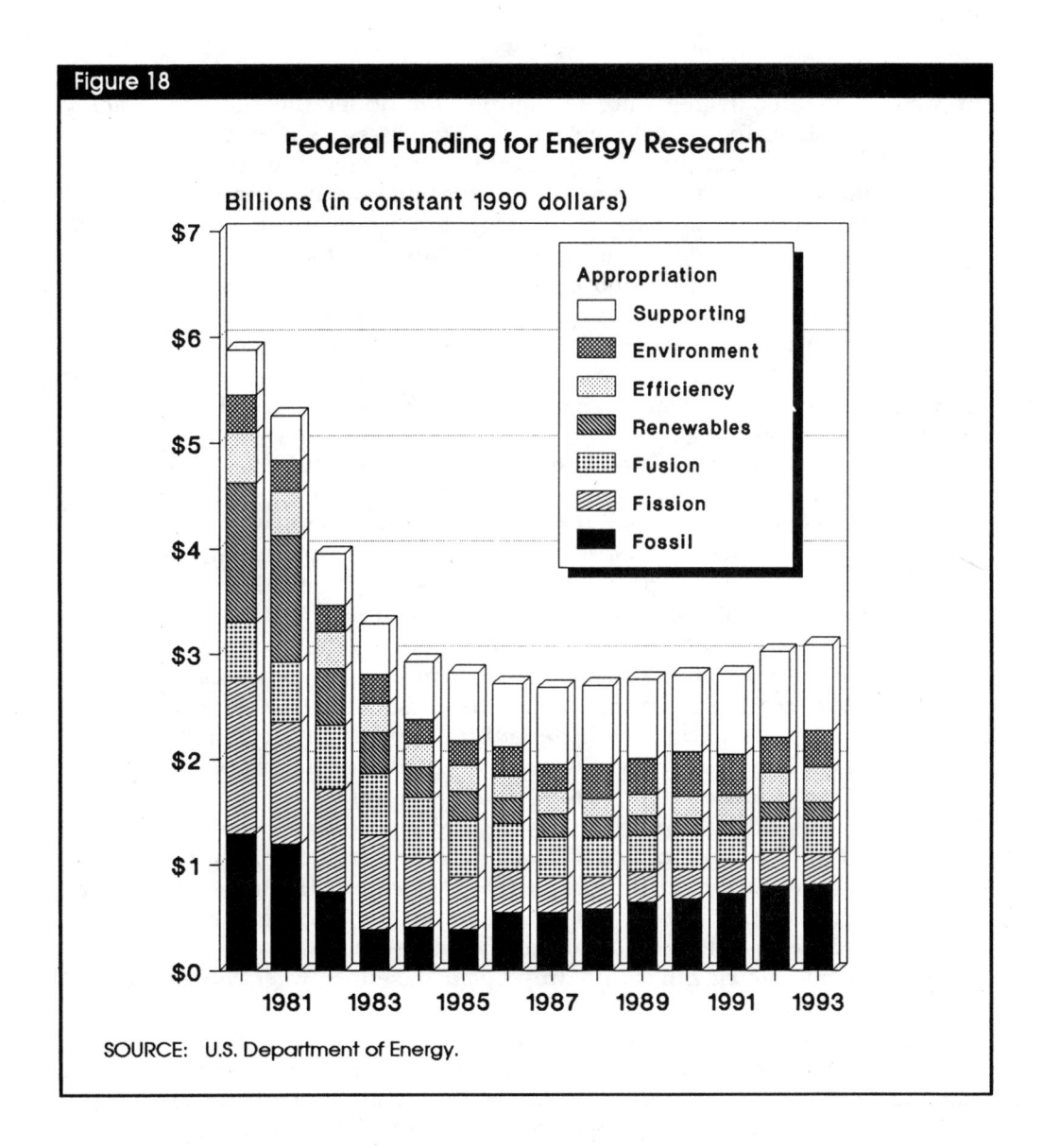

Billions (in constant 1990 dollars)

Appropriation:
- Supporting
- Environment
- Efficiency
- Renewables
- Fusion
- Fission
- Fossil

SOURCE: U.S. Department of Energy.

and photovoltaics, would provide roughly 4 percent of the total. Coal's share in 2030 would remain dominant at 49 percent; nuclear power would provide 22 percent; natural gas, 6 percent; and oil, less than 2 percent.[196]

To put the National Energy Strategy in perspective, three other forward-looking energy analyses are worth comparing. One is the 1992 study by ICF for the Edison Electric Institute on future generating options and greenhouse gas controls.[197] Another is the joint study by the American Gas Association, The Alliance to Save Energy and the Solar Energy Industries Association.[198] The third is a 1991 study carried out by a group of energy researchers at The Alliance to Save Energy, the American Council for an Energy-Efficient Economy, the Natural Resources Defense Council and the Union of Concerned Scientists.[199]

Each of these analyses concludes that the nation could diversify its generating supply to a greater degree than is planned for under the National Energy Strategy—*and* hold the electricity sector's CO_2 emissions at or below present levels. Where they disagree is on the cost of such control measures.

ICF makes its projections of future generating supplies through 2015. It assumes, like the National Energy Strategy, that government incentives and cost-effective demand-side management programs could hold the increase in electricity demand to slightly more than 50 percent over the period. It also assumes, like the National Energy Strategy, that renewables would account for 21 percent of the nation's generating mix—except that renewables would achieve this level of penetration in 2015 instead of 2030. Moreover, it finds that the electricity sector could stabilize its CO_2 emissions in 2015 at 1990 levels by raising the use of natural gas to 26 percent of the nation's fuel mix, while cutting the use of coal to 34 percent. (Nuclear power's share would be 18 percent.) The net cost of these measures is estimated at $13 billion annually by 2015, raising electricity prices 5 to 10 percent above baseline projections.

In *An Alternative Energy Future,* the joint study led by the American Gas Association, the forecast hinges on the elimination of subsidies for carbon-rich fuels, greater use of high-efficiency gas turbines (by industrial cogenerators as well as utilities) and aggressive pursuit of energy efficiency options. With the institution of these market-based measures, electricity demand growth would be held to 5 percent through 2010, instead of rising 50 percent through 2015, as forecast in the National Energy Strategy and in the ICF study. The fuel mix projections are much more consistent by comparison. In *An Alternative Energy Future*, renewables would account for nearly 17 percent of the nation's generating mix in 2010, natural gas's share would rise to 26 percent, and the fractions provided by nuclear power and coal would drop to 15 percent and 39 percent, respectively. Perhaps the most important difference in this study is that CO_2 emissions in the electricity sector would fall 13 percent *below* 1990 levels, rather than merely stabilizing at those levels, and the nation's electric bill would drop *below* baseline projections instead of rising above them.

An Alternative Energy Future incorporates many assumptions from another report by four groups that advocate greater use of efficiency and renewables. This 1991 study, *America's Energy Choices*, finds that implementation of cost-effective, market-based incentive programs could keep electricity demand essentially stable for the first three decades of the 21st century—following a 15 percent spurt of growth during the 1990s. Consequently, electricity demand in 2030 might be at only 60 percent of the level projected in the National Energy Strategy. Moreover, this study projects that renewables' share of the generating mix could reach 40 percent in 2030, almost double what the National Energy Strategy forecasts. Coal's share, at 53 percent, would constitute a larger percentage of the fuel mix than in the other analyses. Yet total kilowatt-hours of generation from coal would be lower because of reduced electricity demand. By the same token, no new nuclear plants would be required to meet the nation's generating requirements.

The most intriguing finding of *America's Energy Choices* concerns the

Figure 19

Projections of Future U.S. Generating Mix

Trillions of kilowatt-hours

- National Energy Strategy (50 percent CO2 increase)
- ICF low-impact scenario (CO2 stabilization)
- Market scenario
- Base
- Climate stabilization

Years: 1991, 2010, 2015, 2030, 2030

Legend: Coal, Nuclear, Oil, Natural gas, Hydro/renewables

SOURCE: *National Energy Strategy*, 1991; *America's Energy Choices*, 1991; and ICF Resources, *Assessment of Greenhouse Gas Emissions Policies on Electric Utilities*, 1992.

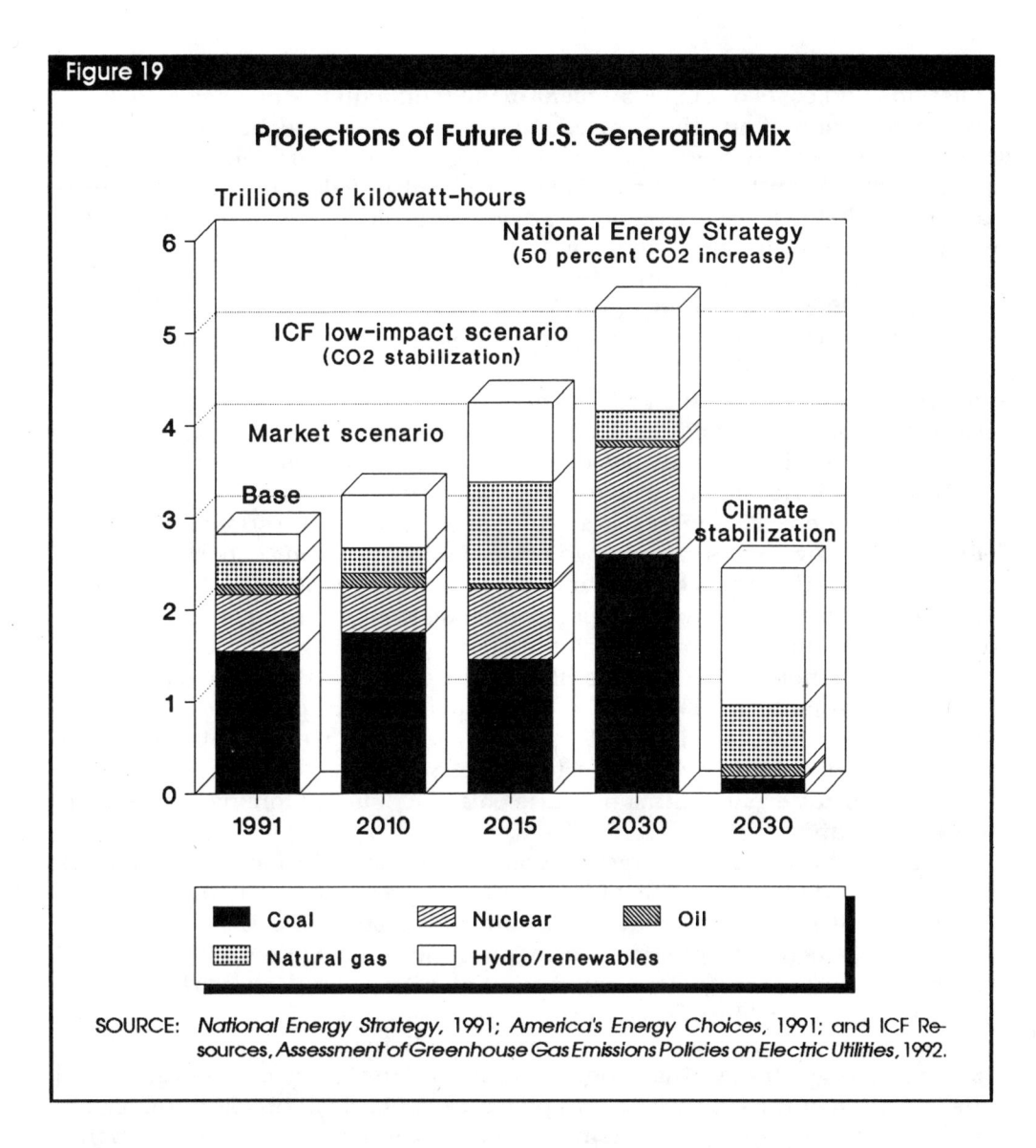

magnitude of savings that might result from a major investment in efficiency and renewables. In a market-based scenario, total energy savings—including those realized outside the electricity sector—would total $3.1 trillion over the next 40 years, compared with added investment costs of just $1.2 trillion. The cumulative net savings, in other words, would be $1.9 trillion over the period (or $600 billion after a 7 percent discount rate is applied). Moreover, the nation's carbon dioxide emissions in 2030 would fall to nearly 30 percent *below* prevailing levels even as the gross national product rises to 2.5 times *above* the present level. The net savings and CO_2 reductions might be even greater, the study finds, if environmental and security costs were incorporated into the market price of fossil fuels.

Such diverse projections of energy supply and demand, trends in CO_2 emissions and costs of CO_2 abatement make it difficult to draw any hard-and-fast conclusions about the future. Nevertheless, several important points emerge from these studies. First and foremost, the potential for future efficiency gains affects the outlook for virtually everything else. If demand-side measures are able to hold electricity demand growth to a minimum (without affecting economic growth similarly), there will be little need for precipitous changes in the fuel mix. Coal can continue to make a sizable contribution, especially as new, more efficient units take over for old, inefficient ones. When and if the utility sector's CO_2 emissions start to grow, however, more fuel switching would be required. Natural gas-fired plants—owned by utilities and industry alike—are first in line to assume a greater role. Indeed, they already are the fastest growing major source of power. Eventually, fuel cells using natural gas or coal-based molten carbonate fuel may emerge as the premier fossil energy sources.

Whether efforts should be made to revive nuclear power and to commercialize renewable energy sources are really second-order questions. If energy efficiency improvements can serve in place of new generating capacity—and do so at less cost—there may be little impetus to invest huge sums in these supply-side alternatives. That said, the specter of fossil energy shortages—of both raw fuel in the ground and storage of combustion byproducts in the air—is bound to keep these noncarbon options alive. The real question is whether both nuclear and solar technologies must be pursued. Energy planners could strive for a future in which breeder reactors and nuclear fusion essentially power the world. Alternatively, they could seek to have photovoltaics and other renewables take over. In any event, neither option is poised for such a dominant role in the foreseeable future.

In a world with many energy options yet no clear solutions, policymakers will be inclined to select a mix of fuels rather than relying exclusively on one. Several studies conclude that the Bush administration's National Energy Strategy is much too conservative in its outlook for energy efficiency and renewables in the early 21st century. Yet with federal support for these programs lagging behind all of the other supply-side technologies, the strategy's forecast could become a self-fulfilling prophecy. Some studies do agree with the National Energy Strategy that a new generation of fission reactors—introduced after the turn of the century—could play a central role in ameliorating global warming. Yet if public skepticism of nuclear power is not assuaged, such a forecast may amount to nothing more than a pipe dream.

Ultimately, investment capital is the most limited resource of all. Accordingly, a least-cost investment strategy will have to ferret out the most affordable choices. It is hard to single out the best options, however, when the full costs of energy are not reflected in the marketplace and the future costs of energy are not completely clear. What is known is that power plants are long-lived, capital-intensive assets. Investment decisions made today will have ramifications long into the future. The final section of this chapter examines ways in which policies to ameliorate the greenhouse effect may transform energy pricing and energy markets. From this discussion emerges a greenhouse gambit for the electric power industry.

Conclusions

The fundamental question now before energy planners is how to complete the transition from an Industrial Age energy policy to one better suited for the coming Electronic Age—and keep global warming in check. Ideally, market forces would set the speed and direction of this change without the need for government intervention. Such an approach, according to market theorists, maximizes the efficiency of change at the lowest possible costs to society. Unfortunately, market reality does not always coincide with market theory. Governments do intervene in markets, consumers act inefficiently at times, and markets cannot fully evaluate the costs of things they do not buy and sell.

Most governments consider energy security matters far too important to leave solely in the invisible hands of the marketplace. The U.S. government, for one, provides more than $25 billion a year in direct energy subsidies, mainly in the form of tax breaks to fossil energy producers.[200] It also lends a military presence in strategic oil regions, underwrites the insurance liability of the nuclear industry and funds billions of dollars a year in energy research and development programs through the Department of Energy. Such activities shift some of the financial burden away from energy producers to taxpayers at large, ostensibly because these actions benefit the economy overall. But this intervention also obscures the real price of energy and "tilts the playing field" on which energy decisions are made.

Consumers, meanwhile, often lack full information about their product choices, which leads to "irrational" purchasing decisions. Even when products' energy costs are labeled, consumers may not look beyond the higher price tag to appreciate the energy savings. Energy suppliers, on the other hand, face fewer capital constraints and are more willing to make long-term investments. The resulting "payback gap" sways the market toward supply-side options. "Split incentives" between the equipment installers and equipment operators often leads to the same result. Installers have less concern for efficiency, because operators end up paying the energy bills. Average-cost pricing of electricity is yet another way in which consumers are lulled into increasing energy consumption, since the potential for incremental energy savings is not measured against the marginal cost of supply.

Finally, there is the matter of environmental externalities. The market has no means of making rational and efficient decisions about environmental assets that are not priced in the marketplace. Pollution-control expenditures—like the tens of billions of dollars a year that electric utilities spend on air pollution—can serve as proxies for environmental costs, since they are passed on to consumers. But the residual costs of pollution must be measured in other ways. Energy-related air pollution problems, for example, are thought to add $40 billion a year in U.S. health care expenses, by some estimates. The Environmental Protection Agency figures that air pollution also leads to at least $2.5 billion in annual crop losses. Energy-related ozone pollution and acid rain problems cause billions of dollars more in damage each year to forests, water bodies and corrodible man-made materials.[201]

The Clean Air Act amendments of 1990 are intended to reduce this damage. Hence, the market price of energy will "internalize" more of the costs of air pollution—or at least the costs of *cleaning up* the pollution. Still, market pricing has a long way to go before it reflects the "full" costs of energy production. The potential for global warming, in particular, is not reflected in the price of fossil fuels and other greenhouse gases, despite the threat they pose to human health, natural ecosystems and urban infrastructure.

Factoring in Externalities

To get a better handle on the "hidden" environmental costs of electricity production, the U.S. Department of Energy and the New York State Energy and Developmental Authority funded a comprehensive study by researchers at Pace University, in White Plains, N.Y., in 1990. (The study followed pioneering research begun in Germany several years earlier.) While conceding the difficulty of assigning exact dollar values for environmental damages related to ozone pollution, acid rain, global warming and the like, the Pace University researchers operated on "the basic tenet that 'a crude approximation' of these damage costs is closer to an accurate accounting for resource costs than is a value of zero."[202]

With this caveat in mind, the researchers calculated how the price of electricity might change if external health and environmental costs were reflected fully in the rates charged to customers. They estimated, for example, that electricity generated from older coal plants would essentially double in price, from 5.8 cents to 11.6 cents per kilowatt-hour. The increase in price from new coal plants outfitted with scrubbers would be only 3.9 cents per kWh, however, because of lower emissions of sulfur and nitrogen oxides. Natural gas plants, which burn cleaner still, would be assessed an increase of only a penny per kilowatt-hour.

No generating source was found to be environmentally benign in the Pace University study. While nuclear plants produce no air pollution, for instance, workers are exposed to radiation inside the reactors, waste disposal and plant decommissioning costs are not fully reflected in electricity rates, and the potential for another Chernobyl-like accident persists. These unrealized costs add up to 2.9 cents per kWh, by the researchers' reckoning, although accident potential alone accounts for 2.3 cents of the total. Those who discount the possibility of another severe nuclear accident may be inclined to discount this cost estimate as well.

Even the renewable technologies are not free of health and environmental hazards, the Pace University researchers concluded. Manufacture of photovoltaic modules commonly involves use of toxic substances such as cadmium and arsenic, which could be released to the environment in the event of fire or improper disposal. Similarly, solar thermal plants use toxic heat transfer fluids. Moreover, the siting of solar plants in arid regions could upset fragile desert ecosystems. Even wind turbines make noise, pose a hazard to flying birds and, to some, are aesthetically displeasing on the landscape. Nevertheless, the renewable technologies fared the best in the Pace University study, with unrealized costs ranging from only 0.1 to 0.4 cents per kWh.

If such hidden costs of power generation were factored into the market price of electricity, the nation's new generating capacity might take a dramatic swing toward renewables. According to the 1990 white paper prepared for the Department of Energy by an interagency task force, renewables could double their expected share of power generation by 2010 if the base price of conventional energy sources were raised by 2 cents per kWh to reflect their unrealized environmental costs and tax benefits. On that basis, renewables could cost-effectively provide up to 40 percent of the new generating capacity brought on-line over the next two decades, the white paper found, although the surcharge on other sources would inflate the price of electricity by 25 to 30 percent.[203]

Since the Pace University study has been completed, many public service commissions have begun their own process of evaluating the environmental costs of electricity generation. In New York State, the commission's integrated resource planning assessments now tack on a surcharge of 1.405 cents per kWh for coal-burning plants operating without scrubbers in densely populated areas. This surcharge equals nearly 25 percent of these plants' total generating costs. Cleaner-burning power plants are penalized proportionately less.[204] At least 27 other states are looking into similar programs, whereby utilities offer an accounting of external environmental costs in their integrated resource planning assessments. (As of 1991, 18 states had mandated that this be done.)[205]

Meanwhile, public utility commissions in at least eight states now run periodic auctions that permit conservation contractors to compete against energy producers to provide future increments of "supply."[206] In some cases, conservation bids get preferential treatment to allow for the unquantified environmental benefits of saving energy. The Bonneville Power Administration, for instance, considers energy saved to be comparable in value to energy produced even if a conservation bid is up to 10 percent more expensive. The State of Wisconsin gives a similar 15 percent bonus to either conservation contractors or builders of nonfossil-burning power plants. California, Colorado, Massachusetts, Nevada, New Jersey and Oregon are among other states developing even more complex methods to evaluate the environmental costs of future power generation.[207]

The import of these regulatory changes is twofold. First, further technological advances and tax breaks are no longer likely to be the only means by which renewables and (to a lesser extent) nuclear power can take advantage in the marketplace; price comparisons incorporating external environmental costs also may narrow the gap with fossil-generated electricity. Second, environmental costs that are internalized and passed through to ratepayers will spur more conservation and efficiency investments, reducing electricity demand overall. These actions, serendipitously, will cause CO_2 emissions to fall in the electricity sector.

Electric utility executives are less than pleased by these developments, however, and for some good reasons. They point out that if environmental costs are to be estimated in market prices, *all* sources of a pollutant should be assessed, not just ones that happen to be subject to state regulation. The effect of putting environmental "adders" on electricity generation alone could even be counterproductive if consumers increase their use of less efficient and more environmentally harmful forms of energy to avoid paying the added fees.

Utility executives also say it is wrong to assume that environmental adders will benefit society when they are grafted on to a system that is already heavily regulated. The purpose of the traditional command and control approach, presumably, has been to reduce pollution to the point where the costs of further reductions outweigh the benefits. Placing environmental adders on top of these regulations may move society farther away from an optimum level of pollution control, rather than closer to it. If regulators want to internalize unpriced environmental costs, utility executives agree that use of adders or a market of tradable emissions permits make it possible. But first, they insist, regulators must abandon the traditional command and control approach.[208]

Assessing Carbon Taxes

New regulatory approaches to pollution control are gaining favor. The new Clean Air Act will allow utilities to buy and sell sulfur dioxide emissions credits beginning in 1995 to combat acid rain, for example. Utilities capable of reducing sulfur emissions at less than the asking price of the credits will be able to sell some credits to other utilities lacking low-cost options, benefiting both parties in the exchange. In the Los Angeles basin, the trading concept already has been extended to include nitrogen oxides and reactive organic compounds—two key ingredients of smog—and a broader range of companies are eligible to participate in the local exchanges, including petroleum refiners and heavy industry.[209]

Meanwhile, back on Capitol Hill, the U.S. House of Representatives has passed national energy legislation to carry forward the trend launched by more than half of the state public utility commissions: creation of economic incentives to level the playing field between renewables and other more environmentally harmful sources of electricity. The House legislation provides a production incentive of 1.5 cents per kWh for renewable generating sources. (The Senate has not acted on this proposal.)[210]

Legislators' next step may be to zero in on the potential global warming costs posed by burning of fossil fuels. A number of state utility commissions already impute such costs in their resource planning assessments. California, for one, assigns a 1-cent-per-kWh adder to coal-fired electricity generation from out-of-state producers.[211] A handful of other states have similar CO_2 adders, although the values they have chosen vary by a factor of 20.[212] While these imputed costs are strictly for resource planning purposes, they may lay the groundwork for an eventual carbon tax on electricity production. A 1-cent-per-kWh fee would be equivalent to charging a $28 tax per (short) ton of carbon. Considering the varying carbon content of fossil fuels, such a tax would raise the delivery price of coal by approximately $18 per ton; oil, by $3.90 a barrel; and natural gas, by 48 cents per thousand cubic feet.

The idea of instituting such a carbon tax is still mainly anathema in the United States, however. Proposed legislation seeking a $28-per-ton carbon tax on fossil fuels has languished in Congress ever since Rep. Fortney Stark (D-Calif.) introduced a bill (HR 1086) in 1991. The Department of Energy has weighed in with its own downbeat assessment, reporting to Congress in 1991 that a carbon tax of $450 per ton would be required to curb the nation's CO_2 emissions 20 percent by 2000. Such a tax would more than double the price of natural gas, heating oil and electricity, DOE estimated, and reduce the gross national product by 1.4 percent, or $95 billion a year.[213]

ICF reached a similar conclusion in its recent analysis of greenhouse gas emissions policies prepared for the Edison Electric Institute. ICF figured that a $100-per-ton tax on utilities' consumption of fossil fuels would make only a modest dent in the industry's projected CO_2 emissions in the years ahead. The industry's CO_2 emissions would increase by 12 percent through 2005, instead of rising 26 percent above 1990 levels, because of the tax. By 2015, the industry's CO_2 emissions would rise 57 percent above 1990 levels, instead of 73 percent, as forecast in a baseline projection that assumes high electricity demand growth and continued heavy reliance on coal.

ICF found that a much higher carbon tax would be needed to hold down utilities' CO_2 emissions because—at $100 per ton—the utilities would find it cheaper to pay the tax and continue using most of their existing coal-fired generating capacity, as opposed to switching to less carbon-intensive fuels. Consequently, carbon tax proceeds would total $87 billion in 2015, with the added levies passed on to ratepayers. Limited fuel switching also would raise annual generation expenses by $19 billion, so that the total economic cost of the carbon tax would be $107 billion in 2015 in this high-impact scenario.[214]

Not all studies have drawn such pessimistic conclusions about the costs related to imposing a carbon tax, however. The Congressional Budget Office has used an econometric model developed by Data Resources Inc. of Lexington, Mass., and found that a $100-per-ton tax on all carbon sources (not just utility fuels) could reduce the nation's CO_2 emissions 6 percent *below* 1988 levels by 2000.[215] Another model developed by Dale Jorgenson and his associates at Harvard University draws an even rosier picture, whereby the nation's carbon emissions would fall 27 percent below 1988 levels by 2000 in response to a $100-per-ton tax.[216] (Each of these models assumed that a $10-per-ton carbon tax would be instituted in 1991 and increased in $10 yearly increments until the $100 level was reached in 2000.)

In the Jorgenson model, the macroeconomic cost of the carbon tax would reduce the size of the year 2000 gross national product by less than 1 percent—or about the width of the pencil line to draw the GNP growth curve. Two assumptions are critical to this model's optimistic results. First, consumers are expected to find numerous cost-effective ways to save energy in response to higher prices. Second, a wide range of noncarbon generating sources are thought to be available to substitute economically for coal. When ICF crafted a similar "low-impact" scenario in its study for the Edison Electric Institute, the annual cost of the carbon tax paid by utilities fell to $72 billion in 2015,

Figure 20

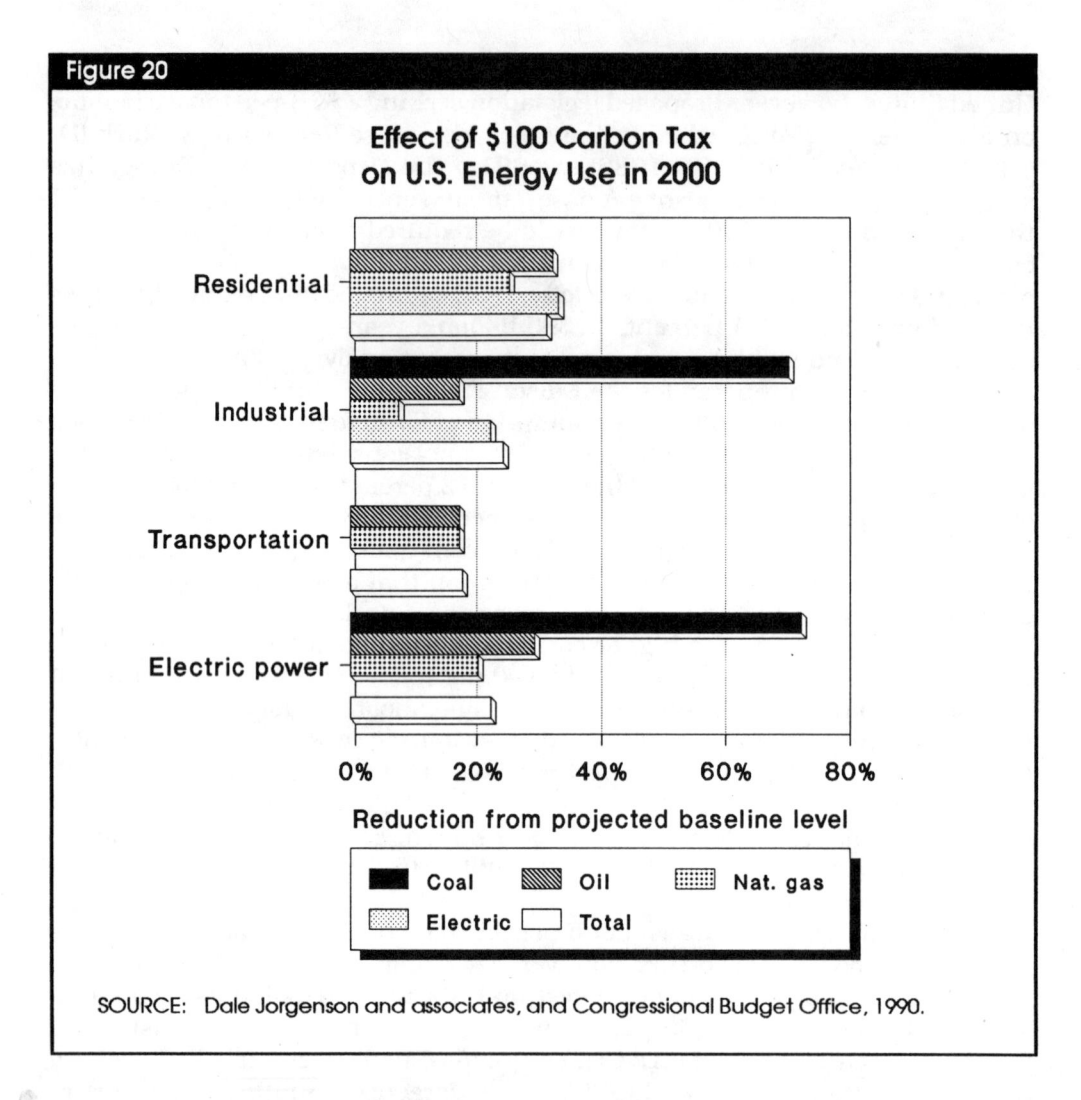

Effect of $100 Carbon Tax on U.S. Energy Use in 2000

[Bar chart showing reduction from projected baseline level (0% to 80%) for Residential, Industrial, Transportation, and Electric power sectors, broken down by Coal, Oil, Nat. gas, Electric, and Total.]

Reduction from projected baseline level

Legend: Coal, Oil, Nat. gas, Electric, Total

SOURCE: Dale Jorgenson and associates, and Congressional Budget Office, 1990.

equal to 1.2 percent yearly GNP loss, and the sector's carbon emissions were projected to drop 14 percent below 1990 levels.[217]

Thus, analyses making favorable assumptions about energy efficiency and alternative energy development suggest that a carbon tax could hold CO_2 emissions at or below 1990 levels at a comparatively small cost to the economy. Nevertheless, ICF recommended against the institution of such a tax, because other means of reducing greenhouse gas emissions would be less costly. These options include capturing methane from coalbeds and landfills, planting trees to sequester carbon, and promoting selective electrification of the commercial and industrial sectors. ICF also noted that putting a carbon tax on utilities' fossil fuel consumption would create a political risk, in that utility commissions might prevent full recovery of costs of the carbon tax through rate increases. Moreover, utilities could not be certain how their customers would respond to tax-related rate increases.

"However," the ICF study concludes, "the above discussion of costs is taken only from the perspective of the utilities and their customers. The [carbon] tax, while a 'cost to the electricity sector,' represents a transfer payment in the context of the U.S. economy."[218] The tens of billions of dollars of revenue generated from a carbon tax, in other words, would not disappear into a black hole. Rather, they would be available for purposes that may cushion the macroeconomic effect of imposing such a tax. This intriguing idea bears further examination.

Europe Moves Forward

In Europe, serious consideration is being given to a carbon-based energy tax that would generate up to $1 trillion in revenue by the turn of the century.[219] The tax would be used to promote energy conservation, alternative energy development and tax reform—since, in principle, the tax would be revenue-neutral. More important in a greenhouse context, the carbon tax would play a central role in the European Community's commitment to stabilizing its CO_2 emissions. At present, the 12 member community nations account for 13 percent of world's carbon emissions from burning fossil fuels.

Several European countries—including Denmark, Finland, the Netherlands, Norway and Sweden—have instituted modest carbon taxes already.[220] Denmark was the most recent addition to this group, when its parliament voted in December 1991 to place a carbon tax of $58 per metric ton on household energy use and a $29-per-ton tax on industrial energy use. (The industrial tax will not go into effect until January 1993.) As has become common practice in such countries with carbon taxes, the Danish government will use proceeds from the tax to promote energy efficiency and redress distortions in the tax code that favor electricity production from central-station producers. Energy-intensive industries will be eligible to receive graduated tax refunds and may receive a full refund if they meet certain criteria for economic and energy efficiency improvements.[221]

Given the high rate of taxation in Scandinavia, it is perhaps not surprising that Norway and Sweden have instituted the largest carbon taxes. Sweden's tax is $150 per metric ton, and the government plans to raise the tax to $192 per ton as of Jan. 1, 1993. (Energy-intensive industries and electric utilities are exempt from this tax, however.) Also in 1993, Sweden plans to raise its tax on household power consumption by 3 percent, which will be offset by a 3 percent reduction in the general value-added tax. The net effect is a fiscal policy that gives households an additional incentive to conserve energy while lowering the purchase price of most goods and services.[222]

The most talked-about carbon tax, however, is the one that has been proposed for the European Community as a whole. As originally proposed, the tax would amount to $3 per barrel of oil, or its energy equivalent in other fuels, beginning in 1993. The tax would then increase $1 a barrel per year until it reaches $10 a barrel in 2000. Renewable energy sources would be exempt from the tax, but nuclear power would not. The European Parliament and each member nation of the EC must approve the tax before it goes into effect.

If divisions within the EC serve as any indication, it may be quite difficult for nations to agree upon the institution of a mutual carbon-based tax. Ireland, Greece, Portugal and Spain strongly oppose the community-wide tax, citing concern for their fragile economies. Denmark, Germany and the Netherlands, on the other hand, strongly support the tax. Belgium and Great Britain, meanwhile, are among other EC nations that remain lukewarm on the tax proposal and want more formal study before its enactment.[223]

As currently envisioned, the EC's energy tax would have two components. One part of the tax would be levied on the amount of energy produced (except for renewable resources, which would be exempt), and the other part would be levied according to the carbon content of the fuel consumed. The percentage of revenues that would be drawn from each component has spawned a vigorous debate within the departments of the European Commission, the EC's governing body. The environment department wants at least three-quarters of the tax to be based on energy consumption as a means of promoting rapid energy efficiency improvements. The energy department wants at least half of the tax to be based on carbon content, however, as a means of promoting a switch to low-carbon energy sources—including nuclear power.[224]

Another struggle within the European Commission has pitted the departments of industry and competition against the departments of environment and energy. The industry department strongly opposed the energy tax until it won exemptions for energy-intensive industries such as steel, chemicals, nonferrous metals, cement, glass and paper. The competition department, which oversees international trade, also vigorously opposed the energy tax until it received assurances that the tax would not be instituted unless and until the European Community's principal trading partners—the United States and Japan—agreed to take similar measures.[225]

Because of these political compromises, the EC appears likely to approve a watered-down version of the energy tax that exempts some industries and provides subsidies for others, while falling well short of its original goal of curbing projected CO_2 emissions in 2000 by 6.5 percent. If no CO_2 control measures are enacted, the EC's emissions are expected to increase 12 percent by 2000. The combination of the energy tax and other energy-saving measures (such as stricter appliance efficiency standards, higher automobile taxes and more stringent speed limits) is supposed to forestall the CO_2 increase.[226]

America Holds Back

The exasperating experience of the European Community in instituting a carbon tax may serve as a parable for the fate of similar initiatives around the world. The lack of support for the tax among the community's poorer members suggests that developing nations also may be reluctant to embrace such a tax. The debate within the European Commission as to how to divide the tax's proceeds illustrates the divergent views of policymakers regarding the promotion of efficiency measures versus supply-side programs, and renewable energy development versus nuclear power development. The spirited opposition of

European business interests to the tax suggests as well that energy-intensive industries will seek exemptions from such a tax unless their competitors are subject to similar tax treatment.

Here is where the United States could take a leadership position. As the world's largest economic power—and the largest emitter of carbon dioxide—the United States would send a powerful message by saying that it supports the institution of a carbon tax domestically and abroad. But the U.S. government continues to have no stomach for such a plan, largely because of concerns about what might happen to the economy.

It is no secret that the United States consumes significantly more energy to produce a dollar's worth of goods and services than most of its trading partners. Japan consumes only half as much energy per unit of economic output, for example, while Western European countries consume about a third less energy. Since the United States requires relatively more energy to keep its economy humming, a carbon tax that increases energy prices would appear likely to further disadvantage American industry in the global marketplace.

It is not simply that Americans are profligate energy consumers, however. The government's longstanding policy of maintaining low energy prices has made it easier for energy-intensive industries to do business in the United States and sell their wares abroad. The nation's geography, climate and infrastructure also must be taken into consideration. The United States is one of only a few countries that spreads over an entire continent; its climate ranges from sub-arctic to sub-tropical. Americans are a highly mobile people as well, traveling great distances to work and play. Moreover, their homes and offices are comparatively spacious, climate-controlled and outfitted with modern (often energy-intensive) conveniences. Given these attributes, the United States may never achieve—or even seek to achieve—the lower level of energy intensity that prevails in other nations.

Perhaps most significant is that government intervention to lower the nation's energy intensity would run counter to the economic principle that "the market knows best." This revered principle holds that the market already has found the most efficient, cost-effective ways to use energy. To force the market to become even more sparing in its energy use would impose net costs on society—not bring about net benefits. In a free market, after all, there is no such thing as a free lunch.

This same economic principle is at the root of economic concerns about addressing global warming. Since the bulk of greenhouse gas emissions are related to energy production, policies to ameliorate global warming would lead invariably to greater conservation and efficiency measures plus expanded use of noncarbon fuels. These options are presumed to be more expensive than maintaining the status quo; otherwise, consumers would be taking advantage of them already.

With this principle in mind, it is easy to see how the costs of greenhouse gas controls add up quickly in most econometric studies. Perhaps the influential model is the product of Alan Manne of Stanford University and Richard Richels of the Electric Power Research Institute.[227] As a flurry of global warming bills appeared in Congress following the greenhouse summer of 1988, these two economists set about the task of calculating what it would cost the United States to reduce its CO_2 emissions 20 percent below existing levels and maintain them at that reduced rate throughout the 21st century. Their preliminary findings, released at a conference in late 1989, quickly became front-page news in *The New York Times.* The headline read: "Staggering Cost Foreseen To Curb Warming of Earth."[228]

Soon corporate executives and even White House economists were citing this seminal work in their own policy evaluations of greenhouse gas controls. The *1990 Economic Report of the President,* in a chapter entitled "The Economy and the Environment," reported that policies to hold CO_2 emissions at 80 percent of the current level might result in up to $3.6 trillion of lost national income during the 21st century.[229] The forecast of such a "staggering" loss—equal to a 5 percent contraction of the gross national product—was taken directly from the Manne-Richels model and hardly represented the kind of outcome on which a president seeks to base an economic program.

The point that may have been lost on those who read news accounts at the time, however, was that the 5-percent-loss estimate was one in a range of figures divined by Manne and Richels and was meant to reflect a highly improbable worst-case scenario. The thinking behind the worst-case scenario goes something like this: Energy services will never be cheaper than they are now, because any future technological innovations must be accompanied by higher energy prices. Fossil fuels will remain the cheapest sources of energy by far during the 21st century, despite a rapidly diminishing supply. And consumers will be very slow to change their energy consumption habits even as their fuel bills rise.

By the middle of the 21st century in this worst-case scenario, insatiable demand for energy has devoured most of the world's proven oil reserves and virtually all of the world's natural gas supplies—leaving only carbon-laden synthetic fuels as commercially viable energy sources. (Noncarbon fuels are presumed to be still more expensive.) These synthetic fuels are made principally from tar sands and oil shale, however, which emit two-thirds more carbon per unit of energy produced than conventional coal. As a result, U.S. CO_2 emissions would be spewing forth at a rate of 15 billion tons a year by the end of the 21st century—almost triple the current rate—were it not for the government's policy of holding CO_2 emissions 20 percent below the current level.

Thus, energy policymakers are left with only one choice in this worst-case scenario: Place a high carbon tax on fossil fuels so that nonfossil energy sources are able to compete and consumers are compelled to conserve. The carbon tax would soar to $400 per ton during the decade of the 2020s, when domestic natural gas supplies are assumed to be nearly exhausted—and the introduction of noncarbon fuel alternatives is still a decade away. Then, as these alternatives begin to come on-stream, the tax would drop to $250 a ton by 2040 and remain at that level for the rest of the century. The resulting 5 percent cost to the economy equals $300 billion a year in today's dollars, or slightly more than the United States now spends on defense.

But consider what happens when a few key assumptions in the Manne-Richels model are changed. The most important of these assumptions concerns the "autonomous rate" of energy-efficiency improvements. In the worst-case, the rate of improvement is set at zero, i.e., higher energy prices must accompany any efficiency improvements. If the autonomous rate of improvement is set at 1 percent a year instead, meaning that 99 cents next year will buy the same amount of energy services as a dollar buys today, the projected cost of the CO_2 controls is cut in half, to $1.8 trillion through the 21st century.[230] With efficiency improvements curbing incremental growth of the fuel supply, a greater portion of the energy mix can be carbon-based without exceeding the CO_2 emissions limit. Having fewer restraints on carbon energy sources in turn reduces the size of the carbon tax—and the macroeconomic penalty as well.

With different assumptions about the development of noncarbon energy sources, the costs of CO_2 abatement fall even further in the Manne-Richels model. If these alternative fuels were to enter the marketplace faster and be one-third cheaper than is supposed in the worst-case scenario, the cost of CO_2 emissions reductions would fall another $1 trillion. Now, instead of costing the U.S. economy $3.6 trillion through the 21st century, the CO_2 control program costs only $800 billion, equal to a 1 percent annual loss of gross national product.[231] Expressed another way, the nation's output of goods and services would be 99 percent as large in any given year over the period—and the annual rate of economic growth would be barely distinguishable from what would have occurred without enactment of a greenhouse plan.

The Energy Gambit

In the final analysis, two questions will largely determine whether energy planners opt for the greenhouse gambit. First, would it be possible to achieve greenhouse gas reductions with little or no adverse effect on the economy? And second, are there larger gains that could benefit the economy in the course of taking the greenhouse gambit?

The cost of CO_2 controls: Estimates of how much it would cost to hold CO_2 emissions at or below prevailing levels vary with assumptions made about the efficacy of the marketplace and the rate of technological innovation. If one assumes that the market is functioning well and that future technological innovations will come only in response to higher energy prices, the cost of CO_2 abatement would appear to be quite high. Manne and Richels offer several cost estimates—ranging from 1 to 5 percent of GNP—based on differing assumptions built into their model. Their "best guess" projection is for a 2.2 percent loss of GNP resulting from a 20 percent cut in CO_2 emissions.[232]

Other studies noted in this chapter have come to similar or more favorable conclusions about the costs of CO_2 abatement. In general, the analyses with lower cost estimates take a bottom-up, engineering approach to the question, rather than a top-down, econometric approach. These studies have identified many cost-effective and environmentally sound ways to provide energy services that apparently are thwarted by market imperfections and government subsidies that favor competing sources. These studies find that a comprehensive program of regulatory reforms, market-based incentives and real-cost pricing of energy would enable the relatively untapped energy sources to come to the fore.

One study by four of the nation's leading energy and environmental groups concludes that very aggressive pursuit of a comprehensive energy reform program over the next four decades could reduce the nation's primary energy demand by 25 percent, increase the renewable energy supply by 400 percent and reduce carbon dioxide emissions by 70 percent, without affecting overall growth in the economy. Better still, projected energy savings of $5 trillion would outweigh added investment costs of $2.7 trillion, providing net energy savings of $2.3 trillion over the period. (The net savings shrinks to $600 billion when a 7 percent discount rate is included.)[233]

This "climate stabilization scenario" raises the possibility that even a huge carbon abatement program might not cost the economy at all. Rather, it would free up capital now spent on energy to invest in other areas. Moreover, the amount of carbon abatement would be sufficient to stabilize the amount of CO_2 in the atmosphere—an accomplishment that no other study has envisioned. (Most other studies seek to cap the amount of CO_2 that is emitted *into* the atmosphere, or to reduce such emissions by up to 20 percent, which would slow but not halt the rise in the atmospheric CO_2.)

Although no other carbon abatement study has drawn conclusions as favorable as those described here, several highly detailed analyses affirm the possibility that emissions could be reduced below current levels and provide a net economic benefit. The congressional Office of Technology Assessment holds out the prospect that "tough" controls on all greenhouse gases—not just carbon dioxide—could bring about a 20 to 35 percent reduction in emissions by 2015 and yield a net annual savings of $20 billion (although the range of estimates in this study runs as high as $150 billion in net annual costs).[234] Similarly, the National Academy of Sciences has concluded that "the United States could reduce its greenhouse gas emissions by between 10 and 40 percent of the 1990 level at very low cost. Some reductions could even be at net savings if the proper policies are implemented."[235] These studies also make the point that costs of emissions controls would be minimized—and ameliorative effects on global warming would be maximized—if a broad range of greenhouse gases and emissions sources were targeted, not just carbon dioxide in the utility sector alone.

Given the wide range of estimates concerning the macroeconomic effects of greenhouse gas abatement, one can easily find a study to suit a particular point of view. The historical record is less subject to interpretation, however. Between 1973 and 1986, the energy intensity of the U.S. economy fell by an average of 2.4 percent per year—and the nation's CO_2 emissions remained stable—despite a large increase in coal-fired electricity production. Over the same period, the economy managed to expand at an annual rate of 2.5 percent, which is an acceptable, if not stellar, rate of growth.[236] If this performance can be replicated in the decades ahead—without the large increase in coal-fired electricity generation—the nation would be virtually assured of reducing its CO_2 emissions while maintaining healthy growth of the economy.

Figure 21

Energy Savings Potential Through 2030

Trillions (present value at 3% discount rate)

Scenario	Investment costs	Energy savings	Net savings
Market	-$1.2	$3.1	$1.9
Environmental	-$2.1	$4.2	$2.1
CO2 Stabilization	-$2.7	$5.0	$2.3

3 scenarios by conservation advocates

Cumulative values: Investment costs, Energy savings, Net savings

SOURCE: *America's Energy Choices,* 1991.

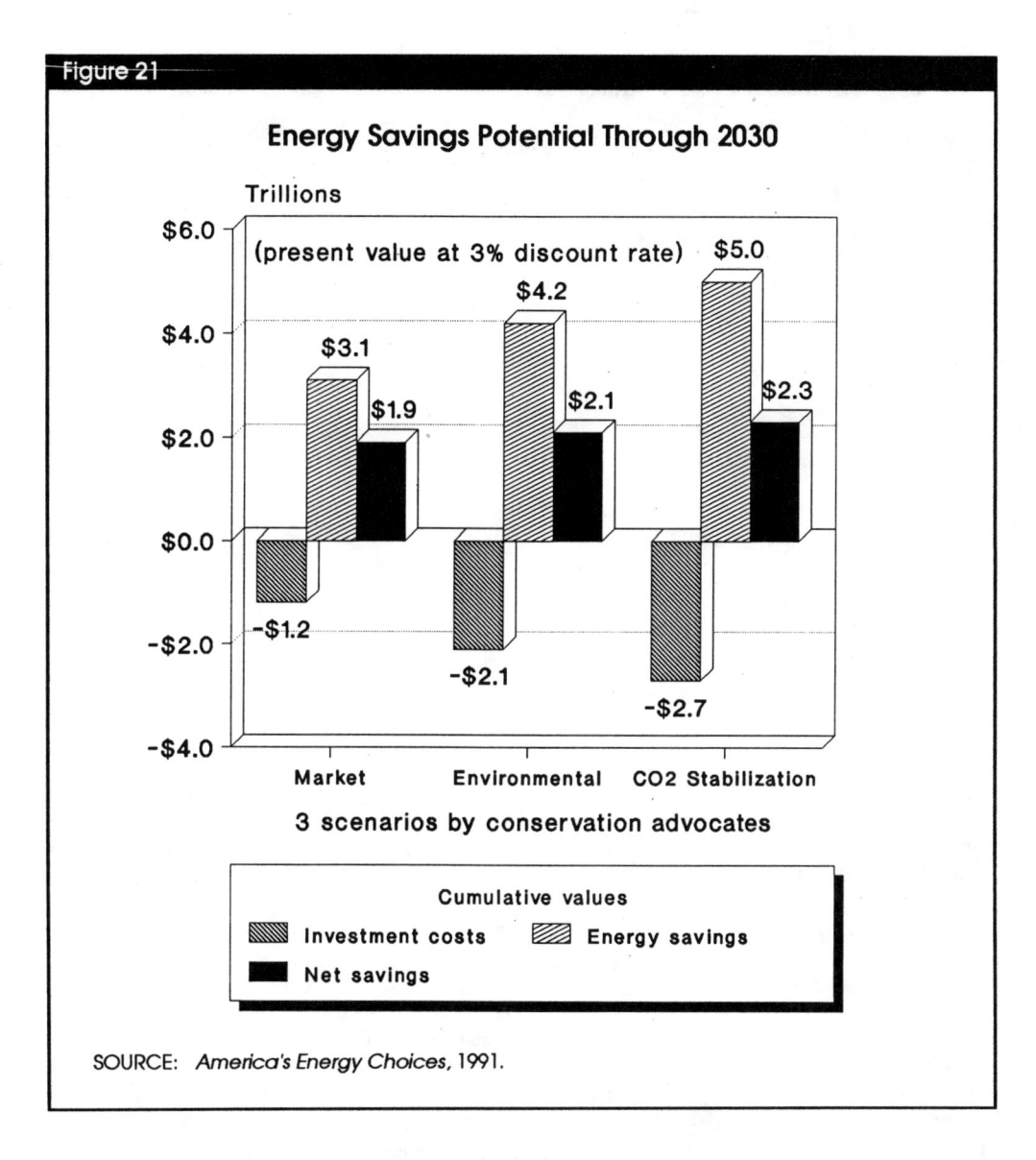

Some argue that the late 1970s and early 1980s was a unique period in which the world was buffeted by two oil price shocks that led the "low-hanging fruit" of energy savings to be plucked. But many energy policy initiatives that were in budding stages then are only now blossoming in full. National appliance efficiency standards did not go into effect until 1987 and 1988, and they will be strengthened in 1993. Three-quarters of the state public utility commissions now use integrated resource planning methods to assess future generating requirements. And nearly two-thirds of the commissions offer financial incentives for utility demand-side management programs, a development that is causing a boom in DSM investments. The states and the Congress also are taking steps toward "leveling the playing field" for renewable energy and

Table 5

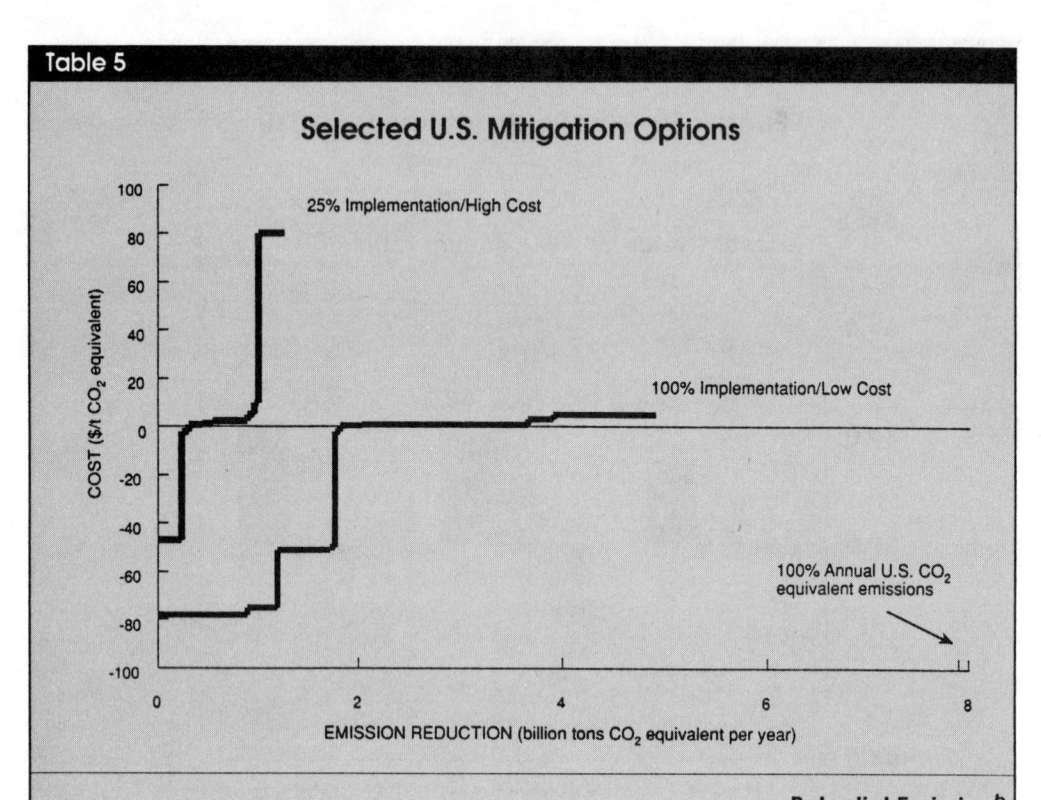

Selected U.S. Mitigation Options

Mitigation Option	Net Implementation Cost[a]	Potential Emission[b] Reduction (metric tons of CO_2 equivalent per year)
Building energy efficiency	Net benefit	900 million
Vehicle efficiency (no fleet change)	Net benefit	300 million
Industrial energy management	Net benefit to low cost	500 million
Transportation system management	Net benefit to low cost	50 million
Power plant heat rate improvements	Net benefit to low cost	50 million
Landfill gas collection	Low cost	200 million
Halocarbon-CFC usage reduction	Low cost	1400 million
Agriculture	Low cost	200 million
Reforestation	Low to moderate cost	200 million
Electricity supply	Low to moderate cost	1000 million

[a] Net benefit = cost less than or equal to zero.
Low cost = cost between $1 and $9 per ton of CO_2 equivalent.
Moderate cost = cost between $10 and $99 per ton of CO_2 equivalent.
High cost = cost of $100 or more per ton of CO_2 equivalent.

[b] This "maximum feasible" potential emission reduction assumes 100 percent implementation of each option in reasonable applications and is an optimistic "upper bound" on emission reductions.

SOURCE: National Academy of Sciences, *Policy Implications of Greenhouse Warming*, 1991.

efficiency investments to make them better able to compete with conventional forms of energy. As these trends continue to unfold, investments in greenhouse gas abatement are likely to appear less "staggering" and become much more affordable.

Benefits of the gambit: Enactment of a strategic plan to address global warming may address U.S. policy concerns beyond energy and the environment. These concerns include fiscal problems at home, competitiveness issues abroad and leadership throughout the world as a new millennium approaches.

Over the last two decades, U.S. energy policy has been driven largely by fears of growing dependence on foreign oil. In the name of national security, President Carter spent billions of dollars in an unsuccessful attempt to develop carbon-based synthetic fuels. President Reagan sanctioned oil drilling in environmentally sensitive regions. And President Bush committed U.S. armed forces to defend the world's access to oil in the volatile Middle East. Pitched battles have been fought in Congress as well over whether to open the Arctic National Wildlife Refuge, whether to raise automobile fuel economy standards and how to divvy up $3 billion in annual government spending on energy research and development.

By comparison, the $15 billion spent to safeguard oil shipments in the Persian Gulf in 1989 and the $30 billion spent in the war to liberate Kuwait in 1990-91 engendered relatively little public debate. The U.S. government found the money and the muster when it perceived a threat to one of the nation's vital interests. If global warming were regarded in the same way, there is little doubt that the government would find the resolve to act once again. The real question is whether the energy policy debate will be transformed in the absence of such a clear and present danger.

The "energy gambit," as defined here, represents a strategic decision by policymakers to devitalize the importance of fossil fuels before global warming leaves them with no other alternative. It is further designed as a "no-regrets" policy, since benefits would accrue even if the climate does not change. The energy gambit would provide a rationale for the government to reduce or eliminate taxpayer subsidies for fossil energy development, for example, saving billions of dollars a year; and it would give the government more discretion in the use of military force, saving lives as well as money.

The apparent losers in the energy gambit are oil and coal interests. But they might be losers anyway if, on the basis of new scientific evidence, a crash program is enacted to develop alternatives—as was the case with CFCs after the discovery of the ozone hole. Ironically, the long-term interests of fossil energy producers could be served by the energy gambit, since a more deliberate approach would be taken to development of alternative fuels and energy-saving technologies. Should CO_2 controls be enacted at some future date, the combination of more availability of noncarbon fuels and slower energy demand growth would make it easier to stay within the emissions limits without requiring wholesale shifts away from oil and coal. In effect, fossil energy producers would secure their own future by acknowledging a larger role for these competing sources, even though their first choice would be a future in which global warming presented no encumbrance.

The energy gambit also could enable the United States to turn its energy largesse to its advantage if it were to sign an international agreement that requires nations to stabilize or reduce their CO_2 emissions. The greatest potential for least-cost energy savings, after all, is in countries that have done the least so far to realize them. Whereas the nations of Western Europe and Japan have already skimmed the cream of their economic churns, the United States (not to mention many formerly planned economies and developing nations) has more of these efficiency gains still to reap. Accordingly, the U.S. should be able to spend fewer dollars on each percentage reduction of greenhouse gas abatement than most of its trading partners. On that basis, the energy gambit seems likely to boost the competitiveness of the United States in the world economy rather than hindering it.

In addition to making the U.S. economy run more efficiently, the energy gambit would create additional jobs in the energy sector even as it shrinks the required size of energy-related investments and frees up capital for other purposes. The American Gas Association, The Alliance to Save Energy and the Solar Energy Industries Association estimate that 215,000 new jobs would be created if market forces were harnessed to promote the natural gas, energy efficiency and renewable energy industries. By contrast, only 44,000 jobs would be lost in the coal industry.[237]

Jobs creation would extend beyond the energy industry, moreover. A narrow focus on energy-related impacts of CO_2 abatement (such as in the Manne-Richels model) fails to see the interaction of capital markets. America's energy industries are in fact extremely capital-intensive, accounting for 11 to 25 percent of all capital investment in the United States in recent years.[238] As reduced growth in energy demand reduces the size of investments in new energy facilities, more capital would be available to invest in new manufacturing plants, machinery and the nation's infrastructure, creating jobs and further bolstering U.S. competitiveness. As an added bonus, the nation could reduce its $450 billion annual energy bill by perhaps $100 billion as a result of investments in energy efficiency, putting more spending money in consumers' pockets.[239]

The energy gambit, employed to its fullest logical extent, also would make use of carbon taxes. Several studies have shown that carbon taxes could provide annual revenues of $80 billion to $140 billion annually.[240] While fossil energy producers may regard such a tax as a loss of income, other sectors of the economy could view it as a transfer of income. Indeed, the macroeconomic effect of the tax hinges much more on how the tax proceeds are used than on the size of the carbon tax itself. (Studies that focus solely on energy-related impacts once again fail to consider this broader effect.)

Recently, researchers at the U.S. Environmental Protection Agency teamed up with analysts at Data Resources Institute to calculate the possible effects of phasing in a carbon tax over 20 years at 5 percent annual increments. In 2010, the carbon tax would reach $38 per ton and provide $80 billion in projected revenue.[241] If the tax proceeds were used to reduce personal income taxes, the short-term macroeconomic loss would be almost as large as the size of the tax itself. If the proceeds were used to reduce the federal budget deficit instead, the

Figure 22

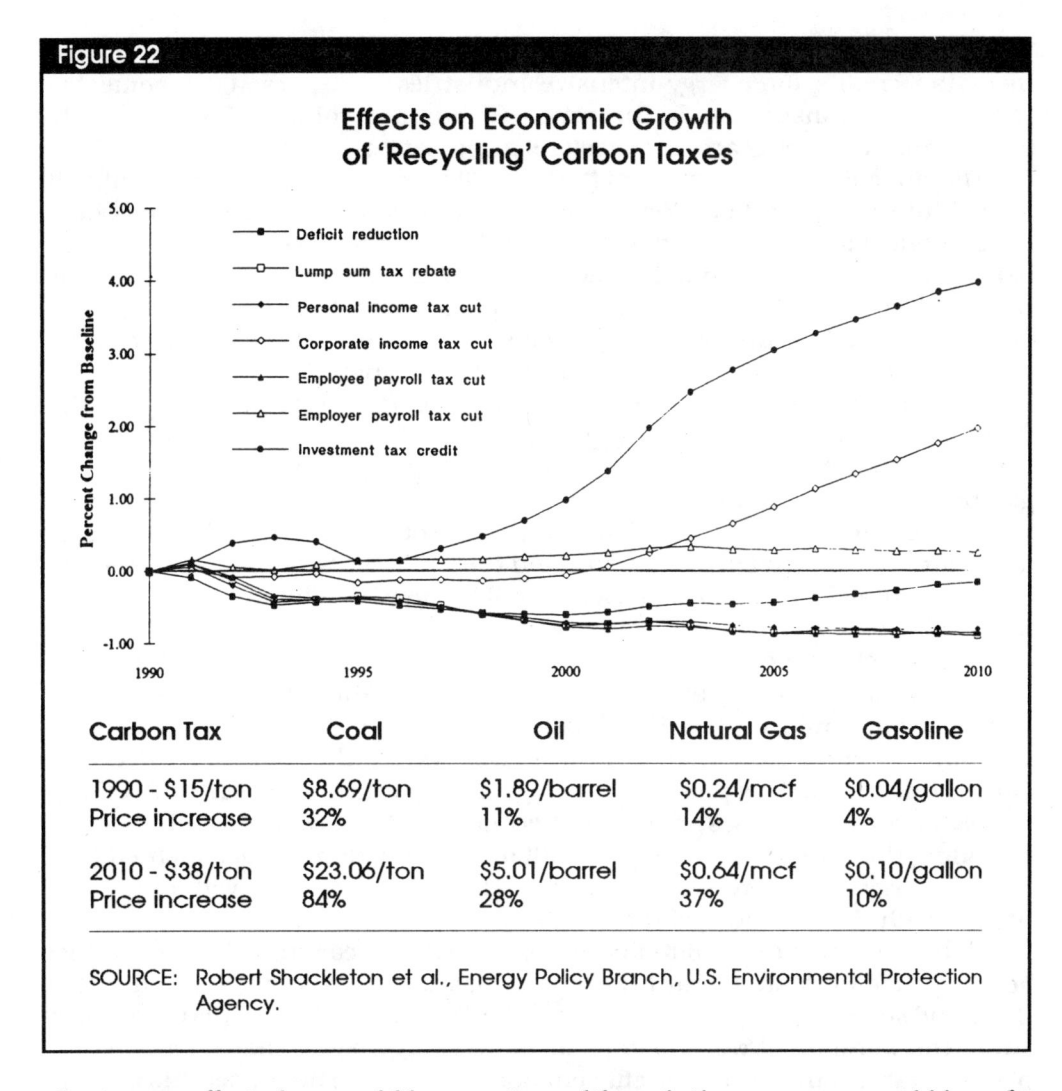

Effects on Economic Growth of 'Recycling' Carbon Taxes

Legend:
- Deficit reduction
- Lump sum tax rebate
- Personal income tax cut
- Corporate income tax cut
- Employee payroll tax cut
- Employer payroll tax cut
- Investment tax credit

Y-axis: Percent Change from Baseline (-1.00 to 5.00)
X-axis: 1990 to 2010

Carbon Tax	Coal	Oil	Natural Gas	Gasoline
1990 - $15/ton	$8.69/ton	$1.89/barrel	$0.24/mcf	$0.04/gallon
Price increase	32%	11%	14%	4%
2010 - $38/ton	$23.06/ton	$5.01/barrel	$0.64/mcf	$0.10/gallon
Price increase	84%	28%	37%	10%

SOURCE: Robert Shackleton et al., Energy Policy Branch, U.S. Environmental Protection Agency.

short-term effect also would be negative, although the economy would benefit in the long run from a decline in interest rates.

Alternatively, the carbon tax proceeds could be "recycled" into the economy as an investment tax credit. While consumption of carbon-intensive goods and services would be reduced, investment in those economic sectors eligible for the tax credit would eliminate any short-term losses and actually enable the economy to grow faster over the long run than if the carbon tax had *not* been enacted. In effect, the carbon tax would bring about an ecological redistribution of the tax burden, from those who emit more CO_2 in the atmosphere as they go about their business, to those who emit less. The investment tax credit, moreover, could ease the financial burden on carbon-intensive industries if they were eligible to receive the credit for investments in energy efficiency improvements and diversification into noncarbon energy sources. Indeed, several European nations with carbon taxes are taking this approach already.

The most politically palatable option may be to offer a combination of investment tax credits for energy-intensive industries and personal income tax reductions for consumers. In this way, DRI found, it would still be possible to achieve reductions in carbon emissions at no net cost to the economy.[242]

The current anti-tax sentiment that prevails in America may make it difficult to institute a carbon tax. Yet when U.S. taxpayers consider that personal income and payroll taxes' share of federal revenues have climbed from 37 percent to 82 percent of total tax revenues over the last 50 years—at the same time that corporate income taxes have dropped from 24 percent to less than 9 percent of the total—taxpayers may decide it is time to shift some of the burden away from themselves toward heavily polluting industries.[243]

The American public is in a mood for change in this 1992 election year. The $400 billion annual federal budget deficit—and $4 trillion debt overall—are constant reminders of the shortcomings of current fiscal policy. The energy gambit, including carbon taxes, offers a fresh approach to this seemingly intractable problem. But it could also be part of something far more significant: a new strategic vision for the nation—and the world—where fiscal, market and environmental forces come together to build a new base for economic activity in the 21st century.

The first chapter of this book discussed how "chaos" may reign in the global climate as climate patterns shift from one equilibrium state to another. A similar theory could apply to the global economy. The global economy today is in a state of flux as the Industrial Age gives way to the new Electronic Age. During this transitional phase, technology, energy and investment are working largely at cross-purposes, amplifying economic disruptions as well as compounding the threat to the globe. Until a strategic plan is implemented that brings energy and environmental priorities into balance, chaos will continue to lurk in both the climate *and* the economy.

In the Industrial Age, nations that acquired and consumed the most fuel generally had the fastest-growing economies, attained the highest standards of living and secured the brightest prospects for the future. Today, this is no longer true. The world must work to implement an energy policy that favors the use of sustainable sources and the efficient use of all fuels. The sooner this is done, the smoother the transition to the new era will be, the more the quality of life will be enhanced and the greater the likelihood that global warming will be held in check.

This is the new energy formula for the Electronic Age. It is the essence of the greenhouse gambit.

Notes

1. Ged R. Davis, "Energy for Planet Earth," *Scientific American*, September 1990.
2. The carbon dioxide content of the atmosphere has increased from 275 parts per million to 356 parts per million over the last 140 years, an increase of more than 30 percent. At present rates of growth in fossil fuel consumption, the atmosphere's CO_2 content is expected to exceed 600 parts per million by the middle of the 21st century. By the 23rd century, as the last of the fossil fuels are exhumed, the CO_2 content could reach 3,000 parts per million. Barring a series of massive volcanic eruptions, a meteor strike or so-called "nuclear winter," it is inconceivable that such an increase in atmospheric CO_2 would have no effect in warming the climate.
3. John P. Holdren, "Energy in Transition," *Scientific American*, September 1990.
4. "50 and 100 Years Ago," *Scientific American*, August 1987.
5. Louis C. Hunter and Lynwood Bryant, *A History of Industrial Power in the United States, 1780-1830, Volume 3: The Transmission of Power*, MIT Press, Cambridge, Mass., 1991.
6. Energy Information Administration, *Annual Energy Review 1990*, U.S. Department of Energy, Washington, D.C., May 1991.
7. ICF Resources, *Assessment of Greenhouse Gas Emissions Policies on the Electric Utility Industry: Costs, Impacts and Opportunities*, prepared for the Edison Electric Institute, Washington, D.C., January 1992.
8. *Ibid.*
9. *Ibid.*
10. Intergovernmental Panel on Climate Change, *Scientific Assessment of Climate Change, Summary and Report*, World Meteorological Organization/United Nations Environment Programme, Cambridge University Press, Cambridge, Mass., 1990.
11. World Resources Institute, *World Resources: 1990-91*, Oxford University Press, New York, N.Y., 1990; and Ralph M. Perhac, *Greenhouse Gases and Global Climate: Options for the Electric Utility Industry*, Electric Power Research Institute, Palo Alto, Calif., undated manuscript.
12. See second citation of note 11, and U.S. Environmental Protection Agency, *Policy Options for Stabilizing Global Climate*, 21P-2003.2, December 1990.
13. See second citation of note 11; World Resources Institute, *World Resources: 1988-89*, Basic Books, New York, N.Y., 1988; and Christopher Flavin, "World Carbon Dioxide Emissions Fall," Worldwatch Insitute press release, Washington, D.C., Dec. 6, 1991.
14. Arnold P. Fickett, Clark W. Gellings and Amory B. Lovins, "Efficient Use of Electricity," *Scientific American*, September 1990.
15. See note 1.
16. Statistics on sectoral energy demand are from the Energy Information Administration, *Monthly Energy Review*, U.S. Department of Energy, May 1992.
17. Mark P. Mills, *Ecowatts: Using Electricity to Save Energy and Cut Greenhouse Gases*, Science Concepts, Chevy Chase, Md., April 1991. According to this analysis, the United States would have produced 4.04 pounds of CO_2 per dollar of gross national product in 1990, rather than 2.34 lbs./$GNP, were it not for improvements in the "carbon dioxide efficiency" of the economy. This analysis calculates that improved automobile efficiency standards accounted for 10

percent of the improvement in the nation's carbon dioxide efficiency and that expansion of nuclear power accounted for 12 percent of the improvement. The analysis suggests that substitution of electricity for less efficient carbon energy applications accounted for much of the remaining improvement.

18. R. Squitieri, O. Yu, C. Roach, "The Coming Boom in Computer Loads," *Public Utilities Fortnightly*, Dec. 25, 1986.
19. Chauncey Starr, "Global Climate Change and the Electric Power Industry," Electric Power Research Institute, July 28, 1988.
20. Energy Research Group, *Carbon Dioxide Reduction Through Electrification of the Industrial and Transportation Sectors*, prepared for the Edison Electric Institute, Washington, D.C., draft April 1989.
21. *Ibid.*
22. See note 6, and *Saving Energy and Reducing CO_2 With Electricity: Estimates of Potential*, EPRI CU-7440, Electric Power Research Institute, Palo Alto, Calif., October 1991.
23. See note 20.
24. See second citation of note 22.
25. "The Greenhouse: New Climate, Same Old Target," *Electrical World*, September 1988.
26. Paul E. Waggoner and Roger R. Revelle, "Summary of the AAAS Panel on Climatic Variability, Climate Change and the Planning and Management of U.S. Water Resources," American Association for the Advancement of Science, Washington, D.C., Sept. 27, 1988.
27. "Taps Run Wild, As Water Use Doubles," *Environment Week*, July 5, 1990.
28. See note 26.
29. *Impact of Abnormal Ultimate Heat Sink Temperatures Upon Power Plant Operations*, American Society of Mechanical Engineers, October 1989.
30. George Lagassa, "Nationwide Effect of Drought Severe But Varied," *ASE*, September 1988.
31. "Utility Concerns Reach Accord to Protect Salmon," *The Wall Street Journal*, March 15, 1991.
32. "SCEcorp Unit, Partners Weigh Desalination Plant," *The Wall Street Journal*, Feb. 15, 1991.
33. James Young, Southern California Edison Co., personal communication, Jan. 9, 1992.
34. Kenneth P. Linder et al., *Potential Impacts of Climate Change on Electric Utilities*, New York State Energy Research and Development Authority, Report 88-2, December 1987.
35. U.S. Environmental Protection Agency, *The Potential Effects of Global Climate Change on the United States*, Report to Congress, Volume 2: National Studies, Washington, D.C., October 1988.
36. Information on nationwide effects is drawn from note 35, and Ken Linder, ICF Inc., "Electricity Demand: How It Might Be," *EPA Journal*, January/February 1989.
37. See note 34, and "How Will Global Warming Affect U.S. Demand?" *Electrical World*, September 1988.
38. See note 34.
39. *Ibid.*
40. See note 7 for all of the information provided in this section.
41. "Summary of 'Assessment of Greenhouse Gas Emissions Policies on the Electric Utility Industry: Costs, Impacts and Opportunities,'" Edison Electric Institute, Washington, D.C., undated manuscript.
42. See note 7.

43. U.S. Congress, Office of Technology Assessment, *Changing by Degrees: Steps to Reduce Greenhouse Gases*, OTA-O-482, U.S. Government Printing Office, Washington, D.C., February 1991.
44. "Dispute Over Ad Campaign Underlies Utility Division on Global Warming," *Electric Utility Week*, June 2, 1991.
45. "Can PR Cool the Greenhouse?" *Science*, June 28, 1991.
46. *Ibid.*
47. Mark DeMichele, "Success Through Environmental Leadership," *Public Utilities Fortnightly*, May 24, 1990.
48. Dennis Wamsted, "SCE Crafts Carbon Dioxide Reduction Plan," *Environment Week*, May 23, 1991.
49. "California Utilities Plan to Reduce Some Emissions," *The Wall Street Journal*, May 22, 1991.
50. David Stipp, "New England Electric Plans Pollution Cuts," *The Wall Street Journal*, Nov. 11, 1991; and "NEES to Seek Proposals for Measures to Offset 'Greenhouse Gas' Emissions," *Electric Utility Week*, Nov. 11, 1991.
51. Howard S. Geller, Eric Hirst, Evan Mills, Arthur H. Rosenfeld, Marc Ross, "Getting America Back on the Energy-Efficiency Track: No-Regrets Policies for Slowing Climate Change," American Council for an Energy-Efficient Economy, Washington, D.C., October 1991.
52. U.S. Department of Energy, *National Energy Strategy: Powerful Ideas for America*, U.S. Government Printing Office, Washington, D.C., February 1991.
53. *Impact of Demand-side Management on Future Customer Electricity Demand: An Update*, CU-6953, Electric Power Research Institute, Palo Alto, Calif., September 1990.
54. *Ibid.*
55. Veronika Rabl, "The Opportunity in DSM," *EPRI Journal*, October/November 1991.
56. See second citation of note 22.
57. Douglas G. Cogan and Susan L. Williams, *Generating Energy Alternatives: 1987 Edition*, Investor Responsibility Research Center, Washington, D.C., 1987.
58. Thomas R. Kuhn, Edison Electric Institute, "Forum: Electrifying Improvements," *Issues in Science and Technology*, Summer 1991.
59. Alfred E. Kahn, "Environmentalists Hijack the Utility Regulators," *The Wall Street Journal*, Aug. 7, 1991.
60. David Moskovitz, "Will Least-Cost Planning Work Without Significant Reform?" National Association of Regulatory Utility Commissioners conference, Aspen, Colo., revised June 10, 1988.
61. See note 57.
62. Mark A. Crawford, "Electricity Crunch Foreseen...Maybe," *Science*, Nov. 18, 1988.
63. Leslie Lamarre, "The Opportunity in DSM," *EPRI Journal*, October/November 1991.
64. See note 14.
65. S. Nadel and J. Jordan, *Do Incentives Work? A Preliminary Evaluation*, draft, American Council for an Energy Efficient Economy, Washington, D.C., October 1991.
66. See note 63.
67. Eric Hirst, Oak Ridge National Laboratory, testimony before the Colorado Public Utilities Commission, printed manuscript, November 1991.
68. *Ibid.*
69. David Stipp, "Utilities Rush to Profit from Selling Less," *The Wall Street Journal*, Nov. 15, 1990.

70. Wil Lepkowski, "Energy Policy," *Chemical & Engineering News*, June 17, 1991.
71. See note 67.
72. Robin Goldwyn Blumenthal, "Con Edison Sees Renewed Vigor For New York," *The Wall Street Journal*, Aug. 2, 1991.
73. See note 69.
74. See note 67.
75. Eric Hirst, Oak Ridge National Laboratory, "Fulfilling the Demand-side Promise," *Public Utilities Fortnightly*, July 1, 1991.
76. See note 69.
77. "Ontario Hydro Says Savings Will Allow It to Defer New Plants," *The Wall Street Journal*, Jan. 17, 1992.
78. See second citation of note 22 and note 55.
79. See note 53.
80. See note 14.
81. Amory B. Lovins and Hunter L. Lovins, "Least-Cost Climatic Stabilization," Rocky Mountain Institute, Snowmass, Colo., printed manuscript, May 17, 1991.
82. Amory B. Lovins, "Energy Strategy: The Road Not Taken?" *Foreign Affairs*, October 1976.
83. Amory B. Lovins in *Meeting the Energy Challenges of the 1990s*, GAO/RCED 91-66, U.S. General Accounting Office, Washington, D.C., March 1991.
84. National Academy of Sciences, *Policy Implications of Global Warming*, Reports of the Synthesis and Mitigation Panels, National Academy Press, Washington, D.C., 1991.
85. See note 51.
86. *Ibid.*
87. *Ibid.*
88. Chauncey Starr, Milton F. Searl and Sy Alpert, "Energy Sources: A Realistic Outlook," *Science*, May 15, 1992.
89. See note 52.
90. See notes 14, 22, 51, 53, 81 and 84.
91. See note 52.
92. U.S. General Accounting Office, "Electricity Supply: Utility Demand-side Management Programs Can Reduce Electricity Use," GAO/RECED-92-13, Washington, D.C., October 1991. This report summarizes statistics of the North American Electric Reliability Council.
93. See note 43.
94. Susan L. Williams and Kevin Porter, *Power Plays: Profiles of America's Leading Independent Renewable Electricity Developers*, Investor Responsibility Research Center, Washington, D.C., 1989.
95. See note 7.
96. *BP Statistical Review of World Energy*, London, England, June 1992.
97. *Ibid.*
98. See note 16.
99. See note 7, and Paul D. Holtberg et al., *Baseline Projection Data Book*, 1992 Edition, Gas Research Institute, Washington, D.C., 1992.
100. See note 7. From the executive summary: "As a result, coal market impacts are less severe, and coal production under stabilization or reduction policies is actually *greater* under the Low Impact Scenario than under the High Impact Scenario, although there is less coal production under both policy cases relative to the base case."
101. See note 96, and Joseph Haggin, "Coal," *Chemical & Engineering News*, June 17, 1991.

102. See note 96.
103. See note 1.
104. See note 43.
105. See note 7.
106. See note 84.
107. William Fulkerson, Roddie R. Judkins and Manoj K. Sanghvi, "Energy from Fossil Fuels," *Scientific American*, September 1990; and U.S. Congress, Office of Technology Assessment, *Energy Technology Choices: Shaping Our Future*, OTA-E-493, U.S. Government Printing Office, Washington, D.C., July 1991.
108. See second citation of note 107.
109. See note 88.
110. See note 33, and "Global Warming Plays Role in Planning for Next Clean Coal Project Competition," *The Energy Report*, Dec. 26, 1988.
111. Dennis Wamsted, "Greenhouse Gas Concerns May Tarnish Luster Of 'Clean' Fluidized Beds," *Environmental Week*, Sept. 5, 1991.
112. R.J. Charlson et al., "Climate Forcing by Anthropogenic Aerosols," *Science*, Jan. 24, 1992.
113. See note 88.
114. *Ibid.*
115. See notes 16, 96 and second citation of note 99.
116. See first citation of note 107, and Robert H. Williams and Eric D. Larson, "Expanding Roles for Gas Turbines in Power Generation," *Electricity: Efficient End-Use and New Generation Technologies and Their Planning Implications*, Thomas B. Johansson, Birgit Bodlund and Robert H. Williams, editors, Lund University Press, Lund, Sweden, 1988.
117. Robert Johnson, "Producers of Natural Gas Face Dilemma as Prices Hit New Lows," *The Wall Street Journal*, Jan. 27, 1992.
118. See note 96.
119. William L. Fisher, "Energy Policy After the Gulf War," *Forum for Applied Research and Public Policy*, Spring 1992. Fisher cites estimates by the U.S. Department of Energy, the National Research Council and the Gas Research Institute.
120. Peter Weber, "Sold on Fuel Cells," *World Watch*, January/February 1992.
121. Amal Kumar Naj, "Clean Fuel Cells Sparking New Interest," *The Wall Street Journal*, March 19, 1992.
122. *Ibid.*, and see second citation of note 107.
123. See note 121.
124. American Gas Association, The Alliance to Save Energy and Solar Energy Industries Association, *An Alternative Energy Future*, Washington, D.C., April 1992.
125. See note 52, and Christopher Flavin and Nicholas Lenssen, "Nuclear Power Industry at a Standstill," Vital Signs Brief No. 4, Worldwatch Institute, Washington, D.C., May 1992.
126. Sherwood H. Smith, "A Case for Reviving the Nuclear Option," *Forum for Applied Research and Public Policy*, Winter 1991. Emission offsets have been updated to reflect nuclear generation in 1991.
127. See note 52.
128. See note 7.
129. Steve Cohn, "Public Not Convinced On Nuclear Comeback," *Forum for Applied Research and Public Policy*, Winter 1991.
130. John Horgan, "The Finance of Fission," *Scientific American*, June 1989.
131. See full citation of note 125; International Atomic Energy Agency *Newsbriefs*, January/February 1991; and Wolf Hafele, "Energy from Nuclear Power," *Scientific American*, September 1990.

132. See second full citation of note 131.
133. "AP600: An Advanced Passive Plant," Westinghouse Corp., undated brochure.
134. Construction is expected to resume at the two Watts Bar units owned by the Tennessee Valley Authority.
135. Neil Bermann, "Uranium Market Likely to Be Depressed by Sales from Commercial, Military Stocks Well Into '90s," *The Wall Street Journal*, Sept. 16, 1991.
136. Stephen C. Stinson, "Nuclear," *Chemical & Engineering News*, June 17, 1991.
137. *Ibid.*
138. Amal Kumar Naj, "Japan Clears GE In Nuclear Project, Value is $1 Billion," *The Wall Street Journal*, May 21, 1991.
139. See note 136.
140. See notes 14, 22, 51, 53, 81 and 84.
141. *Ibid.*, and Nicholas Lenssen, *Nuclear Waste: The Problem That Won't Go Away*, Worldwatch Paper #106, Worldwatch Institute, Washington, D.C., December 1991.
142. *Ibid.*
143. "Measure Would Exempt Yucca Mountain from Requirements for Nevada Permits," *Environment Reporter*, Sept. 13, 1991; and Paul Slovic, James H. Flynn and Mark Layman, "Perceived Risk, Trust, and the Politics of Nuclear Waste," *Science*, Dec. 13, 1991.
144. See second citation of note 143, and Richard Monastersky, "Yucca Site: A Conclusion and Controversy," *Science News*, Oct. 26, 1991.
145. David Stipp, "Yankee Atomic Spotlights Massive Costs Needed to Decommission Nuclear Plants," *The Wall Street Journal*, June 2, 1992; and Cynthia Pollock Shea, "Breaking Up Is Hard to Do," *World Watch*, July/August 1989. Shea also is the author of *Decommissioning: Nuclear Power's Missing Link*, Worldwatch Paper #69.
146. Phil Hill, "Germany Shuts Down Two New Nukes," *Environmental Action*, November/December 1989.
147. Michael W. Golay and Neil E. Todreas, "Advanced Light-Water Reactors," *Scientific American*, April 1990.
148. Jonathan Weil, "House Passes Energy Bill to Overhaul Utility Regulation, Promote Efficiency," *The Wall Street Journal*, May 29, 1992.
149. William F. Malec, "TVA Re-Examines the Nuclear Option," *Forum for Applied Research and Public Policy*, Winter 1991.
150. See note 148.
151. Bill Keepin and Gregory Kats, "Greenhouse Warming: Comparative Analysis of Nuclear and Efficiency Abatement Strategies," *Energy Policy*, December 1988.
152. See note 1.
153. *Ibid.*
154. See note 88.
155. See note 16.
156. Christopher Flavin and Nicholas Lenssen, *Beyond the Petroleum Age: Designing a Solar Economy*, Worldwatch Paper #100, Worldwatch Institute, Washington, D.C., December 1990.
157. See note 92.
158. See note 94.
159. See note 83.
160. "Photovoltaics—Clean Energy Now & For the Future," *Rocky Mountain Institute Newsletter*, Spring 1991.
161. Richard Curry, "Production of PV Modules Worldwide Rises 14% to Record 48 MW During 1990," *Photovoltaic Insider's Report*, February 1991.
162. Richard Curry, "Dow Chemical Acquires Minority Equity Interest in AstroPower," *Photovoltaic Insider's Report*, December 1991.

163. Richard Curry, "Texas Instruments Says 10% Nowhere Near Maximum Efficiency of Its Spheral Solar Module," *Photovoltaics Insider's Report*, November 1991.
164. Richard Curry, "DOE Envisions Accelerated PV Commercialization With 1,400 MW Installed by 2000," *Photovoltaics Insider's Report*, November 1991.
165. American Wind Energy Association, "Tehachapi Now Leads World in Wind Generating Capacity," *Wind Energy Weekly*, May 4, 1992.
166. Carl J. Weinberg and Robert H. Williams, "Energy from the Sun," *Scientific American*, September 1990.
167. American Wind Energy Association, "Northern States Power Plans 10-MW Windfarm in Minnesota," *Wind Energy Weekly*, Oct. 7, 1991.
168. See note 156.
169. "Wind-Power Venture Set With Iowa-Illinois Gas Unit," *The Wall Street Journal*, Nov. 13, 1991.
170. Thomas W. Lippman, "Future of Wind Power Gets a Lift," *The Washington Post*, Nov. 17, 1991.
171. David Stipp, "'Wind Farms'" May Energize the Midwest," *The Wall Street Journal*, Sept. 6, 1991.
172. See second citation of note 11 and note 88.
173. See note 94.
174. Patrick Lee, "Top Solar Power Firm Cuts Work Force in Half," *Los Angeles Times*, July 6, 1991.
175. See note 33, and Kimberly Dozier, "California Utilities Plan Solar Two," *The Energy Daily*, August 30, 1991.
176. See note 94, and Scott Spiewak, "Geothermal Heats Up," *Cogeneration & Resource Recovery*, September/October 1991.
177. "A Big Chill Grip Geothermal Energy," *Science*, Nov. 22, 1991.
178. See note 94.
179. Matthew L. Wald, "Mining Deep Underground for Energy," *The New York Times*, Nov. 3, 1991.
180. See note 94.
181. See note 166.
182. See note 88.
183. See note 11.
184. See note 16.
185. See note 43.
186. *Ibid.*, and see notes 88 and 156, which cites figures from the World Energy Conference.
187. See notes 11 and 43.
188. See note 107.
189. See note 156.
190. James R. Chiles, "Tomorrow's Energy Today," *Audubon*, January 1990. In his calculation, Zweibel assumes that each photovoltaic module is 12 percent efficient and that half of the required land area is set aside for tracking and mounting and to allow for maintenance of the modules.
191. See note 160.
192. See note 156.
193. National Renewable Energy Laboratory, Idaho National Engineering Laboratory, Los Alamos National Laboratory, Oak Ridge National Laboratory, and Sandia National Laboratories, *The Potential for Renewable Energy, An Interagency White Paper*, SERI/TP-260-3674, Golden, Colo., March 1990.
194. Mary Anne Gozewski and Tom McCord, "DOE Seeks Biggest Fossil Energy Budget Since 1982; Gas Gets Biggest Increase," *The Energy Report*, Feb. 3, 1992.

195. See note 52.
196. *Ibid.*
197. See note 7.
198. See note 124.
199. The Alliance to Save Energy, American Council for an Energy-Efficient Economy, Natural Resources Defense Council, Union of Concerned Scientists, and the Tellus Institute, *America's Energy Choices: Investing in a Strong Economy and a Clean Environment,*" Washington, D.C., 1991.
200. Harold M. Hubbard, "The Real Cost of Energy," *Scientific American,* April 1991; and H.R. Heede, R.E. Morgan, and S. Ridley, *The Hidden Costs of Energy,* Center for Renewable Resources, Washington, D.C., 1985.
201. *Ibid.,* and Sandra Postel, "Toward A New 'Eco'-Nomics," *World Watch,* September/October 1990.
202. Pace University Center for Environmental Legal Studies, *Environmental Costs of Electricity,* Oceana Publications Inc., New York, 1990.
203. See note 193.
204. See first citation of note 200.
205. *Ibid.,* and "States Considering 'Social Costs' When Planning for Electric Utilities," *Environment Reporter,* Oct. 25, 1991.
206. See note 83.
207. See first citation of note 200.
208. See, for example, Paul L. Joskow, "Dealing With Environmental Externalities: Let's Do It Right," Issues and Trends Briefing Paper #61, Edison Electric Institute, Washington, D.C., 1992.
209. Richard W. Stevenson, "Trying a Market Approach to Smog," *The New York Times,* March 25, 1992.
210. "House Passes Energy Bill Offering Incentives Senate Version Lacks," *International Solar Energy Intelligence Report,* June 1, 1992.
211. "Western Fuels Group Argues Against Carbon Dioxide Penalty in California," *The Energy Report,* July 22, 1991.
212. See note 208.
213. U.S. Department of Energy, *Limiting Net Greenhouse Gas Emissions in the United States: Report to Congress,* Washington, D.C., December 1991.
214. See note 7.
215. U.S. Congress, Congressional Budget Office, *Carbon Charges as a Response to Global Warming: The Effects of Taxing Fossil Fuels,* Washington, D.C., August 1990. The Congressional Budget Office conducted sensitivity analyses on a total of three econometric models and concluded that a 1 to 3 percent loss in gross national product might result of using carbon taxes to achieve a 20 percent reduction in the nation's CO_2 emissions.
216. Dale W. Jorgenson and Peter J. Wilcoxen, "Reducing U.S. Carbon Dioxide Emissions: The Cost of Different Goals," CSIA Discusion Paper 91-9, Kennedy School of Government, Harvard University, Cambridge, Mass., October 1991.
217. See note 7.
218. Ibid.
219. James Tanner, "Carbon Tax to Limit Use of Fossil Fuels Becomes Embroiled in Global Politics," *The Wall Street Journal,* June 9, 1992.
220. See note 156.
221. "Denmark Passes Carbon Tax," *Business and the Environment,* Jan. 24, 1992.
222. "Sweden Takes Action on Carbon Dioxide," *Global Environmental Change Report,* May 1992.

223. "Proposed Energy Tax Hits Major Snags in Run-Up to Joint Ministers' Meeting," *International Environment Reporter*, Dec. 4, 1991.
224. Debora MacKenzie, "Europe Haggles Over Carbon Taxes," *New Scientist*, Oct. 5, 1991.
225. Martin Du Bois, "EC May Fall Short of Goal in Cutting Rise in Emissions," *The Wall Street Journal*, May 12, 1992.
226. *Ibid.*, and Marlise Simons, "Europe Proposes Taxes on Fuels Tied to Warming," *The New York Times*, May 14, 1992.
227. For details, see Alan S. Manne and Richard G. Richels, "Global CO_2 Emission Reductions," *Energy Journal*, August 1990.
228. Peter Passell, "Staggering Cost Is Foreseen To Curb Warming of Earth," *The New York Times*, Nov. 19, 1989.
229. *1990 Economic Report of the President*, U.S. Government Printing Office, Washington, D.C., February 1990.
230. See note 227, and Mary Beth Zimmerman, "Assessing the Costs of Climate Change Policies: The Uses and Limits of the Model," The Alliance to Save Energy, Washington, D.C., manuscript dated April 10, 1990.
231. See notes 227 and 228.
232. See note 227, and Bob Davis, "In Rio, They're Eyeing Greenhouse Two-Step," *The Wall Street Journal*, April 20, 1992.
233. See note 199.
234. See note 43.
235. See note 84.
236. See notes 199, 216 and second citation of full citation of note 232. Dale Jorgenson of Harvard University and Peter Wilcoxen of the University of Texas calculate that the U.S. economy would have grown at an annual rate of 3.5 percent from 1974 through 1985, instead of 2.5 percent annually, had it not been for the oil price shocks of 1973 and 1979.
237. See note 124.
238. See note 92.
239. *Ibid.*, and see note 51. The technical potential for energy savings is drawn from studies by The Alliance to Save Energy and the Congressional Budget Office.
240. See notes 7, 199, 213, 215 and 216. ICF calculates that a $100 per ton tax levied on carbon fuels used by electric utilities could generate $87 billion in revenue by 2015. Four energy and environmental groups estimate that a carbon tax of $92 per ton levied on all sources could generate $140 billion annually.
241. Robert Shackleton et al., "The Efficiency Value of Carbon Tax Revenues," U.S. Environmental Protection Agency, Energy Policy Branch, Washington, D.C., draft manuscript, June 2, 1992. Carbon taxes recycled through an investment tax credit would increase baseline growth of the gross national product by 0.2 percent a year and reduce carbon emissions below the baseline forecast by 3 to 12 percent in 2010.
242. *Ibid.*, and William K. Stevens, "New Studies Predict Profits In Heading Off Warming," *The New York Times*, March 17, 1992.
243. See note 199, which cites the president's fiscal year 1992 budget proposal.

Box 5-A

1a. According to the Energy Information Administration, the United States imported 35 percent of the oil it consumed in 1973. Preliminary figures for 1991 indicate that imported oil's share rose to 45 percent of the total.
2a. *National Energy Strategy: Powerful Ideas for America,* U.S. Government Printing Office, Washington, D.C., February 1991.
3a. U.S. Congress, Office of Technology Assessment, Changing by Degrees: Steps to Reduce Greenhouse Gases, OTA-O-482, U.S. Government Printing Office, Washington, D.C., February 1991.
4a. See note 2a. With implementation of the National Energy Strategy, the nation's oil imports would be estimated to fall by 40 percent in 2010, before rising to 60 percent by 2030, according to the document's authors.

Box 5-B

1b. "The Greenhouse: New Climate, Same Old Target," *Electrical World,* September 1988.
2b. Bill Paul, "Some Utilities in Northeast Cut Voltage; Heat Drought Strain Others Nationwide," *The Wall Street Journal,* Aug. 5, 1988.
3b. Arthur Fisher, "Global Warming: Playing Dice with the Earth's Climate," *Popular Science,* August 1989.
4b. Scott Kilman and Bill Richards, "Drought Eases, but Too Late to Undo Food-Price Impact," *The Wall Street Journal,* July 29, 1988.
5b. George Lagassa, "Nationwide Effect of Drought Severe But Varied," *ASE,* September 1988.

Box 5-C

1c. Amory B. Lovins in *Meeting the Energy Challenges of the 1990s,* Report GAO/RCED 91-66, U.S. General Accounting Office, Washington, D.C., March 1991.
2c. Rodd Aubrey, "Radio Wave Bulbs May Last 20 Years," Associated Press wire story, June 1, 1992.
3c. *Impact of Demand-side Management on Future Customer Electricity Demand: An Update,* CU-6953, Electric Power Research Institute, Palo Alto, Calif., September 1990.
4c. Howard S. Geller, Eric Hirst, Evan Mills, Arthur H. Rosenfeld, Marc Ross, "Getting America Back on the Energy-Efficiency Track: No-Regrets Policies for Slowing Climate Change," American Council for an Energy-Efficient Economy, Washington, D.C., October 1991.
5c. *Ibid.*
6c. Alex Wilson and John Morrill, *Consumer Guide to Home Energy Savings,* American Council for an Energy-Efficient Economy, Washington, D.C., 1991.
7c. *Ibid.*
8c. See note 4c.
9c. *Ibid.*
10c. See note 3c.
11c. See note 6c.
12c. Roger S. Carlsmith et al., *Energy Efficiency: How Far Can We Go?,* ORNL/TM-11441, Oak Ridge National Laboratory, Oak Ridge, Tenn., January 1990.

13c. See note 3c.
14c. See note 3c. Depending on the application, EPRI estimates that high-efficiency motors reduce electricity consumption by 3 to 10 percent, and additional savings of up to 35 percent are obtainable from variable-speed drives. See note 15c below for other estimates.
15c. Rick Bevington and Arthur H. Rosenfeld, "Energy for Buildings and Homes," *Scientific American*, September 1990.
16c. *Saving Energy and Reducing CO_2 With Electricity: Estimates of Potential*, EPRI CU-7440, Electric Power Research Institute, Palo Alto, Calif., October 1991.

Box 5-D

1d. Leslie Lamarre, "The Opportunity in DSM," *EPRI Journal*, October/November 1991.
2d. Amory B. Lovins in *Meeting the Energy Challenges of the 1990s*, GAO/RCED 91-66, U.S. General Accounting Office, Washington, D.C., March 1991.
3d. U.S. Congress, Office of Technology Assessment, *Changing by Degrees: Steps to Reduce Greenhouse Gases*, OTA-O-482, U.S. Government Printing Office, Washington, D.C., February 1991; and Michael Shepard, "How to Improve Energy Efficiency," *Issues in Science and Technology*, Summer 1991.
4d. See note 2d.
5d. Jeffrey Taylor, "New Rules Harness Power of Free Market To Curb Air Pollution," *The Wall Street Journal*, April 14, 1992
6d. See note 2d and "1990 Competitek Forum," *Rocky Mountain Institute Newsletter*, Fall/Winter 1990.
7d. Peter Passell, "Greenhouse Gamblers," *The New York Times*, June 19, 1991.
8d. See note 2d.

Box 5-E

1e. U.S. Department of Energy, *A Systems Study for the Removal, Recovery and Disposal of Carbon Dioxide from Fossil Fuel Power Plants in the U.S.*, DOE/CH/00016-2, prepared by Brookhaven National Laboratory for the Office of Energy Research, Washington, D.C., December 1984.
2e. C.A. Hendriks, et al., "Technology and Cost of Recovery and Storage of Carbon Dioxide from an Integrated Gasifier Combined Cycle Plant," University of Utrecht, Utrecht, The Netherlands, draft manuscript dated April 18, 1990.
3e. Robert H. Williams, "Hydrogen from Coal with Gas and Oil Well Sequestering of the Recovered CO_2," Princeton University Center for Energy and Environmental Studies, Princeton, N.J., draft manuscript dated June 16, 1990.
4e. Chauncey Starr, Milton F. Searl and Sy Alpert, "Energy Sources: A Realistic Outlook," *Science*, May 15, 1992.

Box 5-F

1f. International Atomic Energy Association, *IAEA Newsbriefs*, January/February 1992; Christopher Flavin and Nicholas Lenssen, "Nuclear Power Industry at a Standstill," Vial Signs Brief No. 4, Worldwatch Institute, Washington, D.C., May 1992; and Wolf Hafele, "Energy from Nuclear Power," *Scientific American*, September 1990.

2f. John F. Fialka, "Atomic Power Plants Are Cause of Concern Amid USSR Crisis," *The Wall Street Journal*, Sept. 9, 1991.
3f. E.S. Browning, "Paris and Bonn Moving to Ally Power Industries," *The Wall Street Journal*, April 4, 1989.
4f. "U.K. Drops Nuclear Plants from Privatization Plan," *The Wall Street Journal*, Nov. 10, 1989.
5f. Matthew L. Wald, "Japan Now Ahead in Nuclear Power, Too," *The New York Times*, Feb. 27, 1990; and "Japan's Nuclear Energy Pace," *The Wall Street Journal*, Oct. 28, 1991.
6f. Jacob M. Schlesinger and Quentin Hardy, "Japan Faces Possible Summer Electricity Crunch," *The Wall Street Journal*, April 4, 1991.
7f. *Ibid.*
8f. See second citation of note 5f, and Jacob M. Schlesinger, "Japan Energy Plan Spurs Public Fission," *The Wall Street Journal*, Jan. 30, 1991.

Box 5-G

1g. Mark Crawford, "U.S. Fusion Program: Struggling to Stay in the Game," *Science*, Dec. 14, 1990.
2g. Christopher Flavin and Nicholas Lenssen, *Beyond the Petroleum Age: Designing a Solar Economy*, Worldwatch Paper 100, Worldwatch Institute, Washington, D.C., December 1990.
3g. Jerry E. Bishop, "Future of Hot Fusion Is Boiling Down to the Behavior of A Few Helium Atoms," *The Wall Street Journal*, Sept. 12, 1990.
4g. Jerry E. Bishop and Ken Wells, "Hot Fusion Test Using New Fuel Shows Promise," *The Wall Street Journal*, Nov. 11, 1991.
5g. Robert W. Conn et al., "The International Thermonuclear Experimental Reactor," *Scientific American*, April 1992.
6g. David P. Hamilton, "The Fusion Community Picks Up the Pieces," *Science*, March 6, 1992.
7g. Robert Pool, "Fusion Breakthrough?" *Science*, April 7, 1989.
8g. Jerry E. Bishop, "Electric Power Research Institute to Pay $12 Million More to Study 'Cold Fusion,'" *The Wall Street Journal*, March 18, 1992.

Chapter 6
Conclusions

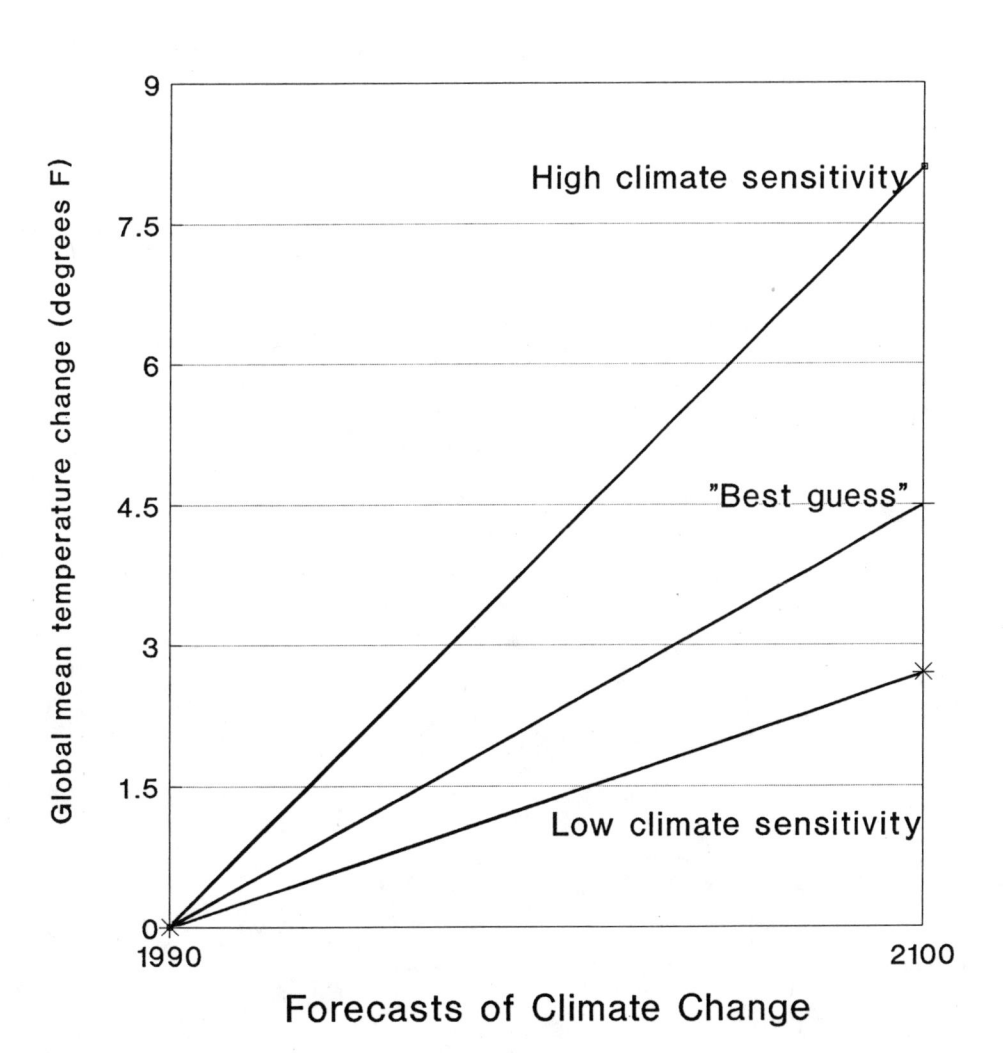

Forecasts of Climate Change

(Chart showing Global mean temperature change (degrees F) from 1990 to 2100, with three lines: "High climate sensitivity" reaching ~8°F, "Best guess" reaching 4.5°F, and "Low climate sensitivity" reaching ~2.7°F.)

Contents of Chapter 6

The Greenhouse Gambit
    The Road from Rio 467
    Passing the Torch 470
    Does Climate Change Matter? 472
    The Regret in 'No Regrets' 475
    The Grand Bargain 478

Notes 484

The Greenhouse Gambit

In chess, a gambit is an opening in which a player risks one or more minor pieces to gain an advantage in position. The greenhouse gambit does much the same. It puts at risk the world's conventional way of doing business—especially its heavy reliance on fossil fuels—in an effort to hold climate change in check. The "minor" pieces in this geopolitical game are huge petroleum and coal companies that constitute the world's largest industry. The "major" pieces are the climate and the Earth itself. In this high-stakes gambit, the world is transformed one way or another, depending on the outcome of the match.

The Road from Rio

The Earth Summit held in Rio de Janeiro in June 1992 constitutes the opening move in the world's greenhouse gambit. Delegates from 153 nations signed a first-ever treaty on climate change. Yet many dismiss the treaty as an overly cautious—even cowardly—attempt to counter the threat posed by global warming. While the treaty is legally binding, it has no teeth. It commits no nation to targets or timetables to stabilize its greenhouse gas emissions. Instead, it endorses the principle that a return to "earlier levels" of emissions by the end of the century would contribute to the long-term objective of stabilizing greenhouse gases in the atmosphere. Industrialized nations will report periodically on their progress toward limiting emissions, "with the aim" of returning to 1990 levels. Developing nations will try to control their emissions as well, without necessarily stabilizing them.

The agreement is a far cry from what many hoped would be achieved at Rio. Four years earlier, when delegates from 48 nations convened in Toronto to discuss the prospect of greenhouse gas controls, they cast a nonbinding vote in favor of a 20 percent reduction in such emissions by 2005. Several countries in Europe subsequently pledged to reach that goal. Sen. Tim Wirth (D-Colo.) introduced legislation requesting the United States to do the same. A flurry of government and industry studies weighed the feasibility of achieving such a goal.

Also in 1988, George Bush was a candidate for president. While not endorsing the 20 percent reduction target, Bush assured voters: "Those who think we're powerless to do anything about the greenhouse effect are forgetting about the White House effect." In the wake of Earth Summit, these words ring with new irony. Of the 118 heads of state who went to Rio, Bush was among the last to say he would attend. He apparently made up his mind only weeks before the summit—and after the language of the climate treaty had been softened to meet his objections. Over a series of preparatory meetings, the United States had made clear it would not go along with the 20 percent reduction target. Then it became a matter of whether America would commit to stabilizing its emissions at 1990 levels by the year 2000. In the end, "the

White House effect" was to make sure that no nation at Rio would be bound to achieve any targets or timetables whatever.

To put the Rio climate agreement in perspective, it is worth recalling that stabilization of greenhouse gases emitted *to* the atmosphere would not bring about a stabilization of greenhouse gases *in* the atmosphere—nor would a 20 percent cut, for that matter. If carbon dioxide emissions were stabilized at the current level—approximately 6 billion metric tons of carbon a year—the atmospheric concentration of CO_2 still would rise to 500 parts per million by the end of the 21st century—a 40 percent increase from today's level of 356 parts per million. A 20 percent reduction in CO_2 emissions would shrink the yearly total to less than 5 billion tons of carbon, yet the atmospheric level of CO_2 would still reach almost 450 parts per million by 2100—almost a 30 percent increase. To maintain atmospheric CO_2 levels where they are today, carbon emissions would in fact have to drop to less than 2 billion tons of carbon a year—a 66 percent reduction. In effect, the world would have to roll back the clock to the level of emissions that prevailed in the early 1950s, when there were only half as many people on the planet and less than a fifth as many cars and generating plants.

With business as usual, on the other hand, the world in 2100 is likely to have more than twice as many people and an economy that is perhaps 10 times as large. Carbon emissions might be spewing forth at a rate of nearly 27 billion tons a year, in which case the concentration of carbon dioxide in the atmosphere would exceed 800 parts per million—well over twice today's level. Then there are the other greenhouse gases to consider. Methane is a far more potent trapper of heat in the atmosphere, with 10 to 20 times the radiative forcing potential of carbon dioxide, and its emissions are growing at a faster rate. With business as usual, the concentration of methane in the atmosphere could rise from 1.7 parts per million today to 4 parts per million by 2100. Yet to achieve atmospheric stabilization, emissions of methane and other heat-trapping gases would have to be held below current levels, instead of allowing them to more than double.

Viewed in this context, the steps taken at the Earth Summit in Rio appear especially tentative. But the significance of the event cannot be dismissed entirely. Rio did manage to bring together delegates from 178 nations for the largest international conference ever held. Counting representatives of the non-governmental organizations who attended the Global Forum, a companion meeting to the summit, 35,000 people participated in all. If nothing else, the sheer size of the gathering underscored common concerns about the state of the global environment and the need for nations to work together in creating new economic development programs that sustain the environment over the long term.

Delegates at the formal meeting in Rio—the United Nations Conference on Environment and Development—also approved more than a climate treaty. Representatives of more than 150 nations approved a Biodiversity Convention, a legally binding treaty that requires signatories to take inventories of plants and wildlife and to enact plans to protect endangered species. (The United States refused to sign this treaty, however, because it also requires signatories to share research, profits and technologies with nations whose genetic resources they use.) By consensus, UNCED delegates endorsed a nonbinding Statement on Forestry Principles and the Rio Declaration on Environment and

Conclusions

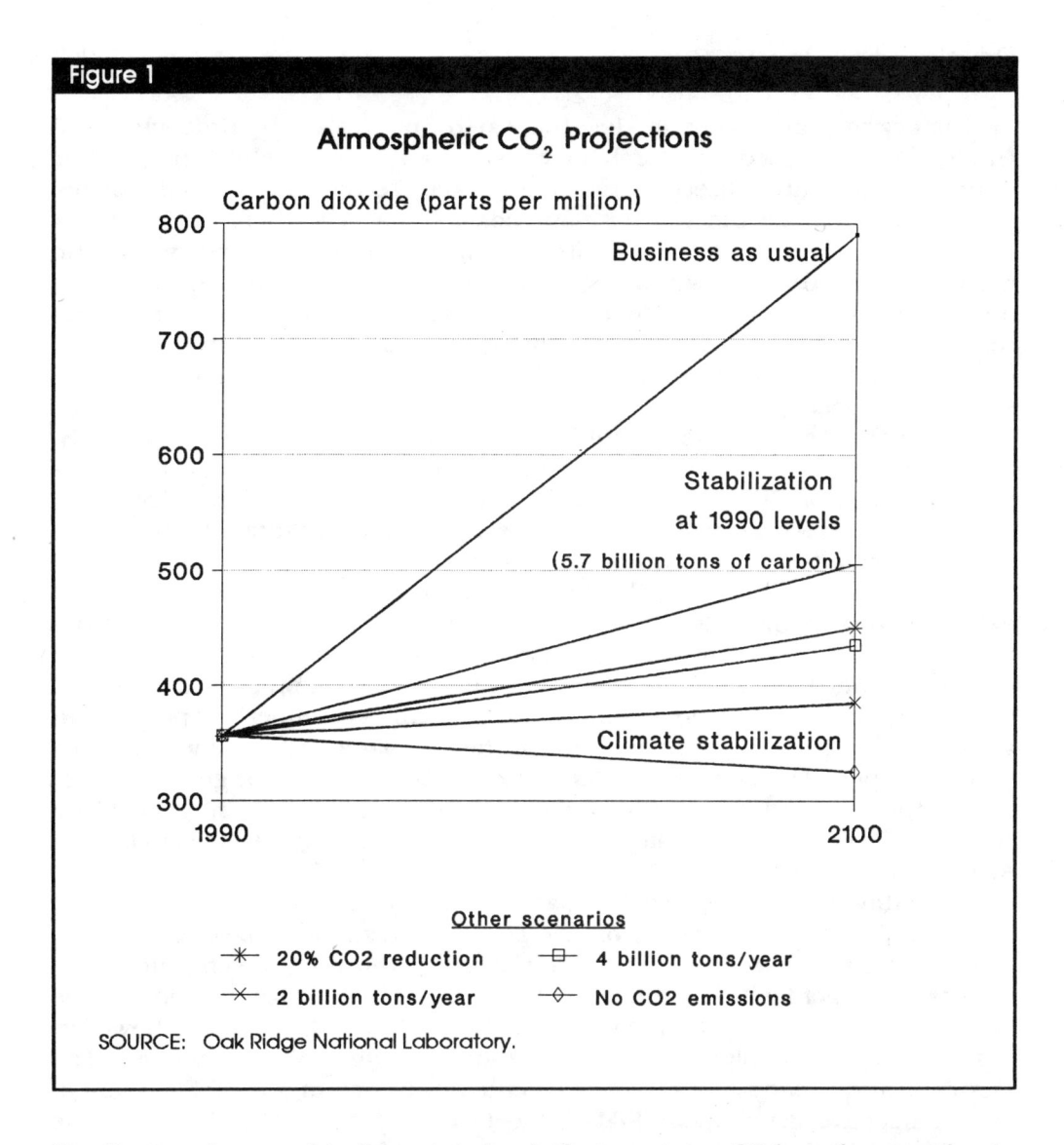

Figure 1

Atmospheric CO_2 Projections

Carbon dioxide (parts per million)

- Business as usual
- Stabilization at 1990 levels (5.7 billion tons of carbon)
- Climate stabilization

1990 — 2100

Other scenarios
- ✽ 20% CO2 reduction
- ☐ 4 billion tons/year
- ✕ 2 billion tons/year
- ◇ No CO2 emissions

SOURCE: Oak Ridge National Laboratory.

Development, a nonbinding statement that contains 27 broad principles to guide environmental policy. Finally, a mammoth 800-page document known as Agenda 21 was approved at UNCED. Agenda 21 is to serve as a blueprint for environmental action through the 21st century, covering a wide range of topics, ranging from pollution to poverty, population and waste management.

Ultimately, Rio may be remembered as the start of a process that changes the way the world thinks about economic growth, emphasizing the importance of maintaining the environmental foundation on which economic development is based. In a more practical sense, Rio has provided a framework in which nations can assess and respond to changing global environmental conditions. And it has led the United Nations to create a Sustainable Development Commission to monitor how countries will comply with promises made at the summit.

Passing the Torch

History may also come to view the Earth Summit as the time when the United States passed the torch of environmental leadership on to other nations. "Today an unprecedented era of peace, freedom and stability makes concerted action on the environment possible as never before," President Bush told convention delegates after he signed the climate treaty, and he promised that "the U.S. will work to carry forward the promise of Rio." Yet other world leaders chided Bush, as the chief executive of the world's one remaining superpower, for not assuming more of a leadership role at the conference. Bush was unrepentant. "I have not come here to apologize," he said in his speech.[1]

While the American president may have been right to claim that U.S. environmental protection laws are "second to none," his statements rang hollow among convention delegates who also were aware that America ranks first in the production of greenhouse gases. Moreover, if the United States claims to be a leader in cleaning up the environment, its presence was greatly overshadowed at the EarthTech exhibition held in conjunction with the Rio summit. Of 400 exhibitors displaying their wares at EarthTech, only 25 were American companies. The Germans and the Japanese had much larger contingents. The disparity at EarthTech and the different style of rhetoric expressed at the Earth Summit symbolizes the trepidation that characterizes the American position on global warming and the commitment of the Germans and Japanese to turn pollution-control and energy efficiency technologies into major growth industries of the 21st century. Loathe to pick winners and losers through anything that smacks of Industrial Policy, the American government declines to lay out such a vision of the future.

For a time after the defeat of Germany and Japan in World War II, the United States had a virtual monopoly on the commericalization of new technology. Spin-offs from defense-related research benefited the aircraft, computer, electronics and machine tool industries. While the U.S. government still devotes nearly two-thirds of its R&D budget to defense-related purposes, however, defense R&D gradually has ceased to stimulate the civilian economy. The Germans and Japanese, meanwhile, spend only 5 percent and 12.5 percent of their respective government R&D budgets on defense. Their priorities lie elsewhere. The Germans spend nearly 15 percent of government R&D funds on industrial development; the United States spends only 0.2 percent. Similarly, the Japanese spend nearly one-quarter of government R&D money on energy technologies; the United States spends less than 4 percent. In terms of gross national product, the United States devotes less than 2 percent of its financial resources to government R&D in nondefense areas, compared with 2.8 percent in Germany and 3.0 percent in Japan.[2]

The Japanese, in particular, have embraced a philosophy that global warming will provide more in the way of new investment opportunities than economic hardship. In 1991, Japan's Ministry of International Trade and Industry drew up "New Earth 21," a five-plank platform that would lead to the "restoration of a green planet" by the end of the 21st century. It also created a new Research Institute of Innovative Technology for the Earth (RITE) to foster the development of new environmental technology. As such, RITE stands as the world's first environmental technology institute.[3]

Because many of the environmental issues RITE plans to address are global in scope, the Japanese trade ministry also has established an International Center for Environmental Technology Transfer. Over the next 10 years, the transfer center expects to train 10,000 people around the globe in the fields of energy conservation, environmental protection and pollution control technology. Not surprisingly, many of the technologies disseminated by the center will be made in Japan. In another symbolic move, the United Nations Environment Program has chosen to build its new International Environmental Technology Center in Japan. Construction will begin in Kansai Science City in the fall of 1992.[4]

Perhaps more than any other nation, Japan has a firmly held belief that new technologies will overcome the world's environmental problems. The government is strongly committed to further development of nuclear power, for instance, as an environmentally suitable generating source. Most of the $400 million earmarked by the government for "global environmental issues" in 1992 will be spent on nuclear energy programs. RITE also is devoting half of its $28 million budget to two innovative carbon sequestration projects. One project is trying to breed a new species of super-algae that would photosynthesize industrial carbon emissions into carbohydrates that can serve as food or fuel. The other anti-global warming project is to develop selective permeable membrane technology that separates carbon dioxide from other gases at a low cost and with high efficiency.[5]

Other nations are more reticent about the ability of technology to solve the world's ecological problems. Yet even in the United States, a number of novel concepts have been put forward as a means of controlling global warming. One idea is to fill barges with iron ore and dump their contents at sea to create massive "blooms" of phytoplankton. The marine life would sequester carbon directly and emit a sulfur compound that stimulates the seeding of clouds. Other proposals include using aircraft, balloons or cannons to cast a pall of dust and soot in the atmosphere, so as to prevent some of the sun's rays from striking the Earth's surface. The National Academy of Sciences has even evaluated a scheme whereby rockets would deploy in orbit 50,000 giant space mirrors to deflect sunlight before it reaches the atmosphere. The academy has concluded like most scientists, however, that so little is known about the potential side-effects of these geoengineering schemes—and so much remains to be discovered about the behavior of the climate—that now is not the time to put these pie-in-the-sky ideas into effect.[6]

Perhaps the most practical, down-to-earth approach to halting the buildup of greenhouse gases is to reduce the rate at which buildings, motor vehicles and industrial machinery consume fossil energy. Such a reduction could be achieved by improving the energy efficiency of capital stock and introducing more low-carbon and noncarbon fuels into the energy supply. The western Europeans and the Japanese have shown that prosperous and competitive economies can be attained while consuming comparatively little energy per unit of gross domestic product. Ultimately, the United States may have to approach, if not match, the level of energy intensity now prevailing in these countries if it is to remain competitive in a world of rising energy costs.

The difficult question is over what time frame this objective should be achieved. In most societies, virtually the entire capital stock is replaced within 50 years—which coincidentally is about the time frame in which most projections of global warming are made. If the United States committed itself to a long-term program of constructing better-insulated buildings, manufacturing more efficient machines and developing more affordable alternative fuels, it might take care of the global warming problem (or at least its contributing role) as the equipment stock turns over.

Many scientists believe, however, that global warming will not afford a lead time of 50 years. Indeed, the amount of greenhouse gases built up over the last 50 years may commit the Earth to at least 1 degree F of additional warming regardless of plans put into effect now. In addition, it is not at all clear that the greenhouse/efficiency goal would be reached if left solely in the invisible hands of the marketplace. On the contrary, the findings of this report suggest that "business as usual" will lead to substantial increases in greenhouse emissions, if present trends are any indication.

To meet the nutritional needs of a burgeoning human and livestock population, the agricultural sector has become highly dependent on energy- and chemical-intensive farming practices since World War II. An even more long-standing tradition in the forestry sector has been to fell virgin timber—a vast carbon storehouse—and then harvest trees on a rotational basis only after the virgin supply is practically wiped out. The auto sector, meanwhile, has stalled on fuel economy gains since the early 1980s and once again is promoting power as a chief inducement to the driving public. Oil companies in the 1990s are reformulating gasolines to make engines "burn cleaner," but the required refining methods consume more crude oil and hence increase CO_2 emissions. Finally, the utility sector is winning consumers over with generally clean and efficient uses of electricity. But the principal generating source is coal—the most abundant and carbon-rich of all fossil fuels. With electricity demand correlated with economic growth (and a virtual necessity in the Electronic Age), reliance on coal is bound to increase without a shift in fuel-generating priorities.

Does Climate Change Matter?

The U.S. government has resisted calls to interfere with these marketplace trends under the economic tenet that the "market knows best" how to allocate the use of resources. To force the market to become more thrifty in the use of energy, for example, or to switch to noncarbon energy sources, is presumed to impose net costs on society; otherwise, the market would be doing these things already. Nevertheless, several prominent studies suggest that a meaningful carbon abatement program might not cost the economy at all. Rather, it could free up capital now spent on energy to invest in other things.

In general, these analyses take a bottom-up, engineering approach to the energy question, instead of a top-down, econometric approach. They find that many cost-effective and environmentally sound ways to provide energy services are apparently thwarted by market imperfections and government subsidies that favor competing sources. These studies also generally find that a comprehensive program of regulatory reforms, market-based incentives and real-cost pricing of energy would enable relatively untapped energy savings and alternative sources to come to the fore.

But even if investments in greenhouse gas controls were to save more money than they cost—a point of contention to be sure—the possibility remains that the money could be put to better use if the costs of global warming turned out to be low and the return on alternative investments was high. Some argue that it is society's duty, in fact, to maximize the return on present-day investments in order to leave an endowment of technology and capital that will improve the quality of life for future generations.

"To me, the great uncertainty is not what our carbon emissions and other greenhouse gases are going to be for the next 75 years...and not what that will do to climate," observes Harvard University professor Thomas Schelling, "but rather, how will people be living? How will they be working? How will they be transporting themselves? What will they be eating?"[7] Turning back the clock to 1900, Schelling wonders, who could have known that future generations would be able to fly across the ocean in jet aircraft, bask in air-conditioned comfort on hot summer days and cook their meals in a matter of minutes in microwave ovens? Schelling contends that technological advancements have made society relatively immune to climate and weather. If the climate changes, does it really matter?

Another university professor, William Nordhaus of Yale University, has examined the economic underpinnings of this argument. He has taken a look at the nation's output of goods and services and concluded that only 3 percent of the nation's economic sectors are "highly sensitive" to climate change; another 10 percent are "somewhat sensitive." That leaves 87 percent of the U.S. gross national product in economic sectors that are only "negligibly affected" by climate.[8]

Nordhaus has also tried to quantify the economic costs imposed on the climate-sensitive sectors if the climate warmed 5 degrees F through 2050, the sea level rose by two feet and rainfall became more spotty in agricultural zones. He figures that greater air conditioning requirements in 2050 would add $490 million to annual utility bills, seawalls to hold back the tide would cost $6.18 billion a year to build and maintain, and the effects on agriculture would be a wash (although his range of estimates is from plus $12 billion to minus $12 billion.) The total quantifiable cost of global warming, therefore, would be $6.67 billion a year, equal to 0.25 percent of the projected gross national product for 2050. A plan to stabilize the nation's CO_2 emissions, meanwhile, is presumed to cost the U.S. economy about 1 to 2 percent in forgone annual income, by Nordhaus's reckoning. Accordingly, the projected benefits of abating global warming—0.25 percent of GNP—would not outweigh the projected costs. Nordhaus's conclusion is that limited, low-cost measures could be taken to reduce greenhouse gas emissions but, on balance, it is better to put up with "modest and gradual" global greenhouse warming than to embark on an ambitious program to combat the problem.[9]

Critics of this analysis believe it contains several fundamental flaws. For starters, one cannot be certain that global warming will be "modest and gradual." Chaos theory suggests that climate change might be impetuous, swift and highly unpredictable. The havoc wrought by extreme events—severe hurricanes, forest fires, drought and the like—would result in far greater monetary damages than is assumed in an orderly transition to a warmer world. Moreover, it is wrong to assume that warming-related losses in "highly sensitive" sectors of the economy would not exceed the 3 percent level of national income they now represent. If domestic food production were threatened, for example, the likely response would not be to let the agriculture sector disappear, with its portion of national income dropping to zero. Rather, the likely response would be to spend even more of the nation's income to maintain production, despite the adverse climatic trend. As food prices increase, agriculture's share of the gross national product would rise instead of fall.

Another criticism of this cost-benefit analysis is that it is highly anthropocentric. Climate change is sure to have impacts on natural habitat and species lacking in commercial value. While these costs are harder to gauge, surely they exist, considering the huge sums spent each year to preserve wetlands, endangered species and the like. Nordhaus acknowledges that climate change is likely to have effects on such things as human health, biological diversity, the quality of life and environmental quality. These intangibles might raise the total cost of global warming to around 1 or 2 percent of global income, by his figuring.[10] Even then, however, the potential costs and benefits of ameliorating global warming might be in the same range. In other words, the economy in 2050 would have about the same value, whether a greenhouse control program is enacted or not. Adding in Schelling's argument that alternative present-day investments could leave a richer legacy for future generations, one might conclude that the greenhouse gambit is simply not worth taking.

One pitfall remains in this thinking, however, and it concerns the basic tool economists use to evaluate future costs and benefits: the "discount rate." Given the choice between having 99 cents today or $1 a year from now, most people would prefer to have the 99 cents today. They know they can invest the 99 cents and increase its value to *more* than $1 a year from now; accordingly, their discount rate of future investments exceeds 1 percent. (A typical discount rate is around 5 percent.) In a similar fashion, the discount rate can be used to assess the present-day value of future damages related to global warming.

If one could determine in 1990, say, that the cost of global warming in 2050 would amount to precisely $186.6 billion, then it would not make economic sense to spend $50 billion in 1990 to keep that amount of damage from occurring. In fact, it would not make sense to spend any more than $10 billion to prevent $186.8 billion in losses in 2050, because $10 billion invested in 1990, earning 5 percent interest annually, would swell to that amount of money in 60 years. Therefore, it makes no difference from an economist's perspective whether to invest $10 billion now to mitigate the future damage or to spend $186.8 billion in 2050 to deal with the problem as it unfolds.

This is where economic analysis may fail policymakers on the global warming question. It provides a rationale for future generations to pay for environmental problems inherited from their forebears and overlooks the possible consequences of delaying action. One must bear in mind that global warming is a moving target. The atmospheric buildup of greenhouse gases will not stop automatically once the doubled-CO_2 threshold is reached—even though this is the point at which most analyses take their "snapshot" of projected costs and benefits. One analysis by the Environmental Protection Agency finds that if industrialized countries wait 20 years to enact greenhouse gas controls (and developing countries wait 45 years)—carrying on with business as usual in the meantime—the result would be a 40 to 50 percent increase in the projected warming commitment by 2050.[11]

Accordingly, a present-value analysis that discounts future environmental damage on the basis of, say, 5 degrees F of projected warming in 2050 may overlook the real stakes involved. A buildup of greenhouse gases that continues unabated midway through the 21st century could commit the climate to 10 degrees F or more of additional warming and hundreds of billions of dollars (perhaps trillions of dollars) of additional environmental damage well into the 22nd century. If the discount-rate analysis applied for the year 2050 fails to appreciate this dynamic of global warming, then the benefits of taking action in the near term to mitigate the problem are undervalued.

The National Academy of Sciences emphasized this important point in its 1991 report, *Policy Implications for Global Warming*:

> Initially, mitigation is likely to reduce real income more than either doing nothing or taking adaptation measures as climatic changes emerge. Ultimately, however, mitigation actions could result in higher real income than waiting and taking adaptation measures. In this scenario, investing in mitigation reduces consumption now, but produces advantages in the future. Expenditures on mitigation options should thus be seen as investments in the future.[12]

This thinking is at the heart of the greenhouse gambit: Whether to make a move in the near term that may involve near-term sacrifice in order to secure a position in the long term that may offer strategic advantage.

The Regret in 'No Regrets'

The nation's energy and environmental policy today continues to relegate the threat posed by global warming as a problem of the future. Yet the more this threat is discounted, the greater becomes its ability to strike. Now it is clear that long before the last pound of fossil energy is exhumed from the earth, the rest of the planet will have run out of room to accommodate the extra carbon dioxide (and other greenhouses gases) without inexorably changing the climate. The greenhouse gambit concedes that some affordable, accessible fossil fuels may never be burned.

The gambit also accepts that real climate insurance may not be purchased at any price, although this is especially true when the carbon abatement program is modest. Because the amount of CO_2 in the atmosphere will continue to grow until the rate of emissions is cut by two-thirds worldwide, stabilization of emissions or even a 20 percent reduction would accomplish very little—especially if the United States acted unilaterally. A 20 percent reduction in U.S. CO_2 emissions would delay a doubling of atmospheric CO_2 by only three years, for example, if other nations failed to follow America's lead.[13] Of course, the present situation is much the reverse of this hypothetical example. Many nations appear willing to stabilize their emissions, yet it is the United States that is refusing to lead.

Until the White House—and to a large extent American industry—are convinced that a climate change insurance policy can be bought that offers good protection at a low annual premium, they have decided that the best global warming policy is one of "no regrets." That is, strategic policy decisions will be made on the basis of economic and environmental criteria other than climate change. If policy decisions yield "free" climate insurance, so much the better. But it would not be prudent, in their view, to enact policies strictly in response to unconfirmed fears about global warming.

The net result is that any greenhouse gas reductions will come about more by coincidence than by design. The United States and most other nations around the world are instituting a ban on chlorofluorocarbons, for example, because of the threat they pose to the Earth's protective ozone layer. In the process, a major source of greenhouse gases is being eliminated, as CFC molecules have thousands of times the heat-absorbing capacity of CO_2 molecules. The end of production of CFCs in the United States by 1995, in fact, will achieve the same reduction in radiative forcing potential as eliminating a billion tons of carbon dioxide emissions—equivalent to roughly one-quarter of the nation's CO_2 emissions.[14]

Remarkably, the latest scientific information suggests that ozone depletion is cooling the upper reaches of the atmosphere and counteracting the warming effect of CFCs. With a no-regrets strategy in place, however, this revelation will not alter the basic decision to phase-out CFCs because they are known to cause another environmental problem. Only now the ban on CFCs will not help much, if at all, in addressing the global warming problem. The bottom line is that the onus will fall more on other greenhouse gases, particularly carbon dioxide, to achieve further emissions reductions.

Fortunately, no-regrets strategies also exist for CO_2. The "Green Lights" program to promote energy-efficient lighting installations, for example, is expected to reduce carbon emissions by 21 million to 55 million tons in the year 2000. Similar programs for "green" buildings, computers and industrial motors are expected to yield another 22 million tons in reductions. Federal regulations that require landfills to capture seeping methane emissions will reduce greenhouse gas emissions by another 20 million to 40 million tons of CO_2 equivalent in 2000. Programs embedded in the National Energy Strategy—including integrated resource planning, expanded use of biofuels and natural gas regulatory reform—are expected to yield savings of 45 million tons of carbon a year in 2000. President Bush's "America the Beautiful" tree-planting program, which aims to plant 1 billion additional trees a year through the

Figure 2

U.S. 'No Regrets' Strategy

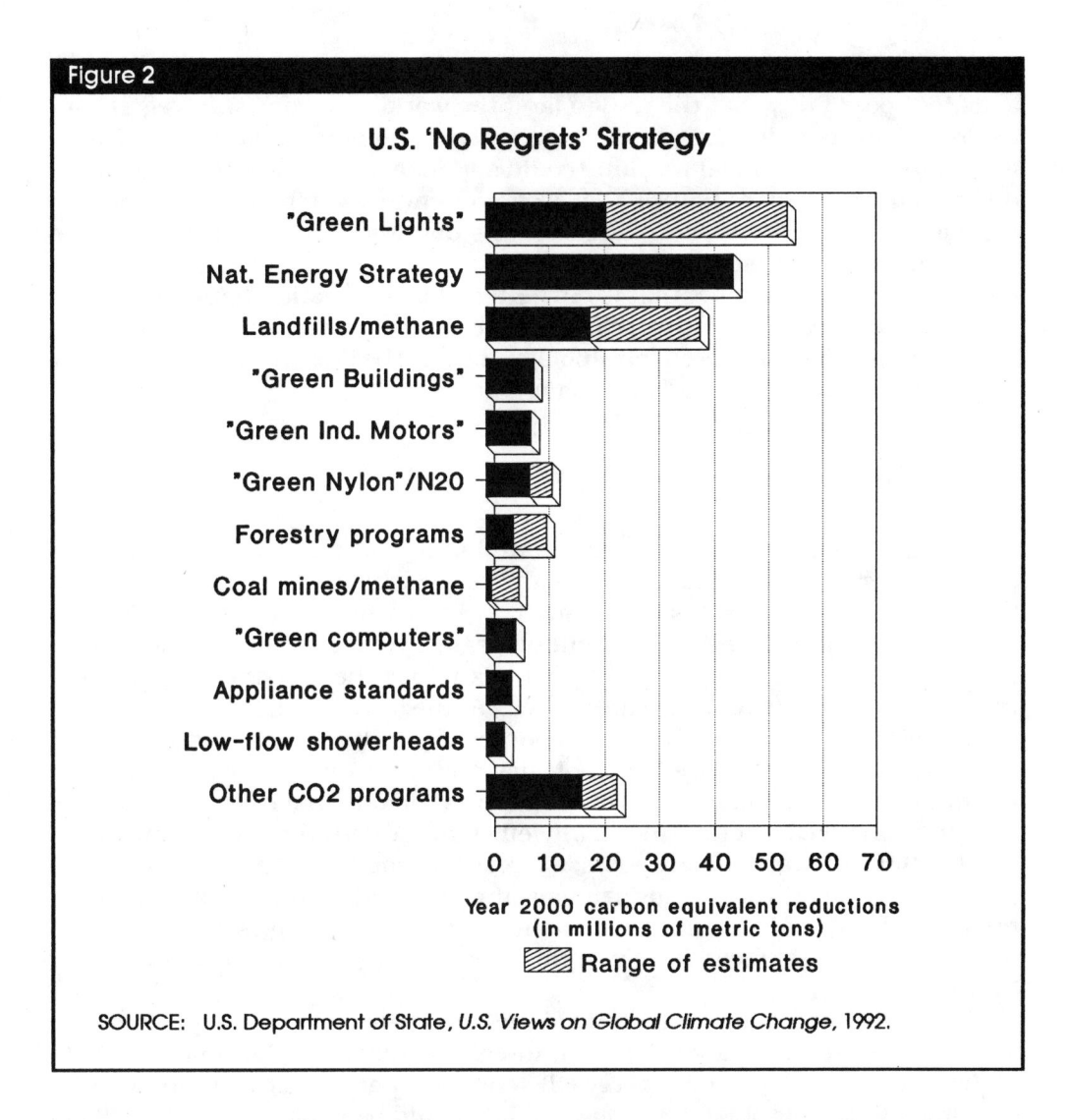

- "Green Lights"
- Nat. Energy Strategy
- Landfills/methane
- "Green Buildings"
- "Green Ind. Motors"
- "Green Nylon"/N2O
- Forestry programs
- Coal mines/methane
- "Green computers"
- Appliance standards
- Low-flow showerheads
- Other CO2 programs

Year 2000 carbon equivalent reductions (in millions of metric tons)

▨ Range of estimates

SOURCE: U.S. Department of State, *U.S. Views on Global Climate Change*, 1992.

1990s, is expected to sequester 5 to 9 million tons of CO_2 from the atmosphere in 2000. All told, federal programs already in effect should hold the nation's CO_2 emissions to a 5 to 8 percent increase above 1990 levels in 2000, while methane, nitrous oxide and CFC emissions are held below 1990 levels. Averaging all of the greenhouse gases together, the increase expected during the 1990s should be only 1.5 to 6.0 percent of CO_2 equivalent (depending on the rate of economic growth)—not far above the emissions stabilization target referred to in the Rio climate treaty.[15]

Still, there are limits to how much climate change insurance this "no-regrets" policy can offer. After the year 2000, the nation's CO_2 emissions are expected to start rising rapidly again because of increased use of coal in power plants, greater consumption of fuel in the nation's transportation sector and rising energy demand associated with growth in the economy overall. In reality, to call such a strategy "no-regrets" is a bit of a misnomer because there remains a distinct possibility that the nation (and the world) someday will "regret" not having taken more immediate and aggressive actions to stem the buildup of greenhouse gases. Global warming could suddenly appear like the bogeyman, meting out punishment until the economy endures a painful mid-course correction to curb its carbon excesses. As one wry observer described this unsettling prospect, such crash measures would be like "trying to install a sprinkler system in a hotel that is currently on fire, or building military forces while you're already under attack, or buying collision insurance after you've crashed your car."[16] In chess terminology, it would be like taking a gambit after your king has already been checkmated.

The Grand Bargain

The greenhouse gambit is another kind of no-regrets strategy. It is more concerned with taking measures to stabilize the climate than it is to preserve the status quo for business. As such, it favors the adoption of relatively aggressive greenhouse gas abatement programs, yet it still adheres to market-based principles. The list of policy options under the greenhouse gambit is varied and it cuts across many industries. At the top of the list is a re-ordering of the government's research and development priorities. Emphasis would be placed on new energy-related technologies destined to capture a large share of the market in a greenhouse world. Examples include fuel cells, photovoltaic systems, long-life batteries and highly efficient industrial motors, appliances and lighting systems. Advanced fission reactors and fusion power also have a place on this list. Yet the government already spends many times more on nuclear-related R&D than on all of these other technologies combined. A reformulated government R&D program would strike a better balance among these options.

The greenhouse gambit also would seek to create a more level playing field for energy sources. To a large extent, integrated resource planning programs at the state level—which put energy efficiency on a par with energy supply—are working toward this objective already. More could be done, however. Billions of dollars could be pared from annual subsidies to fossil energy developers. Vendors of nuclear power could be asked to shoulder more of the costs of nuclear plant insurance and radioactive waste disposal. More fundamentally, the environmental costs of energy production and consumption could be incorporated into market pricing. A carbon tax could serve as a proxy for these external costs, with the revenues recycled as an investment tax credit into an economy that is long on consumption and short on capital formation. Finally, preferential lending could be directed toward the construction of more energy-efficient homes and buildings to reflect their lower operating costs, while a "feebate" system could be used to help overcome the higher capital costs generally associated with these structures. (A feebate assesses a fee on those who purchase inefficient capital stock and passes it on as a rebate to those who purchase more efficient stock.)

A feebate system could play a role in the auto sector as well. Similarly, a carbon tax could be used to stimulate demand for fuel-economical cars as gasoline prices rise at the pump. (Because some of the revenues generated would likely be used to reduce personal income taxes, low-income households could have their income taxes reduced proportionally to offset higher energy prices.) Implicit subsidies such as free parking and freeways built without tolls also could be reduced or eliminated. Finally, government research and development programs could direct more support for expedited development of electric vehicles, bullet trains and other advanced transit systems.

The agriculture and forestry sectors also would have a vital role in the greenhouse gambit. One important objective would be to take highly erodible cropland out of production and plant trees instead. Utilities in search of low-cost carbon offsets might pay for such land-use conversions, and forest products companies could pay farmers to harvest the trees once they mature. Trees raised in this fashion would accomplish several valuable goals at once: Environmental impacts associated with growing (often surplus) crops would be reduced. Carbon sequestration in soils and timber would be greatly enhanced. The long-term supply of timber would increase, taking pressure off the national forests. And greater quantities of long-lived, affordable wood products would be available to serve in place of more energy-intensive materials. New responsibilities also would accompany these new opportunities, namely, applying low-input, sustainable agricultural methods and New Forestry management techniques to maintain harvest yields and to make the land better prepared for possible stresses of climate change. In the laboratory, biotechnology research could be directed toward the same ends.

None of these policy measures would require the use of heavy-handed command and control regulations. All would rely on market forces. While market forces alone cannot solve the world's environmental problems, without them, nothing else will.

With market-based reforms sweeping through the governments of formerly communist nations, so, too, is market reform catching on in western economies as a new approach to sustaining the environment. One sign of the times is the newly created Business Council for Sustainable Development, which features an impresssive list of chief executive officers from 48 companies, including industrial giants such as Aluminum Co. of America, Browning-Ferris, Chevron, ConAgra, Dow Chemical, Du Pont and Northern Telecom, as well as the corporate chiefs of ABB Asea Brown Boveri Ltd., Ciba-Geigy, Kyrocera, Mitsubishi, Nissan Motor, Nippon Steel and Royal Dutch/Shell Group. These business leaders have endorsed the principle that environmental costs should be reflected in the market prices of goods and services. What is more, they support the levy of a value-added tax on electricity and gasoline, the abolition of subsidies for fossil energy and the removal of other tax breaks that spur profligate energy use. In a new book by Stephan Schmidheiny, the business council's director, this new business philosophy toward the environment is spelled out. According to *Changing Course: A Global Business Perspective on Development and the Environment*:

> Progress toward sustainable development makes good business sense because it can create competitive advantages and new opportunities. But it requires far-reaching shifts in corporate attitudes and new ways of doing business. To move from vision to reality demands strong leadership from the top, sustained commitment throughout the organization, and an ability to translate challenge into opportunities.[17]

The United Nations Conference on Trade and Development also released a landmark study in 1992 that concludes the best way to combat the threat of global warming is to enlist the support of private markets. In *Trading Entitlements to Control Carbon Emissions: A Practical Proposal to Combat Global Warming*, a global Environmental Protection Agency is proposed that would issue pollution allowances that would be traded among polluters and others. Exchange-based and over-the-counter futures and options markets also would be established for the allowances, as has been done already for sulfur dioxide allowances under the U.S. Clean Air Act.[18]

From this trading concept emerges the "Grand Bargain" to combat global warming—creating market relationships between rich and poor nations to help preserve the environment. While continuing tension between the North and the South was much in evidence at the recent Earth Summit, there was mutual agreement that the world's most vexing environmental problems stem from a growing population base that seeks to live better yet lacks the resources to do so. The unresolved question is how to make the resources available.

The United Nations figures that an eventual transfer of $125 billion a year might be required from industrialized nations in order for chronic environmental problems of developing nations to be overcome. Some in the West believe this transfer amounts to environmental blackmail, a new way to soak the rich.[19] Others in the North are more sympathetic to the plight of the environment in the South but insist they lack the resources to invest in development projects for which they get nothing in return. Ultimately, investment capital is the most scarce resource on Earth.

The Grand Bargain addresses these issues. Blackmail or not, the fact remains that the world's population is expected to nearly double by the middle of the 21st century, with 90 percent of the growth centered in the Third World. Over the same period, per-capita income may nearly triple, so that the world economy is almost six times larger than it is today. While developing nations account for only one-quarter of the world's greenhouse gas emissions at present, the growing population imbalance combined with continuing economic development suggests that the Third World is likely to account for at least half of a much larger pool of emissions by the middle of the next century.

Accordingly, any serious program to combat global warming is destined to fail if developing nations do not take part—a point often made by parties in the West who feel they are being singled out for greenhouse gas controls. But those in Third World also make the point that industrialized nations want developing nations to "do as I say, not as I do" when it comes to modernizing their economies. While past development efforts involved quantum leaps in energy consumption, future development must find alternative means, they are told.

Figure 3

Unconstrained Carbon Emissions

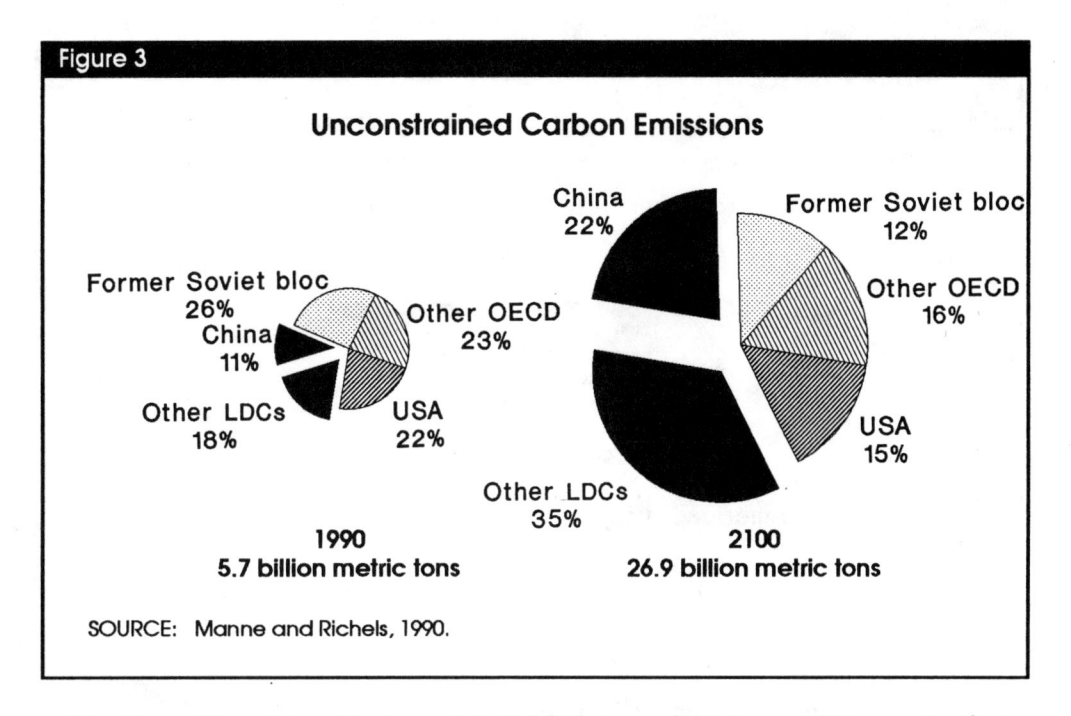

1990
5.7 billion metric tons
- Former Soviet bloc 26%
- China 11%
- Other LDCs 18%
- USA 22%
- Other OECD 23%

2100
26.9 billion metric tons
- China 22%
- Former Soviet bloc 12%
- Other OECD 16%
- USA 15%
- Other LDCs 35%

SOURCE: Manne and Richels, 1990.

The Grand Bargain sidesteps this ticklish issue but does not ignore the fact that a two-thirds reduction in annual global carbon emissions will be required eventually to stabilize the amount of CO_2 in the atmosphere at present levels. On a global per-capita basis, then, emissions must be held to 0.38 tons per person a year, compared with today's average of 1.15 tons. (The more the population grows, however, the lower this per-capita figure must become.)

Considering that per-capita carbon emissions in the United States are the highest in the world, at 5.37 tons, it is probably fanciful to think that Americans can so alter their lifestyles, infrastructure and fuel-use patterns that they will be able to reduce per-capita emissions by 14 times over the next few decades. Yet in Africa, per-capita carbon emissions amount to only 0.29 tons a year—25 percent below the climate stabilization target. In the Far East, the per-capita average is only 0.26 tons, and in India, it is only 0.21 tons.[20]

The United States and other industrialized nations, faced with the prospect of Draconian carbon emissions reductions in their own countries, could decide to invest in other nations where per-capita emissions are low but the potential for sizable increases is high. If an American utility were to find that it costs less to finance the installation of high-efficiency (or noncarbon) generating equipment in Mexico, for example, than to retrofit its own equipment in the United States, the Grand Bargain would permit this least-cost transaction to occur, benefitting both parties in the exchange. By the same token, if European food producers were to conclude that it costs less to eliminate methane emissions from flooded rice paddies in Asia than to drastically reduce energy and fertilizer inputs in their fields in Europe, this exchange could occur as well. The list of possible market-based transactions is endless—and may involve formerly Soviet-bloc nations as well. Eastern Europeans could be paid to shutter old,

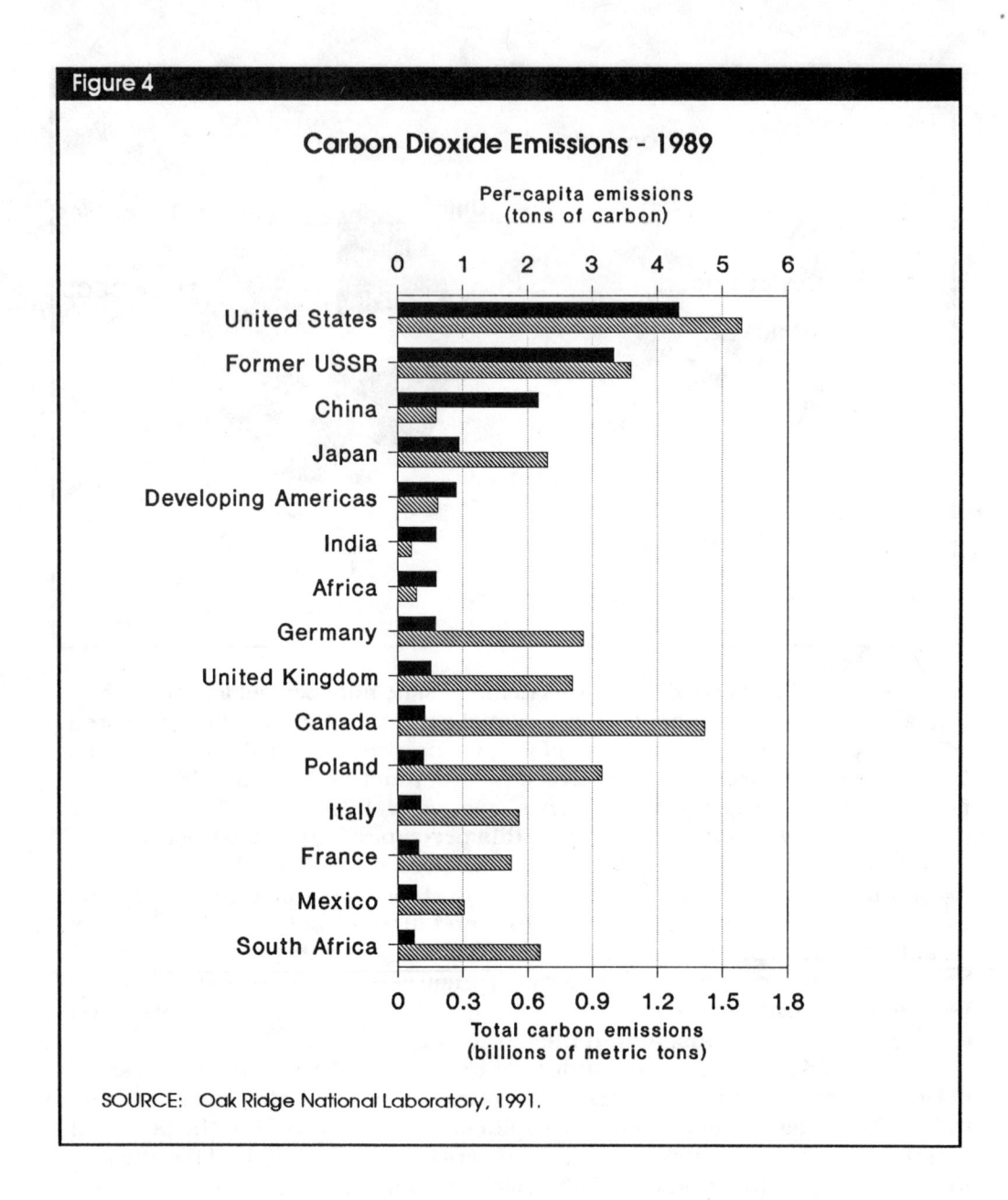

Figure 4

Carbon Dioxide Emissions - 1989

Per-capita emissions (tons of carbon) [top axis: 0–6]
Total carbon emissions (billions of metric tons) [bottom axis: 0–1.8]

Categories (top to bottom): United States, Former USSR, China, Japan, Developing Americas, India, Africa, Germany, United Kingdom, Canada, Poland, Italy, France, Mexico, South Africa.

SOURCE: Oak Ridge National Laboratory, 1991.

highly polluting factories and erect modern, efficient facilities with western technology. Russians could be paid to plug leaks in their creaky natural gas pipeline system and raise their sales volume to the West. Brazilians could be paid to preserve their rainforests and use the proceeds to develop sustainable agroforestry and extractive reserves programs.

Overseeing this whole process would be the global Environmental Protection Agency—or perhaps the U.N. Sustainable Development Commission established at the recent Earth Summit. The international agency would issue a fixed number of greenhouse emissions allowances for each country, taking into account its population and state of economic development, and then reduce the number of permits over a period of years (or more likely decades) until the climate stabilization target is achieved. The Chicago Board of Trade and the American Stock Exchange already have expressed interest in running an exchange to promote market trading of international carbon dioxide allowances.[21]

Much remains to be done before such trading could begin, of course. Appropriate monitoring systems must be put in place. A labyrinth of national environmental laws must be organized into a coordinated international system. Perhaps most important, citizens and governments of the world must become convinced that global warming presents a sufficient threat to make the Grand Bargain worth all of the effort.

If the climate does begin to spin dangerously out of control, one can safely assume that the political resolve *will* be found to address the problem. The question is whether it makes sense to wait and see what the climate has in store. Ameliorative measures that appear moderate, practical and affordable if given sufficient lead time become tough, unworkable and excessively priced if adopted in a crash course. With only eight years before a new millennium begins, businesses, investors and government planners may ask themselves when is the appropriate time to press forward with new technologies and new ways of doing business that enhance our global environmental security as well as our economic well-being.

For those who favor the greenhouse gambit, the time has already arrived.

Notes

1. Rose Gutfeld and John Harwood, "President's Clumsy Handling of Earth Summit Results in a Public Relations Disaster for Him," *The Wall Street Journal*, June 15, 1992.
2. "Gaining New Ground: Technology Priorities for America's Future," Council on Competitiveness, Washington, D.C., April 1991. The percentages listed are for 1989.
3. Frederick S. Myers, "Japan Bids for Leadership in Clean Inudstry," *Science*, May 22, 1992; and Jacob M. Schlesinger, "In Japan, Environment Means an Opportunity for New Technologies," *The Wall Street Journal*, June 3, 1992.
4. *Ibid.*
5. Frederick S. Myers, "A Technical Fix for the Greenhouse," *Science*, May 22, 1992.
6. National Academy of Sciences, *Policy Implications of Global Warming*, Reports of the Synthesis and Mitigation Panels, National Academy Press, Washington, D.C., 1991.
7. Dennis Wamsted, "Climate Change Policies Skewed, Says Schelling," *Environment Week*, Dec. 13, 1990.
8. William Nordhaus, "To Slow or Not to Slow: The Economics of the Greenhouse Effect," paper presented at the annual meetings of the American Association for the Advancement of Science, New Orleans, La., February 1990.
9. *Ibid.*
10. *Ibid.*, and Dennis Wamsted, "Costly CO_2 Reductions Not Justified By Expected Global Warming Damage," *Environment Week*, Feb. 22, 1990.
11. U.S. Environmental Protection Agency, *Policy Options for Stabilizing Global Climate*, 21P-2003.2, December 1990.
12. See note 1.
13. See note 11.
14. U.S. Deppartment of Energy, National Energy Strategy: Powerful Ideas for America, U.S. Government Printing Office, Washington, D.C., February 1991.
15. U.S. Department of State, "U.S. Views on Global Climate Change," undated manuscript. "U.S. Views" was prepared in April 1992 in preparation the United Nations Conference on Environment and Development in Rio de Janeiro.
16. Amory B. Lovins and Hunter L. Lovins, "Least-Cost Climatic Stabilization," Rocky Mountain Institute, Snowmass, Colo., printed manuscript, May 17, 1991.
17. Stephan Schmidheiny, *Changing Course: A Global Business Perspective on Development and the Environment*, MIT Press, Cambridge, Mass., 1992.
18. Jeffrey Taylor, "Global Market In Pollution Rights Proposed by U.N.," *The Wall Street Journal*, Jan. 31, 1992.
19. See, for example, Patricia Adams, "Rio Agenda: Soak the West's Taxpayers," *The Wall Street Journal*, June 3, 1992; and George Melloan, "Flying Down to Rio, for Fun and Profit," *The Wall Street Journal*, May 11, 1992.
20. Carbon Dioxide Information Analysis Center, *Trends '91: A Compendium of Data on Global Change*, Oak Ridge National Laboratory, Oak Ridge, Tenn., December 1991.
21. See note 18.

About IRRC's Environmental Information Service

Investors and corporations are being profoundly affected by the public's growing environmental awareness and government's increased environmental regulation. In response to this growing interest in the environment, IRRC has developed an Environmental Information Service that provides some of the most comprehensive information available on corporate environmental performance, trends and activities.

The Environmental Information Service is designed to help institutional investors integrate information on corporate environmental liabilities, risks, performance and opportunities into their portfolio management practices. The service is also increasingly utilized by corporations themselves and consultants to analyze and benchmark their progress in reducing adverse environmental impacts and costs.

Subscribers to the Environmental Information Service receive a number of benefits:

- IRRC's three-volume *Corporate Environmental Profiles Directory* lets you quickly and accurately compare the environmental performance records and potential liabilities of companies in the Standard & Poor's 500. With a profile of each company, the directory contains information on a wide variety of compliance and performance measures—such as toxic chemical releases, oil and chemical spills, hazardous waste cleanup sites and compliance penalties—as well as information on company environmental policies, initiatives and achievements. With more than 1,500 pages of valuable data and information, and updated annually, the *Corporate Environmental Profiles Directory* is the premier information source for anyone interested in corporate environmental performance.

- Six issues of *Investor's Environmental Report* keep you up-to-date on the environmental issues and trends most important to the business and investment community. The newsletter is filled with articles, regular columns and in-depth features on issues such as regulatory and judicial actions affecting business, the performance of environmental investment funds, and trends in corporate environmental practices.

- Special studies help you understand major national and global environmental issues with significant consequences for business. Recent and forthcoming studies examine how companies are finding profitable uses for trash in the post-consumer waste stream, the role of corporations in tropical deforestation and in efforts to develop sustainable forestry practices, and trends in the disclosure of environmental information in corporate security filings.

For additional information on IRRC's Environmental Information Service, please write to IRRC at 1755 Massachusetts Avenue, N.W., Washington, DC 20036, or call (202) 234-7500.

About the Author

Doug Cogan is Manager of Global Issues for IRRC's Environmental Information Service. He is a graduate of Williams College with a degree in political economics. He has been with IRRC for 10 years. Mr. Cogan is co-author of two editions of *Generating Energy Alternatives: Demand-side Management and Renewable Energy at America's Electric Utilities* and the first edition of *Power Plays: Profiles of America's Leading Renewable Electricity Developers.* His first book on a global environmental topic is *Stones in a Glass House: CFCs and Ozone Depletion.* He is now editing a forthcoming IRRC report on the role of corporations in tropical deforestation and sustainable forestry management practices. Mr. Cogan has written numerous articles and short reports on corporate environmental issues and shareholders' growing interest in environment-related business opportunities, liabilities and management practices.

Other Environment and Energy Publications from IRRC

Trash to Cash: New Business Opportunities in the Post-Consumer Waste Stream

The nation's solid waste crisis is changing the way industry does business and is providing new market opportunities. This report analyzes the ability of the glass, aluminum, steel, paper and plastics industries to reclaim their products from the municipal solid waste stream and manufacture new products. The report also features emerging companies that offer recycling, composting and other non-burn disposal options, as well as information on better established waste-to-energy companies.

July 1991/317 pp.   ISBN 0-931035-84-8 $35

The Greenhouse Effect: Investment Implications and Opportunities

On October 4, 1989, IRRC and the World Resources Institute sponsored an investor forum on global warming. This report is a compilation of edited remarks by the 18 featured speakers from corporations, investing institutions, academia and the government. The forum's morning session presented the science of global warming, its potential impacts and government responses to it, and the implications and opportunities for industry. The afternoon session of panel discussions focused on three industries likely to be affected by global climate change: electric power, transportation and real estate.

February 1990/ 166 pp.   ISBN 0-931035-42-2 $35

Power Plays: Profiles of America's Leading Renewable Electricity Developers

This study analyzes the current status of seven renewable electric power technologies—biomass (including the burgeoning waste-to-energy industry), geothermal, wind, hydro, photovoltaics, solar thermal and ocean—and provides in-depth profiles of 104 non-utility companies spearheading the development of these technologies. The report includes data on project locations, cost, financing techniques, energy production and utility contracts.

July 1989/456 pp./Hardcover   ISBN 0-931035-33-33 $100

Tropical Deforestation

This book examines the role of corporations in tropical deforestation, as well as the viability of economic alternatives to forest destruction such as ecotourism,

sustainable forestry and agriculture. A unique feature of this report will be a description of grass-roots economic development projects with investment potential that minimize the destruction of tropical forest ecosystems and preserve indigenous lands and biodiversity.
Available December 1992   ISBN 0-931035-85-6 $45

Stones in a Glass House: CFCs and Ozone Depletion

This report traces the development of the multi-billion-dollar CFC and halon industries, as well as the decade-long debate over the impact of chemicals on the ozone layer. Now that international agreements and U.S. regulations will eliminate the use of the chemicals, the report examines the prospects for new substitute chemicals and new business opportunities for the major players.
1988/147 pp.   ISBN 0-931035-27-9 $35

Generating Energy Alternatives: Demand-side Management, Renewable Energy at America's Electric Utilities

This extensive survey of 123 electric utilities contains comprehensive information on utilities' demand-side management programs, marketing strategies, non-conventional generating technologies and third-party power purchases. A 50-page industry analysis highlights emerging trends and compares the survey results with the 1983 edition.
1987/358 pp.   ISBN 0-931035-16-3 $100

Investors Environmental Report

Published bi-monthly, *Investors Environmental Report* provides updates and in-depth articles about pertinent environmental issues affecting corporations and reports on how corporations and investors are responding to environmental pressures and opportunities. It also analyzes the environmental regulatory issues affecting various industries.
       ISSN 1055-2154 $150 per year

News for Investors

This monthly newsletter keeps readers up-to-date on developments affecting corporate social policy in such areas as the environment, tobacco, investment in South Africa, Northern Ireland, animal testing, energy, military production and equal employment opportunity. It also provides complete descriptions of all social policy shareholder resolutions submitted for inclusion in corporations' proxy statements. During the annual meeting season, monthly checklists show the disposition of resolutions at a glance.
       ISSN 1053-5470 $250 per year